Coherence, Amplification,
and Quantum Effects in
Semiconductor Lasers

# Coherence, Amplification, and Quantum Effects in Semiconductor Lasers

**EDITED BY**
**YOSHIHISA YAMAMOTO**
Basic Research Laboratories
Nippon Telegraph and Telephone Corporation
Musashino-Shi, Tokyo 180, Japan

A Wiley-Interscience Publication
John Wiley & Sons, Inc.
New York / Chichester / Brisbane / Toronto / Singapore

# Wiley Series in Pure and Applied Optics

The Wiley Series in Pure and Applied Optics publishes outstanding books in the field of optics. The nature of these books may be basic ("pure" optics) or practical ("applied" optics). The books are directed towards one or more of the following audiences: researchers in university, government, or industrial laboratories; practioners of optics in industry; or graduate-level courses in universities. The emphasis is on the quality of the book and its importance to the discipline of optics.

In recognition of the importance of preserving what has been written, it is a policy of John Wiley & Sons, Inc. to have books of enduring value published in the United States printed on acid-free paper, and we exert our best efforts to that end.

*Library of Congress Cataloging-in-Publication Data*:

Coherence, amplification, and quantum effects in semiconductor lasers
/edited by Yoshihisa Yamamoto.
     p. cm. -- (Wiley series on pure and applied optics, ISSN 0277-2493)
    "A Wiley-Interscience publication."

    Includes bibliographical references.
    ISBN 0-471-51249-4
    1. Semiconductor lasers.  I. Yamamoto, Yoshihisa.  II. Series.
TA1700.C64 1990
621.36'6--dc20                      90-12608
                                            CIP

Printed in the United States of America

10 9 8 7 6 5 4 3 2 1

# Contributors

E. BERGLIND, Department of Microwave Engineering, The Royal Institute of Technology, Stockholm, Sweden

GUNNAR BJÖRK, Department of Microwave Engineering, The Royal Institute of Technology, Stockholm, Sweden

FEDERICO CAPASSO, AT&T Bell Laboratories, Murray Hill, New Jersey

T. E. DARCIE, AT&T Bell Laboratories, Crawford Hill, Holmdel, New Jersey

CHARLES H. HENRY, AT&T Bell Laboratories, Murray Hill, New Jersey

TERENCE G. HODGKINSON, British Telecom Research Laboratories, Martlesham Heath, Ipswich, Suffolk, England

KAZUHIRO IGETA, NTT Basic Research Laboratories, Nippon Telegraph and Telephone Corporation, Musashino-shi, Tokyo, Japan

ROBERT M. JOPSON, AT&T Bell Laboratories, Crawford Hill, Holmdel, New Jersey

ANDERS KARLSSON, Department of Microwave Engineering, The Royal Institute of Technology, Stockholm, Sweden

SOICHI KOBAYASHI, Phototonic Integration Research, Inc., Columbus, Ohio

PAO-LO LIU, Department of Electrical and Computer Engineering, State University of New York at Buffalo, Buffalo, New York

SUSUMU MACHIDA, NTT Basic Research Laboratories, Nippon Telegraph and Telephone Corporation, Musashino-shi, Tokyo, Japan

TAKAAKI MUKAI, NTT Basic Research Laboratories, Nippon Telegraph and Telephone Corporation, Musashino-shi, Tokyo, Japan

K. NAKAGAWA, Graduate School at Nagatsuta, Tokyo Institute of Technology, Nagatsuta, Midori-ku, Yokohama, Kanagawa, Japan

OLLE NILSSON, Department of Microwave Engineering, The Royal Institute of Technology Stockholm, Sweden

MOTOICHI OHTSU, Graduate School at Nagatsuta, Tokyo Institute of Technology, Nagatsuta, Midori-ku, Yokohama, Kanagawa, Japan

TADASHI SAITOH,   NTT Basic Research Laboratories, Nippon Telegraph and Telephone Corporation, Musashino-shi, Tokyo, Japan

BAHAA E. A. SALEH,   Department of Electrical and Computer Engineering, University of Wisconsin, Madison, Wisconsin

MALVIN C. TEICH,   Columbia Radiation Laboratory, Columbia University, New York, New York

YOSHIHISA YAMAMOTO,   NTT Basic Research Laboratories, Nippon Telegraph and Telephone Corporation, Musashino-shi, Tokyo, Japan

YUZO YOSHIKUNI,   NTT Optoelectronics Laboratories, Morinosato, Atsugi-shi, Kanagawa, Japan

# Contents

# Preface

This book contains 12 review chapters, written by leading scientists in the field, on both classical and quantum aspects of semiconductor lasers and related applications. The first five chapters deal mainly with the classical noise and modulation characteristics of semiconductor lasers. Spectral linewidth broadening and frequency modulation of semiconductor lasers have been extensively studied during the last 10 years because of their potential technological importance for coherent communications and laser spectroscopy. The next three chapters deal with optical signal amplification in semiconductor lasers. Nearly all have been concerned with linear amplifiers operating below the oscillation threshold, but nonlinear injection-locked oscillators operating above the oscillation threshold have also attracted attention. The final four chapters deal with the quantum-mechanical (nonclassical) aspects of semiconductor lasers. A squeezed state with quantum noise reduced to below the standard shot noise limit can be generated and spontaneous emission can either be inhibited or enhanced in microcavity semiconductor lasers.

This book describes the state of the art in this rapidly evolving field. It is hoped that these review articles will help students and active researchers who are interested in the physics and applications of these aspects of semiconductor lasers. It is a pleasure for me to thank all the contributors and the publisher for the fruitful collaboration. Finally, I wish to express my thanks to Minako Goh, who typed the manuscripts carefully within a tight schedule.

YOSHIHISA YAMAMOTO

*Tokyo, Japan*
*January 1991*

Coherence, Amplification,
and Quantum Effects in
Semiconductor Lasers

# 1

# Introduction

YOSHIHISA YAMAMOTO
<br>*NTT Basic Research Laboratories, Musashino-shi, Tokyo, Japan*

The quantum-mechanical theory of laser amplifiers was developed by Shimoda, Takahashi and Townes [1] in the early days of masers before lasers were actually demonstrated. The density matrix master equation method pioneered by them was successfully extended by Scully and Lamb, Jr. [2] to describe the spectral and photon statistical properties of laser oscillators. Haken worked out the quantum-mechanical Langevin equation method to describe the nonlinear dynamics of lasers [3]. The problem is often reduced to the Fokker–Planck equation, by which analytic solutions are obtained without introducing a linearization approximation. The two methods correspond to the Schrödinger picture and the Heisenberg picture in quantum mechanics, and so they should lead to the same conclusion. This point was thoroughly studied by Lax and Louisell [4] and Louisell [5].

A laser is a nonequilibrium open system with an infinite number of degrees of freedom. Macroscopic coherence is established by the balance between the system's stabilizing force (gain saturation) and fluctuating forces (pump noise, vacuum field fluctuations, and dipole fluctuations) from reservoirs. This is a common feature of quantum coherence established in the second-order phase transition.

The quantum theory of lasers can be reduced to these simple equations by preserving only a few degrees of freedom as "systems" and eliminating all the other degrees of freedom as "reservoirs." In this respect, the fluctuation–dissipation theorem studied by Senitzky [6] plays a key role in the theory.

Quantum states of light have been a central issue in quantum optics. A coherent state of light was extensively studied by Glauber [7]. A coherent state is an eigenstate of the photon annihilation operator and has equal

*Coherence, Amplification, and Quantum Effects in Semiconductor Lasers*, Edited by Yoshihisa Yamamoto.
<br>ISBN  0-471-512494  © 1991 John Wiley & Sons, Inc.

amounts of quantum noise in each of the two quadrature components (amplitudes of the $\cos \omega t$ and $\sin \omega t$ components or electric field and magnetic field) to satisfy the minimum uncertainty product. A squeezed state has less quantum noise than a coherent state in one quadrature component but more noise in the other to satisfy the minimum uncertainty product. The mathematical foundation of squeezed states was established by Takahashi [8], Stoler [9], and Yuen [10]. Redistribution of the quantum noise between photon number and phase instead of between two quadrature components is also possible; such a state, called a *number-phase squeezed state*, was studied by Jackiw [11].

These states are not only mathematical basis states for quantum optics calculations; they can be generated in real experiments. Glauber demonstrated that a coherent state can be generated by a classical oscillating current [7]. As a more realistic example, one may ask whether an "oscillating dipole current" in a laser medium produces a coherent state. In a laser oscillator, spontaneous emission with random phase couples to a lasing field and diffuses the phase according to the Schawlow–Townes formula [12]. Except for this point, a laser far above the threshold generates a sequence of states that are not too far from a coherent state, but the physics behind this is much more complicated than the case for radiation produced by an oscillating current studied by Glauber.

In most lasers pumped at well above the threshold, intrinsic quantum noise is so small that it is masked by various extrinsic noises and cannot be measured in real experiments. A semiconductor laser is definitely an exceptional case. Because of its small size and stable operation, quantum noise is much larger than the various extrinsic noises and is observable at any pump rate. In this respect, the quantum theory of lasers can only be tested using a semiconductor laser. Such efforts were pioneered by Armstrong and Smith [13] and Hinkley and Freed [14] soon after the invention of the semiconductor laser.

The interest in semiconductor laser noise has been rekindled recently, because a semiconductor laser is used as a practical light source in advanced optical communications (see Chapter 6) and in laser spectroscopy (see Chapter 5). The quantum noise emerges as the limiting factor in such practical applications. One important discovery obtained in this line of research is that the frequency modulation noise (spectral linewidth) of a semiconductor laser is much larger than the Schawlow–Townes limit because of amplitude-phase coupling via excited carriers (see Chapters 2–4). This amplitude–phase coupling also determines the direct frequency modulation characteristics of a semiconductor laser. Another important discovery is that amplitude modulation noise (photon statistics) of a semiconductor laser is subject to longitudinal-mode partition noise, even though the lasing mode has a much higher intensity than nonlasing modes (see Chapter 10).

Another line of research is related to the recent successful development of quantum optics. One remarkable difference distinguishing the semiconductor laser from all other lasers is that a semiconductor laser is pumped by injection current supplied via an electrical circuit. Optical pumping is a Poisson point process, and pump noise is shot-noise-limited. Electrical pumping is not necessarily a Poisson point process due to Coulomb interaction, and the pump noise for a semiconductor laser can actually be reduced to below the shot-noise limit. As a result of this pump noise suppression, the output of a semiconductor laser is a number-phase squeezed state rather than a coherent state (see Chapters 11 and 12).

The other important noise source in lasers is spontaneous emission. Spontaneous emission is caused by "vacuum fluctuation" of electromagnetic field as well as "radiation reaction." Therefore, spontaneous emission is either inhibited or enhanced if vacuum fluctuation is controlled by means of cavity walls, as was originally predicted by Purcell [15] and demonstrated by Drexhage [16]. Spontaneous emission can be controlled in microcavity semiconductor lasers, which leads to a reduction of the threshold current and quantum noise (see Chapter 13).

The noise added by laser amplifiers was also studied soon after the invention of lasers by Haus and Mullen [17] and Heffner [18]. The generalized uncertainty principle for simultaneous measurement governs the fundamental limit on noise figure of laser amplifiers.

The semiconductor laser has also recently attracted interest for application to direct optical signal amplification. The performance of linear amplifiers operating below the lasing threshold and injection-locked oscillators operating above the lasing threshold was found to be close to the quantum limit (see Chapters 7–9).

The standard quantum limits and information theory aspects in the preparation, measurement, and amplification of light have recently been reviewed by Yamamoto and Haus [19]. The semiconductor laser is a unique device that requires quantum mechanics for its performance to be correctly understood and also offers a real experimental test of the quantum theory of light.

## REFERENCES

1. K. Shimoda, H. Takahashi, and C. H. Townes, *J. Phys. Soc. Jpn.* **12**, 686 (1957).
2. M. O. Scully and W. E. Lamb, Jr., *Phys. Rev.* **159**, 208 (1967); **166**, 246 (1968); **179**, 368 (1969).
3. H. Haken, *Encyclopedia of Physics* 25/2c, S. Flugg, ed., Springer-Verlag, 1970.
4. M. Lax and W. H. Louisell, *Phys. Rev.* **185**, 568 (1969) and references cited therein.

5. W. H. Louisell, *Quantum Statistical Properties of Radiation*, Wiley, New York, 1974.

6. I. R. Senitzky, *Phys. Rev.* **A3**, 421 (1970); *Phys. Rev.* **A6**, 1175 (1972); *Phys. Rev. Lett.* **31**, 955 (1973).

7. R. Glauber, Phys. Rev. **131**, 2766 (1963).

8. H. Takahashi, in *Advances in Communication Systems*, A. V. Barakrishnan, ed., Academic Press, New York, 1965.

9. D. Stoler, *Phys. Rev.* **D4**, 1925 (1971).

10. H. P. Yuen, *Phys. Rev.* **A13**, 2226 (1976).

11. R. Jackiw, *J. Math. Phys.* **9**, 339 (1968).

12. A. Schawlow and C. H. Townes, *Phys. Rev.* **112**, 1940 (1958).

13. J. Armstrong and A. W. Smith, *Phys. Rev.* **140**, A155 (1965).

14. E. D. Hinkley and C. Freed, *Phys. Rev. Lett.* **23**, 277 (1969).

15. E. M. Purcell, *Phys. Rev.* **69**, 681 (1946).

16. K. H. Drexhage, *Progress in Optics*, Vol. 12, E. Wolf, ed., North Holland, New York, 1974.

17. H. A. Haus and J. A. Mullen, *Phys. Rev.* **128**, 2407 (1962).

18. H. Heffner, *Proc. IRE* **50**, 1604 (1962).

19. Y. Yamamoto and H. A. Haus, *Rev. Mod. Phys.* **58**, 1001 (1986).

# 2

# Line Broadening of Semiconductor Lasers

CHARLES H. HENRY

*AT & T Bell Laboratories, Murray Hill, New Jersey*

## 2.1. INTRODUCTION

### 2.1.1. Historical Background

One of the most interesting aspects of a single-mode laser is the spectral purity of the lasing mode. The linewidth and lineshape were of interest from the beginning of laser physics. In their first paper proposing the laser, Schawlow and Townes [1] predicted that the line would be Lorentzian in shape and that it would narrow inversely with power. They also gave a formula for the linewidth. The experimental demonstrations of lasers in the early 1960s stimulated a great amount of additional theoretical work on how to understand the laser and its linewidth.

It was shown by Lax [2] that the treatment by Schawlow and Townes was appropriate for description of a laser below threshold. Above threshold, amplitude fluctuations of the laser field are stabilized and this leads to a factor of 2 reduction in linewidth. The remaining field fluctuations can be thought of as phase fluctuations. Thus the line broadening results from phase noise. The detailed change in linewidth through threshold was calculated by Hempstead and Lax [3]. The extreme narrowness of gas lasers made verification of these predictions very difficult. However, Gerhardt et al. [4] succeeded by using a 500-m folded interferometer and by operating the HeNe laser at microwatt power levels. They quantitatively verified the Schawlow–Townes formula below threshold and the additional narrowing predicted by Hempstead and Lax.

The development of single-mode semiconductor lasers operating at room temperature in the early 1980s renewed activity in the study of

*Coherence, Amplification, and Quantum Effects in Semiconductor Lasers*, Edited by Yoshihisa Yamamoto.
ISBN 0-471-512494 © 1991 John Wiley & Sons, Inc.

linewidth and phase noise. Unlike the gas- or solid-state lasers, the linewidth of semiconductor lasers is broad and readily measured. Conventional semiconductor lasers have linewidth power products of order 50–100 MHz mW. The first careful measurements of the linewidth of semiconductor lasers were made by Fleming and Mooradian [5]. They found that the lineshape was Lorentzian and that the linewidth decreased inversely with facet power, in agreement with theoretical expectations, however, they also found that the linewidth was 50 times wider than predicted by the modified Schawlow–Townes formula. This startling result was explained by the author [6] as due primarily to the change in laser mode frequency with changes in gain. When a laser is above threshold, it responds to fluctuations in spontaneous emission, which change the laser intensity, with changes in gain. The changes in gain are a feedback mechanism that damps out intensity fluctuations. This is precisely the effect that reduced the Schawlow–Townes linewidth formula by 2 times when the laser is above threshold. However, associated with the gain change is a refractive index change, which alters the cavity mode frequency and leads to additional phase fluctuations. To account for this effect, the modified Schawlow–Townes formula must be corrected by a factor of $1 + \alpha^2$, where $\alpha = \Delta n' / \Delta n''$, is the ratio of changes in the real and imaginary parts of the mode refractive index. This parameter is referred to as the linewidth enhancement factor [6]. It is of order 4–7 for conventional semiconductor lasers, leading to sizable increases in linewidth. This same type of correction was given by Lax [2] for gas- or solid-state lasers, when the cavity resonance is tuned away from the center of the gain spectrum. Normally, this detuning is small and we are not aware of its experimental verification. In semiconductor lasers, however, the laser line occurs in the low-energy tail of the optical absorption edge. Detuning is quite large in this case, and large values of $\alpha$ result.

The modification of the linewidth formula by $1 + \alpha^2$ was not the only change necessary for description of the broadening of semiconductor laser lines. Diano et al [7] showed that the Lorentzian lineshape is only an approximation. The actual lineshape has additional peaks in the Lorentzian tail, separated from the line center by the relaxation oscillation frequency. Typically, these peaks are present only at the intensities of about 1% of the line center intensity. The lineshape structure originates from the same mechanism as linewidth enhancement: the feedback existing within the laser to damp out intensity fluctuations. Intensity fluctuations are removed by gain changes occurring during relaxation oscillations that bring the laser back to its steady-state intensity. These relaxation oscillations induce oscillating phase changes that result in the observed additional peak in the power spectrum [8].

While the linewidth of a conventional semiconductor lasers is too broad for many applications, it can be reduced enormously by optical feedback. Wyatt and Devlin [9] showed that linewidths as small as 10 kHz are achievable by forming an external cavity laser. In this device, originally developed by Fleming and Mooradian [10], light is coupled from the laser through an antireflection (AR)-coated facet to a long external passive section consisting of a collimating lens, a beam path about 10–20 cm long, and a wavelength-selective reflector in the form of a diffraction grating. It can be shown that the linewidth narrows as the square of the length of the passive section [11]. The external cavity laser can be tuned coarsely by rotating the diffraction grating and finely by changing the length of the external cavity with a piezoelectric transducer.

These lasers have been extremely valuable for laboratory demonstrations; however, they are bulky and the commercial advantage of the semiconductor laser in the form of a small and rugged chip is lost. To retain these features while still obtaining acceptable linewidths is the goal of much current work on semiconductor lasers. One approach is to form an integrated optic cavity. Olsson et al. [12] have shown that more compact narrow-linewidth lasers can be made by coupling the laser to a passive high-$Q$ resonator in the form of a resonant optical reflector (ROR) [13]. In this way Ackerman et al. [14] recently achieved a linewidth of 7 kHz for a hybrid semiconductor laser with a passive resonator section only 5 mm long.

Another approach to reduced linewidth is to make a monolithic laser with improved properties. It has been shown experimentally by Ogasawara et al. [15], Koch et al. [16], and Westbrook et al. [17] that multiple quantum well (MQW) lasers have $\alpha$ parameters that are about half those of similar lasers with bulk active layers. The MQW lasers also have less loss than conventional lasers. Consequently, smaller linewidths are obtainable than with conventional lasers. To take advantage of reduced loss, the length of the laser cavity is increased by making a distributed feedback (DFB) laser with a long, weakly coupled grating [18]. Another monolithic approach is to add a passive section ended with an external Bragg reflector [19]. These approaches have resulted in demonstrations of monolithic lasers with linewidths of less than 1 MHz by Ogita et al. [18] and Kano et al. [19].

### 2.1.2. Themes

In this chapter, we will review the current understanding of phase noise and line broadening of semiconductor lasers. While making contact with experiment, our main goal is to present a theoretical description of these phenomena. Much of the discussion is based on the papers of the author and collaborators. The author does not mean to imply by this that these approaches to the description of lasers are more important than those of others, but wishes to write about what he knows well.

The difficulty in treating lasers is that both the wave and particle properties of light are required for a complete description. The subject of this chapter is phase noise and phase is a wave property. However, electron–hole recombination occurring in spontaneous and stimulated emission is a particle phenomenon; one photon is generated for each electron–hole recombination. Only the wave aspects of light can be treated classically, while the quantum theory of radiation can be used to treat both particle and wave phenomena. For this reason, rigorous treatments of lasers have traditionally been by means of the quantum theory of radiation. However, a considerable price is paid for this rigor. The theories involve manipulation of time-dependent noncommuting operators, a procedure that most semiconductor laser specialists, including the author, are not comfortable with. The quantum theory of radiation was originally developed for description of modes in a lossless, closed cavity and it is not as easily applied to cavities having loss or to open resonators as the classical theory is. For these reasons, we will develop a description of the lasers in terms of classical quantities (carrier number, phase, and intensity) obeying rate equations. Fluctuation phenomena will be introduced by adding classical Langevin noise sources to these equations. The noise sources are quantum in origin, but the quantum description appears only in evaluation of the diffusion coefficients (correlation functions) of these noise sources. Our goal is thus to present a nearly self-contained classical description of quantum noise adequate to treat nearly all observed phenomena associated with the line broadening of single-mode semiconductor lasers.

The linewidth of semiconductor lasers can vary by more than three orders of magnitude in going from DFB lasers to external cavity lasers. A theory of spontaneous emission in lasers, developed by the author [11] and applied to lasers with external feedback by Kazarinov and the author [20] and to DFB lasers by Kojima and Kyuma [21], shows how this comes about. Spontaneous emission is emitted in all directions and over a large range of frequencies; however, the coupling of spontaneous emission into the lasing mode is governed by a Green's function, which is determined by the laser structure and which can vary enormously from one type of laser to another. We will apply this theory to discuss line broadening in Fabry–Perot (FP) lasers, DFB lasers, and lasers with external feedback. In discussing external feedback, we will derive a relation of chirp and linewidth established by Kazarinov et al. [13] and confirmed by Olsson et al. [22].

The material parameters on which linewidth is dependent are $\alpha$, the spontaneous emission factor $n_{sp}$, which relates the spontaneous emission rate to gain, and the waveguide loss $\gamma$. We will discuss their origin and magnitudes and indicate why $\alpha$ and $\gamma$ are reduced in quantum well lasers.

The understanding of the line broadening of semiconductor lasers is not a closed subject. The narrowest linewidths are achieved by operating the lasers at high power. The theories predict that the linewidth will narrow inversely with power. Actually, this is only approximately true.

Measured linewidths narrow to a minimum "linewidth floor" and then broaden at higher powers. This phenomena is not understood at present. We will end this chapter by reviewing the experimental studies of the linewidth floor.

### 2.1.3. Classical and Quantum Descriptions of Laser Light

Before proceeding with a classical description, we would like to discuss how the laser light is to be pictured and what are the limitations of a classical description. Consider a laser with a spatial mode $\Phi(\mathbf{x})$. The optical field in the laser (i.e., transverse electric field), which is real, can be expressed as

$$E(\mathbf{x}, t) = B[\beta(t)\Phi(\mathbf{x}) + \beta(t)^*\Phi(\mathbf{x})] \tag{2.1}$$

where $\beta(t)$ is a complex wave amplitude oscillating at the cavity frequency.

$$\beta(t) \sim e^{-i\omega_0 t} \tag{2.2}$$

and $\omega_0$ is the angular frequency of the mode. Thus, the wave amplitude undergoes harmonic motion. The energy of the mode is proportional to $\beta\beta^*$. It is convenient to choose the real factor $B$ so that this energy is given by $\hbar\omega_0 I(t)$, where

$$I(t) = \beta(t)^*\beta(t) \tag{2.3}$$

We will refer to $I(t)$ as the *lightwave intensity*.

In the classical description of radiation, the field amplitude has a definite value which can be represented as a point in the complex $\beta$ plane of Figure 2.1. This is no longer the case in the quantum theory of radiation. In the quantum-mechanical treatment of the same problem, the field is again expressed by Eq. (2.1), but this $\beta$ and $\beta^*$ replaced by the destruction and creation operators $b$ and $b^\dagger$. These operators obey the

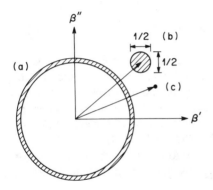

**Figure 2.1.** Distribution of complex field values $\beta$ for (a) a photon-number state; (b) a coherent state; (c) a classical field, which is represented by a point. After Walls [26].

boson commutation relations $[b, b^\dagger] = 1$ and $[b, b] = [b^\dagger, b^\dagger] = 0$. Formally, the quantum theory of radiation is identical to the quantum treatment of the harmonic oscillator. The real and imaginary parts of the field amplitude are analogous to position and momentum operators. They no longer have definite values and their distributions depend on the state of the field (wave function of the harmonic oscillator). Whether the quantum and classical descriptions of the wave field resemble one another depends on what this state is. The photon number states are energy eigenstates of the harmonic oscillator with energies $P\hbar\omega_0$, where $P$ is the photon number. A photon number state corresponds to a thin ring in Figure 2.1 and does not at all resemble classical behavior. The phase is distributed over all possible values. However, for the case of the harmonic oscillator, it is possible to construct packets, which do not have definite energy, but which have relatively well defined values of phase and amplitude. The packets have a fixed Gaussian spread and undergo harmonic motion. A good discussion of the quantum-mechanical harmonic oscillator and these Gaussian packets is given by Bohm [23]. The spread of the packet is consistent with a minimum uncertainty associated with the commutation relations of $b$ and $b^\dagger$. It corresponds to

$$\Delta\beta' = \Delta\beta'' = \tfrac{1}{2} \tag{2.4}$$

where $\beta'$ and $\beta''$ refer to the real and imaginary values of $\beta$. These packets are the "coherent states," first investigated by Glauber [24] and Sudarshan [25] as a description of radiation. The coherent state is a small circle of diameter $\tfrac{1}{2}$ in Figure 2.1. To put this size into perspective, note that a typical laser with facet power of 1 mW has $I \approx 4 \ 10^4$, so that the point is at a distance of 200 from the origin. Thus, coherent states, within the bounds of the uncertainty principle, approach classical wave motion.

We can think of laser optical field as being in a coherent state that is changing in time. It grows or diminishes in amplitude due to gain or loss in the cavity and is altered in a more chaotic manner due to spontaneous emission that adds new contributions to the field of the lasing mode. These changes can be expressed by writing $\beta$ as

$$\beta(t) = I(t)^{1/2} e^{-i\phi(t) - i\omega_0 t} \tag{2.5}$$

where $\phi(t)$ is the instantaneous phase of the packet center and $I(t)$ is the lightwave intensity of the packet center. With appropriate choice of the normalization constant $B$ in Eq. (2.1), $I(t)$ is equal to the instantaneous mean photon number $P(t)$. However, *such a quasicoherent state does not have a definite energy. An energy measurement would yield a Poisson distribution of photons with mean number $P(t) = I(t)$* [24].

This intuitive description was put on a rigorous basis for free radiation by Glauber [24], who showed that any distribution of radiation can be described as a mixture of coherent states having a probability distribution $P(\beta, \beta^*, t)$. In the quantum theory of radiation, all measurable properties of the field are calculated as expectation values of normally ordered operators, for which the $b^\dagger$'s are to the left of the $b$'s. Such operators have expectation values of zero for the vacuum state. Glauber showed that the average of any normally ordered operator $M(b, b^\dagger)$ can be calculated as a classical average of $M(\beta, \beta^*)$ weighted by $P(\beta, \beta^*, t)$. Thus, any measurable property of the radiation could be calculated as an average in the associated classical description. Application of this result immediately shows that

$$\langle I \rangle = \langle \beta^* \beta \rangle = \langle b^\dagger b \rangle = \langle P \rangle \qquad (2.6)$$

so that in general, the average lightwave intensity is equal to the average photon number.

This quantum–classical correspondence was extended to the case of the laser by Lax and Louisell [27] and Lax [28]. They showed that just as in the case of free radiation, measurable properties of laser radiation can be represented as classical average of $M(\beta, \beta^*)$ weighted by $P(\beta, \beta^*, N, t)$, which now also depends on the carrier number $N$ and that $P(\beta, \beta^*, N, t)$ satisfies a Fokker–Planck equation. They found the "drift vectors" and "diffusion coefficients" of this equation. Associated with the Fokker–Planck equation is a set of Langevin rate equations that use the same drift vectors and diffusion coefficients. The form of these equations is given in Section 2.2.2.1.

The classical description is adequate for determining all properties of the semiconductor laser, provided that either the spread in $\beta$ determined by the uncertainty principle is masked by fluctuations in $\beta$ associated with spontaneous emission or that the laser approaches a coherent state when these fluctuations are small. The latter happens when the laser is far above threshold. One exception to this description has been found by Yamamoto et al. [29]. They find no measurable discrepancy of phase noise, but they do find that photon number noise will become sub-Poissonian for a laser with high internal quantum efficiency at levels of about 10 times threshold. Machida et al. [30] have experimentally verified this prediction. In this case, the laser optical field approaches a squeezed state in which, compared with the coherent state, photon-number fluctuations are reduced at the expense of increased phase fluctuations. Such a field is not describable in the semiclassical picture of Glauber and Lax. For example, the mean-square fluctuations of lightwave intensity would become negative.

In summary, we can think of laser light as classical terms. The field is described as a classical spatial mode $B\Phi(\mathbf{x})$ with a time-dependent complex amplitude $\beta(t)$. This instantaneous classical field corresponds to a point in Figure 2.1 with amplitude $I^{1/2}$ and phase $\phi$. The quantum theory of radiation results in only a small modification of this picture in which the instantaneous field is a coherent state, which is not point, but a small circle in Figure 2.1, having uncertainties in phase and amplitude that are consistent with the Heisenberg principle. The greatest difference between the classical and quantum descriptions is that in a classical description the field has a definite energy $P(t)\hbar\omega_0 = I(t)\hbar\omega_0$, while the quantum description leads to a Poisson energy distribution for the coherent state about an average energy $I(t)\hbar\omega_0$. The classical picture, with the addition of a Poissonian distribution of photon number, appears to be adequate for describing of almost all dynamic processes of semiconductor lasers, such as relaxation oscillations, mode partition noise, phase noise, line broadening, and coherent detection. One important exception is the observation of sub-Poissonian statistics of photon number fluctuations, specifically, squeezed states generated by running an efficient AlGaAs laser far above threshold with a constant current source [30].

## 2.2.  FLUCTUATIONS IN SEMICONDUCTOR LASERS

### 2.2.1.  Langevin Rate Equations for the Field Amplitude and Carrier Number

The rate equation for the wave field amplitude $\beta$ is derived in Section 2.3. Here we merely write down the equation and give an intuitive justification for the various terms:

$$\dot{\beta} = \left[ -i\omega_0 + \frac{\Delta G}{2}(1 - i\alpha) \right]\beta + F_\beta(t) \qquad (2.7)$$

The first term on the right-hand side ensures that at threshold, $\beta$ oscillates at the mode angular frequency $\omega_0$. The term $\Delta G/2$ is the net gain (per second), the difference of gain and loss in the cavity. In the absence of other terms, $I(t)$ will increase as $e^{\Delta G t}$ due to this term. In a laser with a uniform active layer, we will show that

$$\Delta G = G - \Gamma = \Delta g v_g \qquad (2.8)$$

where $\Delta G$ is the difference of gain $G = g v_g$ and loss $\Gamma$, where $g$ is the gain per centimeter and $v_g$ is the group velocity. The term with $\alpha$ reflects the fact that as the gain changes the mode angular frequency changes by

$$\Delta\omega_0 = \tfrac{1}{2}\alpha\,\Delta G \qquad (2.9)$$

The last term accounts for random changes in $\beta$ due to spontaneous emission. The role of spontaneous emission during a short time $\Delta t$ is illustrated in Figure 2.2. During $\Delta t$, spontaneous emission adds additional

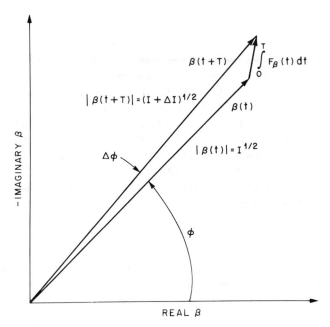

**Figure 2.2.** Phasor diagram of the complex optical field amplitude $\beta$ showing the change due to spontaneous emission in a small time $\Delta T$.

contributions to $\beta$ that change both its phase and its amplitude. While the changes in $\beta$ that are due to $\Delta G$ are regular, the changes caused by spontaneous emission are chaotic. The changes are simulated by adding a random "Langevin force" $F_\beta$ to the rate equation for $\beta$. Like $\beta$, $F_\beta$ is a complex quantity.

The Langevin forces are useful in specifying the different contributions to noise that occur in a system. The semiconductor laser is completely specified by adding an equation for the carrier number $N$:

$$\dot{N} = C - GI - S + F_N(t) \qquad (2.10)$$

where $C$ is the current in carriers per second, $GI$ is the rate of stimulated emission given by the product of gain and lightwave intensity, and $S$ combines both the rates of spontaneous emission into all modes and nonradiative recombination. We add a Langevin force $F_N(t)$ to take into account fluctuations in these rates. For example, the rate of spontaneous emission is not a smooth process; the minority carriers recombine and are generated one at a time resulting in Poisson-distributed generation and recombination events. The rates of $GI$, $C$, and $S$ are to be thought of as average rates, and the fluctuations in these rates are incorporated in contributions to $F_N(t)$.

### 2.2.2.  Langevin Methods

In this section we collect a number of properties of Langevin rate equations that will be useful in our study of semiconductor lasers. They will be presented without proof. For a general discussion of Langevin rate equations, the reviews of Lax [31, 32] are recommended. The author has learned much of the little he knows about fluctuation theory from reading the papers of M. Lax and from personal discussions with him. Langevin rate equations describe the fluctuations of a system. The various sources of noise are added as random Langevin forces. The equations are solved to determine how the system variables fluctuate. By solving the equations many times, one can find a distribution of the system variables ($\beta$ and $N$ in the preceding example). This has been done by computer for the semiconductor laser by Miller [33]. Often we are interested in only the mean squares of the variables and these can be found analytically.

#### 2.2.2.1.  Diffusion Coefficients. Consider a system with several variables $a_i$ obeying a set of Langevin rate equations

$$\dot{a}_i = A_i + F_i(t) \tag{2.11}$$

with drift vectors $A_i$ and forces $F_i(t)$. We will assume that the average of $F_i$ is zero:

$$\langle F_i(t) \rangle = 0 \tag{2.12}$$

If $F_i$ had a nonzero average, it would be included in the drift vector $A_i$. We will assume that the system is "Markoffian," that is, that the random forces have no memory and hence correlations of these forces are delta functions in time:

$$\langle F_i(t) F_j(t') \rangle = 2D_{ij}\delta(t - t') \tag{2.13}$$

where the $D_{ij}$ are the "diffusion coefficients." The Markoffian assumption is reasonable for a semiconductor laser, where steady emission processes are thought to last no longer than a carrier scattering time, about $10^{-13}$ s, a time that is very short compared to any dynamical process of the laser.

The origin of the names "drift" and "diffusion" have to do with the role these coefficients play in the Fokker–Planck equation. When the system noise sources are Markoffian- and Gaussian-distributed, the probability function $P(\mathbf{a}, t)$, determining the distributions of variables obeys a Fokker–Planck equation:

$$\frac{\partial P}{\partial t} = -\sum_i \frac{\partial}{\partial a_i}(A_i P) + \sum_i \sum_i \frac{\partial}{\partial a_i}\frac{\partial}{\partial a_j}(D_{ij}P) \tag{2.14}$$

The Fokker–Planck equation is a form of diffusion equation. The diffusion coefficients $D_{ij}$ promote a spread of the probability distribution $P(\mathbf{a}, t)$, while the drift vectors $A_i$ counteract the spreading and bring the system variables back to their steady-state values. A derivation of the Fokker–Planck equation from the Langevin rate equations with Gaussian Langevin forces is given in Ref. 32, Section 5D.

**2.2.2.2. Determination of Diffusion Coefficients.** The diffusion coefficients must be determined in order to specify the noise sources. This can be done by a variety of methods. The diffusion coefficients are symmetric, so for the laser, which is described by three variables, six diffusion coefficients must be found.

For generation and recombination processes, where the variables change by integers, the diffusion coefficients can be written down by inspection. In this case the drift vectors can be written as a difference of generation and recombination rates, total rate in minus total rate out:

$$A_i = G_i - R_i \tag{2.15}$$

and the diffusion coefficients are given by

$$2D_{ii} = G_i + R_i \tag{2.16}$$

$$2D_{ij} = -(G_{ij} + R_{ij}) \tag{2.17}$$

where $R_{ij}$ and $G_{ij}$ are the contributions to these rates for which the variables $a_i$ and $a_j$ both change. These relations are easily established through a derivation of the Fokker–Planck equation for generation and recombination processes [32, Section 5B].

The diffusion coefficients can also be found if there is independent knowledge about averages of the fluctuations of the variables. It can be shown by calculating averages of the Langevin rate equation [Eq. (2.11)] and using the Markoffian property of the Langevin forces [Eq. (2.13)] [34; 32, Section 6D] that in the limit $\Delta t \to 0$

$$2D_{ij} = \frac{\langle \Delta a_i \, \Delta a_j \rangle}{\Delta t} \tag{2.18}$$

It is often useful to transform the Langevin equations from one set of variables to another. In the transformation $a_i' = a_i'(\mathbf{a}, t)$, Lax [32, Section 6F] has shown that the new drift vectors and diffusion coefficients are

$$A_i' = \frac{\partial a_i'}{\partial t} + \sum_k \frac{\partial a_i'}{\partial a_k} A_k + \sum_k \sum_l \frac{\partial^2 a_i'}{\partial a_k \, \partial a_l} D_{kl} \tag{2.19}$$

$$D_{ij}' = \sum_{kl} \frac{\partial a_i'}{\partial a_k} \frac{\partial a_j'}{\partial a_l} D_{kl} \tag{2.20}$$

We will make use of this change of variables to express the field amplitude in terms of intensity and phase.

**2.2.2.3. Correlations in Time and Frequency.** The Langevin rate equations may not be linear. Often the nonlinearities are dealt with by solving the nonlinear equations to determine the steady-state operating point and then approximating the equations as linear in calculating the fluctuations about the steady state. For such a "quasilinear system," the Langevin approach is particularly useful in computing the spectrum of fluctuations [31]. Consider variables defined as deviations from the steady state with average values of zero. In steady-state operation, the correlation of two variables $a_i(t + \tau)$ and $a_j(t)$ does not depend on $t$:

$$\phi_{ij}(\tau) = \left\langle a_i(t + \tau)a_j(t) \right\rangle = \left\langle a_i(\tau)a_j(0) \right\rangle \qquad (2.21)$$

This is known as "stationarity." Furthermore, it follows from time reversal of the microscopic equations the system and stationarity that $\phi_{ij}(\tau)$ is even in $\tau$ and symmetric in $i$ and $j$ [35, Section 119]

$$\phi_{ij}(\tau) = \phi_{ji}(\tau) = \phi_{ij}(-\tau) \qquad (2.22)$$

Relations (2.21) and (2.22) greatly simplify the correlations among the frequency components of the variables $a_i(\Omega)$, that are defined by Fourier transform

$$a_i(\Omega) = (2\pi)^{-1/2} \int_{-\infty}^{\infty} a_i(t)e^{i\Omega t} \, d\tau \qquad (2.23)$$

Using the last three equations [Eqs. (2.21)–(2.23)], we find that when $\Omega$ and $\Omega'$ are both positive (or both negative)

$$\left\langle a_i(\Omega)a_j(\Omega')^* \right\rangle = (a_i a_j)_\Omega \delta(\Omega - \Omega') \qquad (2.24)$$

where

$$(a_i a_j)_\Omega = \int_{-\infty}^{\infty} \phi_{ij}(\tau)e^{i\Omega\tau} \, d\tau = 2\int_0^\infty \phi_{ij}(\tau)\cos \Omega\tau \, d\tau \qquad (2.25)$$

and

$$\left\langle a_i(\Omega)a_j(\Omega') \right\rangle = 0 \qquad (2.26)$$

Equation (2.25) is referred to as the *Wiener–Khinchin theorem*. As $\phi_{ij}(\tau)$ is even in $\tau$, it follows that $(a_i a_j)_\Omega$ is real and even in $\Omega$. Taking the inverse transform of Eq. (2.25) and setting $\tau = 0$, we find that the average $\left\langle a_i a_j \right\rangle$

is an integral of a "spectral density" over frequency

$$\langle a_i a_j \rangle = \int_{-\infty}^{\infty} (a_i a_j)_\Omega \, df = 2 \int_0^\infty (a_i a_j)_\Omega \, df \qquad (2.27)$$

where the spectral density of $a_i a_j$ is $(a_i a_j)_\Omega$ [35, Section 118].

The preceding relations hold also for the Langevin forces. If we substitute $F_i(\Omega)$ and $F_j(\Omega')^*$ into Eq. (2.24) and then evaluate the spectral density [Eq. (2.25)] using the delta function correlation for Langevin forces [Eq. (2.13)], we find

$$\langle F_i(\Omega) F_j(\Omega')^* \rangle = 2 D_{ij} \delta(\Omega - \Omega') \qquad (2.28)$$

Thus, the same diffusion coefficients describe correlations among both the Langevin forces and the Fourier components of these forces.

**2.2.2.4. Gaussian Random Variables.** The Langevin method is convenient for the calculations of correlation functions among random variables and spectral densities. However, to determine the power spectrum of a laser, the complete distribution of the laser phase change $\Delta\phi(t)$ is needed. Fortunately, we can argue that $\Delta\phi(t)$ is a "Gaussian random variable," one with a Gaussian probability distribution.

Let $x$ be a Gaussian random variable. Its probability distribution is

$$P(x) = (2\pi \langle x^2 \rangle)^{-1/2} \exp\left(-\frac{x^2}{2\langle x^2 \rangle}\right) \qquad (2.29)$$

The entire distribution is determined by the second moment $\langle x^2 \rangle$. It can be shown by direct calculation of the integral $\int_{-\infty}^{\infty} P(x) e^{ix} \, dx$ that

$$\langle e^{ix} \rangle = e^{-1/2 \langle x^2 \rangle} \qquad (2.30)$$

If there are two independent Gaussian random variables $x$ and $y$, with $\langle xy \rangle = 0$, then it can be shown that the sum $s = x + y$ is also a Gaussian random variable. This can be proved by transforming the joint probability distribution $P(x)P(y)$ to new variables that are the sum and difference of $x$ and $y$ and then eliminating the difference variable by integration.

An extension of this argument is that a linear combination of independent Gaussian random variables is a Gaussian random variable. In the quasilinear approximation with Gaussian Langevin forces $F_i$, the solutions $a_i$ are such linear combinations of $F_i(t)$. This is a good approximation for a semiconductor laser. The noise results from electron–hole recombination and spontaneous emission. The recombination events are Poisson-distributed, but the recombination rates are so large ($10^{16}$ s$^{-1}$ mA$^{-1}$), that in any time interval of interest, the number of recombinations is sufficiently large that the Poisson distribution is close to a Gaussian. We will see in Section 2.5 that spontaneous emission results from a sum of contributions from throughout the laser. Such fluctuations, resulting from a linear combination of many independent processes, tend to be Gaussian-distributed according to the central limit theorem [36].

### 2.2.3.  Diffusion Coefficients of a Semiconductor Laser

The conventional derivations of the Langevin rate equations begin with a microscopic model of the laser and treat both the radiation and the carriers quantum-mechanically. This has been done both by Haag [37] and Cohen [38] for the semiconductor laser and by Lax for lasers of isolated atoms [28]. The latter derivation is particularly attractive because a transformation to the classical description is made. All the derivations are long and involved. Here, we will try to shorten this procedure by going immediately to the classical macroscopic rate equations, Eqs. (2.7) and (2.10). It remains only to determine the diffusion coefficients for the Langevin forces in these equations. We do this by a variety of general arguments. The results are justified by their agreement with the work of Lax [28].

We will begin by determining the diffusion coefficients associated with the two wave field amplitudes $\beta$ and $\beta^*$. According to Eq. (2.18), $D_{\beta\beta}$ will be proportional to the average of the product of $\Delta\beta\,\Delta\beta$ occurring in a short time. Figure 2.2 illustrates this change. It is clear that $\Delta\beta$ can take any angle and will average to zero. Therefore

$$2D_{\beta\beta} = 2D_{\beta^*\beta^*} = 0 \qquad (2.31)$$

Equation (2.18) also implies that $D_{\beta\beta^*}$ is real. We will write it as

$$2D_{\beta\beta^*} = R \qquad (2.32)$$

It will be shown in Eq. (2.33) that $R$ is the average rate of spontaneous emission into the lasing mode.

In treating the laser above threshold, it is usually more convenient to express the wave field in terms of phase $\phi$ and lightwave intensity $I$, two real variables, using Eq. (2.5). We can use Eq. (2.19) to transform the Langevin equations for $\beta$ and $\beta^*$ to the new variables, giving

$$\dot{I} = \Delta G I + R + F_I(t) \qquad (2.33)$$

$$\dot{\phi} = \frac{\alpha\,\Delta G}{2} + F_\phi(t) \qquad (2.34)$$

We see that after this transformation, $R$ appears as the spontaneous emission rate. The diffusion coefficients are found by applying transformation (2.20) and using Eqs. (2.31) and (2.32). The result is

$$2D_{II} = 2RI, \qquad 2D_{I\phi} = 0, \qquad 2D_{\phi\phi} = \frac{R}{2I} \qquad (2.35)$$

These diffusion coefficients can be understood by noting that the force causing fluctuations in $I$ will be given by the product of $2I^{1/2}$ and that the component of $F_\beta$ parallel to $I$. Similarly, the force causing the fluctuations in $\phi$ will be given by the component of $F_\beta$ perpendicular to $\beta$ divided by $I^{1/2}$. The lack of correlation of these two components of $F_\beta$ leads to $D_{I\phi} = 0$. Only these diffusion coefficients are needed to derive the linewidth formula in Section 2.3.

The fluctuations in carrier number result from the processes of generation and recombination. The diffusion coefficient $D_{NN}$ can be found from the sum of the rates in and the rates out in the equation for $N$ [Eq. (2.10)]. According to the quantum theory of radiation, the downward rate of optical transitions is $R(P + 1)$, so that if the rate of spontaneous emission is $R$, the rate of stimulated emission is $RP = RI$. The upward rate representing absorption of radiation is $AI$. The gain $G$ is the difference between the rates of emission and absorption

$$G = R - A \qquad (2.36)$$

Another contribution to the rate in (2.10) is the current $C$. Yamamoto and Machida [39] argue that current is not an independently fluctuating quantity, but one that is determined by the source voltage, junction voltage, and series resistance. Following them, we will not consider this term as contributing to $2D_{NN}$. The sum of rates in and rates out then determines $2D_{NN}$ according to (2.16):

$$2D_{NN} = (R + A)I + S \qquad (2.37)$$

Since the spontaneous and stimulated emission processes do not depend on the phase of the wave field, we will assume

$$2D_{N\phi} = 0 \qquad (2.38)$$

We can determine $2D_{NI}$ from the requirement that energy is conserved in lightwave intensity fluctuations. If during a short time, $I$ increases by $\Delta I$, the field energy will have increased by $\Delta I \hbar \omega$. This energy comes from downward transitions of the carriers between levels separated by $\hbar \omega$. Thus, the minority carrier number changes by $\Delta N = -\Delta I$ and the correlation of these two processes results in $\langle \Delta N \Delta I \rangle = -\langle \Delta I^2 \rangle$. From the

**Table 2.1   Diffusion Coefficients of a Semiconductor Laser**

$$2D_{\beta\beta^*} = R$$
$$2D_{\beta\beta} = 2D_{\beta^*\beta^*} = 0$$
$$2D_{II} = 2RI$$
$$2D_{IN} = -2RI$$
$$2D_{NN} = S + (2R - G)I$$
$$2D_{\phi\phi} = \frac{R}{2I}$$
$$2D_{N\phi} = 2D_{I\phi} = 0$$

relation between diffusion coefficients and correlation functions [Eq. (2.18)], we conclude that

$$2D_{NI} = -2D_{II} = -2RI \tag{2.39}$$

This relation can also be established by relating the diffusion coefficients of lightwave intensity and photon number [40, 41]. The diffusion coefficients are summarized in Table 2.1.

## 2.3.   SPECTRA OF SINGLE-MODE LASERS

### 2.3.1.   Measurement of the Laser Power Spectrum

Consider a real random variable $E(t)$ with $\langle E(t) \rangle = 0$. We have in mind the wave field of a laser. We can write $E(t)$ as a Fourier integral over frequency components $E_\omega$:

$$E(t) = (2\pi)^{-1/2} \int_{-\infty}^{\infty} E_\omega e^{-i\omega t} \, d\omega = (2\pi)^{-1/2} \int_{0}^{\infty} E_\omega e^{-i\omega t} \, d\omega + \text{c.c.} \tag{2.40}$$

where "c.c." represents the complex conjugate. The correlation function

$$\langle E_\omega E_{\omega'}^* \rangle = (E^2)_\omega \delta(\omega - \omega') \tag{2.41}$$

where $(E^2)_\omega$ is the spectral density [Eq. (2.24)].

The general procedure for measuring $(E^2)_\omega$ is described by Lax [32, Section 4B]. The field $E(t)$ is passed through a narrowband filter and then into a square-law detector. This is followed by time-averaging. This procedure is diagramed in Figure 2.3a. The filter response $K(\omega)$ is shown in

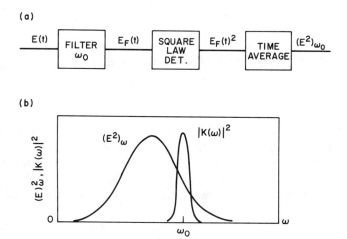

**Figure 2.3.** Diagram of measurement of the optical power spectrum: (a) schematic of the measurement process; (b) lineshape of the filter and the laser power spectrum.

Figure 2.3b. After the signal passes through the filter it is given by

$$E_F(t) = (2\pi)^{-1/2} \int_0^\infty K(\omega) E_\omega e^{-i\omega t} \, d\omega + \text{c.c.} \qquad (2.42)$$

provided that the filtering procedure is carried out for a time long compared to the filter response time $t_f \approx 1/\Delta f$, where $\Delta f$ is the bandwidth of the filter. The time average of the output of the square-law detector is given by

$$\overline{E_F(t)^2} = \frac{2}{2\pi} \int_0^\infty d\omega \int_0^\infty d\omega' K(\omega) K(\omega')^* \overline{E_\omega E_{\omega'}^*} e^{-i(\omega - \omega')t} \qquad (2.43)$$

where terms at twice the optical frequency have been dropped.

To simplify this equation, we need to replace the time average $\overline{E_\omega E_{\omega'}}$ by an ensemble average $\langle E_\omega E_{\omega'} \rangle$. The basic assumption of statistical physics is that sufficiently long time averages are equal to ensemble averages [35, Section 1]. The characteristic time for fluctuations of $E(t)$ is the correlation time $t_{corr}$, the decay time of $\langle E(t)E(0) \rangle$. We expect that for time averages long compared to $t_{corr}$, the two averages are equal. The time $t_{corr}$ is approximately given by the reciprocal of the width of the spectral density. It is clear from Figure 2.3b, that in order to carry out a high resolution measurement of the spectral density, the time of measurement must be long compared to the filter response time and hence much longer than $t_{corr}$, so that the two averages should be equal. Making a change to the ensemble average and using Eq. (2.41), Eq. (2.43) becomes

$$\overline{E_F(t)^2} = 2\int_0^\infty (E^2)_\omega |K(\omega)|^2 \, df \approx 2(E^2)_{\omega_0}\int_0^\infty |K(\omega)|^2 \, df \quad (2.44)$$

The last expression follows because $|K(\omega)|^2$ is a function peaked at $\omega_0$ and spectrally narrow compared to $(E^2)_\omega$ (see Figure 2.3$b$).

To measure the spectral density (or power spectrum) of a semiconductor laser, a tunable optical filter is required. One means of doing this is to use a scanning Fabry–Perot interferometer [5]. However, the preferred method, especially for narrow-linewidth lasers, is to shift $E(t)$ from optical frequencies to radio frequencies and then use a spectrum analyzer as a tunable filter.

The down-shifting of the optical signal is the same procedure as in heterodyne detection. The laser field $E(t)$ is added to the field of a reference laser $E_R(t)$ and the sum of the fields is detected. The signals are normally passed through single-mode optical fibers and added together by means of a 3-dB coupler. The frequency difference of the two lasers must be within the range of the spectrum analyzer. If we write the two laser fields as

$$E_i(t) = I_i(t)^{1/2} e^{-i\phi_i(t) - i\omega_i t} + \text{c.c.}, \qquad i = 1, 2 \quad (2.45)$$

where $i = 1$ and $i = 2$ represent the signal and reference lasers, respectively. Then the detected signal at the difference frequency will be

$$S_{12}(t) \sim I_1(t)^{1/2} I_2(t)^{1/2} e^{-i\phi_1(t) + i\phi_2(t) - i(\omega_1 - \omega_2)t} + \text{c.c.} \quad (2.46)$$

If the reference laser is monochromatic, $\Delta I_2 = \Delta\phi_2 = 0$, $S_{12}(t)$ will be an exact down-shifted replica of $E_1(t)$. For lasers above threshold, fluctuations in intensity are small and can be neglected. Then the fluctuations in $S(t)$ are due to the sum of the phase fluctuations of the two lasers.

The method of down-shifting is also employed in the spectrum analyzer to simulate a tunable filter. The signal $S_{12}(t)$ is multiplied by a monochromatic radio-frequency (RF) signal that is slowly swept in frequency. The resulting signal is then passed through a narrowband amplifier at a fixed frequency and the squared envelope of the filtered signal is displayed. The output of the spectrum analyzer is the spectral density of $S_{12}(t)$, appropriately down-shifted in frequency.

A popular way of measuring the power spectrum of lasers is the self-heterodyne method [42] illustrated in Figure 2.4. In this case $E_1$ and $E_2$ are signals from the same laser. One of the signals is shifted in frequency with an acoustooptic frequency shifter and then delayed by passing it through a coil of optical fiber. The two signals are then added

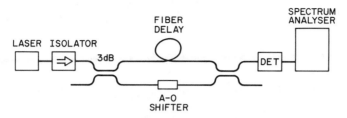

**Figure 2.4.** Schematic diagram of the self-heterodyne method.

together and detected. If the delay is long compared to the laser correlation time, $\Delta\phi_1(t)$ and $\Delta\phi_2(t)$ in Eq. (2.46) are independent. In this case $\langle\Delta\phi_1(t)^2\rangle = \langle\Delta\phi_2(t)^2\rangle$ and this results in the power spectrum of $S(t)$ being twice as broad for a single laser [see Eq. (2.59), below].

Linewidth measurements require that the laser be well isolated from reflections (Figure 2.4). About 60 dB of isolation is needed, and intentional misalignments are made to ensure that surface reflections of the lens and isolator do not return to the laser. The effect of distant reflections is to set up an external cavity coupled to the laser (see Section 2.5.4.3 and Figure 2.21, later). The laser can then oscillate on one of a number of closely spaced external cavity modes. The feedback can narrow the laser line. On the other hand, vibrations and air currents will modulate the mode frequencies and cause mode hopping, leading to broadening and structuring of the laser power spectrum.

### 2.3.2. Power Spectrum of a Laser below Threshold

The optical wave field emitted from a laser is given by

$$E(t) \sim \beta(t)e^{-i\omega_0 t} + \text{c.c.} = (2\pi)^{-1/2}\int_0^\infty \beta_\omega e^{-i\omega t}\,d\omega + \text{c.c.} \quad (2.47)$$

Thus $E_\omega$ is proportional to $\beta_\omega$. The spectral density [Eq. (2.41)] may be calculated directly from evaluation of $\langle\beta_\omega\beta_{\omega'}^*\rangle$. For a laser below threshold, $-\Delta G$ is large and fluctuations in $\Delta G$ due to changes in carrier number $N$ or lightwave intensity $I$ can be neglected. In this case, $\beta(t)$ is completely determined by the Langevin rate equation [Eq. (2.7)]. We can calculate $\beta_\omega$ by transforming this equation to frequency components

$$\beta_\omega = \frac{F_\beta(\omega)}{i[\omega_0 + (\alpha\,\Delta G/2) - \omega] + (\Delta G/2)} \quad (2.48)$$

Substitution into the equation for the spectral density [Eq. (2.41)] and using

$$\langle F_\beta(\omega) F_\beta(\omega')^* \rangle = R\delta(\omega - \omega') \tag{2.49}$$

which follows from Eqs. (2.28) and (2.32), we find

$$(E^2)_\omega \sim \frac{R}{[\omega - (\alpha/2)\,\Delta G - \omega_0]^2 + (\Delta G/2)^2} \tag{2.50}$$

The power spectrum is Lorentzian in shape. For steady-state operation, it follows from the rate equation for $I$ [Eq. (2.33)] that $\Delta G = -(R/I)$. Consequently the linewidth of $(E^2)_\omega$ is given by

$$\Delta\nu = \frac{R}{2\pi I} \tag{2.51}$$

This equation is the Schawlow–Townes linewidth formula. Since the lightwave intensity $I$ is proportional to facet power $P_0$, $\Delta\nu$ narrows inversely with $P_0$.


### 2.3.3. Power Spectrum of a Laser above Threshold in the Low-Frequency Approximation

In this section, we will calculate the laser linewidth and lineshape in the low-frequency approximation and neglecting gain saturation. A more exact calculation will be made in Section 2.4. Above threshold, we can no longer neglect the fluctuations in $\Delta G$. As the laser current $C$ is raised above threshold, the gain $G$ increases to a steady-state value where $G \approx \Gamma$, and the lightwave intensity $I$ increases to a steady-state value for which the rate of stimulated emission $GI$ equals the current above threshold $C - S$. In describing the fluctuations in intensity about the average value of $I$, it is more convenient to describe the field in variables $I$, $\phi$ by means of the Langevin rate equations, Eqs. (2.33), (2.34), and (2.10) for $I$, $\phi$ and $N$.

Solutions of the rate equations for $I$ and $N$ are characterized by rapidly damped relaxation oscillations with a decay rate of order $10^9$ s$^{-1}$. The frequency components of the Langevin forces that are responsible for line broadening are lower than this rate. The laser tends to adiabatically follow perturbations occurring at rates small compared to the relaxation oscillation decay rate. The adiabatic response can be calculated by neglecting the time derivatives $\dot{I}$ and $\dot{N}$ in the rate equations for $I$ and $N$, Eqs. (2.33) and (2.10). Neglecting $\dot{I}$ and dividing the rate equation for $\dot{I}$ by $I$, we have

$$\Delta G(t) = -\frac{R}{I} - \frac{F_I(t)}{I} \tag{2.52}$$

The laser gain adjusts itself in this manner to maintain nearly steady-state intensity. We will take $I$ in Eq. (2.52) to the equal to the steady-state intensity.

If we substitute $\Delta G(t)$ into the Langevin equation for [Eq. $\phi$ (2.34)], we have

$$\dot{\phi} = -\frac{R}{I} - \frac{\alpha F_I(t)}{2I} + F_\phi(t) \qquad (2.53)$$

The term $R/I$ is a constant and results only in a small frequency shift. We will neglect this term. The phase shift in time $t$ is

$$\Delta\phi(t) = \int_0^t F_\phi(t)\, dt - \frac{\alpha}{2I}\int_0^t F_I(t)\, dt \qquad (2.54)$$

We can use this equation to compute the mean-square phase change by squaring this equation, averaging, and noting that the Langevin forces are delta-function-correlated [Eq. (2.13)] with diffusion coefficients given by Eq. (2.35) we find,

$$\langle \Delta\phi^2(t) \rangle = \frac{R}{2I}(1 + \alpha^2)t \qquad (2.55)$$

The linear increase in $\langle\langle\Delta\phi^2(t)\rangle$ with $t$ can be thought of as phase diffusion. The phase change, which is driven by Langevin forces [see Eq. (2.53)] with no restoring force (drift vector), undergoes Brownian motion, causing $\langle\Delta\phi^2\rangle$ to increase linearly in time.

We can use this result to compute the spectral density $(E^2)_\omega$. The optical field of the laser is given by Eq. (2.45) as

$$E(t) \sim e^{-i\phi(t)-i\omega_0 t} + \text{c.c.}$$

where the time dependence of $I(t)$ is neglected. Then $(E^2)_\omega$ is given by the Wiener–Khinchin theorem [Eq. (2.25)]

$$(E^2)_\omega \sim \int_0^\infty \langle E(t)E(0)\rangle \cos\omega t\, dt \qquad (2.56)$$

where

$$\langle E(t)E(0)\rangle \sim \langle e^{-i\Delta\phi(t)}\rangle e^{-i\omega_0 t} + \text{c.c.} \qquad (2.57)$$

We have already argued in Section 2.2.2.4 that the Langevin forces $F_\beta(t)$ are Gaussian random variables. If $I$ is regarded as a constant, then $F_I(t)$ and $F_\phi(t)$ are proportional to the components of $F_\beta$ parallel and perpendicular to the field and are also Gaussian random variables. Equation (2.54) shows that $\Delta\phi(t)$ is sum of these forces and therefore a Gaussian variable. If $\Delta\phi(t)$ is a Gaussian random variable, it follows that [Eq. (2.30)] $\langle e^{-i\Delta\phi} \rangle = e^{-1/2\langle\Delta\phi^2\rangle}$ and

$$\langle E(t)E(0) \rangle \sim e^{-1/2\langle\Delta\phi^2\rangle} \cos \omega_0 t \tag{2.58}$$

We see that since $\langle \Delta\phi \rangle^2 \sim t$ [Eq. (2.55)], the correlation function $\langle E(t)E(0) \rangle$ exponentially decays. The decay results in the Lorentzian broadening of the laser line in the same way that an atomic emission lines is lifetime-broadened.

The spectral density is given by Eqs. (2.56) and (2.58) as

$$(E^2)_\omega \sim \int_0^\infty e^{-1/2\langle\Delta\phi(t)^2\rangle} \cos(\omega - \omega_0)t \, dt \tag{2.59}$$

where $\langle \Delta\phi^2(t) \rangle$ is given by Eq. (2.55). We have neglected the term at the sum of the optical frequencies in writing Eq. (2.59). This integral is readily evaluated; resulting in a Lorentzian centered at $\omega_0$ having a linewidth given by

$$\Delta\nu = \frac{R(1 + \alpha^2)}{4\pi I} \tag{2.60}$$

This formula differs in two ways compared to the linewidth formula for a laser below threshold [Eq. (2.51)]. The stabilization of amplitude fluctuations reduced $\Delta\nu$ twofold. However, the gain fluctuations, brought about by this stabilization process, induce additional phase fluctuations, which increase the linewidth by $(1 + \alpha^2)$.

This process is illustrated in Figure 2.5. Suppose initially $\beta$ has the steady-state amplitude $I^{1/2}$ and then a spontaneous emission event alters $\beta$ by $\Delta\beta$. The phase changes instantly by $\Delta\phi_1$. This will be followed by another change $\Delta\phi_2$ that is brought about by the relaxation oscillations that bring $I$ back to the steady-state intensity. During the return to steady state, there will be a gain change

$$\Delta g(t) = -2\frac{\omega}{c} \Delta n''(t) \tag{2.61}$$

where $\Delta n''$ is the imaginary change in refractive index. Accompanying this change is a change in the real refractive index given by

$$\Delta n' = \alpha \, \Delta n'' \tag{2.62}$$

We can take Eq. (2.62) as the definition of $\alpha$. The propagation constant

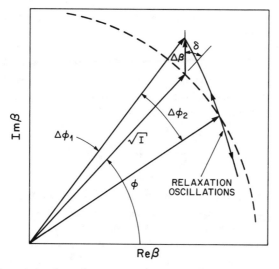

**Figure 2.5.** Changes in phase due to spontaneous emission: $\Delta\phi_1$ occurs instantly; $\Delta\phi_2$ occurs after relaxation oscillations return $I$ to the steady-state value.

$k = (\omega/c)n'$ has a constant value for each mode of the laser, so when $n'$ changes, the mode frequency $\omega$ will also change by

$$\Delta\omega = -\frac{\omega}{c}\,\Delta n' v_{\mathrm{g}} \tag{2.63}$$

The delayed phase change $\Delta\phi_2$ is given by $\int \Delta\omega\, dt$. Using the last three equations, we find

$$\Delta\phi_2 = \frac{\alpha}{2}\int \Delta G\, dt \tag{2.64}$$

This leads to the second term in Eq. (2.54) where $\Delta G$ is given by Eq. (2.52).

### 2.3.4. Evaluation of $R$ and $I$

To make use of linewidth formulas, Eqs. (2.51) and (2.60), we need to express $R$ and $I$ in more readily measured parameters. We can relate $I$ to the total output power from the two ends $2P_0$ and to the current above threshold $C - C_{\mathrm{th}}$ by relations that follow from conservation of energy:

$$\frac{2P_0}{h\nu} = (g - \gamma)v_{\mathrm{g}}I \tag{2.65}$$

where $g$ and $\gamma$ are the gain and loss per unit length in the laser, and

$$C - C_{\text{th}} = g v_g I \tag{2.66}$$

where $C - C_{\text{th}}$ is the electron current above threshold, excluding leakage around the active stripe. These general relations hold for both DFB and FP lasers. For FP lasers, we can relate $g - \gamma$ to the facet reflectivity by

$$g - \gamma = \frac{1}{L} \ln \frac{1}{r_1 r_2} \equiv \gamma_m \tag{2.67}$$

where $r_1$ and $r_2$ are the field reflectivities of the two ends.

We also need to relate the spontaneous emission rate $R$ to other laser parameters. In the absence of cavity losses ($\gamma = \gamma_m = 0$, $r_1 r_2 = 1$), the carrier and photon populations will come into equilibrium. This equilibrium is characterized by no change in free energy when photons are absorbed or generated. In this case, the chemical potential of the photons is equal to the difference in the chemical potentials (Fermi levels) of the conduction band and valence band electrons. This difference is $eV$, where $V$ is the bias voltage at the junction. The equilibrium photon number $P_e$ for an ideal Bose gas is given by Landau and Lifshitz [35, Section 55]

$$P_e = \left[ \exp\left( \frac{\hbar \omega_0 - eV}{kT} \right) - 1 \right]^{-1} \tag{2.68}$$

In the steady state, the rate of spontaneous and stimulated emission must equal the rate of the absorption

$$R(P_e + 1) = AP_e \tag{2.69}$$

Combining the last two equations shows that

$$R = A \exp\left( \frac{eV - \hbar \omega_0}{kT} \right) \tag{2.70}$$

and that $R$ is related to the gain per second is $G = R - A$ by

$$R = G n_{\text{sp}} \tag{2.71}$$

where

$$n_{\text{sp}} = \left[ 1 - \exp\left( \frac{\hbar \omega_0 - eV}{kT} \right) \right]^{-1} \tag{2.72}$$

Note that as $\hbar\omega_0$ goes from less than $eV$ to more than $eV$, $G$ changes sign, $n_{sp}$ goes to infinity and changes sign, but $R$ [Eq. (2.70)] varies smoothly and remains positive. Relation (2.71) states that the rate of spontaneous emission into a closed cavity depends solely on the gain $g$ and parameters $v_g$, $eV$, $kT$, and $\hbar\omega_0$.

If we assume that the relation of $R$ and $G$ derived for a lossless cavity is approximately correct, then by means of Eq. (2.65) relating $I$ and $P_0$ and Eq. (2.71) relating R and $gv_g$, the formula for the linewidth of a laser above threshold [Eq. (2.60)] becomes

$$\Delta\nu = \frac{g(g-\gamma)v_g^2 n_{sp}\hbar\nu(1+\alpha^2)}{8\pi P_0} \tag{2.73}$$

where $g-\gamma=\gamma_m$ for an FP laser. This formula was approximately verified by a study of Welford and Moorodian [43] who measured the linewidth versus power at different temperatures for GaAs transverse junction lasers. Their data are shown in Figure 2.6. The linewidth versus $1/P_0$ has a linear dependence that can be described by a linewidth power product. These authors independently estimated all parameters entering Eq. (2.73) and then compared the calculated linewidth power product with their measured values. They found agreement within 20% at three temperatures. The decrease in linewidth with temperature is attributed to both the decrease in $\alpha$ and $n_{sp}$.

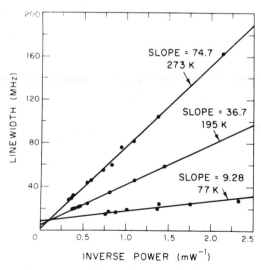

**Figure 2.6.**   AlGaAs laser linewidth versus inverse power at three temperatures. After Welford and Mooradian [43].

The data in Figure 2.6 only approximate the $1/P_0$ dependence of Eq. (2.73). Exact $1/P_0$ behavior requires that the straight lines of data pass through the origin. This is nearly the case for the highest-temperature data, but the intercept at $P_0^{-1} = 0$ becomes larger as the temperature decreases. This is an example of the "linewidth floor." We will discuss this phenomena in Section 2.7.

Another useful expression for the linewidth can be obtained by elimination of $I$ from the linewidth formula [Eq. (2.60)] by relating it to the current above threshold [Eq. (2.66)]

$$\Delta \nu = \frac{g^2 v_g^2 n_{sp}(1 + \alpha^2)}{4\pi(C - C_{th})} \tag{2.74}$$

This expression is useful in comparing lasers with different end reflectivities. Increasing the end reflectivity alters both $g$ and $P_0$ and makes Eq. (2.73) difficult to apply.

Table 2.2 presents a numerical evaluation of the linewidth power product $\Delta \nu \, P_0$ [Eq. (2.73)] and the linewidth current product $\Delta \nu(C - C_{th})e$ [Eq. (2.74)] as different parameters are changed. A value of $n_{sp} = 1.6$ is used for all cases. The first case represents a conventional long-wavelength communications laser with $\alpha = 6$ and $\gamma = 15$ cm$^{-1}$. The subsequent lines represent how the linewidth–power and linewidth–current products are reduced as $\alpha$ and $\gamma$ are reduced to values expected for a quantum well laser. To take advantage of this decreased internal loss, the length of the laser and the facet reflectivities are increased. These changes reduce the linewidth power product from 51 MHz mW to 170 kHz mW. The last line decreases $\gamma$ to 0.1 cm$^{-1}$ and $\alpha$ to 1 as might be expected in a gas- or solid-state laser with high reflecting facets. In this case the linewidth–power product is decreased to 0.3 kHz mW. These examples illustrate that the internal loss $\gamma$ and the linewidth enhancement factor $\alpha$ are the most important parameters controlling the laser linewidth. They each enter quadratically into the formulas for $\Delta \nu$.

**Table 2.2  Calculation of Linewidth–Power and Linewidth–Current Products**

| $\alpha$ | $L$ ($\mu$m) | $r_1 r_2$ | $\gamma$ (cm$^{-1}$) | $g$ (cm$^{-1}$) | $\Delta \nu \, P_0$ (MHz mW) | $\Delta \nu(C - C_{th})$ (MHz mA) |
|---|---|---|---|---|---|---|
| 6 | 250 | 0.30 | 15 | 73 | 51 | 169 |
| 3 | " | " | " | " | 16.1 | 61 |
| " | " | " | 5 | 53 | 11.7 | 32.4 |
| " | 1000 | " | " | 17.0 | 0.94 | 3.33 |
| " | 3000 | " | " | 9.0 | 0.17 | 0.93 |
| " | 1000 | 0.95 | " | 5.5 | 0.013 | 0.35 |
| 1 | " | " | 0.1 | 0.52 | 0.0003 | 0.0009 |

## 2.4. PHASE NOISE SPECTRUM AND LINESHAPE STRUCTURE

### 2.4.1. Frequency Spectra of Fluctuations about the Steady State

The laser linewidth formula [Eq. (2.60)] was derived under the assumption that we need consider only the low-frequency response of the laser to sources of phase and intensity noise. In this section, we include the high-frequency response of the laser and calculate the laser power spectrum and phase noise spectrum. We also take into account the alteration in gain due to lightwave intensity changes as well as carrier number changes. This treatment closely follows [44].

We solve the rate equations of the laser without the former adiabatic approximation but with the equations linearized to describe small oscillations about the steady state. The equations governing the deviations in $\phi$, $I$, and $N$ are obtained by expanding $I$ and $N$ as

$$I(t) = I + i(t) \qquad N(t) = N + n(t) \tag{2.75}$$

and expanding $S$, $G$, and $\alpha \, \Delta G$ as

$$S(t) = S + S_N n(t) \tag{2.76}$$

$$G(t) = G + G_N n(t) - G_I i(t) \tag{2.77}$$

$$\alpha \, \Delta G = \alpha \, G_N n(t) \tag{2.78}$$

which when substituted into the Langevin rate equations [Eqs. (2.10), (2.33), (2.34)] result in the linear equations

$$\dot{\phi} = \frac{\alpha}{2} G_N n + F_\phi(t) \tag{2.79}$$

$$\dot{i} = G_N I n - \Gamma_I i + F_I(t) \tag{2.80}$$

$$\dot{n} = -\Gamma_N n - G i + F_N(t) \tag{2.81}$$

where the damping coefficients $\Gamma_I \equiv G_I I + R/I$ and $\Gamma_N \equiv G_N I + S_N$. Equations (2.76)–(2.78) account for the change in spontaneous emission rate with carrier number, the change in gain with carrier number and lightwave intensity, and the change in mode frequency with carrier number. The change in gain with lightwave intensity (gain saturation) is necessary to account for the large damping of relaxation oscillations that is observed in index-guided lasers and the increase in damping with lightwave intensity [44, 45]. The lack of a contribution of gain saturation to $\alpha \, \Delta G$ will be discussed in Section 2.5.4.2.

The rate equations [Eqs. (2.79)–(2.81)] are linear and can be immediately solved by Fourier analysis. Fourier-transforming all time-dependent quantities [Eq. (2.23)], the rate equations become

$$(-j\Omega + \Gamma_N)n(\Omega) + Gi(\Omega) = F_N(\Omega) \tag{2.82}$$

$$- G_N In(\Omega) + (-j\Omega + \Gamma_I)i(\Omega) = F_I(\Omega) \tag{2.83}$$

$$- j\Omega\phi(\Omega) - \frac{\alpha}{2}G_N n(\Omega) = F_\phi(\Omega) \tag{2.84}$$

The solutions are

$$n(\Omega) = \frac{(\Gamma_I - j\Omega)F_N(\Omega) - GF_I(\Omega)}{\Delta} \tag{2.85}$$

$$i(\Omega) = \frac{G_N IF_N(\Omega) + (\Gamma_N - j\Omega)F_I(\Omega)}{\Delta} \tag{2.86}$$

$$\phi(\Omega) = -\frac{F_\phi(\Omega)}{j\Omega} - \frac{\alpha G_N n(\Omega)}{2j\Omega} \tag{2.87}$$

where

$$\Delta = GG_N I + \Gamma_N \Gamma_I - j\Omega(\Gamma_N + \Gamma_I) - \Omega^2 \tag{2.88}$$

The spectral densities of these variables are given by Eqs. (2.24) and (2.28):

$$(n^2)_\Omega = \frac{2(G^2 D_{II} - 2G\Gamma_I D_{IN} + (\Gamma_I^2 + \Omega^2)D_{NN})}{|\Delta|^2} \tag{2.89}$$

$$(i^2)_\Omega = \frac{2((\Gamma_N^2 + \Omega^2)D_I + 2\Gamma_N G_N ID_{NI} + G_N^2 I^2 D_{NN})}{|\Delta|^2} \tag{2.90}$$

$$(\phi^2)_\Omega = \frac{2D_{\phi\phi}}{\Omega^2} + \frac{\alpha^2 G_N^2 (n^2)_\Omega}{4\Omega^2} \tag{2.91}$$

where the diffusion coefficients are given in Table 2.1. Note that the spectral density for $\phi$ is proportional to $\Omega^{-2}$, this in part justifies the low-frequency approximation made in the last section in deriving the linewidth formula.

The phase change $\Delta\phi(t) = \phi(t) - \phi(0)$

$$\Delta\phi(t) = (2\pi)^{-1/2}\int_{-\infty}^{\infty}\phi(\Omega)[\exp(-j\Omega t) - 1]\,d\Omega \tag{2.92}$$

Using this relation and the delta function correlation of variables of different frequency [Eq. (2.24)], we can express the mean-square phase change as

$$\langle \Delta\phi(t)^2 \rangle = \frac{1}{\pi} \int_{-\infty}^{\infty} (\phi^2)_\Omega [1 - \cos(\Omega t)] \, d\Omega \qquad (2.93)$$

This integral can be done by contour integration [44]. The poles at $\Omega = 0$ contribute the terms linear in $t$ to $\langle \Delta\phi(t)^2 \rangle$. The poles associated with the zeros of $|\Delta|^2$ contribute damped relation oscillations with damping rate

$$\Gamma = \tfrac{1}{2}(\Gamma_N + \Gamma_S) \qquad (2.94)$$

an angular frequency

$$\Omega = \left( GG_N I + \Gamma_N \Gamma_I - \Gamma^2 \right)^{1/2} \qquad (2.95)$$

These parameters are associated with the transient behavior of the laser. Evaluating the contour integral and using Table 2.1 to eliminate the diffusion coefficients results in

$$\Delta\phi(t)^2 = \frac{R}{2I} \left\{ (1 + \alpha^2 A)t \right.$$
$$+ \frac{\alpha^2 A[\cos 3\delta - \exp(-\Gamma t)\cos(\Omega t - 3\delta)]}{2\Gamma \cos \delta}$$
$$+ \left. \frac{\alpha^2 B[\cos \delta - \exp(-\Gamma t)\cos(\Omega t - 3\delta)]}{2\Gamma \cos \delta} \right\} \qquad (2.96)$$

where $A$ and $B$ are given by

$$A = \frac{((1 + \Gamma_I)/G)^2 + \Gamma_I^2 S/G^2 RI}{(1 + (\Gamma_N \Gamma_I / GG_N I))^2} \qquad (2.97)$$

$$B = \frac{G_N(S + 2RI - GI)}{2GR(1 + (\Gamma_N \Gamma_I / GG_N I))} \qquad (2.98)$$

and

$$\cos \delta = \frac{\Omega}{(\Omega^2 + \Gamma^2)^{1/2}} \qquad (2.99)$$

The first term in Eq. (2.96) shows that $A$ is a correction to $\alpha^2$ in the linewidth formula. The term $A$ is associated primarily with gain saturation. The bracketed term in the numerator of the expression for $B$ is $2D_{NN}$, hence $B$ is associated with the Langevin force $F_N(t)$. Numerical evaluation [44] shows that $A \approx 1$ and $B$ is negligible.

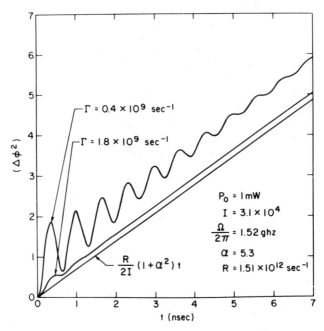

**Figure 2.7.** Mean-square phase change versus time for small damping; realistic damping and for the adiabatic approximation, which leads to a linear dependence on time. After Henry [44].

Figure 2.7 is a plot of the calculated mean-square phase change as a function of time for a GaAs laser. The lower curves exhibit the linear increase of $\langle \Delta\phi^2 \rangle$ with time that results from calculating phase fluctuations in the adiabatic (low frequency) approximation [Eq. (2.55)]. The other two curves are evaluations of Eq. (2.96). They show that superimposed on this linear dependence on time is a contribution of the damped relaxation oscillations. The upper curve, which corresponds to weak damping of relaxation oscillations, obtained by setting the gain saturation parameter to zero, has a significant contribution. The middle curve, with more realistic damping of relaxation oscillations, has only a small oscillatory contribution to $\langle \Delta\phi^2(t) \rangle$.

An experimental measurement of the mean-square phase change $\langle \Delta\phi^2(t) \rangle$ was made by Eichen and Melman [46]. They passed the laser light through a Michelson interferometer and measured the fringe visibility of the output. The detected signal is given by

$$\langle [E(t) + E(t + \tau)]^2 \rangle = 2[\langle E^2 \rangle + \langle E(t)E(t + \tau) \rangle]$$

$$\sim \left(1 + e^{-1/2\langle \Delta\phi^2(\tau) \rangle} \cos \omega_0 \tau \right) \qquad (2.100)$$

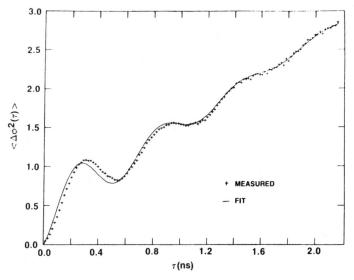

**Figure 2.8.** Direct measurement of $\langle \Delta \phi^2 \rangle$ and comparison with theory of Eq. (2.96). After Eichen and Melman [46].

where $\omega_0 \tau$ is the phase difference associated with the difference in delay in the two arms $\tau$. The last expression is valid when amplitude fluctuations are neglected and Eq. (2.58) is used to evaluate the correlation functions. Thus, they could directly measure $\langle \Delta \phi^2(t) \rangle$. Their data are given in Figure 2.8 along with an excellent fit using Eq. (2.96) with $A = 1$ and $B = 0$.

For realistic damping of relaxation oscillations, the oscillatory term contributes very little to the linewidth. We will see in the next section that the oscillations do contribute significantly to the tails of the lineshape.

It is interesting to compare the phase fluctuations shown in Figure 2.7 with the phase uncertainties due to quantum effects. Referring to Figure 2.1, the phase uncertainty for a coherent state is $(2I^{1/2})^{-1}$. Hence

$$\langle \Delta \phi^2 \rangle = \frac{1}{4I} \approx 10^{-5}$$

in Figure 2.6, where $I = 3 \times 10^4$ (for 1 mW facet power). This is negligible and confirms that the classical treatment is adequate for almost all considerations of phase noise in semiconductor lasers.

### 2.4.2. Laser Lineshape

The power spectrum can be calculated by substituting the expression for $\langle \Delta \phi^2(t) \rangle$ [Eq. (2.96)] into the spectral density [Eq. (2.59)]. This integral

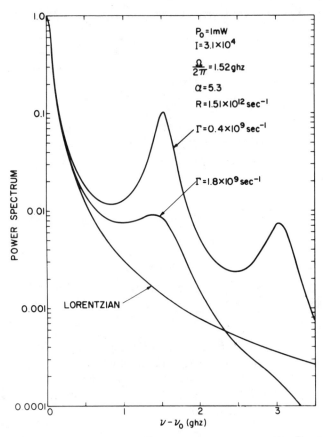

**Figure 2.9.** Power spectrum of the laser line for the three functions $\langle \Delta\phi^2 \rangle$ of Figure 2.7. After Henry [44].

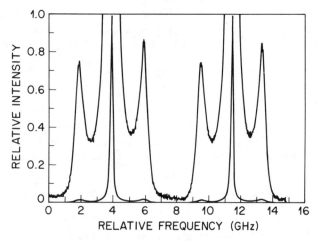

**Figure 2.10.** Fabry–Perot scan of a laser line showing the slightly asymmetric side peaks separated from the main peak by the relaxation oscillation frequency. After Mooradian [48].

must be done numerically with a fast Fourier transform. The results are shown in Figure 2.9 for the three calculations of $\langle \Delta\phi(t)^2 \rangle$ given in Figure 2.7. We see that the tail of the lineshape becomes structured with additional side peaks separated from the central peak by multiples of the relaxation oscillation frequency. The peaks are of low intensity for lasers with highly damped relaxation oscillations. The physical origin of the sidebands is that the laser responds to changes in lightwave intensity with relaxation oscillation that frequency-modulate the laser mode.

The sidebands were first observed by Diano et al. [7]. Vahala et al. [47] also observed them and noted that the peaks on either side of the stop band differ by about 20% in intensity. They explained this as due to a correlation of phase and intensity fluctuations. Intensity fluctuations have been neglected in the computation of the line shape by means of Eq. (2.59). The data of Mooradian [48] are shown in Figure 2.10.

## 2.5. DEPENDENCE OF LINEWIDTH ON CAVITY STRUCTURE

Our discussions of lasers and spontaneous emission thus far were deficient in several respects. The Langevin rate equation for the wave field amplitude $\beta$ [Eq. (2.7)] was presented with physical justification, but without derivation. The diffusion coefficient $2D_{\beta\beta^*} = R$ determining the rate of spontaneous emission into the mode was set equal to $Gn_{sp}$ [Eq. (2.71)], a relation that is exact only for a lossless cavity. In this section, we derive the Langevin rate equation for $\beta$ and an expression for the diffusion coefficient $2D_{\beta\beta^*} = R$ that is valid for lasers of fairly general geometry. This will allow us to rigorously describe the linewidth and phase noise of the lasers of current interest: FP lasers, DFB lasers, and lasers having passive feedback.

### 2.5.1. Solution of the Wave Equation with Spontaneous Emission

We will describe the optical field of the laser in terms of its Fourier components $E_\omega(\mathbf{x})$:

$$E(\mathbf{x}, t) = \int_0^\infty E_\omega(\mathbf{x}) e^{-i\omega t}\, d\omega + \text{c.c.} \tag{2.101}$$

We will assume that $E_\omega$ satisfies a scalar wave equation. This is a good approximation to the vector wave equation for the TE modes and when polarization effects are not of interest.

$$\left[ \nabla^2 + \frac{\omega^2}{c^2}\varepsilon_\omega(\mathbf{x}) \right] E_\omega(\mathbf{x}) = F_\omega(\mathbf{x}) \tag{2.102}$$

Normally the right-hand side of the scalar wave equation is zero. However, to include spontaneous emission we will add a source term $F_\omega(\mathbf{x})$ in analogy with the Langevin rate equation. From the classical point of view, in addition to the induced polarization resulting from $E_\omega(\mathbf{x})$ acting on the charges, which contributes to $\varepsilon_\omega(\mathbf{x})E(\mathbf{x})$, there will be a spontaneous polarization associated with the random motion of the charges, which contributes $F_\omega(\mathbf{x})$ [49].

We can solve Eq. (2.102) by first finding the Green's function $G_\omega(\mathbf{x}, \mathbf{x}')$ satisfying

$$\left[\nabla^2 + \frac{\omega^2}{c^2}\varepsilon_\omega(\mathbf{x})\right]G_\omega(\mathbf{x}, \mathbf{x}') = \delta(\mathbf{x} - \mathbf{x}') \tag{2.103}$$

The field is related to the source by

$$E_\omega(\mathbf{x}) = \int d\mathbf{x}' \, G_\omega(\mathbf{x}, \mathbf{x}') F_\omega(\mathbf{x}') \tag{2.104}$$

We assume that the semiconductor laser is index-guided and sufficiently uniform to be described by a complete set of transverse modes $\phi_n(\mathbf{x})$. For brevity, we will use $\mathbf{x} = (x, z)$, where $x$ represents both transverse coordinates. If we write the dielectric function $\varepsilon_\omega(\mathbf{x})$ as

$$\varepsilon_\omega(\mathbf{x}) = \varepsilon_\omega'(x) + i\varepsilon_\omega(x)'' + \Delta\varepsilon_\omega(x, z) \tag{2.105}$$

where $\Delta\varepsilon_\omega(x, z)$ describes the oscillatory part of the dielectric function associated with a grating in the case of a DFB laser, then the transverse modes satisfy

$$\left[\frac{\partial^2}{\partial x^2} + \frac{\omega^2}{c^2}\varepsilon_\omega'(x)\right]\phi_n(x) = k_n'^2\phi_n(x) \tag{2.106}$$

The transverse modes can be taken as real and orthonormal

$$\int \phi_n\phi_n \, dx = (\phi_n\phi_m) = \delta_{nm} \tag{2.107}$$

The completeness relation is

$$\sum_n \phi_n(x)\phi_n(x') = \delta(x - x') \tag{2.108}$$

We can expand Green's function as

$$G_\omega(\mathbf{x}, \mathbf{x}') = \sum_n g_n(z, z')\phi_n(x)\phi_n(x') \qquad (2.109)$$

Substitution of Green's function [Eq. (2.109)] into its wave equation [Eq. (2.103)] and using Eqs. (2.105)–(2.108) reduces the Green's function equation to

$$\left[\frac{d^2}{dz^2} + k_n^2(z)\right]g_n(z, z') = \delta(z - z') \qquad (2.110)$$

where

$$k_n^2(z) = k_n'^2 + \frac{\omega^2}{c^2}\langle \phi_n | i\varepsilon_\omega''(x) + \Delta\varepsilon_\omega(x, z) | \phi_n \rangle \qquad (2.111)$$

The solution of Eq. (2.110) is discussed at length by Morse and Feshbach [50]. It is a one-dimensional Green's function

$$g_n(z, z') = \frac{Z_{n+}(z_>)Z_{n-}(z_<)}{W_n} \qquad (2.112)$$

where $Z_{n+}$ and $Z_{n-}$ are functions satisfying the homogeneous part of Eq. (2.110) and the boundary conditions at positive and negative $z$, respectively, and $z_>$ and $z_<$ are the greater and lesser values of $z$ and $z'$. This is illustrated in Figure 2.11 for the case of a Fabry–Perot cavity.

The Wronskian $W_n$ is defined by

$$W_n = Z_{n+}'Z_{n-} - Z_{n+}Z_{n-}' \qquad (2.113)$$

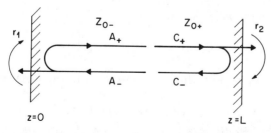

**Figure 2.11.** Diagram of a Fabry–Perot laser cavity showing the solutions $Z_{0-}$ and $Z_{0+}$ satisfying the boundary conditions at each end; $C_\pm$ and $A_\pm$ are the amplitudes of the waves propagating in the positive and negative directions.

where $Z'_{n\pm} = dZ_{n\pm}/dz$. For equations such as Eq. (2.110), which lack a first derivative, *the Wronskion has the remarkable property of being independent of z* [50].

The Wronkskian is a function of the complex propagation constant which depends on $\omega$ and $G = gv_g$. At threshold of the $n$th mode $\omega = \omega_n$, $G = G_n$ and $W_n(\omega_n, G_n) = 0$. In what follows, we will take the lasing mode to be the fundamental mode with threshold values $\omega_0, G_0$ unless stated otherwise. Near threshold

$$W \approx \frac{\partial W}{\partial \omega}(\omega - \omega_0) + \frac{\partial W}{\partial G}\Delta G \tag{2.114}$$

where $\Delta G = G - G_0 = G - \Gamma_0$. For both FP and DFB lasers, the propagation constant near threshold is a function of $\omega - \omega_0 - i(\Delta G/2)(1 - i\alpha)$ and Eq. (2.114) becomes

$$W_0 \approx \frac{dW_0}{d\omega}\left(\omega - \omega_0 - i\frac{\Delta G}{2}(1 - i\alpha)\right) \tag{2.115}$$

A different form of $W$ near threshold for a laser with passive feedback is discussed in Section 2.5.4.

When $W_0 = 0$, it follows from Eq. (2.113) that $Z_{0+}(z) \sim Z_{0-}(z)$. It is convenient to define the normalization of $Z_{0+}(z)$ and $Z_{0-}(z)$ so that for the lasing mode at threshold

$$Z_{0+}(z) = Z_{0-}(z) \equiv Z_0(z) \tag{2.116}$$

Evaluating the optical field [Eq. (2.104)] and keeping only the contribution of the lasing mode, we have

$$E_\omega(\mathbf{x}) = \frac{\phi_0(x)Z_0(z)(\phi_0 Z_0 F_\omega(\mathbf{x}))}{(dW_0/d\omega)[\omega - \omega_0 - i(\Delta G/2)(1 - i\alpha)]} \tag{2.117}$$

where $(\phi_0 Z_0 F(\mathbf{x})) \equiv \int dx \int_0^L dz\, \phi_0 Z_0 F_\omega(\mathbf{x})$.

If we express the field as

$$E_\omega(\mathbf{x}) = B\beta_\omega \phi_0(x)Z_0(z) \tag{2.118}$$

where $B$ is the same normalization contrast as in Eq. (2.1), we have

$$\beta_\omega = \frac{(\phi_0(x)Z_0(z)F_\omega(x))}{B(dW_0/d\omega)[\omega - \omega_0 - i(\Delta G/2)(1 - i\alpha)]}. \tag{2.119}$$

We see that Eqs. (2.101) and (2.118) lead to the same form for the optical field as Eq. (2.1) if

$$\beta(t) = \int_0^\infty \beta_\omega e^{-i\omega t} \, d\omega \qquad (2.120)$$

Multiplying Eq. (2.119) by the bracketed term in the denominator and $e^{-i\omega t}$ and integrating over $\omega$ from 0 to $\infty$, we obtain the rate equation for $\beta(t)$:

$$\dot{\beta} = \left[ -i\omega_0 + \frac{\Delta G}{2}(1 - i\alpha) \right] \beta + F_\beta(t) \qquad (2.121)$$

the form assumed earlier [Eq. (2.7)] for the Langevin rate equation for $\beta$. The Langevin force is

$$F_\beta(t) = \int_0^\infty F_\beta(\omega) e^{-i\omega t} \, d\omega \qquad (2.122)$$

and

$$F_\beta(t) = \frac{-i\left( \phi_0 Z_0 \int_{\omega_0 - \Delta\omega}^{\omega_0 + \Delta\omega} F_\omega(\mathbf{x}) e^{-i\omega t} \, d\omega \right)}{B(dW_0/d\omega)} \qquad (2.123)$$

The limits $\omega_0 \pm \Delta\omega$ have been added. They have no effect because only frequency components of $F_\beta(t)$ near the mode frequency $\omega_0$ influence the laser.

The normalization constant $B$ depends on how we define $I = |\beta|^2$. We will take it to be the number of photons (energy/$\hbar\omega$) in the active part of the laser. In Appendix 2.A, we show that this condition leads to

$$B^2 = \frac{2\pi\hbar\omega_0}{n_0' n_g \left( |Z_0|^2 \right)} \qquad (2.124)$$

where $(|Z_0|^2) = \int_0^L |Z_0|^2 \, dz$, $L$ is the active length of the laser, and $n_g = c/v_g$ is the group index of the fundamental mode.

The diffusion coefficient $2D_{\beta\beta^*} = R$ can be determined by calculating the average $\langle F_\beta(t) F_\beta(t)^* \rangle$ using Eq. (2.122). This average depends on the $\langle F_\omega(\mathbf{x}) F_\omega(\mathbf{x}')^* \rangle$, where $F_\omega(\mathbf{x})$ is the Fourier transform of the time-dependent quantity $F(\mathbf{x}, t)$ that is the source of spontaneous emission. Spontaneous emission is thought to be coherent for no longer than an electron-scattering time, about $10^{-13}$ s. During this time, electrons with thermal energy at room temperature travel only about 0.01 $\mu$m. This distance is sufficiently small that we will assume $F(\mathbf{x}, t)$ is spatially uncorrelated:

$$\langle F(\mathbf{x}, t) F(\mathbf{x}', t') \rangle \sim \delta(\mathbf{x} - \mathbf{x}') \tag{2.125}$$

We will also assume that $F(\mathbf{x}, t)$ obeys stationarity [Eq. (2.21)], that is, $\langle F(\mathbf{x}, t + \tau) F(\mathbf{x}, t) \rangle$ is independent of $t$. As discussed in Section 2.2.2 [Eq. (2.26)], it follows from stationarity that

$$\langle F_\omega(\mathbf{x}) F_{\omega'}(\mathbf{x}') \rangle = 0 \tag{2.126}$$

where $\omega$ and $\omega'$ are both positive or negative and that $\langle F_\omega(\mathbf{x}) F_\omega(\mathbf{x})^* \rangle \sim \delta(\omega - \omega')$; therefore

$$\langle F_\omega(\mathbf{x}) F_{\omega'}(\mathbf{x})^* \rangle = 2 D_{FF^*}(\mathbf{x}, \omega) \delta(\mathbf{x} - \mathbf{x}') \delta(\omega - \omega') \tag{2.127}$$

From the discussion following Eq. (2.26), it also follows that $2 D_{FF^*}(\mathbf{x}, \omega)$ is real and an even function of $\omega$.

We can evaluate $\langle F_\beta(t) F_\beta(t')^* \rangle$ by making use of Eqs. (2.122)–(2.125). This is done in Appendix 2.B, where it is shown that

$$\langle F_\beta(t) F_\beta(t')^* \rangle = R \delta(t - t') \tag{2.128}$$

and that for a laser with uniform active region

$$R = \frac{\left( |Z_0|^2 \right)^2 \left( \phi_0^2 2 D_{FF^*}(\mathbf{x}, \omega_0) \right) n_0' n_g}{\hbar \omega_0 |dW_0/d\omega|^2} \tag{2.129}$$

For a lossless cavity, we established in Section 2.3.4 that $R = g_0 v_g n_{sp}$, where $n_{sp}$ is the negative of the equilibrium photon number [Eq. (2.72). We can use this result to establish a general expression for $2 D_{FF^*}(\mathbf{x}, \omega)$. In Section 2.5.2, we calculate $|Z_0|^2$ and $dW_0/d\omega$ for a general Fabry–Perot cavity. For a lossless cavity of length $L$, these quantities are $(|Z_0|^2) = 2L$ and $|dW_0/d\omega|^2 = (4k_0 L/v_g)^2$. Using these results and $R = g_0 v_g n_{sp}$, we find

$$\left( \phi_0^2 2 D_{FF^*}(\mathbf{x}, \omega_0) \right) = 4\hbar \frac{\omega^3}{c^3} n_0' g_0 n_{sp} = 4\hbar \frac{\omega_0^3}{c^3} n_{sp} \left( \phi_0^2 n' g \right) \tag{2.130}$$

The equality of the first and last expressions in Eq. (2.130) must hold regardless of the functional form of $\phi_0$. For a lossless cavity all the modes approaches threshold as $g$, which is negative, approaches zero and Eq.

(2.130) must then hold for any of these modes. This can only be satisfied if

$$2D_{FF^*}(\mathbf{x}, \omega) = \frac{4\hbar\omega^3 n_{sp}}{c^3} gn' = \frac{4\hbar\omega^4}{c^4}\varepsilon''_\omega P_e \qquad (2.131)$$

The last equality follows from $-(\omega/c)\varepsilon''_\omega = n'g$ and $P_e = -n_{sp}$ is the equilibrium photon number. A different derivation of Eq. (2.131) was presented in ref. [11].

An expression equivalent to Eq. (2.131) was derived by Landau and Lifshitz [49] using quantum mechanics in their discussion of thermal fluctuations of the electromagnetic field. Our expressions agree if we set $V = 0$ and change $P_e$ to $P_e + \frac{1}{2}$ to include vacuum fluctuations, which are beyond the scope of our discussion.

Equation (2.131) can be thought of as a manifestation of the fluctuation–dissipation theorem. In equilibrium, a cavity will have an average photon number $P_e$. This equilibrium photon number is a balance between dissipation controlled by $\varepsilon''_\omega$ and spontaneous emission determined by $2D_{FF^*}(\mathbf{x}, \omega)$. The spontaneous emission results in field fluctuations.

The general expression for $D_{FF^*}$ [Eq. (2.131)] can be used to obtain a more convenient expression for the spontaneous rate [Eq. (2.129)]:

$$R = G_0 n_{sp} F_R \qquad (2.132)$$

where $G_0 = g_0 v_g$ and

$$F_R = \left[\frac{2(|Z_0|^2)k_0}{v_g |dW_0/d\omega|}\right]^2 = \left[\frac{(|Z_0|^2)}{|dW_0/dk^2|}\right]^2 = \left[\frac{(|Z_0|^2)}{|(Z_0^2)|}\right]^2 \qquad (2.133)$$

The last equality can be derived by expanding the axial Green's function as a series of eigenfunctions of the axial wave equation. It was presented earlier by Wang et al. [51] as a special case of a more general result of Arnaud [52]. In the case of a lossless cavity, the axial mode $Z_0$ will be real and according to the last expression in (2.133), $F_R$ goes to 1. The reality of $Z_0$ follows the Hermitean nature of the axial eigenvalue equation for a cavity with closed ends.

We have arrived at a most satisfactory description of spontaneous emission *Spontaneous emission occurs in all directions and over a broad range of optical frequencies. It is described by $D_{FF^*}$ [Eq. (2.131)], which depends only on the material parameters $\varepsilon''_\omega(x), \omega, eV, kT$. The amount of spontaneous emission coupled into the lasing mode is determined by Green's function, which depends on laser geometry. It is given by the $g_0 v_g n_{sp} F_R$ [Eqs. (2.132), (2.133).*

### 2.5.2.   Spontaneous emission in a Fabry–Perot laser

As a first example of the Green function method, we treat the FP laser with arbitrary facet reflectivities. Consider an FP laser of length $L$ and having complex reflectivities $\tilde{r}_1 = r_1 e^{i\theta_1}$ and $\tilde{r}_2 = r_2 e^{i\theta_2}$ at $z = 0$ and $z = L$. The functions $Z_{0-}$ and $Z_{0+}$, solutions satisfying boundary conditions at $z = 0$ and $z = L$, respectively, are given by

$$Z_{0-} = \tilde{r}_1 \exp(ikz) + \exp(-ikz) \tag{2.134}$$

$$Z_{0+} = \tilde{r}_1 \exp(ikz) + \tilde{r}_1 \tilde{r}_2 \exp(2ikL)\exp(-ikz) \tag{2.135}$$

where $k$ is a complex propagation constant. Then

$$k = k_0 - \frac{i}{2}(g - g_0)(1 - i\alpha) + \frac{\omega - \omega_0}{v_g} \tag{2.136}$$

Evaluation of the formula for the Wronskian [Eq. (2.113)] results in

$$W_0 = 2ik\tilde{r}_1\left[1 - \tilde{r}_1\tilde{r}_2 \exp(2ikL)\right] \tag{2.137}$$

Note that $W_0$ is independent of $z$.

The lasing condition is that $W_0 = 0$. We see that it leads to the familiar conditions that the round-trip gain is unity and that the round-trip phase is a multiple of $2\pi$.

$$\tilde{r}_1\tilde{r}_2 \exp(2ik_0 L) = 1 \tag{2.138}$$

where $k_0 = k_0' - i(g_0 - \gamma)/2$. The change in propagation constant for deviations from threshold is

$$\Delta k = \frac{\Delta\omega}{v_g} - \frac{i\,\Delta g}{2}(1 - i\alpha) \tag{2.139}$$

and $dW_0/d\omega$, evaluated at threshold, is

$$\frac{dW_0}{d\omega} = \frac{4k_0 L\tilde{r}_1}{v_g} \tag{2.140}$$

The normalization of $Z_{0+}$ and $Z_{0-}$ has been chosen so that at threshold $Z_{0-} = Z_{0+} = Z_0$, where

$$Z_0 = \tilde{r}_1 e^{ik_0 z} + e^{-ik_0 z} \tag{2.141}$$

If we neglect the small contribution of the oscillatory term, the integral

$\int_0^L |Z_0|^2 \, dz = (|Z_0|^2)$ is given by

$$(|Z_0|^2) = \frac{r_1(r_1 + r_2)[\exp((g_0 - \gamma)L) - 1]}{g_0 - \gamma} = \frac{r_1(r_1 + r_2)(1 - r_1 r_2)L}{-r_1 r_2 \ln(r_1 r_2)}$$

(2.142)

Finally, the rate of spontaneous emission is given by $F_R G_0 n_{sp}$, [Eqs. (2.132), (2.133)], where

$$F_R = \left[ \frac{(r_1 + r_2)(1 - r_1 r_2)}{2 r_1 r_2 \ln(r_1 r_2)} \right]^2$$

(2.143)

The enhancement to the spontaneous emission rate $F_R$ is plotted in Figure 2.12 for the case of equal facet reflectivities $r_1 = r_2 = (R_m)^{1/2}$, where $R_m$ is the facet power reflectivity. We see that cleaved facets, where $R_m \approx 0.3$, $F_R$ is only 1.13. However, in the case of low reflecting facets, $F_R$ is large. For example, for $R_m = 0.01$, $F_R = 4.62$.

The main reason for this enhancement is the single-pass amplification of spontaneous emission that becomes quite large for low facet reflectivities. The resonant amplification during multiple round trips is determined by the resonant denominator in Eq. (2.119). This denominator gives rise to the factor, multiplying $\beta$ in the rate equation (2.121). In contrast, the nonresonant single-pass amplification contributes to the numerator [to $(|Z_0|^2)$ which increases $R$]. As some justification for this assertion, con-

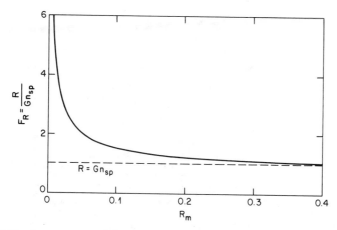

**Figure 2.12.** Spontaneous emission enhancement factor $F_R$ for a Fabry–Perot laser cavity.

sider a laser with small facet reflectivity and $r_1 = r_2 = r$. In this case

$$F_R \cong \frac{1}{r^2(\ln r^2)^2} = \frac{A}{[(g - \gamma)L]^2}$$

where $A = \exp[(g - \gamma)L]$ is the single-pass amplification.

The only difference between the description of the laser in this section and in Section 2.3 is the enhancement of spontaneous emission by $F_R$. Therefore, the laser linewidth, which is proportional to the rate of spontaneous emission, will also be increased by $F_R$. This result was found independently by Ujihara [3], the author [11], Arnaud [52], and Bjork and Nilsson [4] by different methods.

### 2.5.3. Distributed Feedback Lasers

The spontaneous emission enhancement factor $F_R$ for DFB lasers, given by the first expression of Eq. (2.133), has been evaluated by Kojima and Kyuma [21] and more recently by Duan et al. [55]. The results of Kojima and Kyuma are reproduced in Figures 2.13 and 2.14 where the spontaneous emission enhancement $F_R$ is plotted against $\kappa L$. Figure 2.13 shows results for the case of DFB lasers having a uniform grating and combinations of nonreflecting and cleaved ends. Figure 2.14 shows results for DFB lasers with two nonreflecting ends and various phase shifts at the center ranging from a uniform grating to a quarter-wave-shifted grating. The

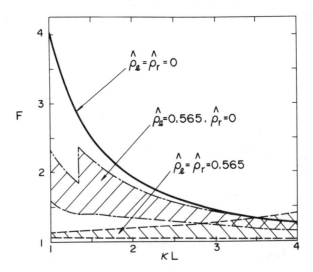

**Figure 2.13.** Spontaneous emission enhancement factor $F_R$ versus $\kappa L$ for distributed feedback lasers with cleaved and perfect AR-coated ends. After Kojima and Kyuma [21].

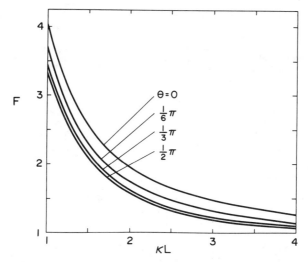

**Figure 2.14.** Spontaneous emission enhancement factor $F_R$ versus $\kappa L$ for distributed feedback lasers with perfect AR-coated ends and a range of phase shifts at the center. After Kojima and Kyuma [21].

formulas of Kojima and Kyuma for $F_R$ are quite long and will not be reproduced here. A different approach to calculating $F_R$, based on evaluation of the last expression in Eq. (2.133), has been given by Wang et al. [51], who found curves quite similar to those of Figure 2.14 for DFB lasers with nonreflecting ends.

In Figures 2.13 and 2.14, $F_R$ is plotted against $\kappa L$, where $\kappa$ is the coupling coefficient of the Bragg reflector and $L$ in the length of the laser. The $1/\kappa$ is the characteristic length for Bragg reflection. The results in Figure 2.14 show that when $\kappa L$ is large and the laser has low end loss, $F_R$ approaches unity, while when $\kappa L$ is small, $F_R$ can be large.

This behavior can be understood by considering how the axial field $Z_0$ changes with $\kappa L$ for a specific example. The field of a DFB laser can be written as

$$Z_0 = C_+(z)e^{ik_0z} + C_-(z)e^{-ik_0z} \qquad (2.144)$$

where $k_0$ is the real propagation constant at the Bragg wavelength. The power flowing to the right and left are proportional to $|C_+|^2$ and $|C_-|^2$, respectively. In Figure 2.15, we plot these powers versus $z$ for $\kappa L = 1$ and $\kappa L = 4$ for a quarter-wave-shifted DFB laser with nonreflecting ends. The mode has been normalized to unity at the center of the cavity. The coefficients $C_\pm$ were calculated by the method of McCall and Platzman [56].

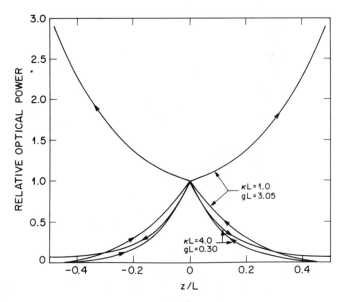

**Figure 2.15.** Power versus axial position in a quarter-wave-shifted laser with perfect AR-coated ends for $\kappa L = 1, 4$.

The last expression for $F_R$ in Eq. (2.133) can be written in terms the $C_\pm$ coefficients as

$$F_R = \left[ \frac{\left( |C_+|^2 + |C_-|^2 \right)}{2|(C_+ C_-)|} \right]^2 \qquad (2.145)$$

where the parentheses denote integration over the length of the laser and the rapidly oscillating contributions to the integrals have been neglected. Figure 2.15 illustrates the for large $\kappa L$, the power flowing in the two directions at each point in the laser is approximately equal $C_+(z) \approx C_-(z)$ and hence the contributions of the numerator and denominator in Eq. (2.145) will be nearly equal. This leads to $F_R$ near unity. Figure 2.15 also shows that for $\kappa L = 1$, the axial field grows as it crosses the cavity is due to amplification needed to overcome end losses. Consequently, $C_+(z)$ and $C_-(z)$ are very different in magnitude near the ends of the cavity, the contribution to the numerator of Eq. (2.145) is greater than the denominator and $F_R$ will be large compared to unity. We conclude that the increase in $F_R$ for small $\kappa_L$ DFB lasers results from single pass amplification. This is the same source of spontaneous emission enhancement that occurs in FP lasers with low reflecting ends.

### 2.5.4. Fabry–Perot Laser with Passive Optical Feedback

**2.5.4.1. Linewidth Reduction.** The Green's function treatment of the FP laser can readily be extended to the important case of passive feedback. Suppose the feedback element is coupled to facet 1. The effect of passive feedback is to render the complex reflectivity of facet 1 frequency-dependent:

$$\tilde{r}_1 \rightarrow r_1(\omega) e^{i\phi(\omega)} \qquad (2.146)$$

For example, setting back the facet reflector by distance $L$ by means of a passive section will make $\phi_1(\omega)$ frequency-dependent according to

$$\frac{d\phi_1}{d\omega} = \tau = \frac{2L}{v_g} \qquad (2.147)$$

where $\tau$ is the reflection delay. Other changes, such as Bragg reflection, will render the modulus of reflectivity $r_1$ frequency-dependent.

The Wronskian of an FP cavity is given by Eq. (2.137). If $\tilde{r}_1$ is independent of frequency, the Wronskian of an FP cavity near threshold is found by differentiating this expression and using Eq. (2.139) for $\Delta k'$.

$$W_0 = \frac{dW_0}{d\omega}\left[\Delta\omega - i\frac{\Delta G}{2}(1 - i\alpha)\right] \qquad (2.148)$$

The frequency dependence of $\tilde{r}_1$ alters $W_0$ to

$$W_0 = \frac{dW_0}{d\omega}\left[(1 + A - iB)\Delta\omega - i\frac{\Delta G}{2}(1 - i\alpha)\right] \qquad (2.149)$$

where

$$A - iB = \frac{1}{i\tau_0}\frac{d\ln(\tilde{r}_1)}{d\omega} = \frac{1}{\tau_0}\frac{d\phi_i}{d\omega} - \frac{i}{\tau_0}\frac{d\ln(r_1)}{d\omega} \qquad (2.150)$$

and $dW_0/d\omega$ is as given by Eq. (2.140).

Substitution of the altered expression for $W_0$ into Eq. (2.119) results in a modification of the rate equation for $\beta$ [Eq. (2.121)]:

$$(1 + A - iB)(\dot{\beta} + i\omega_0\beta) = \frac{\Delta G}{2}(1 - i\alpha) + F_\beta(t) \qquad (2.151)$$

where the right side is unchanged.

This equation can be transformed to intensity and phase variables using Eqs. (2.5) and (2.19) as before. The right sides of the new equations are unchanged from Eqs. (2.33) and (2.34):

$$(1 + A)\dot{I} - 2BI\dot{\phi} = \Delta G I + R + F_1(t) \qquad (2.152)$$

$$(1 + A)\dot{\phi} + \frac{B}{2}\frac{\dot{I}}{I} = \frac{\alpha \Delta G}{2} + F_\phi(t) \qquad (2.153)$$

Laser line broadening results from low-frequency components of the Langevin forces. We can neglect $\dot{I}$ in calculating the response to these components as we did in Section 2.3.3. Doing this, eliminating $\Delta G$ and dropping the constant $R$, which results in only a frequency shift, we find

$$\dot{\phi} = \frac{F_\phi(t) - (\alpha/2I)F_I(t)}{(1 + A + \alpha B)} \qquad (2.154)$$

Comparing with Eq. (2.53), we see that the effect of passive feedback reduces phase fluctuations by

$$F_C = 1 + A + \alpha B \qquad (2.155)$$

The linewidth is proportional to $\langle \Delta\phi^2 \rangle$ and will be reduced by $F_C^2$:

$$\Delta\nu = \frac{\Delta\nu_0}{F_C^2} \qquad (2.156)$$

where $\Delta\nu_0$ is the linewidth that the laser would have without feedback and having a frequency-independent facet reflectivity $\tilde{r}_1(\omega_0)$.

This equation was first presented by Patzak et al. [57] and later derived independently by Kazarinov and the author [20] in the manner presented here. We also showed that the parameters $A$ and $\alpha B$ have a simple graphical interpretation. In general, two conditions determine laser threshold: gain = loss and round-trip phase = $2\pi N$. These two conditions can be plotted in the plane of gain g, versus optical frequency. This is illustrated in Figures 2.16 and 2.17 for an FP laser and an external Bragg reflector (EBR) laser. There are a number of curves of constant phase corresponding to different values $N$. The operating point of the laser corresponds to the lowest point of intersection of the two curves, where both conditions are satisfied.

The parameters $A$ and $\alpha B$ are related to the slopes of the "loss" and "constant phase" curves at the operating point. For a Fabry–Perot laser with constant reflectivities, the slope of the "loss" curve is zero and the slope of the curve of "constant phase" is $d\omega/dG = 2/\alpha$ (Figure 2.16). For a more general case, such as illustrated in Figure 2.17 for a Bragg reflector, both slopes are nonzero. It is shown in Appendix 2.C that with optical feedback slope of the "constant phase" increases to $2/\alpha(1 + A)$

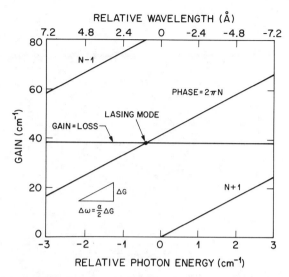

**Figure 2.16.** The two conditions "gain = loss" and "round-trip phase = $2\pi N$" for a Fabry–Perot laser plotted as gain versus photon energy. We refer to the two conditions as the "loss" and "phase curves." After Kazarinov and Henry [20].

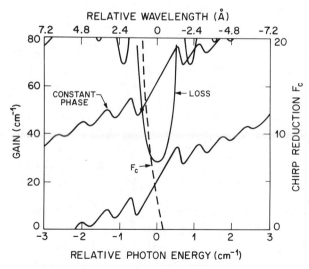

**Figure 2.17.** "Loss" and "phase" curves for a Bragg reflector laser. The Bragg reflector has a length of 5 mm and $\kappa L = 1$. After Kazarinov and Henry [20].

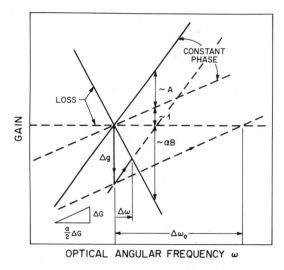

**Figure 2.18.** The two feedback parameters $A$ and $B$ determined from the slopes of the "loss" and "phase" curves at the operating point. The dashed curves show the slopes for a FP laser without feedback. The diagram also shows the slopes with and without feedback. After Kazarinov and Henry [20].

and the downward slope of the "loss" curve becomes $2/\alpha(\alpha B)$. This is illustrated in Figure 18.

**2.5.4.2. Relation of Adiabatic Chirp and Linewidth Reduction.** Chirp is a frequency shift of the laser resulting from a change in light intensity. We will restrict this discussion to chirp occurring at frequencies that are small compared the relaxation oscillation damping rate. In this regime, the laser adiabatically follows any intensity change with a change in frequency. This phenomenon is used to frequency-modulate semiconductor lasers. We will show here that in the case of optical feedback, the laser chirp is reduced by $F_C$ [Eq. (2.155)], the square root of the factor that describes linewidth reduction [Eq. (2.156)]. Hereafter, we will call $F_C$ the *chirp reduction factor*.

Adiabatic chirp results from two effects. The first is gain saturation with increasing optical intensity, the same phenomenon that causes the strong damping of relaxation oscillations. This was expressed earlier [Eq. (2.77)] as

$$\Delta G = -G_I \Delta I \tag{2.157}$$

where $G_I$ is positive. The second is that the change in refractive index with gain saturation is smaller than that occurring with the gain change due to a change in carrier number. In the discussion given here, for simplicity, we will assume that the refractive index change associated with gain saturation is negligible.

The physical origin of gain saturation is not settled. Yamada and Suematsu [58], Kazarinov et al. [59], and Agrawal [60] have attributed it to spectral hole burning, the depletion of levels generating stimulated emission. Recently, this has been disputed by Kessler and Ippen [60], who attribute it to carrier heating and Su [61], who attributes it to standing waves that spatially modulate the carrier density, setting up a grating that increases loss at the laser wavelength.

We can use the laser rate equations with optical feedback [Eqs. (2.152), (2.153)] to describe adiabatic chirp by neglecting the Langevin forces and $R$, setting $\dot{I} = 0$, replacing $\phi$ with the optical frequency chirp $\Delta\omega$. We will write the gain change $\Delta G$ in Eq. (2.152) as

$$\Delta G = G_N \Delta N - G_I \Delta I \qquad (2.158)$$

The optical frequency change associated with the gain change in Eq. (2.153) is assumed due only to $\Delta N$; hence

$$\frac{\alpha \Delta G}{2} = \frac{\alpha G_N \Delta N}{2} \qquad (2.159)$$

The rate equations [Eqs. (2.152), (2.153)] become

$$-2B \Delta\omega = G_N \Delta N - G_I \Delta I \qquad (2.160)$$

$$(1 + A) \Delta\omega = \frac{\alpha G_N}{2} \Delta N \qquad (2.161)$$

Eliminating $G_N \Delta N$, we find

$$\Delta\omega = \frac{(\alpha/2)G_I \Delta I}{1 + A + \alpha B} = \frac{\Delta\omega_0}{F_C} \qquad (2.162)$$

where $\Delta\omega_0 = -(\alpha/2)G_I \Delta I$ is the adiabatic chirp without feedback and $F_C$ is the chirp reduction factor [Eq. (2.155)].

Adiabatic chirp and its reduction by optical feedback is graphically described in Figure 2.18. The dashed lines show changes in the laser without feedback. Gain saturation reduces gain by $-G_I \Delta I/v_g$. There is no refractive index or round-trip phase change accompanying this change. Consequently, curves of "constant phase" move vertically downward by this amount. The new operating point occurs at the intersection of the "constant phase" and "loss" curves and corresponds to an optical frequency shift $\Delta\omega$. In a laser without feedback, the frequency shift $\Delta\omega_0$ corresponds to the shift of the mode associated with the gain change

$G_N \Delta N$ necessary to bring the gain equal to loss. With optical feedback, the shift in optical frequency is reduced because, with $B > 0$, loss drops with increasing frequency, so less gain change is required. In addition, with $A > 0$, the mode requires a smaller frequency change to maintain constant round-trip phase when the refractive index is altered by a change $\Delta N$.

The equations for chirp and linewidth reduction [Eqs. (2.162), (2.156)] imply that optical feedback reduces chirp by the square root of linewidth reduction. This was elegantly verified by Olsson et al. [22]. Using a silicon chip Bragg reflector (SCBR) laser [63], in which optical feedback is produced by a Bragg reflector mode with waveguides with $Si_3N_4$ core and $SiO_2$ cladding layers deposited on silicon, they accurately measured the chirp and linewidth as the operating point was changed. Their results are shown in Figure 2.19 along with a gain versus optical frequency diagram showing the operating point. The operating point could be changed by varying the temperature. Because of the difference in temperature dependence of the refractive indices of silica and semiconductor materials (modes in semiconductor material shift at about 1 Å $°C^{-1}$ and those in silica at about $\frac{1}{8}$ Å $°C^{-1}$) the phase curve moves up with increasing temperature, tuning the operating point. Linewidth and chirp reduction occur primarily on the long-wavelength side of the Bragg reflection band, where $\alpha B$ is large. Changing the operating point from the center to the

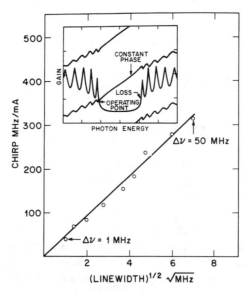

**Figure 2.19.** Chirp versus square root of linewidth for a silicon chip Bragg reflector (SCBR) laser and an inset showing the "loss" and "phase" curves for this laser. After Olsson et al. [22].

long-wavelength side of the Bragg reflection band resulted in a change of 50 in the linewidth $\Delta\nu$ and about 7 in chirp. A plot of chirp versus square root of linewidth yielded a linear relation as expected.

**2.5.4.3. External Cavity Lasers.** Wyatt [64] demonstrated that laser linewidths of less than 1 kHz can be achieved by a semiconductor laser with a long passive section. This is achieved by AR coating one of the laser facets and using a microscope objective to collimate the beam and send it to a reflector about 10–15 cm away. Figure 2.20 illustrates this type of laser made by Olsson and van der Ziel [65] along with a power spectrum obtained combining by the fields of two such lasers in a single-mode optical fiber and displaying the detected beat signal on a spectrum analyzer.

A long cavity will have many closely spaced modes. Stable single-mode operation is achieved by using a diffraction grating to reduce the reflection band to about 5–10 Å. In addition an etalon is added to the cavity to reduce the peak reflection to 1–2 Å. This filtering, together with the small spontaneous emission rate in each longitudinal mode and optical nonlinearities [66, 58], tends to greatly suppress all but one mode. This mode can

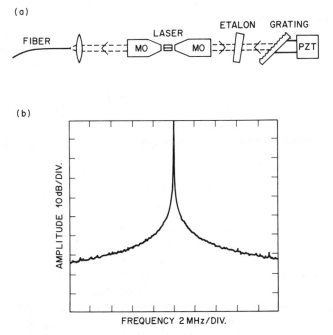

**Figure 2.20.** Diagram of an external grating laser and a spectrum of the beat of two such laser as measured with a spectrum analyser. After Olsson and Van der Ziel [65].

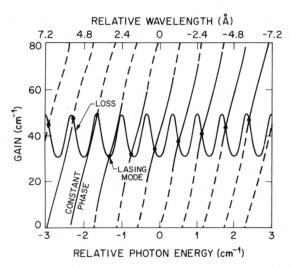

**Figure 2.21.** "Loss" and "phase" curves for a laser with feedback from a reflector set back by 5 mm. The oscillations of the loss curve result from an interface field reflectivity of 0.1. After Kazarinov and Henry [20].

be tuned by mounting the grating reflector on a piezoelectric drive and changing the length of the cavity by about half a wavelength.

The gain–optical frequency diagram for an external cavity reflector laser is shown in Figure 2.21. The oscillations in the loss curve are due to beating between the residual AR-coated facet reflection and the distant reflection. This oscillation will be small in lasers with very good AR coatings and efficient optical feedback. Lasing takes place near the peak reflectivity where the loss is minimum. The optical feedback is dominated by the steep slope of the "constant phase" curve, and the chirp reduction will be given by

$$F_C = 1 + A = 1 + \frac{1}{\tau_0} \frac{d\phi_1}{d\omega} = 1 + \frac{L_1 v_{g0}}{L_0 v_{g1}} \qquad (2.163)$$

The linewidth will be reduced by the square of $F_C$. Note that if feedback to the AR coated laser is small, $r_1$ will be reduced and that will increase spontaneous emission and linewidth as illustrated in Figure 2.12.

Low feedback also leads to an instability known as "coherence collapse" [67–70]. In the case of low feedback, the laser can be viewed as self-locked to the field coming back from the distant reflector. Coherence collapse is an instability in which fluctuations cause the laser to jump out of the self-locked state [69]. This may be caused by either spontaneous emission noise or the chaos associated with the nonlinear dynamics of a laser with

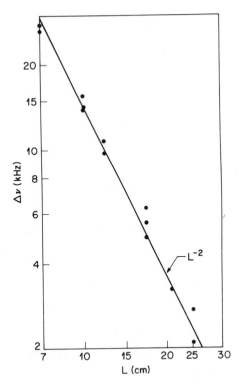

**Figure 2.22.** Linewidth versus length for an external cavity laser. After Linke and Pollock [71].

feedback. Coherence collapse is prevented by strong optical feedback, an AR-coated facet, and wavelength selective feedback.

For external cavities long compared to the length of the active section, the linewidth should diminish as $L_1^{-2}$ [Eqs. (2.156), (2.163)]. This has been verified by Linke and Pollack [71]. Their data are shown in Figure 2.22. The linewidths of external cavity lasers should narrow inversely with power, in the same manner as conventional lasers, this was verified by Olsson and van der Ziel [65]. Their data are shown in Figure 2.23.

### 2.5.4.4.   Resonant Optical Reflector Laser.

The external cavity laser is capable of achieving low-kilohertz linewidths. This is achieved at the expense of a massive mechanically rigid external cavity made with bulk optics and more suitable for laboratory experiments than practical applications. Furthermore, the cavities are subject to microphonics. Olsson and van der Ziel [65] found frequency variations of 1 MHz occurring on a time scale of 50 ms caused mainly by microphonics.

Bulk optics can be avoided by coupling the laser to a long waveguide with a narrowband reflector at one end; however a long waveguide, such as $L_1 \approx 15$ cm (Figure 2.22), is required to achieve low-kilohertz linewidths. To reduce the length, the waveguide can be replaced by an integrated

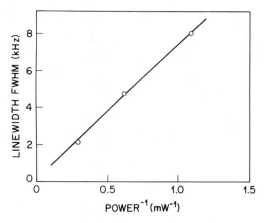

**Figure 2.23.** Linewidth versus inverse power for an external-cavity laser. After Olsson and Van der Ziel [65].

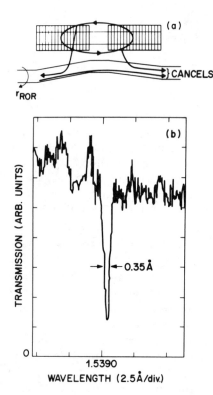

**Figure 2.24.** Transmission spectrum of a resonant optical reflector and diagram of the device. The downward dip corresponds to a reflection resonance. After Olsson et al. [12].

optic resonator. In this way, a compact narrow-linewidth hybrid laser can be achieved. Recently Ackerman et al. [14] have reported such a laser made with a 250-$\mu$m-long amplifying section butt-coupled to a 5-mm-long resonant optical reflector chip and having a linewidth of 7 kHz at 1.32-$\mu$m wavelength.

The invention of the resonant optical reflector (ROR) by Kazarinov et al. [13] was stimulated by experiments by Olsson et al. [63] showing that narrow Bragg reflectors appreciably reduced the lasers linewidth and the understanding that this line narrowing was due to the $B$ term, the rapid increase in reflectivity with optical frequency.

What was needed for extreme line narrowing was to couple the laser to a very narrow reflector. Such a spectrally narrow feature can be achieved by forming an integrated optic cavity with two Bragg reflectors shown in Figure 2.24. Such a, cavity will have a mode at the center of the Bragg reflection spectrum if the space between reflectors is properly chosen so that the wave makes a $\pi/2$ (quarter-wave) phase change relative to the grating lines when crossing the center of the cavity. This is the quarter-wave-shifted cavity used in DFB lasers. However, such a cavity exhibits a transmission resonance, not the desired reflection resonance. If side-by-side coupling instead of butt-coupling is employed, as shown in Figure 2.24, a reflection resonance is seen at the input of the waveguide that couples to the resonator. Away from resonance little power is reflected. At resonance, as shown in Figure 2.24, energy is stored in the cavity. This energy couples back to give a reflected wave and to cancel the transmitted wave, as shown.

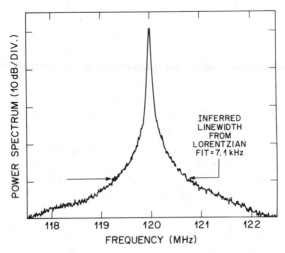

**Figure 2.25.** Power spectrum of a resonant optical reflector laser. After Ackerman et al. [14].

Figure 2.24 shows a measured transmission spectrum of an ROR taken by Olsson et al. [12]. These authors achieved an ROR laser having a linewidth of 135 kHz at 5 mW. More recent work by Ackerman et al. [14], in which many lasers were tested, resulted in linewidths as low as 26 kHz at 1.55 $\mu$m with a 2.5-mm-long passive chip and 7 kHz at 1.32 $\mu$m with a 5-mm-long passive chip (see Figure 2.25). The typical laser powers were 3–6 mW in these experiments. The linewidth that is achievable by this approach decreases as the square of the loss in the resonator. It was estimated that the average waveguide loss of the ROR, including the grating, is about 1 dB cm$^{-1}$.

## 2.6. BULK PROPERTIES

The material properties determining laser linewidth are the spontaneous emission factor $n_{sp}$, the linewidth enhancement factor $\alpha$, and the waveguide loss $\gamma$. They are the subject of this section. Fortuitously, both $\alpha$ and $\gamma$ can be reduced in quantum well lasers.

### 2.6.1. Absorption, gain and emission spectra

The spectra of spontaneous emission r, gain g, and absorption a are shown in Figure 2.26 as a function of photon energy $\hbar\omega$ [72, 73]. These quantities are measured in cm$^{-1}$. The associated rates per second, $R$, $G$, and $A$, are the related by $G = g v_g$, and so on. The curves were determined by measuring the relative spontaneous emission spectrum r. The absorption spectrum a was then determined by use of the Einstein relation [Eq. (2.70)]. The absolute values of a and r were determined from the measured absorption coefficient far above the bandgap. The gain g is related to r by $r/g = n_{sp} = [1 - \exp(\hbar\omega_0 - eV)/kT)]^{-1}$. The bias $eV$ was then adjusted so that g had a maximum at the lasing energy $\hbar\omega_0$.

Inspection of Figure 2.26 shows that $n_{sp} = r/g$ goes to 1 for energies several $kT$ less than the laser photon energy and this ratio approaches $\infty$ at $\hbar\omega = eV$, where $g = 0$. At $\hbar\omega_0$, $n_{sp}$ was measured to be 1.6 in Figure 2.26, which is data on 1.3-$\mu$m InGaAsP material. An earlier and similar experiment on GaAs yielded $n_{sp} = 2.6$ [74]. The parameter $n_{sp}$ tends to be small when the level of inversion is high. We can expect it to be between 1 and 2 for the relatively short lasers used in optical communications and also for lasers with small mode occupation factors $\Gamma_{act}$ of the active layer.

### 2.6.2. Linewidth Enhancement Factor

The linewidth enhancement factor (2.62) $\alpha = \Delta n'/\Delta n''$ is the ratio of the changes in the real and imaginary parts of the refractive index with change

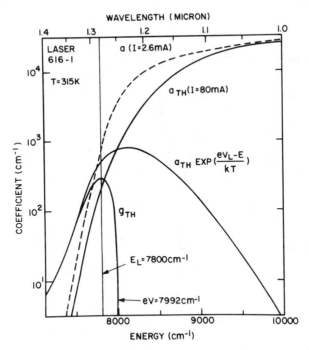

**Figure 2.26.** Absorption, emission, and gain spectra versus photon energy for a 1.3-$\mu$m InGaAs laser. The spontaneous emission spectrum $r$ is measured. The absorption spectra $a$ and the gain spectra $g$ were deduced by means of the Einstein relations for a semiconductor laser [Eqs. (2.68)–(2.72)]. The terms $E_L$ and $eV_L$ refer to the laser photon and bias energy at threshold. After Henry et al. [72].

in carrier number. Above threshold these changes should be determined at constant light intensity.

The spectra of $\Delta n'$ and $\Delta n''$ in going from low bias up to threshold are shown in Figure 2.27 for a GaAs buried heterostructure laser [74]. The curve for $\Delta n''$ was measured by first finding the curve for gain change $\Delta g$. This was done by using the Einstein relations to analyze spontaneous emission spectra to determine the curves for $g$ at low bias and at threshold. Then $\Delta n''$ was converted from $\Delta g$ by means of Eq. (2.61). The real refractive index change $\Delta n'$ was determined from $\Delta n''$ by use of the Kramers–Kronig relations [35, Section 22]. These relations give an exact relation between $\Delta \varepsilon'$ and $\Delta \varepsilon''$, the changes in the real and imaginary parts of the dielectric function. Using the approximation that $\Delta \varepsilon' + i \Delta \varepsilon'' \cong 2n'(\Delta n' + i \Delta n')$, we see that to the extent that $n'$ is real and nearly constant, the relations also apply to $\Delta n'$ and $\Delta n''$. A value of $\alpha = 6.2$ at $\hbar \omega_0$ was found from these experiments. This value included an estimate of the contribution to the refractive index by free carriers.

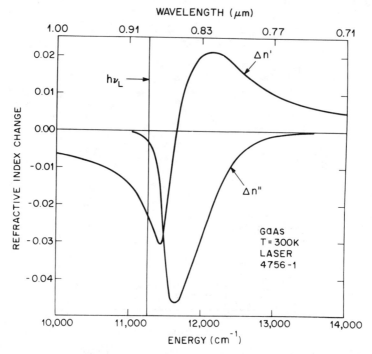

**Figure 2.27.** Real and imaginary changes in refractive index that occur when a GaAs laser is taken from low current up to threshold. The imaginary change was determined from the increase in gain (and decrease in net absorption). The real change was then found by a Kramers–Kronig transformation. After Henry et al. [74].

This contribution can be calculated from the Drude theory

$$\Delta n' = \frac{2\pi e^2}{\omega^2 n'} \left( \frac{1}{m_e} + \frac{1}{m_v} \right) \frac{\Delta N}{V} \qquad (2.164)$$

where $e$ is the electron change, $\Delta N/V$ is the change in carrier density, $m_e$ is the effective mass for the conduction band, and

$$m_v = \frac{m_l^{3/2} + m_h^{3/2}}{m_l^{1/2} + m_h^{1/2}} \qquad (2.165)$$

where $m_l$ and $m_h$ are the light and heavy masses of the valence band [75]. At optical communications wavelengths, 1.3–1.6 $\mu$m, Eq. (2.164) contributes significantly to $\alpha$ (see Figure 2.28, discussed below).

There is a substantial literature devoted to measurements of $\alpha$. A good summary of these data has been given by Osinski and Buus [76]. Their report indicates that most of the values for index guided lasers (without

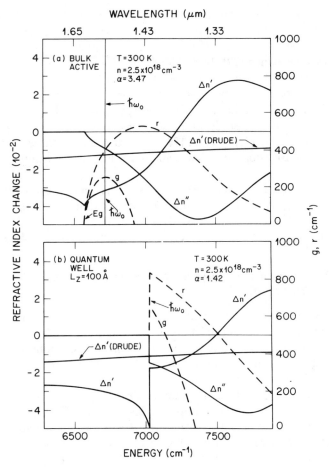

**Figure 2.28.** Calculated changes in the real and imaginary parts of the refractive index, the free carrier contribution to the real change in refractive index, the gain, and the spontaneous emission spectra for (a) a bulk laser and (b) a quantum well laser.

quantum wells) are in the range $5 \pm 1$ for AlGaAs lasers and $6 \pm 2$ for InGaAsP lasers.

## 2.6.3. Waveguide Loss

Our discussion of the linewidth formula in Section 2.3.4 and Table 2.2 pointed out the crucial importance of low waveguide loss in achieving narrow-linewidth lasers. The loss originates from absorption by carriers excited within the bands and scattering by waveguide imperfections. We will briefly discuss the dominant mechanism, which is intervalence band absorption.

Near the top of the valence band, the states resemble the $P_{3/2}$ and $P_{1/2}$ states of an atom that are split by spin–orbit interaction. The additional effect of the periodic potential is to divide the upper $P_{3/2}$ state into light and heavy hole bands [77]. Intervalence band absorption originates from optical transitions between the upper light and heavy hole bands and the split-off $P_{1/2}$ band. At laser photon energies, the heavy hole–split-off band transition is dominant [78]. The absorption is directly proportional to the carrier density. A study of intervalence band absorption in GaAs, InGaAs, and InP by Henry et al. [78] found that for all three crystals, the absorption for $10^{18}$ holes cm$^{-3}$ was about 13 cm$^{-1}$ at 1.3 $\mu$m and 25 cm$^{-1}$ at 1.6 $\mu$m.

In a typical buried heterostructure laser, the mode occupation of the active layer $\Gamma_{act}$ is about 0.25 and the carrier density is approximately $2.5 \times 10^{18}$ cm$^{-1}$ (see Figure 2.28). The waveguide loss, which is weighted by $\Gamma_{act}$ and $N/V$, will be about 8 cm$^{-1}$ at 1.3 $\mu$m and 16 cm$^{-1}$ at 1.6 $\mu$m.

### 2.6.4. Reduction of the Linewidth Enhancement Factor and Loss in Quantum Well Lasers

There have been many theoretical calculations showing that the $\alpha$ parameter is substantially reduced in multiple quantum well lasers [79–82] compared to bulk active layer lasers. Ohtoshi and Chinone [82] predict that $\alpha$ will be even smaller in strained quantum well lasers than in conventional quantum well lasers.

There are also several experimental confirmations of the reduction of $\alpha$ in quantum well lasers compared to bulk lasers. For InGaAsP lasers, Koch et al. [16] report that in similar structures $\alpha \approx 6$ for bulk active layers and $\alpha \approx 3$ for InGaAs MQW lasers. Westbrook et al. [17] find $\alpha \approx 5$–6 for bulk active layers and $\alpha \approx 2$–3 for MQW InGaAsP lasers.

Figure 2.28 presents the results of a simple calculation that shows how this reduction comes about. The values for $r$, $g$, $\Delta n'$, and $\Delta n''$ are calculated for bulk and quantum well heterostructures with a quantum well width of 100 Å. The details of this calculation are given in Appendix 2.D. The main difference between bulk and quantum well lasers is that the optical density of states increases as $A(\hbar\omega - E_g)^{1/2}$ in bulk lasers, while in quantum well lasers it rises abruptly from zero to $A(E_1)^{1/2}$ for $\hbar\omega > E_1 + E_g$, where $E_1$ is the sum of energies of the first quantum well level for each band. Thus, the onset of the absorption edge is much more abrupt for quantum well lasers. Figure 2.28 shows that there is an abrupt rise of gain at $E_g + E_1$ and this is also the energy of peak gain. The effect of this sudden rise in gain is to cause a discontinuity in $\Delta n''$. A Kramers–Kronig calculation of $\Delta n'$ shows that $\Delta n'$ abruptly decreases at this discontinuity. Comparing Figures 2.28a and 2.28b, we see that for the same carrier density, $g$ is 2.6 times as large for the multiple-quantum well (MQW) laser. Both the large value of $g$ (and $\Delta n''$) and the discontinuous drop in

$\Delta n'$ at the absorption edge contribute to reducing $\alpha$ in MQW lasers compared to bulk lasers. In these calculations, $\alpha$ decreases from 3.47 in bulk to 1.42 in MQW structures.

Our calculated values of $\alpha$ as well as published calculations of $\alpha$ tend to be less than the measured values. One reason for this is that the theoretical models do not take into account the Urbach tail that broadens absorption edges [82]. This broadening causes lasing to take place in the tail, and this movement of the lasing energy into the tail tends to increase $\alpha$.

The results of Figure 2.28 also show that intervalence band absorption will be reduced in quantum well lasers. Since, in our example, the gain achieved in quantum well lasers is 2.6 times as great as in bulk active layer lasers at the same carrier density, $\Gamma_{act}$ in the QW case can be less than that in the bulk case by this factor. The intervalence band absorption will be reduced by the same factor provided that this absorption is negligible outside of the active layer.

The reduction of active-layer thickness in MQW lasers compared to bulk lasers will also allow the stripes to be wider and still result in single mode operation. This should also reduce the side wall roughness loss [84, 85], a significant loss, but one that is less than the intervalence band absorption loss.

The above remarks may help explain the remarkable low values of internal loss $\gamma$ that have been reported in single QW lasers. These low loss values are deduced by measuring the differential quantum efficiencies of the lasers versus length. Wang et al. [86] report a loss of 2.5 cm$^{-1}$ in GaAs/AlGaAs graded-index single QW lasers, and Temkin et al. [87] report losses as low as 2.5 cm$^{-1}$ in InGaAs/InP graded-index single QW lasers with 2–3-$\mu$m-wide stripe widths.

## 2.7. LINEWIDTH FLOOR

### 2.7.1. Background

The theory of laser linewidth presented above gives a good description of lasers at low power. However, since the linewidth narrows inversely with power, there is great motivation to try to make the linewidth narrower by running the laser at as high power as possible. Such experiments reveal additional broadening mechanisms that result in a minimum "linewidth floor" as power is increased.

The first observation of a linewidth floor came from the studies of Mooradian and coworkers [48] of AlGaAs lasers. The data of Welford and Mooradian [43], Figure 2.6 shows that the linewidth narrows in a manner that depends linearly of inverse power. As mentioned earlier, Welford and Mooradian found that this narrowing was consistent with the linewidth formula [Eq. (2.73)]. However, contrary to the linewidth formula, the extrapolated linewidth does not go to zero at infinite power, but approaches a finite value, which increases as the temperature is lowered. These extrapolated linewidth floors are in the range of a few megahertz. An additional measurement at 1.6 K, gave an extrapolated linewidth floor of about 30 MHz [48].

Mooradian has suggested that the linewidth floor is due to carrier density fluctuations [48]. Carrier number fluctuations will alter the refraction index and hence the laser frequency; however, they also alter the gain, and this leads·to feedback mechanisms in a laser above threshold that suppresses such fluctuations. Mooradian points out that fluctuations of trapped carriers may be important because they could contribute to refractive index changes and frequency fluctuations without affecting the gain.

The linewidth floor was encountered in more recent experiments on DFB and monolithic DBR InGaAsP bulk and quantum well lasers [88–91], where it is observed that with increasing power the linewidth reduces to a minimum value and then increases again. A striking example of this are the data of Koch and Koren [88] shown in Figure 2.29.

Yasaka et al. [92] have demonstrated an association of the linewidth floor and axial spatial hole burning in DFB lasers. The spatial distribution

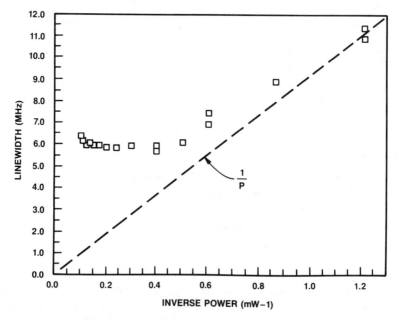

**Figure 2.29.** Linewidth versus inverse optical power of a InGaAs/InP Bragg reflector laser. After Koch and Koren [89].

of intensity in DFB lasers tends to be more nonuniform than in FP lasers. These nonuniformities result in variations in stimulated emission that then cause variations in the carrier density, gain, and Bragg wavelength. Spatial hole burning can be offset by adjusting the current distribution so that more current goes to regions of greater stimulated emission. They did this by using a three-part contact in which the ratio of currents to the center and ends was varied. With uniform current, the laser 300 $\mu$m long, had a linewidth floor of 20 MHz. With separate contacts, linewidths of 15 MHz were obtained with no sign of a linewidth floor.

Kikuchi [93] has measured the frequency noise spectrum as well as the power spectrum of a 1.3-$\mu$m DFB laser. The frequency noise spectrum was measured by passing the laser through a frequency discriminator consisting of a 1-GHz bandwidth FP interferometer. He finds that spectrum of FM noise has a spectral density with two distinct contributions:

$$S_F(f) = \frac{C}{P} + \frac{K}{f} \qquad (2.166)$$

where $f$ is frequency, $P$ is optical power, and $C$ and $K$ are constants. The

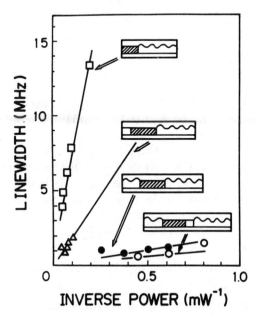

**Figure 2.30.** Linewidth versus inverse power of four butt-jointed InGaAs/InP Bragg reflector lasers of different geometry. After Kano et al. [19].

first term is white noise with a flat spectral density that decreases inversely with power. The second contribution is that of $1/f$ noise that was found to be independent of power. Kikuchi concludes that the FM noise resulting in the linewidth floor has a $1/f$ type of spectrum in the 1–100-kHz range and that it is independent of power. He also remarks that this low-frequency noise will not be very detrimental to high-bit-rate coherent optical communications systems employing differential phase shift-keying.

### 2.7.2.  Narrow-Linewidth InGaAsP Lasers

In spite of mechanisms leading to residual linewidths at high power, remarkably narrow linewidth lasers have been recently reported. Kano et al. [19] have fabricated a variety of monolithic DBR lasers having passive Bragg reflectors. The lowest linewidth achieved was 580 kHz in a structure having a Bragg reflector at each end and a passive section to increase the length of the device. Their results are shown in Figure 2.30.

Ogita et al. [18] have achieved a linewidth of 830 kHz in an 1200-$\mu$m-long InGaAsP DFB laser operating at 30 mW. The laser had buried facets resulting in low reflectivity at each end. Such lasers normally require a quarter-wave shift at the center to operate single mode. To avoid spatial hole burning, instead three phase shifts were added to the grating, which has the effect of making the mode more uniform along the length of the laser.

### 2.8.  CONCLUSIONS

A theoretical description of semiconductor lasers has been presented in which the optical field is described classically by Langevin rate equations. The quantum noise is introduced by appropriate determinations of the diffusion coefficients describing the correlations of the Langevin forces. The role of laser structure was found by adding a source of spontaneous emission to the wave equation and then using the Green function method to find the rate of spontaneous emission into the mode and the altered form of the rate equations for the case of passive optical feedback. The theory provides an adequate description of nearly all aspects of line broadening and phase noise. These include:

1. The linewidth and lineshape of a laser below threshold.
2. The linewidth and power dependence of lasers above threshold.
3. The structure in the lineshape due to relaxation oscillations.
4. Phase diffusion described by the increase in the mean-square phase with time.

5. Corrections to the linewidth formula due to increased spontaneous emission in open cavities such as FP lasers with low facet reflectivity or DFB lasers with low $\kappa L$.

6. The reduction in linewidth and adiabatic chirp by external feedback.

The main inadequacy of theory at present is in accounting for the linewidth floor. This appears to be associated with deviations of the laser from ideal behavior at high power. The experimental association of a $1/f$ component of frequency noise with the linewidth floor greatly restricts the possible mechanisms for this phenomenon. In view of the great interest in this subject, the linewidth floor mechanism is likely to be delineated in the near future.

These studies point out that low waveguide loss is of utmost importance in the fabrication of narrow-linewidth lasers. In the case of passive feedback, low loss in the passive section is necessary to have a long section or to have a high-$Q$ passive resonator. In uniform lasers, low loss is necessary to reduce gain that enters quadratically in determining the linewidth. The current achievements in low-loss glass waveguides and the low losses achieved in single quantum well lasers make it likely that there will be continued reductions in the linewidths of hybrid and monolithic lasers.

## ACKNOWLEDGMENT

The author is grateful to D. A. Ackerman, R. F. Kazarinov, T. L. Koch, C.-Y. Kuo, M. Lax, O. Nilsson, and N. A. Olsson for stimulating discussions.

## APPENDIX 2.A   CALCULATION OF NORMALIZATION CONSTANT $B$

The normalization constant $B$ is determined by equating the energy in the active portion of the laser to $I\hbar\omega_0$, where $I = |\beta|^2$:

$$I\hbar\omega_0 = \frac{1}{8\pi} \int_0^L \left[ \frac{d}{d\omega}(\varepsilon_\omega\omega)E^2 + H^2 \right] dz \qquad (2.\text{A1})$$

where

$$E(z,t) = B\beta Z_0(z)\phi_0(x) + \text{c.c.} \qquad (2.\text{A2})$$

We can write $Z_0(z)$ as the sum of forward and backward propagating parts:

$$Z_0(z) = A_+(z)e^{ik_0z} + A_-(z)e^{-ik_0z} \tag{2.A3}$$

For a TE field in the $x$ direction, the transverse $H$ field in the $y$ direction is similar to Eq. (2.A2), but with $Z_0(z)$ replaced by $Z_{0H} = (c/i\omega)(d/dz)Z_0$:

$$Z_{0H}(z) = n_0'\left[A_+(z)e^{ik_0z} - A_-(z)e^{-ik_0z}\right] \tag{2.A4}$$

When oscillating terms are neglected, we find $(|Z_0|^2) \equiv \int_0^L |Z_0|^2\, dz = (|Z_{0H}|^2)$.

Evaluating Eq. (2.A1) using Eqs. (2.A2)–(2.A4) and $c/v_g = n_g = n_0' + \omega\, dn_0'/d\omega$, we find

$$B^2 = \frac{2\pi\hbar\omega_0}{n_0' n_g (|Z_0|^2)} \tag{2.A5}$$

## APPENDIX 2.B    CALCULATION OF $\langle F_\beta(t)F_\beta(t)^*\rangle$

The correlation function of $F_\beta(t)$ and $F_\beta(t')^*$ can be obtained by using Eqs. (2.123) and (2.127):

$$\langle F_\beta(t)F_\beta(t')^*\rangle = \frac{\int_{\omega_0-\Delta\omega}^{\omega_0+\Delta\omega}\left(\phi_0^2|Z_0|^2 2D_{FF^*}(\mathbf{x},\omega)\right)e^{-i\omega(t-t')}\,d\omega}{B^2|dW_0/d\omega|^2} \tag{2.B1}$$

where the parenthesis indicates integration over both $x$ and $z$. We assume that $2D_{FF^*}(\mathbf{x},\omega)$ is slowly varying with $\omega$ and remove it from the integral. We also take $\Delta\omega$ to be sufficiently large that the remaining integral

$$\int_{\omega_0-\Delta\omega}^{\omega_0+\Delta\omega} e^{-i\omega(t-t')}\,d\omega \approx 2\pi\delta(t-t') \tag{2.B2}$$

For a uniform active layer, the double integral over $x$ and $z$ can be separated. With these changes and the elimination of $B^2$ by use of Eq. (2.A5) the correlation function [Eq. (2.B1)] becomes

$$\langle F_\beta(t)F_\beta(t')^*\rangle = R\delta(t-t') \tag{2.B3}$$

where $R$ is given by Eq. (2.129).

## APPENDIX 2.C   SLOPES OF THE "LOSS" AND "CONSTANT PHASE" CURVES AT THE OPERATING POINT

The two conditions determining laser threshold are gain equal loss

$$g = \gamma + \frac{1}{L} \ln \frac{1}{r_1(\omega)} + \frac{1}{L} \ln \frac{1}{r_2} \tag{2.C1}$$

and round-trip phase equals $2\pi N$

$$\phi_1(\omega) + \phi_2 + 2kL - \alpha gL = 0 \tag{2.C2}$$

The slope of the phase curve $dG/d\omega$, where $G = gv_g$, is found by differentiating Eq. (2.C2) with respect to $\omega$, using $dk/d\omega = 1/v_g$ and $\tau_0 = 2L/v_g$. It is given by

$$\frac{dG}{d\omega} = \frac{2}{\alpha} \left( 1 + \frac{1}{\tau_0} \frac{d\phi_1}{d\omega} \right) = \frac{2}{\alpha} (1 + A) \tag{2.C3}$$

The increase in slope of the phase curve relative to that of the FP laser without feedback is $1 + A$.

The slope of the "loss" curve is given by differentiating Eq. (2.C1):

$$\frac{dG}{d\omega} = -\frac{v_g}{L} \frac{d(\ln r_1)}{d\omega} = -\frac{2}{\alpha} \frac{\alpha}{\tau_0} \frac{d(\ln r_1)}{d\omega} = -\frac{2}{\alpha} (\alpha B) \tag{2.C4}$$

Thus, the negative slope of the loss curve is $\alpha B$ times the slope of the phase curve of a FP laser without feedback (see Figure 2.18).

## APPENDIX 2.D   SIMPLE TWO-BAND MODEL OF A LASER

We present here a simple and conventional model of a laser in which the energy bands are approximated by a two-band model and the quantum well energy levels are replaced by those of an infinite square well. We will take the conduction band mass as $m_c$ and the valence band mass as $m_v = (m_l^{3/2} + m_h^{3/2})^{2/3}$, where $m_l$ and $m_h$ are the light and heavy hole masses, respectively.

The quantum well energies are given by $\varepsilon_{0r} N^2$, where

$$\varepsilon_{0r} = \frac{\hbar^2}{2m_r} \left( \frac{\pi}{L_z} \right)^2 \tag{2.D1}$$

where $L_z$ is the width of the quantum well and $r$ is $c$ or $v$, for the conduction and valence quantum well levels. The bulk density of states of the bands is given by

$$\rho_{\text{bulk}} = \left(\frac{2}{\pi}\right)^{1/2} \left(\frac{m_r}{\pi \hbar^2}\right)^{3/2} \varepsilon^{1/2} = A_r \varepsilon^{1/2} \tag{2.D2}$$

where $\varepsilon$ is the energy of the carrier relative to the band edge. For a quantum well, the density of states is modified into a series of steps. At each step the density of states rises to the value for $\rho_{\text{bulk}}$:

$$\rho_{\text{QW}_r} = A_r \varepsilon_{0r}^{1/2} \text{ in} \left(\frac{\varepsilon}{\varepsilon_{0r}}\right)^{1/2} \tag{2.D3}$$

where $\text{in}(x)$ is $x$ truncated to an integer. The Fermi levels $\mu_r$ are adjusted to the values necessary to specify the carrier densities given by

$$n, p = \int_0^\infty \left[\exp\left(\frac{\varepsilon - \mu_r}{kT}\right) + 1\right]^{-1} \rho \, d\varepsilon \tag{2.D4}$$

where $\mu_r$ are measured relative to the band edges. We approximate the absorption coefficient $a_0$ for an empty conduction band and a full valence band as

$$a_{0_{\text{bulk}}} = A_{\text{abs}} (\hbar\omega - E_g)^{1/2} \tag{2.D5}$$

where $A_{\text{abs}} = 16000 \text{ cm}^{-1}/(0.1 \text{ eV})^{1/2}$ was determined by the value of the absorption coefficient 0.1 eV above the band gap. In the quantum well case

$$a_{0\text{QW}} = A_{\text{abs}} (\varepsilon_{0c} + \varepsilon_{0v})^{1/2} \text{ in} \left(\frac{\hbar\omega - E_g}{\varepsilon_{0c} + \varepsilon_{0v}}\right)^{1/2} \tag{2.D6}$$

The level occupation values for an electron in the conduction band $f_e = (e^{\varepsilon - \mu_c/kT} + 1)^{-1}$ and hole in the valence band is given by $f_h = (e^{\varepsilon - \mu_v/kT} + 1)^{-1}$. The absorption coefficient of upward transitions is

$$a(\hbar\omega) = a_0(\hbar\omega)(1 - f_e)(1 - f_h) \tag{2.D7}$$

The coefficient of spontaneous emission is given by

$$r(\hbar\omega) = a_0(\hbar\omega) f_e f_h \tag{2.D8}$$

The gain is given by

$$g(\hbar\omega) = r - a = a_0(\hbar\omega)(f_e + f_h - 1) \qquad (2.D9)$$

In calculating $f_e$ and $f_h$ in the preceding formulas

$$\varepsilon_c = (\hbar\omega - E_g)(m_{red}/m_c)$$

and $\varepsilon_v = (\hbar\omega - E_g)(m_{red}/m_v)$, where $1/m_{red} = 1/m_c + 1/m_v$. The curves in Figure 2.28 were calculated using $m_c = 0.07m_0$, $m_h = 0.55m_0$, and $m_1 = 0.078m_0$.

## REFERENCES

1. A. L. Schawlow and C. H. Townes, *Phys. Rev.* **112**, 1940–1949 (1958).

2. M. Lax, *Phys. Rev.* **160**, 290–307 (1967).

3. R. D. Hempstead and M. Lax, *Phys. Rev.* **161**, 350–366 (1967).

4. H. Gerhardt, H. Welling, and A. Guttner, *Z. Physik* **253**, 113–126 (1972).

5. M. W. Fleming and A. Mooradian, *Appl. Phys. Lett.* **38**, 511 (1981).

6. C. H. Henry, *IEEE J. Quantum Electron.* **QE-18**, 259–264 (1982).

7. B. Diano, P. Spano, M. Tamburrini, and S. Piazzolla, *IEEE J. Quantum Electron.* **QE-19**, 266–269 (1983).

8. K. Vahala and A. Yariv, *IEEE J. Quantum Electron.* **QE-19**, 1102–1109 (1983).

9. R. Wyatt and W. J. Devlin, *Electron Lett.* **19**, 110–112 (1983).

10. M. W. Flemming and A. Mooradian, *IEEE J. Quantum Electron.* **QE-7**, 44–59 (1981).

11. C. H. Henry, *J. Lightwave Technol.* **4**, 288–297 (March 1986).

12. N. A. Olsson, C. H. Henry, R. F. Kazarinov, H. J. Lee, and B. H. Johnson, *Appl. Phys. Lett.* **51**, 1141–1142 (1987).

13. R. F. Kazarinov, C. H. Henry, and N. A. Olsson, *IEEE J. Quantum Electron.* **QE-23**, 1419–1425 (1987).

14. D. A. Ackerman, M. I. Dahbura, C.-Y. Kuo, Y. Shani, C. H. Henry, R. C. Kislter, and R. F. Kazarinov, "Compact hybrid resonant-optical reflector lasers with very narrow linewidths," paper presented at Integrated Photonics Research Conference, Hilton Head, SC, Mar. 26–28, 1990.

15. N. Ogasawara, R. Itoh, and R. Morita, *Jpn. J. Appl. Phys.* **L24**, L519–L521 (1985).

16. T. L. Koch, V. Koren, and B. I. Miller, "High performance tunable 1.6 $\mu$m InGaAs/InGaAsP multiple-quantum-well distributed Bragg reflector laser," paper presented at 11th IEEE International Semiconductor Laser Conference, Boston, MA, Aug. 29–Sept. 1, 1988, Paper J-1.

17. L. D. Westbrook, D. M. Cooper, and P. C. Spurdens, "Measurements of linewidth enhancement, gain and spontaneous emission in InGaAs quantum well lasers with InGaAsP barriers," paper presented at 11th IEEE Interna-

tional Semiconductor Laser Conference, Boston, MA, Aug. 29–Sept. 1, 1988, Paper P-1.

18. S. Ogita, Y. Kataki, M. Matsuda, Y. Kuwahara, and H. Ishikawa, *Electron. Lett.* **25**, 629–630 (1989).

19. F. Kano, Y. Tohmori, Y. Kondo, M. Nakao, M. Fukuda, and K. Oe, *Electron. Lett.* **25**, 709–710 (1989).

20. R. F. Kazarinov and C. H. Henry, *IEEE J. Quantum Electron.* **QE-23**, 1401–1409 (1987).

21. K. Kojima and K. Kyuma, *Jpn. J. Appl. Phys.* **27**, L1721–L1723 (1988).

22. N. A. Olsson, C. H. Henry, R. F. Kazarinov, H. J. Lee, and B. H. Johnson, *Appl. Phys. Lett.* **41**, 92–93 (1987).

23. D. Bohm, *Quantum Theory*, Prentice-Hall, Englewood Cliffs, NJ, 1951, Chapter 13.

24. R. J. Glauber, *Phys. Rev.* **131**, 2766–2788 (1963).

25. E. C. Sudarshan, *Phys. Rev. Lett.* **10**, 277 (1963).

26. D. F. Walls, *Nature* **306**, 141–146 (1983).

27. M. Lax and W. H. Louisell, *IEEE J. Quantum Electron.* **QE-3**, 47–58 (1967).

28. M. Lax, *Phys. Rev.* **157**, 213–231 (1967).

29. Y. Yamamoto, S. Machida, and O. Nilsson, *Phys. Rev.* **A34**, 4025–4042 (1986).

30. S. Machida, Y. Yamamoto, and Y. Itoya, *Phys. Rev. Lett.* **58**, 1000–1004 (1987).

31. M. Lax, *Rev. Mod. Phys.* **32**, 25–64 (1960).

32. M. Lax, "Fluctuations and coherence phenomena in classical and quantum physics," *Bradneis University Summer Institute in Theoretical Physics 1966*, Vol. 2, M. Chretian, E. P. Gross, and S. Desar, eds., Gordon and Breach, New York, 1966.

33. S. E. Miller, *IEEE J. Quantum Electron.* **QE-22**, 16–19 (1986).

34. M. Lax, *Rev. Mod. Phys.* **38**, 541–566 (1966).

35. L. D. Landau and E. M. Lifshitz, *Statistical Physics*, 1st ed., Addison Wesley, Reading, MA, 1958.

36. J. M. Wozencraft and J. M. Jacobs, *Principles of Communications Engineering*, Wiley, New York, 1956, Chapter 2.

37. H. Haag, *Phys. Rev.* **184**, 338 (1969).

38. J. S. Cohen, "Langevin equations for semiconductor lasers," Phillips Research Laboratories Nat. Lab. Technical note 195/88.

39. Y. Yamamoto and S. Machida, *Phys. Rev.* **A35**, 5114–5130 (1987).

40. M. Lax, *IEEE J. Quantum Electron.* **QE-3**, 37–46 (1967).

41. C. H. Henry, *J. Lightwave Technol* **LT-4**, p. 298 (1986).

42. T. Okoshi, K. Kikucki, and A. Nayayama, *Electron. Lett.* **28**, 1011–1012 (1985).

43. D. Welford and A. Mooradian, *Appl. Phys. Lett.* **40**, 865–867 (1982).

44. C. H. Henry, *IEEE J. Quantum Electron.* **QE-19**, 1391–1397 (1983).

45. R. S. Tucker, *IEEE J. Lightwave Technol.* **LT-3**, 1180–1192 (1985).

46. E. Eichen and P. Melman, *Electron. Lett.* **20**, 826–828 (1985).

47. K. Vahala, Ch. Harder, and A. Yariv, *Appl. Phys. Lett.* **42**, 211–213 (1983).

48. A. Mooradian, *Physics Today*, May, 1985, pp. 43–48.

49. L. D. Landau and E. M. Lifshitz, *Electrodynamics of Continuous Media*, Pergamon Press, New York, 1960, Section 88.

50. P. M. Morse and H. Feshbach, *Methods of Theoretical Physics*, McGraw-Hill, New York, 1953, Chapter 7.

51. J. Wang, N. Schumk, and K. Petermann, *Electron. Lett.* **23**, 715–717 (1989).

52. J. Arnaud, *Opt. Quantum Electron.* **18**, 335–343 (1986).

53. K. Ujihara, *IEEE J. Quantum Electron.* **QE-20**, 814–818 (1984).

54. G. Bjork and O. Nilsson, *IEEE J. Quantum Electron.* **QE-23**, 1303–1313 (1987).

55. G. Duan, P. Gallian and G. Debarge, *IEEE J. Quant. Electron.* **26**, 32–44 (1990).

56. S. L. McCall and P. M. Platzman, *IEEE J. Quantum Electron.* **FBQE-21**, 1899–1904 (1985).

57. E. Patzak, A. Sugimura, S. Saito, T. Mukai, and H. Oleson, *Electron. Lett.* **19**, 1026–1027 (1983).

58. M. Yamada and Y. Suematzu, *IEEE J. Quantum Electron.* **QE-15**, 743 (1979).

59. R. F. Kazarinov, C. H. Henry, and R. A. Logan, *J. Appl. Phys.* **53**, 4631–4644 (1982).

60. G. P. Agrawal, *J. Appl. Phys.* **63**, 1232–1235 (1988).

61. M. P. Kesler and E. P. Ippen, *Appl. Phys. Lett.* **51**, 1765–1767 (1987).

62. C. B. Su, *Electron. Lett.* **24**, 320–371 (1988).

63. N. A. Olsson, C. H. Henry, R. F. Kazarinov, H. J. Lee, K. J. Orlowsky, B. H. Johnson, R. E. Scotti, D. A. Ackerman, and D. J. Anthony, *IEEE J. Quantum Electron.* **24**, 143–147 (1988).

64. R. Wyatt, *Electron. Lett.* **21**, 658–659 (1985).

65. N. A. Olsson and J. P. van der Ziel, *J. Lightwave Technol.* **LT-5**, 510–515 (1987).

66. A. P. Bogotov, P. G. Elisseev, and B. N. Sverdlov, *IEEE J. Quantum Electron.* **QE-11**, 510–515 (1975).

67. D. Lenstra, B. H. Verbeek, and A. J. den Boef, *IEEE J. Quantum Electron.* **QE-21**, 674–679 (1985).

68. H. Temkin, N. A. Olsson, J. H. Abeles, R. A. Logan, and M. B. Panish, *IEEE J. Quantum Electron.* **QE-22**, 286–293 (Feb. 1986).

69. C. H. Henry and R. F. Kazarinov, *IEEE J. Quantum Electron.* **QE-22**, 294–301 (Feb. 1986).

70. F. Morgensen, H. Olesen, and G. Jacobsen, *IEEE J. Quantum Electron.* **QE-21**, 1152–1156 (1985).

71. R. A. Linke and K. J. Pollock, "Linewidth vs. length dependence for an external cavity laser," in *Proceedings of the 10th IEEE International Semiconductor Laser Conference*, Kanazawa, 1986, pp. 118–119.

72. C. H. Henry, R. A. Logan, H. Temkin, and F. R. Merritt, *IEEE J. Quantum Electron.* **QE-19**, 941–946 (1983).

73. C. H. Henry, "Spectral properties of semiconductor lasers," in *Semiconductors and Semimetals*, Vol. 22, Part B, W. T Tsang, editor, Academic Press, New York, 1985.

74. C. H. Henry, R. A. Logan and K. A. Bertness, *J. Appl. Phys.* **52**, 4457–4461 (1981).

75. W. G. Spitzer and H. Y. Fan, *Phys. Rev.* **196**, 882–890 (1957).

76. M. Osinski and J. Buus, *IEEE J. Quantum Electron.* **QE-23**, 9–29 (1987).

77. E. O. Kane, *J. Phys. Chem. Solids*, **1**, 249–261 (1989).

78. C. H. Henry, R. A. Logan, F. R. Merritt, and J. P. Luongo, *IEEE J. Quantum Electron.* **QE-19**, 947–952 (1983).

79. M. G. Burt, *Electron. Lett.* **20**, 27–28 (1984).

80. Y. Anakawa and A. Yariv, *IEEE J. Quantum Electron.* **QE-21**, 1666–1674 (1985).

81. L. D. Westbrook and M. J. Adams, *IEEE Proc.* **135**, Part J., 223–225 (1988).

82. T. Ohtoshi and N. Chinone, *IEEE Photonics Technol. Lett.* **1**, 117–118 (1989).

83. J. D. Dow and D. Redfield, *Phys. Rev.* **B5**, 594–610 (1972).

84. D. Marcuse, *Bell Syst. Tech. J.* **48**, 3187 (1969).

85. C. H. Henry, R. A. Logan and F. R. Merritt, *IEEE J. Quantum Electron.* **QE-17**, 2196–2204 (1989).

86. C. A. Wang, H. K. Choi, and M. K. Connors, *IEEE Photonics Technol. Lett.* **1**, 351–352 (1989).

87. H. Temkin, T. Tanbun-Ek, R. A. Logan, J. A. Lewis, and N. K. Dutta, *Appl. Phys. Lett* **56**, 1222–1224 (1990).

88. T. L. Koch, U. Koren, and B. I. Miller, *Appl. Phys. Lett.* **53**, 1036–1038 (1988).

89. T. L. Koch and U. Koren, *J. Lightwave Technol.* (in press).

90. K. Kobayashi and I. Mito, "Progress in narrow-linewidth tunable laser sources," OFC/IOOC'87, paper WC1, Reno, NV, 1987.

91. C. Zah, C. Caneau, S. G. Menocal, P. S. D. Lin, A. S. Gozdz, F. Faviee, and T. P. Lee, *Electron. Lett.* **24**, 94–96 (1988).

92. H. Yasaka, M. Fukuda, and T. I. Kegami, *Electron. Lett.* **24**, 760–761 (1988).

93. K. Kikuchi, *IEEE J. Quantum Electron.* **QE-25**, 684–688 (1989).

# 3

# Modulation and Noise Spectra of Complicated Laser Structures

OLLE NILSSON, ANDERS KARLSSON, AND EILERT BERGLIND
*Department of Microwave Engineering, Royal Institute of Technology, Stockholm, Sweden*

## 3.1. INTRODUCTION

In coherent optical communication the requirements on the lasers regarding their tuning, modulation, and noise properties are severe and often mutually contradictory. There are several questions of interest for the system designer which, to our knowledge, have not yet been fully answered. Two examples of such questions are as follows. It is well known that the linewidth, that is, the low-frequency FM noise, can be greatly reduced by the use of a long external cavity. What about the higher frequency FM noise? One should expect that the noise becomes large as one approaches a modulation frequency corresponding to the round-trip time in the cavity and also that it may be significant at the relaxation frequency, which is lower in a long cavity laser. To what extent, if any, is the tunability of, for example, a multielectrode DFB laser accompanied by increased FM noise?

In order to find answers to questions such as these we will formulate a theory allowing for high modulation frequencies and several active laser regions with different material properties. The treatment will be strictly small signal in terms of modulation, but the stationary state is assumed to be known from a full nonlinear analysis. Noise sources are introduced with spectral densities and correlation properties satisfying the fluctuation–dissipation theorem. It turns out to be possible to arrive at a formal solution within certain restrictions. However, the reduction of this solution to a tractable form is not trivial except in simple cases. We will point out

*Coherence, Amplification, and Quantum Effects in Semiconductor Lasers*, Edited by Yoshihisa Yamamoto.
ISBN 0-471-512494 © 1991 John Wiley & Sons, Inc.

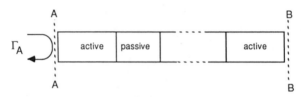

**Figure 3.1.** Model of a laser structure: $\Gamma_A$ is the reflection coefficient in plane $A$; to the left of plane $A$ and to the right of plane $B$ there are no reflections.

some of the difficulties, relate previous theories to our treatment, and give some new results for cases where reliable and self-consistent approximations can be made.

Finally we would like to point out that this chapter is neither a review of the subject nor a monograph of a completed study. Rather, it is a presentation of an ongoing and yet unfinished work, and we hope that some of the problems encountered will serve as a challenge for other workers in the field.

## 3.2. FORMAL STRUCTURE OF THE THEORY

### 3.2.1. Linearized Multiport Description

Consider a laser structure made up by active regions, gratings, mirrors, waveguiding sections, and other components, as depicted schematically in Figure 3.1. We will assume that the laser operates in one single transverse and longitudinal mode. To the left of plane $A$ and to the right of plane $B$ there are no reflections. The approach is to calculate the field distribution in the structure and the reflection coefficient, $\Gamma_A$, in plane $A$ as functions of frequency for fixed values of material parameters, that is, fixed values of gain and refractive indices obtained from a simultaneous solution of the carrier-rate equations and the oscillation condition $1/\Gamma_A = 0$ in the absence of noise and modulation. The influence of modulation and noise, such as fluctuations in carrier numbers, will be described in terms of small equivalent source currents inserted at appropriate locations in the structure in a self-consistent way using a small signal linearization.

The equivalent total admittance in plane $A$ can be written as

$$Y = G_L + Y_A \qquad (3.1)$$

where $G_L$ is the equivalent load conductance and $Y_A$ the active admittance as seen looking into the laser structure. Expressed in the reflection coefficient $\Gamma_A$, we get

$$Y = G_L \frac{2}{1 + \Gamma_A} \tag{3.2}$$

where the reflexion coefficient is defined as

$$\Gamma_A \equiv \frac{V_{A,\text{out}}}{V_{A,\text{in}}} = \frac{G_L - Y_A}{G_L + Y_A} = \frac{2G_L - Y}{Y} \tag{3.3}$$

At free-running oscillation we have

$$Y(\omega_0) = 0 \tag{3.4}$$

Since the reflection coefficient can be determined as a function of frequency, $Y$ can be considered as a known function of frequency.

If there are no source currents in the active regions, but only an incoming voltage wave $V_{A,\text{in}}$, such as those stemming from zero-point fluctuations, we will have for the total voltage and current in plane $A$

$$Y(\omega)V_A(\omega) = I_A(\omega) \tag{3.5}$$

but if we use the definition of $\Gamma_A$ and Eq. (3.2) we obtain

$$I_A(\omega) = 2G_L V_{A,\text{in}} \tag{3.6}$$

which can be seen as a mere definition of the current source describing an incoming voltage wave. Generally the outgoing voltage will be given by

$$V_{A,\text{out}} = V_A - V_{A,\text{in}} = V_A - \frac{I_A}{2G_L} \tag{3.7}$$

Let us emphasize here the significance of Eq. (3.7). When we make measurements on the laser output field, we get $V_{A,\text{out}}$, the equivalent voltage for the outgoing field, but standard laser theories will give us $V_A$, the equivalent voltage for the internal field. But by using Eq. (3.7) we can subtract the incoming vacuum fluctuation $V_{A,\text{in}}$, which is known from the fluctuation–dissipation theorem [1–3].

To find the voltage amplitude $V_A$ in the presence of small source currents in the active regions and at the outputs, we need a multiport description. We therefore divide the laser in regions so small that the field can be considered as constant within each region. We will also need the voltage amplitudes $V_i$ at the locations of the source currents $I_i$. Therefore, noting that the multiport is linear as defined, we write

$$Y \cdot V_A = I_A + \sum_i H_{Ai} I_i + H_{AB} I_B$$

$$Y \cdot V_i = H_{iA} I_A + \sum_j H_{ij} I_j + H_{iB} I_B \qquad (3.8)$$

$$Y \cdot V_B = H_{BA} I_A + \sum_i H_{Bi} I_i + H_{BB} I_B$$

where assuming reciprocity

$$H_{Ai} = H_{iA}, \ldots \qquad (3.9)$$

Here, of course, both $Y$ and the transfer functions $H$ are functions of frequency. There are several possible ways to calculate the transfer functions $H$, and we will return to this problem in Section 3.2.2 and later. Another problem related to this is that the number of regions is very large —for a typical laser on the order of $10^4$ regions! An important question is therefore when and to what extent many regions can be lumped into one without introducing significant errors. This will also be discussed later.

### 3.2.2. Determination of the Transfer Functions $H(\omega)$

The transfer functions $H_{Ai}$ and so on can, of course, be determined by introducing source currents $I_{A,i,B}$ at the individual locations and then calculating the resulting voltage distributions. For example, $H_{iA}$ (and $H_{Ai}$ due to reciprocity) is obtained from $V_i/V_A$ with all currents except $I_A$ equal to zero. Instead of using this direct approach one may study how the admittance $Y_A \equiv I_A/V_A$ changes when we connect a small additional conductance $\delta G_i$ at the location $i$ such that $I_i$ becomes $-V_i \cdot \delta G_i$. One then finds that

$$H_{Ai}^2 = \frac{\partial Y_A}{\partial G_i} \qquad (3.10)$$

$$H_{ii} = H_{Ai}^2 - \frac{1}{2} \cdot \frac{Y_A}{H_{Ai}^2} \cdot \frac{\partial^2 Y_A}{\partial G_i^2} \qquad (3.11)$$

$$H_{ij} = H_{Ai} H_{Aj} - \frac{1}{2} \cdot \frac{Y_A}{H_{Ai} H_{Aj}} \cdot \frac{\partial^2 Y_A}{\partial G_i \partial G_j} \qquad (3.12)$$

These quantities can be readily found from a transfer matrix analysis of the laser structure. They are particularly suitable if one wants to approximate the real laser with a lumped model as is done below in some simple cases.

### 3.2.3. The Modulated Field

So far it has been understood that we have been dealing with monochromatic fields. Rather than treating the modulated fields as a superposition of such fields, we write them as

$$V_{A,i,B}(t) = Re\{V_{A,i,B0}e^{j\omega_0 t + c_{A,i,B}(t)}\} \tag{3.13}$$

where

$$c(t) = a(t) + j\varphi(t) \tag{3.14}$$

describes the amplitude and phase modulation at the various locations. Here $a(t)$ and $\varphi(t)$ are real and will be treated as small quantities except for a very slowly varying part of $\varphi(t)$ expressing the phase diffusion common to all fields in the laser. Therefore it is always true that

$$|c_i(t) - c_A(t)| \ll 1, |c_i(t) - c_B(t)| \ll 1, |c_i(t) - c_j(t)| \ll 1 \tag{3.15}$$

and that we can take the small source currents to be of the form

$$I_{A,i,B}(t) = Re\{I_{A,i,B\Omega}e^{j(\omega_0 + \Omega)t + c_{A,i,B}(t)}\} \tag{3.16}$$

where

$$\Omega = \omega - \omega_0 \tag{3.17}$$

and where only the phase diffusion part in $c(t)$ is significant. Equations (3.8) can then be approximated as

$$\frac{1}{2}V_{A0}Y_\Delta(\Omega) \cdot (a_{A\Omega} + j\varphi_{A\Omega}) = I_{A\Omega} + \sum_i H_{\Delta Ai}(\Omega)I_{i\Omega} + H_{\Delta AB}(\Omega)I_{B\Omega}$$

$$\frac{1}{2}V_{i0}Y_\Delta(\Omega) \cdot (a_{i\Omega} + j\varphi_{i\Omega}) = H_{\Delta Ai}(\Omega)I_{A\Omega} + \sum_j H_{\Delta ij}(\Omega)I_{j\Omega}$$

$$+ H_{\Delta Bi}(\Omega)I_{B\Omega} \tag{3.18}$$

$$\frac{1}{2}V_{B0}Y_\Delta(\Omega) \cdot (a_{B\Omega} + j\varphi_{B\Omega}) = H_{\Delta AB}(\Omega)I_{A\Omega} + \sum_i H_{\Delta Bi}(\Omega)I_{i\Omega}$$

$$+ H_{\Delta BB}(\Omega)I_{B\Omega}$$

and the corresponding equations with $\Omega$ replaced by $-\Omega$. We also used Eq. (3.4) and the notation

$$Y_\Delta(\Omega) = Y(\omega), \qquad H_\Delta(\Omega) = H(\omega) \tag{3.19}$$

$$a(t), \varphi(t) = Re\{a_\Omega, \varphi_\Omega e^{j\Omega t}\} \tag{3.20}$$

Note that positive and negative frequency components will be coupled, as will be shown below, and that $Y_\Delta(-\Omega)$ does not equal $Y_\Delta^*(\Omega)$ in the general case. Before calculating the source currents we treat the carrier-rate equations.

### 3.2.4.  Carrier-Rate Equations

The carrier-rate equation for region $i$ can be written

$$\frac{dN_i}{dt} = J_i - R_i - \frac{P_{si}}{\hbar\omega} - D_i\left(N_i - \frac{N_{i-1} + N_{i+1}}{2}\right) - \Gamma_{Ri} - \Gamma_{si} \quad (3.21)$$

where $N_i$ is the number of carriers, $J_i$ is the number of injected carriers per unit time possibly including noise, $R_i$ is the spontaneous recombination rate, $P_{si}$ is the net generated power through stimulated emission–absorption, $D_i$ is a constant expressing the coupling of neighboring regions due to carrier diffusion (the noise associated with the diffusion process has been neglected), $\Gamma_{Ri}$ is the recombination shot noise fluctuation, and finally $\Gamma_{si}$ is the carrier fluctuation accompanying the stimulated emission–absorption. The noise sources will be discussed in Sections 3.2.6–3.2.7. The steady-state solution to Eq. (3.21) can be assumed as known; for example, it can be solved iteratively with the help of the transfer matrix method [4, 5]. To find a small signal equation we define

$$s_{Ni} = \frac{1}{g_{si}} \cdot \frac{\partial g_{si}}{\partial N_i} \quad (3.22)$$

$$s_{Pi} = \frac{|V_i|^2}{g_{si}} \cdot \frac{\partial g_{si}}{\partial |V_i|^2} \quad (3.23)$$

where $g_{si}$ is the gain constant, $s_{Ni}$ expresses the differential gain and $s_{Pi}$ expresses the nonlinear gain. We also need the relative intensity variation

$$\frac{\Delta|V_i|^2}{|V_i|^2} = 2a_i(t) \quad (3.24)$$

Using

$$\Delta N_i(t) = Re\{\Delta N_{i\Omega} e^{j\Omega t}\} \quad (3.25)$$

we can do a small signal Fourier decomposition of the rate equation, finding

$$\left(j\Omega + \frac{P_{si}}{\hbar\omega} s_{Ni} + \frac{dR_i}{dN_i}\right)\Delta N_{i\Omega} = \Delta J_{i\Omega} - 2a_\Omega \frac{P_{si}}{\hbar\omega}(1 + s_{Pi}) - \Gamma_{Ri\Omega} - \Gamma_{si\Omega}$$

$$- D_i\left(\Delta N_{i\Omega} - \frac{\Delta N_{i-1\Omega} + \Delta N_{i+1\Omega}}{2}\right) \quad (3.26)$$

The diffusion term in Eq. (3.26) is significant in the case of spatial hole burning on a scale of the optical standing-wave pattern. We have introduced it in Eqs. (3.21) and (3.26) only for completeness, and we have not made any explicit use of the diffusion term in the paragraphs to follow.

### 3.2.5. Source Currents Due to Fluctuation in Carrier Number and Nonlinear Gain

A fluctuation in carrier number $\Delta N_i$ gives rise to a fluctuation in complex conductivity that we account for by a source current

$$I_{Ni}(t) = Re\left\{ V_{i0}e^{j\omega_0 t + c_i(t)} \cdot \Delta N_i(t) \cdot \frac{2P_{si}}{|V_{i0}|^2}s_{Ni} \cdot (1 - j\alpha_i)\right\} \quad (3.27)$$

where

$$\alpha = \frac{\partial\varepsilon'}{\partial N}\bigg/\frac{\partial\varepsilon''}{\partial N} \quad (3.28)$$

is the linewidth enhancement factor which is usually negative in semiconductor lasers. With Eq. (3.25) one obtains for the current to be used in Eq. (3.18):

$$I_{Ni\Omega} = \frac{V_{i0}P_{si}}{|V_{i0}|^2} \cdot s_{Ni} \cdot (1 - j\alpha_i)\Delta N_{i\Omega} \quad (3.29)$$

The source current for the nonlinear gain is similarly obtained as

$$I_{Pi\Omega} = \frac{V_{i0}P_{si}}{|V_{i0}|^2} \cdot s_{Pi}2a_{i\Omega} \quad (3.30)$$

Here we have neglected the nonlinearity in $\varepsilon'$.

### 3.2.6. Noise Currents Due to Dipole Fluctuations and Optical Losses

Every stimulated recombination or excitation event is accompanied by noise stemming from the fluctuating dipoles. The noise current for the dipole fluctuation $I_{Di}$ is assumed to have a white, Gaussian spectrum. The noise sources can be assumed as spatially uncorrelated, and the single-sided power spectral density of $I_D$ is [6]

$$S_{fDi} = 2\hbar\omega\frac{2P_{si}}{|V_{i0}|^2}(2n_{spi} - 1) \quad (3.31)$$

where $n_{\mathrm{sp}\,i}$ is the spontaneous emission factor for region $i$. We also have a noise current $I_{ri}$, stemming from other losses than stimulated absorption, such as free carrier absorption and scattering losses. It has a power spectral density given by

$$S_{fri} = 2\hbar\omega \frac{2P_{ri}}{|V_{i0}|^2} \tag{3.32}$$

where $P_{ri}$ is the total loss mentioned above.

### 3.2.7. Noise Sources in the Carrier-Rate Equation

Three noise sources enter the carrier-rate equation. The first is the pump noise from the current injection, which can be directly included in $J_i$. This may or may not be suppressed depending on the pumping mechanism of the laser [7]. The second is the recombination shot noise $\Gamma_{Ri}$, the power spectral density of which is

$$S_{fRi} = 2R_i \tag{3.33}$$

where $R_i$ is the total recombination. Finally we have the dipole fluctuation $\Gamma_{si}$ that must enter the rate equation since every stimulated emission or absorption event also involves a carrier. It therefore must hold that

$$\hbar\omega\Gamma_{si}(t) = \langle V_i(t) \cdot I_{Di}(t) \rangle \tag{3.34}$$

where the average is taken over several light periods. If the in-phase component of the fluctuation is defined to have the phase of $V_i$ one finds

$$\Gamma_{si\Omega} = \frac{|V_{i0}|}{2\hbar\omega} [I_{Di\Omega} + I^*_{Di-\Omega}] \tag{3.35}$$

### 3.2.8. Noise Currents Due to the Load and to the Far End of the Laser

Even in the absence of deterministic input signals we will have fluctuations due to the zero-point fluctuations (represented earlier by $V_{A,\,\mathrm{in}}$). Thus the noise contribution to $I_A$ will have a spectral density given by

$$S_{fA} = 2\hbar\omega G_L = 2\hbar\omega \frac{2P_A}{|V_A|^2} \tag{3.36}$$

where $P_A$ is the output power at facet $A$. Similarly we have a noise contribution to $I_B$ with the spectral density

$$S_{fB} = 2\hbar\omega \frac{2P_B}{|V_B|^2} \tag{3.37}$$

where $P_B$ is the output power at facet $B$.

### 3.2.9. How to Obtain Noise and Modulation Properties

When all noise currents are given we can form the total noise current to be used in Eq. (3.18) as

$$I_{i\Omega} = I_{Ni\Omega} + I_{Pi\Omega} + I_{Di\Omega} + I_{ri\Omega} \tag{3.38}$$

Here $I_{Ni\Omega}$ is given by Eq. (3.29), $I_{Pi\Omega}$ by Eq. (3.30), $I_{Di\Omega}$ is the dipole fluctuation current defined by Eq. (3.31), and finally $I_{ri\Omega}$ is the noise current stemming from optical losses defined by Eq. (3.32). With this the theory is formally complete. Once we know the admittance $Y$ and the transfer functions $H$ we should, in principle, by solving the full system of equations Eq. (3.18) and using the results of Sections 3.2.3 through 3.2.8, be able to find the AM and FM modulation for arbitrary time-varying small-signal pump current distributions as well as the AM and FM noise spectra up to very high frequencies. In the next section we will discuss some applications of the formalism and some of the difficulties encountered in reducing the large set of equations to a tractable form.

## 3.3. APPLICATIONS AND COMPARISON WITH EARLIER RESULTS

### 3.3.1. Low-Frequency FM Noise

In the low-frequency limit we may use

$$Y_\Delta(\Omega) = \Omega \cdot \frac{\partial Y_\Delta}{\partial\Omega} \tag{3.39}$$

so that

$$Y_\Delta(-\Omega) = -Y_\Delta(\Omega) \tag{3.40}$$

and $H_\Delta(\Omega) \approx H_\Delta(0)$. We may also choose a reference plane such that

$$\frac{\partial Y_\Delta}{\partial\Omega} = j\frac{\partial B_\Delta}{\partial\Omega} \tag{3.41}$$

where $B_\Delta$ is the imaginary part of the admittance $Y_\Delta$. Using this in the first of Eqs. (3.18), again taking $V_{A0}$ as real, and adding the complex conjugate of the corresponding equation for $-\Omega$, one obtains

$$j\Omega \frac{\partial B_\Delta}{\partial \Omega} V_{A0} a_{A\Omega} = I_{A\Omega} + I^*_{A-\Omega} + \sum_i \left[ H_{\Delta Ai}(0) I_{i\Omega} + H^*_{\Delta Ai}(0) I^*_{i-\Omega} \right]$$
$$+ H_{\Delta AB}(0) I_{B\Omega} + H^*_{\Delta AB}(0) I^*_{B-\Omega} \qquad (3.42)$$

Taking the difference instead, one gets

$$-\Omega \frac{\partial B_\Delta}{\partial \Omega} V_{A0} \varphi_{A\Omega} = I_{A\Omega} - I^*_{A-\Omega} + \sum_i \left[ H_{\Delta Ai}(0) I_{i\Omega} - H^*_{\Delta Ai}(0) I^*_{i-\Omega} \right]$$
$$+ H_{\Delta AB}(0) I_{B\Omega} - H^*_{\Delta AB}(0) I^*_{B-\Omega} \qquad (3.43)$$

Since the amplitude fluctuations stay finite in the low frequency limit, the right-hand side of Eq. (3.42) must vanish in this limit, and it is possible to express the contribution from the $\Delta N_i$ values in all the other primary noise sources. If we neglect nonlinear gain ($s_p = 0$) and assume that

$$V_{i0} = H_{\Delta Ai}(0) \cdot V_{A0} \qquad (3.44)$$

and that by Eq. (3.25)

$$\Delta N_{i\Omega} = \Delta N^*_{i-\Omega} \qquad (3.45)$$

we find for Eq. (3.43)

$$-\Omega \frac{\partial B_\Delta}{\partial \Omega} V_{A0} \varphi_{A\Omega} = [I_\Omega - I^*_{-\Omega}] + j[I_\Omega + I^*_{-\Omega}] \cdot \frac{\Sigma_i \Delta N_{i\Omega} k_i [b_i + \alpha_i]}{\Sigma_i \Delta N_{i\Omega} k_i [1 - \alpha_i b_i]}$$
$$(3.46)$$

where

$$I_\Omega \equiv I_{A\Omega} + \sum_i H_{Ai}(0) [I_{D\Omega} + I_{P\Omega} + I_{r\Omega}] + H_{AB\Omega}(0) I_{B\Omega} \qquad (3.47)$$

$$k_i \equiv \left[ H_{\Delta Ai}(0)^2 + H^*_{\Delta Ai}(0)^2 \right] \cdot \frac{P_{si}}{|V_{i0}|^2} \cdot s_{Ni} \qquad (3.48)$$

$$jb_i \equiv -\frac{H_{\Delta Ai}(0)^2 - H^*_{\Delta Ai}(0)^2}{H_{\Delta Ai}(0)^2 + H^*_{\Delta Ai}(0)^2} \qquad (3.49)$$

has been used. Using Eqs. (3.46)–(3.49) we can calculate the low-frequency FM noise and hence the linewidth. In order to compare with earlier results [6] and others, which were obtained under the assumption of constant $n_{sp}$ and $\alpha$, we make the same assumption here. If, furthermore, $b_i$ also is constant $= b$, the quotient between the sums in Eq. (3.49) simply becomes $(b + \alpha)/(1 - \alpha b)$. Denoting this factor by $\alpha_T$, we obtain the linewidth as

$$\Delta\nu = \frac{h\nu}{\pi} \cdot \frac{n_{sp}P_s}{P_A^2} \cdot \left(\frac{G_L}{\partial B_\Delta/\partial\Omega}\right)^2 \cdot (1 + \alpha_T^2) \qquad (3.50)$$

Formally this agrees exactly with Ref. 6, if we identify our $b$ with $\rho_n$ of Ref. 6 and take $\rho$ in Ref. 6 as zero (due to our present choice of reference plane). Indeed, if we assume from the very outset that we can lump the active region into one section (which has been implicitly assumed in most earlier treatments) and use Eq. (3.10) to calculate $H$, we find that our $b$ is identical with $\rho_n$ of Ref. 6. The factor $\xi$ in Ref. 6 describing pump noise suppression, was somewhat erroneously and unnecessarily introduced and should be taken as unity. (Since no assumption of pump noise suppression —entering only through $\Delta N$ in Eq. (3.46)—has been made here, we can conclude that it does not affect the linewidth. This may also be inferred from physical grounds since the gain clamping will keep the carrier density constant for injection current modulation giving no low-frequency FM response. However, for inhomogeneous $\alpha$, pump noise does affect the linewidth as was pointed out by Arnaud [8], see also Section 3.3.3.) The assumption that $b_i$ is constant is strictly true for a FP laser with highly reflecting mirrors. It does not hold for a FP laser with small reflections or for a DFB laser, however. How this affects the validity of Eq. (3.50) remains to be investigated. Note, however, in any case, that $\alpha_T$ does not always equal the material $\alpha$, a situation that can arise, such as in lasers with external reflexions or in a DBR laser with passive Bragg reflectors. It can also be directly proved that the results in Ref. 6 include those of Ujihara [9], Henry [10], Arnaud [11], Wang et al. [12], and others. In fact, one can use the method of Arnaud [11] to prove directly that

$$\left|\frac{\partial B_\Delta}{\partial\Omega}\right|_{\rho=0} = \frac{1}{V_{A0}^2} \cdot \left|\int_{Vol}\left(\frac{\partial(\omega\varepsilon)}{\partial\omega}E^2 - \frac{\partial(\omega\mu)}{\partial\omega}H^2\right)dV\right| \qquad (3.51)$$

which shows that Petermann's and Arnaud's $K$ factors are automatically included in Ref. 6 and hence also in the present formulation.

### 3.3.2.  Short Fabry–Perot Laser with Highly Reflecting Mirrors

We consider a Fabry–Perot (FP) laser where the round-trip-time $\tau$ is short ($\Omega\tau \ll 1$), so that $Y_\Delta(\Omega) \approx \Omega(\partial Y/\partial\Omega)$ and $H(\Omega) \approx H(0)$. In this case the field distribution in the laser can be regarded to vary uniformly throughout the laser, that is, $c_A = c_i = c_B$. We do not need the full system of equations, but only the top equation, which becomes, after replacing the summation by an integral

$$\frac{1}{2}V_{A0}\Omega\frac{\partial Y}{\partial \Omega}c_{A\Omega} = I_{A\Omega} + I_{B\Omega} + \int_0^L \frac{V(x)}{V_{A0}}I_{x\Omega}(x)\,dx \qquad (3.52)$$

Here $V(x)/V_{A0}$ is real in the limit of large end mirror reflectivity. Furthermore, if carrier diffusion is large enough to prohibit spatial hole burning on the scale of optical standing-wave pattern, one can use a single rate equation for the entire active region replacing the quantities $N_i$ and so on in Eq. (3.21) and the following equations with the corresponding quantities for the entire region and also deleting the diffusion term.

If one introduces a small additional conductance $\delta G/L$ per unit length in the active region and calculates $H_{iA}$ using Eq. (3.10), the result is unity. Note, however, that such a lumping can be done only if $V(x)/V_{A0}$ is real, that is, for large mirror reflectivities. The resulting equations can then be written

$$\left(j\Omega + \frac{P_s}{\hbar\omega}s_N + \frac{dR}{dN}\right)\Delta N_\Omega = \Delta J_\Omega - 2a_\Omega\frac{P_s}{\hbar\omega}(1 + s_P) - \Gamma_{R\Omega}$$

$$- \frac{V_{A0}}{2\hbar\omega}[I_{D\Omega} + I_{D-\Omega}^*] \qquad (3.53)$$

$$j\Omega\frac{\partial B_\Delta}{\partial \Omega}V_{A0}a_{A\Omega} = \frac{2P_s s_N}{V_{A0}}\Delta N_\Omega + [I_{D\Omega} + I_{D-\Omega}^*] + [I_{P\Omega} + I_{P-\Omega}^*]$$

$$+ [I_{0\Omega} + I_{0-\Omega}^*] \qquad (3.54)$$

$$j\Omega\frac{\partial B_\Delta}{\partial \Omega}V_{A0}\varphi_{A\Omega} = -\alpha\frac{2P_s s_N}{V_{A0}}\Delta N_\Omega - j[I_{D\Omega} - I_{D-\Omega}^*] - j[I_{0\Omega} - I_{0-\Omega}^*]$$

$$(3.55)$$

where

$$I_{0\Omega} = I_{A\Omega} + I_{B\Omega} + I_{r\Omega} \qquad (3.56)$$

is due to all losses, external and internal. Note that $I_P$ does not enter the phase equation [Eq. (3.55)]. By comparing Eqs. (3.53)–(3.56) with the quantum-mechanical Langevin equations for laser noise [2, 7], we find that the equations describing the laser noise are exactly the same (cf. Ref. 7, Eqs. A1–A3, and identify the terms), apart from our inclusion of nonlinear gain and that $\alpha$ as defined here is negative.

To illustrate the effect of nonlinear gain, we can model the gain saturation of the material gain simply as

$$g = A(N - N_0)\frac{1}{1 + \varepsilon S} \qquad (3.57)$$

where $S$ is the number of photons in the cavity and $\varepsilon$ is a measure of the strength of the nonlinear gain. Using this model we get for $s_p$ and $\alpha$ from Eqs. (3.23) and (3.28)

$$s_p = -\frac{\varepsilon S}{1 + \varepsilon S} \tag{3.58}$$

$$\alpha = \alpha_0(1 + \varepsilon S) \tag{3.59}$$

To facilitate comparison with Ref. 7, we use the same parameter values as given in Figure 3 of Ref. 7; $\omega_0/Q_e = 4 \cdot 10^{11}\text{s}^{-1}$, $\omega_0/Q_0 = 10^{11}\text{s}^{-1}$, $n_{sp} = 2$, $A_0^2 = 10^5 R$, $\alpha = 2$ and $\tau_{sp} = 2$ ns, which translates to the notation used here as $\partial B_A/\partial\Omega = 2G_L(Q_e/\omega_0)$, $P_s/\hbar\omega = 10^5 R(\omega_0/Q_e + \omega_0/Q_0)$, $P_r/\hbar\omega = 10^5 R(\omega_0/Q_0)$, $P_A/\hbar\omega = 10^5 R(\omega_0/Q_e)$, $dR/dN = \tau_{sp}^{-1}$, $P_s s_N/\hbar\omega = n_{sp}(\tau_{sp}R)^{-1}$, $n_{sp} = 2$, and $\alpha = -2$. Using Eqs. (3.53)–(3.59) and (3.7) we have calculated the external amplitude and frequency noise spectra as a function of pumping parameter $R = I/I_{th} - 1$ [not to be confused with the recombination rate in Eq. (3.21)] with and without nonlinear gain with pump noise completely suppressed ($\Delta J_\Omega = 0$).

For every parameter value $R$, the same current is used (and the threshold current is the same) both with and without nonlinear gain. Note that the recombination rate will be higher and that the output power will be lower when nonlinear gain is included. The total gain is fixed by the cavity losses so that higher carrier densities are required to maintain oscillation for higher output powers. We have used $\varepsilon = 10^{-7}$ in the calculations, which might overestimate the nonlinear gain, but is chosen so that the effect is clearly seen.

We see in Figure 3.2 that the main effect of nonlinear gain is to reduce the relaxation peak. In Figure 3.2 we also have internal cavity losses, which limits the amount of amplitude squeezing available. If pump noise were suppressed and there were no nonlinear gain and no internal losses apart from carrier recombination, it would be possible to obtain an arbitrary amount of amplitude squeezing just by increasing the pumping. This is so simply because the number of stimulated emitted photons increases compared to the number of spontaneously emitted (which is fixed owing to gain clamping). However, when we have nonlinear gain, the spontaneous emission rate also increases with pumping (since a higher total inversion is required), thus giving a limit to the amount of amplitude squeezing available—for this particular case, using the simple model [Eq. (3.57)], the limit is $0.5 \cdot 10^5 \varepsilon$ in the units of Figure 3.2.

The external frequency noise is plotted in Figure 3.3. We see that the relaxation peak is reduced when nonlinear gain is included. We also see the usual inverse output power variation of the frequency noise, but that the frequency noise (and the linewidth) is rebroadened due to nonlinear gain for high pumping rates. Also the partial unclamping of the carrier

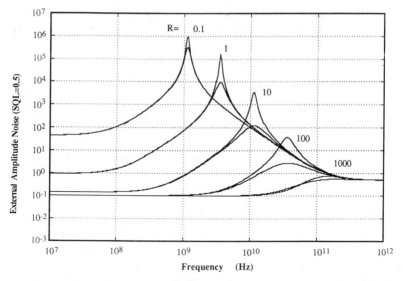

**Figure 3.2.** External amplitude noise (single-sided) (standard quantum limit = 0.5, corresponding to the shot-noise limit at a photodetector) for varying pump parameter, with pump noise completely suppressed. For numerical parameters see Ref. 7 (Figure 3) and above; also note that in Ref. 7 angular frequency is used on the $x$ axis, not frequency as here. The curves with lower relaxation peak are with nonlinear gain included.

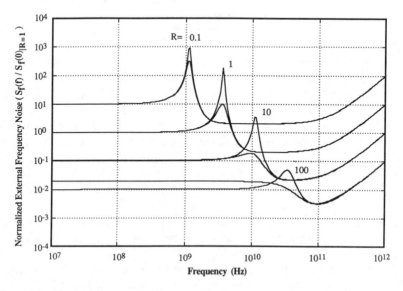

**Figure 3.3.** External frequency noise normalized to the low-frequency value of the frequency noise without nonlinear gain and with $R = 1$. Numerical parameters as in Figure 3.2; the curves with lower relaxation peak are with nonlinear gain included.

density due to nonlinear gain gives an injection current FM response even for low frequencies, implying that the laser linewidth will be affected by pump noise and recombination noise. For the case of constant spectral density of the pump noise, $s_p$ proportional to $S$, and all other parameters unchanged, this would yield a power independent contribution to the linewidth, as discussed by Agrawal and Roy [13].

### 3.3.3. Lasers with Inhomogeneous $\alpha$ Factor

The FM noise properties of lasers with inhomogeneous $\alpha$ factors have been discussed by Arnaud [14, 15]. In Ref. 14, he treats a laser with a transversally inhomogeneous $\alpha$, so that all layers with different $\alpha$ sense the same optical field. It is shown that the resulting linewidth enhancement will be given by an average of $1 + \alpha^2$ over the layers, weighted with respect to dissipated power (or injected current). Another important result is that pump noise increases the linewidth if $\alpha$ is inhomogeneous. These results can also be obtained in the present formalism by taking the transfer functions $H$ as unity and taking the field fluctuations in all sections to be the same. In Ref. 15 he treats a laser with a variable $\alpha$ in the longitudinal direction and for simplicity considers only a ring laser with some nonreciprocal material inserted so that only waves propagating in one direction are lasing. The result for the linewidth [14, Eq. 42] is again an average over the varying $1 + \alpha^2$, but here with respect to the reciprocal of the power gain. Although the present formalism is not directly applicable to the ring laser case, it seems that Arnaud's results can also be derived provided a reinterpretation of the equations are made.

The FM modulation properties of lasers with inhomogeneous $\alpha$ have been discussed by Nilsson and Yamamoto [16] and others, and in the introduction it was suggested that the tunability of multisection lasers might be associated with an increased FM noise. To some extent this must be true at least because any noise current in a tuning electrode will introduce FM noise in proportion to the tunability. Even if the current externally fed into the tuning electrode can easily be made virtually noise-free by use of a series resistance [7], there remains the noise due to spontaneous and nonradiative processes. Therefore there is an equivalent shot-noise modulation current with a spectral density

$$S_I = 2eI_{\text{th}} \tag{3.60}$$

where $I_{\text{th}} = eR$ can be called the *threshold current* of the tuning electrode, $e$ is the electric charge, and $R$ is the spontaneous recombination rate. If the modulation efficiency is expressed as $df/dI_m$ [Hz/A], the resulting FM spectral density becomes

$$S_f = 2eI_{\text{th}} \cdot \left( \frac{df}{dI_m} \right)^2 \tag{3.61}$$

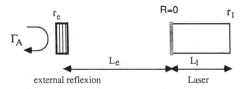

**Figure 3.4.** Extended-cavity laser. Numerical parameters are $r_1, r_{e0} \approx 0.57$, $L_1 = 300 \ \mu m$, which give for the corresponding round-trip time $\tau_1 \approx 6$ ps; $L_e$, which gives $\tau_e$, is used as a variable parameter. A typical value in an experimental setup is $L_e = 3$ cm.

and the contribution $\Delta \nu'$ to the linewidth becomes [6]

$$\Delta \nu' = \frac{1}{4\pi} S_{d\phi/dt}(0) = \pi S_f(0) = 2\pi e I_{\text{th}} \cdot \left( \frac{df}{dI_m} \right)^2 \qquad (3.62)$$

Using the typical values $I_{\text{th}} = 20$ mA and $df/dI_m = 1$ GHz/mA one obtains $\Delta \nu' = 20$ kHz, which is a very modest contribution since such lasers usually have a linewidth of megahertz order.

### 3.3.4.  Short FP Laser with a Long External Cavity

To present some new results we will show how the frequency noise of a short, AR-coated FP laser with an extended passive cavity can be calculated, and show results from such a calculation. The cavity configuration is shown in Figure 3.4.

For simplicity we assume a perfect AR coating at the left laser facet. The external reflection is assumed to be a frequency-dependent mirror (grating), which we model by

$$r_e = r_{e0} \cdot \exp \left[ -\left( \frac{\Omega}{2\pi f_0} \right)^2 \right] \qquad (3.63)$$

Here $f_0$, equal to 30 GHz, gives the spectral width of the grating, the numerical value chosen using typical values for lasers with external grating reflectors. We choose to calculate the external noise spectrum as measured from the grating side because the frequency dependence of the $H$ functions are simpler to handle in this case. It should be noted, though, that the high-frequency noise spectra are expected to be at least slightly different at the left and right outputs. The source admittance to be used can be calculated, giving

$$Y_\Delta(\Omega) = 2G_L \frac{Q_e}{\omega_0 \tau_1} \left( \frac{r_{e0}}{r_e} - e^{-j\Omega(\tau_e + \tau_1)} \right) \cdot \frac{1 + r_e}{e^{-j\Omega(\tau_e + \tau_1)} + r_{e0}} \qquad (3.64)$$

where

$$\frac{\omega_0}{Q_e} = \frac{1 - r_e^2}{r_e \tau_1} \tag{3.65}$$

gives the external $Q$ value for output to the left for the solitary laser if it had the external mirror at the left facet. Looking at Eq. (3.64) for low frequencies, we can see the expected linewidth improvement $[(\tau_1 + \tau_e)/\tau_1]^2$ as given by Henry [10]. The transfer function $H_{iA}$ can be calculated using Eq. (3.38). To simplify the calculation one can assume that a relative gain change corresponds to the same relative change in conductance.

Using for the laser the parameters of Ref. 7 (Figure 3) and Section 3.3.2 above, apart from the external $Q$ value, which is given from Eq. (3.65) and that we also have output coupling from the right laser facet, we have calculated the frequency noise, and the results are shown in Figure 3.5. From the curves we can apart from the apparent linewidth reduction also see that the relaxation peak is damped. This damping can be inferred from the ordinary rate equations by scaling the photon number, the photon

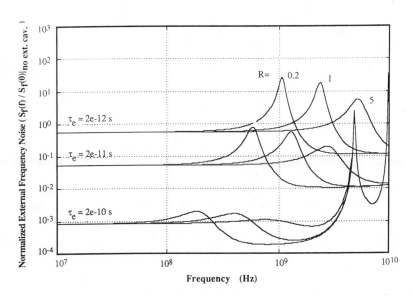

**Figure 3.5.** Frequency spectra normalized to the low-frequency noise spectra for $\tau_e = 0$. The spectra are plotted for three different pump parameters, and for three different lengths indicating both the shift of the relaxation peak with pumping and with passive cavity length. Parameters from Ref. 7 (Figure 3) and Section 3.3.2 (this chapter), apart from the external $Q$ values which are defined from the corresponding reflectances and round-trip times. Note the twofold normalization, to the cavity without any extended passive section and to yield the same low-frequency spectra irrespective of pump level.

lifetime, and the gain. The damping is further enhanced because the noise at high $\Omega$ is seen at a positive real, and hence lossy, part in the admittance function. A significant noise peak, however, is seen at a frequency corresponding to the next mode. The height of this peak may not be accurately given by this small signal analysis, but it is clear that the peak cannot be neglected, and that it warrants attention from the users of this type of laser.

### 3.3.5.   Possibilities of Reducing the General Case to a Tractable Form

Since one is seldom interested in microscopic details of the field pattern in a laser, it should be possible to make use of the fact that the field functions are quasiperiodic over many optical wavelengths to simplify the problem. A natural way to proceed would be to split the active region(s) into a number of lumped regions using Eq. (3.10)–(3.12) to determine the $H$ functions. Such a procedure, however, will introduce errors for the following reasons: The field patterns at the frequencies $\omega_0$ and $\omega_0 \pm \Omega$ do not coincide in the general case. Instead of products of the form $H_{A1}(0) \cdot H_{A1}(\Omega)$ which will appear in terms such as $H_{A1}(\Omega) \cdot I_{N1}$, one should therefore use overlap integrals. Another and perhaps more serious source of error is that such a simple lumping of regions seems to lead to a model that is internally inconsistent if the complex field amplitude distribution, $V(x)/V(0)$, is not real everywhere as it is in, for example, FP lasers with highly reflecting mirrors. This relates to the fact that the power generated in an active region is determined by the integral of the absolute value of the field squared while $H_{A1}$ is determined by the integral of the field squared. To obtain amplitude fluctuations consistent with energy conservation, it seems that one has to introduce at least one more degree of freedom in modeling an extended active region. A simple but critical test case where the correct result can be readily calculated directly is a laser with an amplifying section outside the cavity. Another test is, of course, that the low-frequency modulation response should agree with that obtained from the nonlinear stationary solution by varying the injection current(s). This solution can be extended to modulation frequencies covering the relaxation oscillation peak as long as the approximation $Y_\Delta(\Omega) \approx \Omega \partial Y/\partial \omega$ holds. Such an approach has been taken by Lang and Yariv [17], who developed an analytical tool to calculate the modulation properties of multicavity lasers.

### 3.4.  CONCLUSIONS

In this chapter we have presented a theory for the modulation and noise spectra of complicated laser structures. The theory has been shown to reproduce several earlier established results, and we have also presented new results for the frequency noise spectrum of a short Fabry–Perot laser with an extended passive cavity. However, we have only briefly discussed how the large set of equations stipulated by the theory could be reduced to a more tractable form in the general case. Some suggestions on this have been made, but a satisfactory solution has not been presented.

## ACKNOWLEDGMENTS

This work was initiated while one of the authors, Olle Nilsson, was a visiting scientist at the NTT Basic Research Laboratories, ECL Musashino-shi, Tokyo 180,Japan, during fall 1988, and he wants to acknowledge both the hospitality of NTT and the many stimulating discussions with Dr. Y. Yamamoto.

## REFERENCES

1. M. Lax, *Phys. Rev. A* **160**, 290–296 (1967).
2. Y. Yamamoto and N. Imoto, *IEEE J. Quantum Electron.* **QE-22**, 2032–2042 (1986).
3. O. Nilsson, Y. Yamamoto, and S. Machida, *IEEE J. Quantum Electron.* **QE-22**, 2043–2052 (1986).
4. G. Björk and O. Nilsson, *J. Lightwave Technol.* **LT-5**, 140–146 (1987).
5. J. Whiteaway, G. H. B. Thompson, A. Collar, and C. Armistead, *IEEE J. Quantum Electron.* **QE-25**, 1261–1279 (1989).
6. G. Björk and O. Nilsson, *IEEE J. Quantum Electron.* **QE-23**, 1303–1312 (1987).
7. Y. Yamamoto, S. Machida, and O. Nilsson, *Phys. Rev. A* **34**, 4025–4042 (1986).
8. J. Arnaud, *Opt. Commun.* **68**, 423–426 (1988).
9. K. Ujihara, *IEEE J. Quantum Electron.* **QE-20**, 814–818 (1984).
10. C. H. Henry, *J. Lightwave. Technol.* **LT-4**, 288–297 (1986).
11. J. Arnaud, *Opt. Quantum Electron.* **18**, 335–343 (1986).
12. J. Wang, N. Shunk, and K. Petermann, *Electron. Lett.* **23**, 715–717 (1987).
13. G. P. Agrawal and R. Roy, *Phys. Rev. A* **137**, 2495–2501 (1988).
14. J. Arnaud, *Opt. Lett.* **13**, 728–730 (1988).
15. J. Arnaud, *IEEE J. Quantum Electron.* **QE-25**, 668–677 (1989).
16. O. Nilsson and Y. Yamamoto, *Appl. Phys. Lett.* **46**, 223–225 (1986).
17. R. Lang and A. Yariv, *IEEE J. Quantum Electron.* **QE-21**, 1683–1688 (1985).

# 4

# Frequency Tunability, Frequency Modulation, and Spectra Linewidth of Complicated Structure Lasers

Y. YOSHIKUNI
*NTT Optoelectronics Laboratories, Morinosato, Atsugi-shi, Kanagawa, Japan*

## 4.1. INTRODUCTION

Increasing interest in coherent communication systems has led to several new requirements for semiconductor laser light sources. Such systems require single-mode light sources with narrow spectral linewidth in the wavelength region for optical communications. Linewidth requirement depends on modulation and detection schemes as well as transmission bit rate.

When used in some simple detection schemes not sensitive to phase noise, conventional long-wavelength single-mode lasers, such as distributed feedback (DFB) [1, 2] and distributed Bragg reflector (DBR) [3] lasers, are sufficient: for example, in 1-Gbit/s (one gigabit per second) transmission frequency shift-keying (FSK)–heterodyne systems with the envelope detection schemes [4] allow up to 90-MHz linewidth [5]. However, to date the linewidth of lasers is not narrow enough for use in some sophisticated modulation and detection schemes, such as phase shift-keying (PSK)–homodyne, which requires 300-kHz linewidth for 1-Gbit/s-transmission systems [5].

Coherent systems need two lasers closely aligned in frequency: typically a few gigahertz, which is much smaller than 1 Å in wavelength. (It should be noted that 1 Å in wavelength corresponds to 17.8 GHz at 1.3 $\mu$m, and corresponds to 12.5 GHz at 1.55 $\mu$m.) Wavelength-tunable lasers are needed to maintain such precise tracking. Wavelength-tunable lasers are

*Coherence, Amplification, and Quantum Effects in Semiconductor Lasers,* Edited by Yoshihisa Yamamoto.
ISBN   0-471-512494   © 1991 John Wiley & Sons, Inc.

also key devices in many attractive systems such as wavelength- or frequency-division multiplexing (WDM–FDM), channel selection in multichannel CATV (cable television) systems, and optical switching in local-area networks. Requirements for wavelength tuning depend on application. Local oscillator tracking requires a tuning range of only a few angstroms, but it must have continuous tuning without mode-hopping. WDM–FDM applications often require a tunability of up to hundreds of angstroms. Continuous tuning is also desirable, as it makes wavelength control much simpler, although it is not essential to this system. Some combination of mode-hopping with continuous tuning may be acceptable. Even in conventional single-mode lasers, wavelength tuning can be achieved by varying the operating temperature, although it takes a long time to set lasing wavelength using this method. It also disables some applications that require tuning speed, and makes wavelength tracking difficult owing to relatively poor controllability. Thus, electrically tunable lasers are desirable.

Direct frequency modulation (FM) characteristics are important, when frequency shift-keying (FSK) or continuous-phase shift-keying (CPFSK) are used. Frequency modulation (FM) is a kind of dynamic wavelength (frequency) tuning. The requirements for FM, however, are very different from those for wavelength tuning. The tuning range required is as small as a few gigahertz, but continuous tuning is inherently indispensable. Frequency modulation response should be constant at all frequencies contained in modulation signal.

Many efforts are being made in the area of conventional single-mode lasers to find a device to meet all requirements. The introduction of long-cavity and multiple quantum well (MQW) structures significantly reduce spectral linewidth to 250 kHz [6]. It seems, however, that conventional single-frequency lasers do not sufficiently satisfy all requirements. Some tuning structure must be introduced especially for wavelength tuning. Thus, lasers with complicated structures were proposed and fabricated. In the following section, we investigate experimental results of such complicated lasers related to coherent communications: spectral linewidth, wavelength tuning, and frequency modulation characteristics.

## 4.2. EXPERIMENTAL RESULTS OF COMPLICATED LASERS

Figure 4.1 shows typical cavity structures of some of the complicated structure lasers that were studied. Main experimental results are also shown. In this section we will see the characteristics of five kinds of lasers: (1) external-cavity semiconductor lasers, (2) distributed Bragg reflector (DBR) lasers with phase control region, (3) coupled cavity lasers, (4) phase-tunable distributed feedback lasers, and (5) multielectrode distributed feedback lasers.

| Cavity Structure | Line-width (Hz) | Tuning (nm) | | FM response (Modulation Frequency) | REF |
|---|---|---|---|---|---|
| | | Total | Continuous | | |
| 1) External-cavity lasers 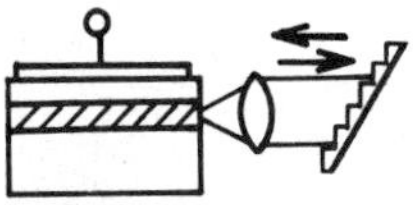 | 2k | – | --- | – | 10 |
| | 800k | 154 | – | – | 12 |
| | 20k | 15 | 15 | — | 13 |
| 2) DBR Lasers with PC regions 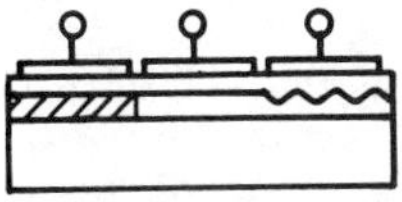 | – | 5.8 | 3.1 | – | 19 |
| | – | 10.0 | 4.4 | – | 21 |
| | 3-25M | 8.0 | 1.4 | 10-15GHz/mA ( 100MHz) | 25 |
| 3) Coupled cavity lasers  | – | 30.0 | – | – | 28 |
| | – | – | – | 0.1-0.2GHz/mA ( 0-6 GHz) | 31 |
| 4) Phase tunable DFB lasers 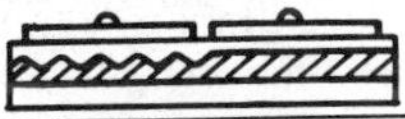 | 30M | --- | – | 13GHz/mA (0.01-300MHz) | 35 |
| | – | 5.0 | – | - | 37 |
|  | 20M | 1.2 | 1.2 | 1.6GHz/mA (<200MHz) | 38 |
| 5) Multi-electrode DFB lasers 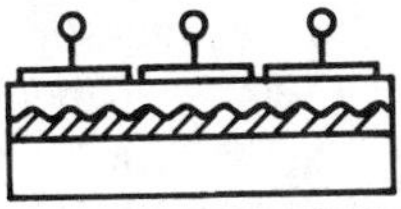 | 30M | 2.4 | 2.4 | - | 44 |
| | – | 1.6 | 0.22 | 1-2GHz/mA (100Hz-100MHz) | 47 |
| | 350k | 1.6 | 1.6 | 0.2GHz/mA (10kHz-10GHz) | 56 |

**Figure 4.1.** Typical cavity structures of some of the complicated structure lasers with main experiment results. Shaded waveguides in the cavity structures represent the active waveguides, while the others represent passive waveguides.

### 4.2.1. External-Cavity Semiconductor Lasers

External-cavity semiconductor lasers have been studied extensively [7, 8] because of their attractive features, such as narrow linewidth and wide wavelength tunability. The external-cavity configuration facilitates the extension of the laser cavity to a few tens of centimeters. This is desirable for narrow-linewidth lasers. Most external-cavity lasers use a grating as a wavelength filter to obtain single longitudinal-mode oscillation. This configuration also enables wavelength tuning in a very wide range by mechanical rotation of the grating.

Because of their long cavity length, longitudinal-mode spacing in external-cavity lasers is small: for example, 1 GHz for a 15-cm cavity. It is much smaller than the resolution of the grating: typically ~ 1 Å. Therefore, in regard to one longitudinal mode, the dispersive character of the grating can be ignored. Thus, the spectral linewidth of external cavity lasers is approximated by Fabry–Perot (FP) lasers with a long passive section. Linewidth $\Delta\nu$ of such lasers is expressed as [9]

$$\Delta\nu = \frac{R}{4\pi I}(1 + \alpha^2)\xi^2 \tag{4.1}$$

where $R$ is the spontaneous emission rate, $I$ is the photon number in the cavity, $\alpha$ is the linewidth enhancement factor, and $\xi$ is the fraction of the effective optical cavity length occupied by the active region. Since $\xi$ is inversely proportional to external-cavity length $L$, linewidth is expected to reduce proportionally to $1/L^2$. This implies that long cavity length in eternal-cavity lasers possibly produces much smaller linewidth than it would in conventional semiconductor lasers.

The $1/L^2$ dependence of linewidth on cavity length $L$ was experimentally observed and reported [10]. The device used was a 1.5-$\mu$m channeled substrate buried heterostructure FP laser with 250-$\mu$m cavity (see inset in Figure 4.2). One facet was coated with antireflection films to reduce facet reflectivity to $\sim 0.1\%$. The light output from the coated facet was collimated with a microscope lens ($NA = 0.85$) and fed back to the laser with a 600-line/mm grating. The laser-to-grating coupling coefficient was estimated to be $\sim 40\%$ per round trip. Such high coupling efficiency produced threshold current as low as 25 mA. Linewidth was measured by the heterodyne beat signal between two identical lasers. This was because the commonly used self-heterodyne method [11] is impractical for high-resolution measurements: it requires 100 km of fiber for a 2-kHz resolution. The 3-dB linewidth was estimated from the linewidth at 40 dB below peak

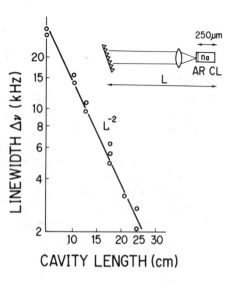

**Figure 4.2.** Measured cavity length dependence of 3-dB spectral linewidth in an external-cavity laser [10]. The inset shows cavity structure. Linewidth was measured by the heterodyne beat signal between two identical lasers. The 3-dB linewidth was estimated from the linewidth at 40 dB below peak power. Output powers were held at 3 mW with bias current of 50 ±3 mA and lasing wavelength was tuned at 1523.5 nm.

power to reduce the effect of center frequency fluctuation. The result is shown in Figure 4.2 for cavity length between 7 and 25 cm, where injection current was adjusted within 50 ± 3mA to keep the output power at 3 mW. The measured linewidth ranged from 27 to 2.5 kHz and was closely approximated by a $1/L^2$ line given by

$$\Delta\nu \, [\text{kHz}] = \frac{4.5 \times 10^3}{P \, [\text{mW}] \cdot L^2 \, [\text{cm}]}$$

with output power $P$ and total cavity length $L$. Minimum linewidth as narrow as 2.5 kHz was obtained with 25-cm cavity.

The external-grating configuration also facilitates a wide range of wavelength tuning by mere mechanical rotation of the grating. A tuning range as large as the whole semiconductor gain width (154 nm) was reported [12]. Wavelength tuning by rotating the grating is discontinuous because of the fixed longitudinal-mode frequency. Continuous frequency tuning within a longitudinal-mode spacing can be performed by changing the cavity length and consequence frequency shift of the longitudinal mode. The continuous frequency tuning range can be expanded if cavity length is synchronously adjusted with the rotation of the grating.

Continuous frequency tuning of 15 and 0.8 nm were reported by mechanical [13] and electrical [14] adjustment of the cavity length with grating rotation. Direct frequency modulation in external-cavity lasers is limited in low-frequency region because of the long round-trip time associated with the long cavity [15].

Despite their many attractive features, problems with long-term stability and mechanical tuning limit application of such lasers. Some trials were made to eliminate these problems using optical fiber as external-cavity [16] and mechanically stabilizing package of external cavity laser [14].

### 4.2.2. Distributed Bragg Reflector (DBR) Lasers with Phase Control Region

Tunable DBR lasers [17, 18] initially achieved 0.9-nm continuous tuning by current injection in the Bragg grating region. Introduction of a phase-tuning region extended the range to 3.1 nm [19]. The frequency tuning in DBR lasers is almost the same as that of external-cavity lasers. The current injection in the Bragg grating of DBR lasers is equivalent to the mechanical rotation of the grating in external-cavity lasers. Cavity length adjustment in DBR lasers is achieved by current injection in the phase control region. The only difference between DBR and external-cavity lasers is in grating dispersion.

Longitudinal-mode spacing in external-cavity lasers is small compared with resolution of the grating. Therefore, the dispersion character of the grating is negligible. On the other hand, longitudinal-mode spacing in DBR lasers is comparable to grating resolution. Therefore, dispersion character of the grating induces considerable phase shift in reflected light from the grating. This phase shift $\Phi_B$ is expressed in complex reflectivity $r_B$ of the Bragg reflector given by [20]

$$r_B = \frac{-j\kappa \sinh \gamma L_B}{\gamma \cosh \gamma L_B + \gamma j\delta \sinh \gamma L_B} \qquad (4.2)$$

$$= |r_B| \exp(j\Phi_B) \qquad (4.3)$$

with

$$\gamma^2 = \kappa^2 + (j\delta)^2 \qquad (4.4)$$

$$\delta = \frac{2\pi n_B}{\lambda} - \frac{\pi}{\Lambda} + \frac{j\alpha_B}{2} \qquad (4.5)$$

where $\kappa$ is the grating coupling coefficient, $L_B$ is the grating length, $n_B$ is the grating refractive index, $\lambda$ is the light wavelength, $\Lambda$ is the grating pitch, and $\alpha_B$ is the loss in grating. For a DBR laser consisting of a Bragg reflector with complex reflectivity $r_B$, phase control region with length of $L_p$, active region with length of $L_a$, and mirror facet with amplitude reflectivity $r_m$, the lasing condition for the lasers is expressed as

$$r_B \cdot r_m \exp 2j(\beta_a L_a + \beta_p L_p) = 1 \qquad (4.6)$$

with

$$\beta_a = \frac{2\pi n_a}{\lambda} - \frac{jg_a}{2} \qquad (4.7)$$

$$\beta_p = \frac{2\pi n_p}{\lambda} - \frac{jg_p}{2} \qquad (4.8)$$

where $n_{a(p)}$ is the refractive index and $g_{a(p)}$ is the power gain in active (phase control) regions. Equation (4.6) produces two equations for round-trip gain and round-trip phase as

$$\Phi_B + 2\left(\frac{2\pi n_a}{\lambda} L_a + \frac{2\pi n_p}{\lambda} L_p\right) = 2m\pi \qquad (4.9)$$

$$-\ln |r_B| \cdot |r_m| = g_a L_a + g_p L_p \qquad (4.10)$$

with any integer number $m$. The resonance frequency of longitudinal modes, that is, possible lasing frequency, is determined by Eq. (4.9), and Eq. (4.10) gives required threshold gain. As the required threshold gain in

Eq. (4.10) becomes minimum at the Bragg wavelength $\lambda_B$ given by

$$\frac{2\pi n_B}{\lambda_B} - \frac{\pi}{\Lambda} = 0 \qquad (4.11)$$

the laser oscillates in the longitudinal mode nearest the Bragg wavelength. Current injection in the phase control region decreases refractive index in the region through the plasma effect in carriers. Consequently, lasing frequency continuously shifts toward shorter wavelength.

If the lasing wavelength shifts away from Bragg frequency, the threshold gain increases according to Eq. (4.10). Further shift in the lasing frequency produces lasing mode change to the next longitudinal mode in the longer wavelength that is nearest to the Bragg wavelength at the time. Thus the current injection in the phase control region causes cyclic wavelength change, consisting of a continuous shift to short wavelength and a mode hop to long wavelength. On the other hand, current injection in the DBR region causes the Bragg wavelength to shift to short wavelength (see e.g., Ref. 21 for the experimental results). Consequently, lasing modes hop to the next shorter wavelength mode. The shift in Bragg wavelength also produces a change in phase shift $\Phi_B$, and consequent continuous wavelength shift to short wavelength (see e.g., Ref. 22 for experiments).

The tuning characteristics for a 1.5-$\mu$m DBR laser [23] are shown in Figure 4.3. The device consists of three current injection regions: a 190-$\mu$m active, an 80-$\mu$m phase control (PC), and a 700-$\mu$m DBR region (see cavity configuration shown in Figure 4.3a). The active region contains two layers of different composition: an active layer that provides lasing gain for 1.5-$\mu$m light and a waveguide layer with 0.23-$\mu$m thickness and 1.3-$\mu$m bandgap composition. The waveguide layer extends to the PC and DBR regions providing good optical coupling between the regions. Both the PC and the DBR regions contain only the waveguide layer which acts as a transparent waveguide for 1.5-$\mu$m lasing light.

Wavelength change by current injection in the DBR region is shown in Figure 4.3b. With increasing current to the DBR region, the lasing wavelength continuously shifts to short wavelength and then hops to the next mode in short wavelength as explained previously. The total discrete tuning range exceeds 5 nm.

Wavelength tuning by current injection in the PC region is shown in Figure 4.3C. With increasing current in the PC region, the lasing wavelength exhibits periodic changes with continuous shift to short wavelength and mode hop to long wavelength. A continuous tuning range of over 1.2 nm is obtained. This range is large enough to cover the gap in the wavelength tuning in Figure 4.3b. Thus, the device demonstrated continuous coverage of a wide tuning range of 5.8 nm (720 GHz) with four-mode

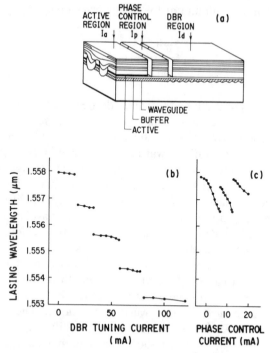

**Figure 4.3.** Wavelength tuning characteristics of a 1.5-$\mu$m DBR laser with a phase tuning region, consisting of three current injection regions: a 190-$\mu$m active, an 80-$\mu$m phase control (PC), and a 700 $\mu$m DBR region [23]. (*a*) configuration; (*b*) wavelength change by current injection in DBR; (*c*) wavelength tuning by current injection in the PC region.

hops, where the currents into the DBR and PC regions were simultaneously controlled by steplike current increase in the DBR region and sawtoothlike current injection in the PC regions. The device showed stable single-mode oscillation at any wavelength in the tuning region with submode suppression of over 30 dB. The output power was kept constant at 2 mW by changing active region current.

Simultaneous adjustment in the DBR and PC region also enables continuous tuning without mode-hopping. If the wavelength shift in longitudinal mode follows the shift in Bragg wavelength, the lasing wavelength shifts continuously with no mode-hopping as in the continuous tuning of external-cavity lasers.

It can be shown from Eqs. (4.2)–(4.11), that the lasing wavelength shifts exactly with the Bragg wavelength, if the refractive index change in the DBR region $\Delta n_{\mathrm{B}}$ and that in the PC region $\Delta n_{\mathrm{p}}$ satisfy a relation given by

$$\Delta n_{\mathrm{B}} = \frac{L_{\mathrm{p}}}{L_{\mathrm{p}} + L_{\mathrm{a}}} \Delta n_{\mathrm{p}} \qquad (4.12)$$

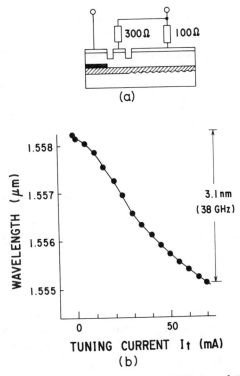

**Figure 4.4.** Continuous wavelength tuning in a 1.5-$\mu$m DBR laser [19] (*a*) Configuration: The tuning currents in the DFB and PC regions are supplied from one current source with constant-current ratio determined by 300- and 100-$\Omega$ resistance. (*b*) wavelength shift versus total tuning current: continuous tuning range is 3.1 nm (38 GHz). (*c*) linewidth change under the continuous wavelength tuning.

In that case, any lasing parameters other than wavelength, including the lasing mode, does not change, because $\delta$ in Eq. (4.4) and consequently $r_B$ in Eq. (4.2) are kept constant.

Continuous wavelength tuning obtained with this scheme is shown in Figure 4.4 [19]. Here, the tuning currents to the DBR and PC region are supplied from one current source through resistances connected to each region to keep the current ratio constant (Figure 4.4*a*). As shown in Figure 4.4*b*, with increase in the tuning current lasing wavelength shifts continuously to short wavelength, producing continuous tuning range as large as 3.1 nm (380 GHz). No mode-hopping is observed, even though the tuning range is about three times larger than longitudinal-mode spacing. The linewidth during the tuning is shown in Figure 4.4*c* [24]. With shift to short wavelength, the linewidth increases monotonically from 23 to 100 MHz. One reason for this linewidth increase is cavity loss increase caused by free carrier absorption by injected carriers in tuning regions. Wave-

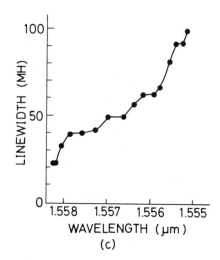

**Figure 4.4.**    *(Continued)*

(c)

length tuning in DBR lasers often causes linewidth broadening [25]. It may cause some problems in practical applications.

The lasing condition in Eq. (4.10) is similar to that for FP lasers with facet reflectivity of $r_B$ and $r_m$. Equation (4.9) is also the same as in FP lasers with active region of $L_a$ and passive region of $L_p + L_{eff}$, if we introduce the effective length of DBR region $L_{eff}$ as

$$2n_p L_{eff} = \frac{\partial \Phi_B}{\partial k} \qquad (4.13)$$

where $k = \omega/c$ is the wavenumber in vacuum. Therefore, linewidths of DBR lasers are expressed in Eq. (4.1) as in external-cavity lasers.

It should be noted, however, that high effective reflectivity $r_B$ does not always means narrow linewidth in DBR lasers in contrast with FP lasers. This is because higher effective reflectivity, that is, strong coupling in grating, also introduces reduction in effective length $L_{eff}$. This enlarges linewidth in spite of increase in $r_B$, especially in short $L_a$ and $L_p$. Therefore, coupling strength, that is, $\kappa$ value, in the grating should be optimized according to the lengths $L_a$ and $L_p$.

Figure 4.5 shows linewidths of five DBR lasers as function of inverse power [26]. The devices are 1.5-$\mu$m butt-jointed DBR lasers. Active region lengths are 200 $\mu$m for samples $a$ and $a'$ and 300 $\mu$m for $b$, $c$, and $d$. Samples $a$, $a'$, and $b$ have one DBR region and one facet mirror, while samples $c$ and $d$ have two DBR regions on both sides. Samples $b$ and $d$ have a phase control region in the cavity, while the other samples do not. Coupling coefficient $\kappa$ is almost same in samples $a$, $b$, $c$, and $d$, but is smaller in sample $a'$.

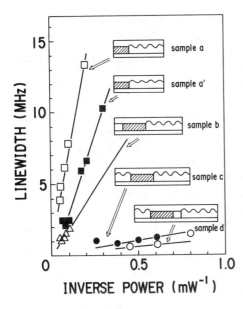

**Figure 4.5.** Linewidths of five DBR lasers as function of inverse power [26]. The devices are 1.5-$\mu$m butt-jointed DBR lasers with active region lengths of 200 $\mu$m (samples $a$, $a'$) and 300 $\mu$m ($b, c, d$). Samples $a$, $a'$, and $b$ have one DBR region and one facet mirror, while samples $c$ and $d$ have two DBR regions on both sides. Samples $b$ and $d$ have phase-control region in cavity while the other samples do not. Coupling coefficient $\kappa$ is almost the same in samples $a$, $b$, $c$, and $d$, but smaller in sample $a'$.

Samples $a$ and $a'$ are almost identical except with respect to coupling coefficient. Linewidth is narrower in the smaller $\kappa$ device, sample $a'$, than in sample $a$: 35 and 50 MHz mW$^{-1}$ in linewidth power product, respectively. This is because reduction of effective length more strongly affects linewidth than increase in effective reflectivity does in these short-cavity devices. Linewidth in sample $b$ is reduced to 20 MHz mW$^{-1}$ by a long cavity length associated with phase tuning region and longer active region. Further reduction of linewidth is produced by introduction of DBR regions on both facets (resp. 2.2 and 1.3 MHz mW$^{-1}$ in samples $c$ and $d$) because of higher reflectivity of Bragg reflector than facet mirror, and extension in cavity length associated with effective length of the grating. Introduction of a phase control region also reduces linewidth in this structure (sample $d$). Linewidth as narrow as 580 kHz is obtained in sample $d$ at 2 mW of output power.

### 4.2.3. Coupled Cavity Lasers

When two or more laser cavities are optically coupled, interference between longitudinal modes of both cavities strongly affects the lasing characteristics. The cavity loss of the lasers changes depending on relative alignment of the longitudinal-mode frequencies of both cavities. In a coupled cavity laser consisting of two laser cavities with different longitudinal-mode spacings, cavity loss should change periodically with wavelength, with wavelength period determined by differences in longitudinal-mode spacing.

If the wavelength period is large compared to gain width of the lasers, the laser should oscillate in a single longitudinal mode at the wavelength of minimum loss. Using this phenomena, a number of single longitudinal-mode coupled cavity lasers were reported. Among these lasers the cleaved coupled cavity ($C^3$) laser was studied extensively because of its stable single-mode nature and relatively simple structure. Because $C^3$ lasers consist of two independent laser cavities, the lasing characteristics strongly depend on the bias condition of both cavities. For operation of $C^3$ lasers, one of two cavities is always biased above its threshold. The other cavities can be operated either above or below threshold.

Figure 4.6 shows typical mode behavior observed in a 1.55-$\mu$m ridge-waveguide $C^3$ laser [27]. Longitudinal-mode variations are shown for pairs of injection current to both sections $I_1$ and $I_2$. The dotted curve represents the lasing threshold of the coupled laser. The thick lines in the figure represent mode boundaries, where a lasing mode changes to the next mode with current change in $I_1$ or $I_2$. The laser oscillates in a single mode between the thick lines. The operation of $C^3$ lasers is classified into three domains. In regions $A$ and $B$ in Figure 4.6, one diode is being biased above its lasing threshold; and the other diode, biased below. In this

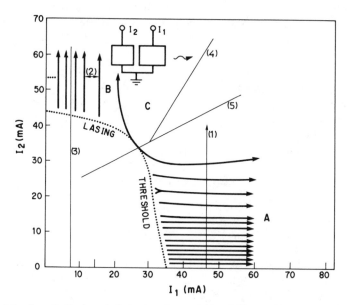

**Figure 4.6.** Longitudinal-mode behavior observed in a 1.55-$\mu$m ridge-waveguide $C^3$ laser for pairs of injection current to sections $I_1$ and $I_2$ [27]. Dotted curve represents the lasing threshold of the coupled laser. Thick lines in the figure represent mode boundary. In regions $A$ and $B$, one diode is biased above its threshold while the other diode is biased below its threshold. In region $C$ both diodes are biased above threshold. Thin lines represent various operations lines discussed in the text.

region, carrier density in the diode biased above threshold is clamped at threshold value because of its laser action. Meanwhile, the other cavity acts as a tunable etalon because current injection in it alters the carrier density and consequently the refractive index. So current injection in the diode below threshold shifts the lasing frequency rapidly with changing oscillation mode in several longitudinal modes.

On the other hand, in region $C$ both diodes are biased above threshold. Since the carrier densities are clamped at their threshold values in both diodes, the laser oscillates in one single mode even when changing the currents $I_1$ and $I_2$. $C^3$ lasers can be used in several operation modes, according to their complex mode behavior. The narrow lines in figure 4.6 show typical operation lines. On operations line 1, one diode is biased at a constant current above threshold and the current to the other one varied. Along with this operations line, lasing frequency can be tuned at several longitudinal modes in a wide frequency range. The tuning range is very large because small wavelength change in longitudinal modes causes mode-hopping and consequently large change in lasing wavelength. It also produces a large tuning rate: specifically, wavelength change for unit current. A tuning range as large as 30 nm and a tuning rate of 2.6 nm/mA were reported along with this operations line [28].

Operations line 2 represents continuous frequency tuning in a narrow frequency range or FM operation [29]. On this operations line the laser oscillates in one single mode with shifting frequency according to the current of the diode below threshold. Operations lines 3–5 represent operation in one single mode under large intensity modulation [30]. According to the distinct regions in $I_1$ and $I_2$, frequency modulation characteristics change drastically.

Figure 4.7 [31] shows small signal FM response of a 1.3-$\mu$m $C^3$ laser. The device is a channeled substrate BH $C^3$ consisting of 140-$\mu$m front and 114-$\mu$m back cavities with a 5-$\mu$m gap between them. The threshold currents for the front and back cavities are 18 and 21 mA. The front cavity is pumped above threshold by DC current (30 mA), and the back cavity is modulated by RF signal with DC bias $I_B$. (Figure 4.7a).

Figure 4.7b shows FM response as a function of DC bias current ($I_B$) for fixed modulation frequency of 100 kHz. For $I_B$ below threshold (21 mA), the FM response oscillates rapidly from positive to negative value. When the back cavity is biased below threshold, it acts as a tunable etalon for the lasing front cavity. A modulation in the current to the back cavity changes the resonance frequency of the etalon, and consequently changes the threshold of the front cavity depending on the relative alignment of the modes of the two cavities. This change in threshold causes changes in the carrier density and consequent modulation in oscillation frequency.

The oscillation in the FM response in Figure 4.7b reflects periodic changes in the relative alignment of the modes of the two cavities. As

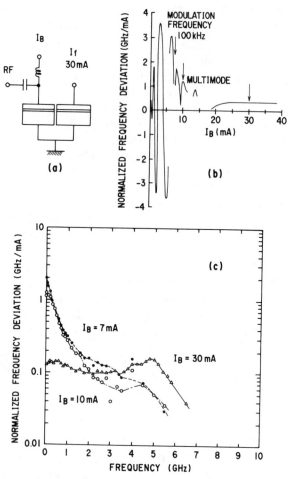

**Figure 4.7.** Small-signal FM response of a 1.3-$\mu$m C$^3$ laser [31]. (*a*) Configuration: The front cavity is pumped above the threshold by DC current (30 mA), and the back cavity is modulated by RF signal with DC bias $I_B$. (*b*) FM response as a function of DC bias current $I_B$, for fixed modulation frequency of 100 kHz. (*c*) Modulation frequency dependence of the FM response for the three bias currents indicated by arrows in Figure 4.7*b*.

shown, when increasing current above threshold, FM response becomes constant at relatively small value because carrier density in the back cavity is also clamped.

Modulation frequency dependence of the FM response is shown in Figure 4.7*c* for the three bias currents, which are indicated by the arrows in Figure 4.7*b*. When DC bias is below threshold ($I_B$ = 7, 10 mA), FM response is large at low frequency and shows rapid roll off above 200 MHz. This rapid rolloff is due to relatively long carrier lifetime in the back cavity below threshold. When DC bias is above threshold ($I_B$ = 30 mA), FM response in the low-frequency region decreases because of the clamped carrier density. Frequency response is roughly constant up to 6 GHz.

### 4.2.4.  Phase-Tunable Distributed Feedback Lasers

If the facet reflectivity is not sufficiently small, the lasing condition of DFB lasers is strongly influenced by the relative grating phase at their facets [32, 33]. That phenomenon has been utilized in the frequency tuning of DFB lasers. An uncorrugated waveguide is placed between a DFB laser and the cleaved facet. Separate electrodes are formed on the DFB region for pumping and the uncorrugated waveguide region for phase control. The relative phase at the reflective facet was effectively controlled by current injection in the phase control region through refractive index change. The operating principle is shown in Figure 4.8a, where threshold condition change with phase tuning is calculated [34]. The calculated device is a DFB laser with an output facet with 0 reflectivity and a back facet with 0.54 field reflectivity. The effect of the phase tuning is approximated by change in phase of the reflective facet $\theta$, which is introduced to the calculation through the complex reflectivity of the facet $\rho_r = 0.54 \exp(i\theta)$. The threshold conditions were calculated for various $\theta$ and are plotted in the $\alpha L$ versus $\Delta \beta L$ plane. Here $\alpha L$ is the required threshold gain, and $\Delta \beta L$ is the propagation constant difference from the Bragg condition. TE $+ 1$ mode represents the DFB mode close to the Bragg wavelength with positive $\Delta \beta L$ (in shorter wavelength) and TE $- 1$ mode represents that with negative $\Delta \beta L$ (in longer wavelength). At $\theta = \pi/2$ (open circles) two DFB modes close to the Bragg wavelength (TE $\pm 1$ modes) have the same threshold gain, indicating both modes likely to lase. Increase in $\theta$ reduces threshold gain for TE $- 1$ mode and increases it for TE $+ 1$ mode. Therefore single mode oscillation likely to occur in TE $- 1$ mode around $\theta = \pi$ (squares). With increasing $\theta$ beyond $3/2\pi$ (triangles), the lowest threshold gain mode switches from TE $- 1$ mode to TE $+ 1$ mode indicating lasing mode likely to change from the longer wavelength mode (TE $- 1$) to the shorter wavelength mode (TE $+ 1$). With further increase in $\theta$, lasing mode switches to TE $- 1$ mode again at $\theta = \pi/2$. Between the mode changes, the lasing mode continuously shifts to larger $\Delta \beta L$ with increasing $\theta$. Thus, increase in $\theta$ rotates the lasing mode on a loop shown by arrows in Figure 4.8a.

Figure 4.8b shows the device structure of a 1.3 $\mu$m phase tunable DFB laser [34]. The device is a DFB–DC–PBH laser with both facets cleaved, consisting of a 350-$\mu$m DFB region and a 250-$\mu$m tuning region. Both regions consist of a 1.3-$\mu$m active layer and a waveguide layer. Corrugation grating is formed only in the DFB region. Current injection in phase control region increases the refractive index in the region and effectively increases $\theta$ of its facet. It changes the lasing mode as shown in Figure 4.8a.

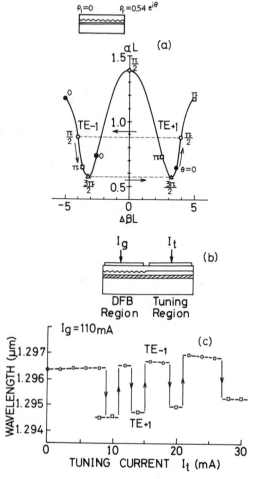

**Figure 4.8.** Wavelength tuning in phase-tunable DFB lasers [34]. (*a*) Calculated threshold gain $\alpha L$ versus propagation constant deviation from Bragg condition $\Delta\beta L$ with various facet phase $\theta$: 0 (dots), $\pi/2$ (open circles), $\pi$ (squares), and $3\pi/2$ (triangles). (*b*) Configuration: The device consists of a 350-$\mu$m DFB region and a 250-$\mu$m tuning region. Both regions consist of a 1.3-$\mu$m active and a waveguide layer. Corrugation grating is formed only in DFB region. (*c*) Observed lasing wavelength change by tuning current in a 1.3-$\mu$m phase-tunable DFB–DC–PBH laser with both facets cleaved.

Figure 4.8*c* shows lasing wavelength change with the tuning current in the device. Current in the DFB region is fixed at 110 mA. With increase in tuning current, the lasing mode periodically changes in longitudinal modes close to the Bragg wavelength in shorter (TE + 1) and longer wavelength (TE − 1) as discussed in Figure 4.8*a*. The facet phase change also produces small continuous wavelength tuning. It is clearly shown in wavelength shift with tuning current around 25 mA. This continuous wavelength tuning was applied to frequency modulation, and flat FM response on modulation frequency was realized [35]. The continuous tuning was also successfully used to compensate for dynamic frequency shift under direct intensity modulation [36].

Phase control also affects the linewidth because it changes cavity loss. The linewidth was measured in the device for various tuning currents. An increase in the tuning current broadened or narrowed the linewidth periodically as a result of periodical change in the cavity loss and also in the single-mode stability. It suggests that the facet phase has an important role in the linewidth of the as-fabricated DFB. When keeping the tuning current at the optimum value, the linewidth reduced linearly with the DFB current. Minimum linewidth of 9.6 MHz was observed [34].

Phase tuning also alters optical gain in the region because the tuning region of the device in Figure 4.8$b$ contains an active layer that provides optical gain for lasing light as well as refractive index change. When carrier density increases in the phase tuning region, it should decrease in the DFB region to keep the total gain balanced with cavity loss. This carrier increase reduces refractive index, and consequently shifts the Bragg wavelength to short wavelength. Using this wavelength shift, a discrete tuning range as large as 5 nm was obtained in a device of the same structure [37]. The device was a 1.5-$\mu$m ridge-waveguide DFB laser consisting of a 200-$\mu$m-long DFB and phase shift regions. Current in the DFB region was increased from 0 to 250 mA with corresponding decrease in current into the phase tuning region. When current in the DFB region is 0, gain was totally provided by the phase tuning region. Therefore the device was operated as a kind of DBR. With increasing current in the DFB region, the operation mode of the laser changes to a hybrid DBR/DFB mode and then changes to a pure DFB mode. This change in the operation mode causes a large shift of the Bragg wavelength. Thus lasing wavelength moves to short wavelength with mode-hopping up to 5 nm.

Bragg wavelength shift, however, partially compensates for continuous shifting in the mode because with current injection in the phase tuning region, the facet phase effect shifts lasing mode to shorter wavelength, while Bragg wavelength shifts to longer wavelength. Consequently, the continuous tuning range in the device shown in Figure 4.8$b$ is small. If the phase tuning region consists of passive waveguide (transparent for lasing light), the Bragg wavelength does not shift with phase tuning, so such devices have larger continuous tuning range. In a 1.5-$\mu$m DFB laser with a passive phase control region consisting of transparent waveguide, a continuous tuning range as large as 1.2 nm was reported [38]. With increase in tuning current, lasing wavelength continuously shifts to short wavelength, and suddenly changes to longer wavelength with mode jump. The continuous tuning range was over 1.2 nm.

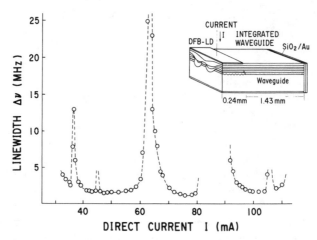

**Figure 4.9.** Spectral linewidth of a DFB external-cavity laser as a function of the injection current [41]. The device (shown in inset) is a 240-$\mu$m-long DFB laser integrated with 1.43-mm-long passive waveguide. The sharp increase in linewidth is due to the mode change at the current.

A passive waveguide can also be used for linewidth reduction of DFB lasers [39, 40], as it enables a long-cavity laser to be produced without causing threshold current increase. Figure 4.9 shows the spectral linewidth of DFB external-cavity lasers as a function of the injection current [41]. The device is a 240-$\mu$m-long DFB laser integrated with a 1.43-nm-long passive waveguide. With increasing injection current, the linewidth changes periodically with a sharp increase for the mode change at some current levels. Linewidth is minimized at the current level just below the point where the lasing wavelength jumped to the next external-cavity mode with longer wavelength. The minimum linewidth obtained is 2 MHz, much narrower than conventional DFB lasers.

### 4.2.5. Multielectrode Distributed Feedback (DFB) Lasers

Distributed feedback lasers' characteristics are strongly affected by spatial distribution in the carrier density along the laser cavity [42]. This phenomenon can be applied to wavelength tuning in DFB lasers, if carrier density distribution is artificially controlled. In multielectrode DFB lasers, carrier density distribution is controlled by partitioned electrodes into two [43–45] or three sections [46, 47].

The carrier density distribution forms corresponding distributions in optical gain and refractive index. Here, we assume that gain and refractive index linearly depend on the carrier density with proportional constants uniform in the cavity. In FP lasers, any change in the refractive index due to the local carrier density has no effect on the lasing frequency. This is because carrier density should be clamped, when averaged over the cavity, at the threshold level and because the lasing frequency is determined only by the averaged refractive index, which should remain constant under the above assumption.

Meanwhile, this effect cannot be canceled in DFB lasers, since lasing frequency shift caused by the index change is nonlinear around the Bragg frequency, and also because cavity loss varies with carrier density distribution. Hence, carrier distribution created by current injection into an electrode causes lasing frequency shift.

The operating principle of the multielectrode DFB is complicated compared to the lasers discussed earlier. Because of its local feedback nature, changes in local gain are always associated with changes in feedback condition. That makes the separate treatment in loss and phase conditions as in DBR lasers impossible. Thus, analysis for DFB lasers should depend on numerical calculations [48, 49].

Multielectrode DFB lasers can be easily analyzed by mentally dividing their cavity into several segments, as in most complicated lasers that have several segments along with their laser cavities—for example, the active, the phase control, and the Bragg regions in DBR lasers. Although the multielectrode DFB has no intrinsic structures in its cavity, carrier density distribution and consequent distribution in gain and refractive index form a structure along the cavity. Assuming that refractive index $n$ and gain coefficient $g$ are uniform in a given section, wave propagation in that section is described by an F matrix [49]. Defining a $z$ axis along the laser cavity, the $z$ dependence of the laser field in a section can be expressed with the field of the forward $E_r(z)$ and the backward $E_s(z)$ propagating waves by

$$E(z) = E_r(z) + E_s(z) \tag{4.14a}$$

$$E_r(z) = R \exp(-j\beta z) \tag{4.14b}$$

$$E_s(z) = S \exp(j\beta z) \tag{4.14c}$$

where $\beta = n\omega/c$ and $\omega$ is the light frequency.

In a section with the grating, the $z$ dependence of the laser field can be written as

$$E(z) = R(z)\exp(-j\beta_0 z) + S(z)\exp(j\beta_0 z) \tag{4.15}$$

where $\beta_0$ is the wave vector for the Bragg condition. A pair of coupled wave equations is given by [50]

$$-R' + (g - j\Delta\beta)R = j\kappa S e^{-j\Omega} \tag{4.16a}$$

$$S' + (g - j\Delta\beta)S = j\kappa R e^{-j\Omega} \tag{4.16b}$$

where $\Delta\beta \simeq \beta - \beta_0$, $g$ is the field gain coefficient, $\kappa$ is the coupling coefficient of the grating, and $\Omega$ is the grating phase. The general solution of the equations are given by

$$E_r(z) = (ae^{\gamma z} + \rho e^{-j\Omega}be^{-\gamma z})e^{-j\beta_0 z} \tag{4.17a}$$

$$E_s(z) = (\rho e^{j\Omega}ae^{\gamma z} + be^{-\gamma z})e^{j\beta_0 z} \tag{4.17b}$$

with arbitary constants $a$ and $b$. Here $\gamma$ satisfies a dispersion relation

$$\gamma^2 = (g - j\Delta\beta)^2 + \kappa^2 \tag{4.18}$$

and

$$\rho = \frac{-\gamma + (g - j\Delta\beta)}{j\kappa} \tag{4.19}$$

Thus, we can write the amplitudes of the forward and backward wave at the right facet $(z = z_i + l)$ of section $i$ in terms of those at the left facet $(z = z_i)$ as [49]

$$\begin{pmatrix} E_r(z_i + l) \\ E_s(z_i + l) \end{pmatrix} = \begin{pmatrix} F_{11} & F_{22} \\ F_{21} & F_{22} \end{pmatrix} \begin{pmatrix} E_r(z_i) \\ E_s(z_i) \end{pmatrix}, \tag{4.20}$$

with

$$F_{11} = \left( \cosh \gamma l + \frac{(g - j\Delta\beta)l}{\gamma l} \sinh \gamma l \right) e^{-j\beta_0 l}$$

$$F_{21} = \frac{j\kappa l}{\gamma l} \sinh \gamma l \, e^{j(\beta_0 l + \Omega)}$$

$$F_{12} = \frac{-j\kappa l}{\gamma l} \sinh \gamma l \, e^{-j(\beta_0 l + \Omega)}$$

$$F_{22} = \left( \cosh \gamma l - \frac{(g - j\Delta\beta)l}{\gamma l} \sinh \gamma l \right) e^{j\beta_0 l}$$

At the section boundaries, $E_{r(s),i}$ in section $i$ and $E_{r(s),i+1}$ in section $i + 1$ satisfy the boundary condition. If refractive index difference between the sections is small enough, the boundary condition is given by $E_{r(s),i} = E_{r(s),i+1}$. Therefore, in a laser with $n$ sections and total length $L$, $E_{r(s)}(L)$

at the right laser facet are related to $E_{r(s)}(0)$ at the left laser facet as

$$\begin{pmatrix} E_r(L) \\ E_S(L) \end{pmatrix} = \prod_{i=1}^{n} F_i \begin{pmatrix} E_r(0) \\ E_S(0) \end{pmatrix}$$

$$= F_l \begin{pmatrix} E_r(0) \\ E_s(0) \end{pmatrix} \tag{4.21}$$

It should be noted that $E_s(L)$ in Eq. (4.21) represents field amplitude of the right-going wave injected to the left laser facet and should be zero for laser operation. Therefore, the laser oscillates only if infinitesimal $E_s(L)$ produces finite $E_s(0)$. This implies threshold condition of the laser expressed by

$$F_l(2, 2) = 0 \tag{4.22}$$

Thus, lasing frequency and required threshold gain are determined by finding pairs of $\omega$ and $g$ which produce $F_l$ satisfying Eq. (4.22).

In multi-electrode DFB lasers, the refractive index deviation $dn$ is created by carrier density change in the active layer. Therefore it follows the optical gain deviation $dg$. The linewidth enhancement factor $\alpha$ [51] relates $dn$ to $ad$ as

$$dn = -\frac{c}{\omega} \alpha \, dg \tag{4.23}$$

We will use $\alpha$ value of 4.0 in the following calculations. ($\alpha$ can vary widely, see [52–54]).

Figure 4.10 shows calculated results for a three electrode DFB laser, where cavity length is 500 $\mu$m, $\kappa$ is 20 cm$^{-1}$, and reflectivity of both facets is 0 [47]. Here we assume the laser consists of three sections and carrier density is uniform in each section. The carrier densities in the side sections are equal but differ from that in the center section. Therefore, the refractive index in the center section differs from that in the side sections by $dn$. The optical gain in the center and side sections also differs by $dg$, which satisfys Eq. 4.23. When the refractive index is uniform (dots: $dn = 0$), the laser mode is symmetrical with respect to the Bragg wavelength, and the two modes nearest the Bragg wavelength are degenerate in the threshold gain. When the carrier density in the center section increases, refractive index in that section decreases, so that lasing wavelength and required threshold gain both change, as indicated by the solid arrows in Figure 4.10. The broken arrows represent center carrier density decrease. Increase (triangles: $dn + 0.002$) in the refractive index causes the shorter wavelength mode to have lower threshold gain than the other modes, so single-mode oscillation occurs in the shorter wavelength mode.

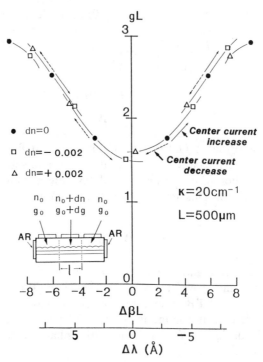

**Figure 4.10.** Calculated threshold gain $gL$ versus propagation constant deviation from Bragg condition $\Delta\beta L$ in a multielectrode DFB laser [47].

On the other hand, decrease in the refractive index (squares: $dn = -0.002$) causes single-mode oscillation in the longer wavelength mode. Change in the threshold gain also causes lasing wavelength shift through the shift of Bragg wavelength. This effect can be calculated also by Eq. (4.23).

Lasing wavelength selection by current ratio control is experimentally observed and demonstrated in Figure 4.11 [47]. The device is a 1.3-$\mu$m three-electrode DFB laser. Reflection from the laser facets is suppressed by an antireflection coating in the front and by a slanted facet in the rear so as to clearly investigate the mode selection by current control. Currents are supplied to three electrodes, as shown in the left side of the figure. The electrodes on the front and rear are connected to each other. Currents applied to the center and side electrodes are adjusted, so that carrier density nonuniformity could be created artificially between the side and the center regions of the laser cavity. The total current is kept constant at 110 mA even though the current to the center is changed. Figure 4.11a shows the lasing spectrum where currents are supplied to the electrodes with the same current density ($I_1 = 37$ mA). In this uniform excitation case, two longitudinal modes on both sides of the Bragg wavelength oscillate. As shown in the figure, single-longitudinal-mode oscilla-

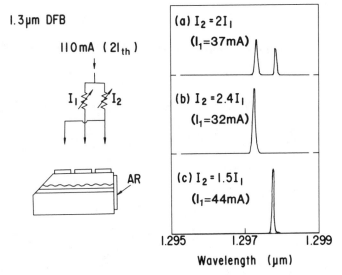

**Figure 4.11.**  Lasing mode selection by current ration adjustment observed in a 1.3-$\mu$m multielectrode DFB laser [47]. Reflection from both facets are suppressed by an AR film and a slanted facet. Lasing spectra are shown for three center electrode currents where total current is held at 110 mA. (*a*) Uniform injection ($I_1 = 37$ mA) produces two-mode oscillation. (*b*) Decrease in the center current ($I_1 = 32$ mA) causes single-mode oscillation in the short-wavelength mode. (*c*) Increase in the center current ($I_1 = 44$ mA) causes single-mode oscillation in the long-wavelength mode.

tion can be obtained by introducing the carrier density nonuniformity along the cavity. The shorter wavelength mode is selected by decreasing the current to the center, which increases refractive index in the center (Figure 4.11*b*; $I_1 = 32$ mA). The longer wavelength mode is selected by increasing it (Figure 4.11*c*; $I_1 = 44$ mA).

Figure 4.12 shows experimental characteristics of a 1.5 $\mu$m two-electrode DFB laser [44]. The cavity length is 200 $\mu$m. An antireflection coat was formed on the front facet. The top electrode is divided into two equal lengths to control the carrier density distribution along the cavity. The wavelength tuning is achieved by adjusting the ratio of the current injected into the front electrode $I_1$ to the total current $I_t$. The lasing wavelength shifts continuously over 2.4 nm (290 GHz) when changing the current ratio from 10 to 60%. In this constant total current condition (110 mA), output power also changes between 1.5 and 7 mW. Continuous wavelength tuning with 21 Å was also reported (under 5-mW constant power) with automatic control of the total current between 80 to 130 mA. The linewidth was typically 30 MHz, the same as conventional DFB lasers. Linewidth of multielectrode DFB lasers is expressed by the same equation as conventional DFB lasers [55], if carrier density nonuniformity is ignored. Therefore, the same methods can be used for linewidth reduction. Long cavity

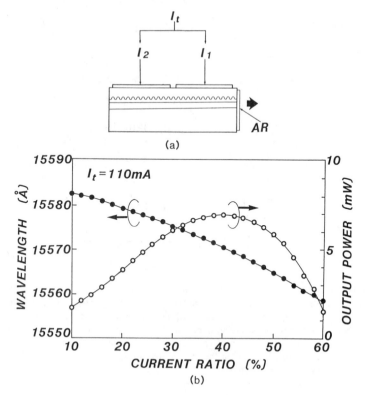

**Figure 4.12.** Observed continuous wavelength tuning in a 1.5-$\mu$m two-electrode DFB laser [44]. The cavity length is 200 $\mu$m. An antireflection coat is formed on the front facet. The wavelength tuning is achieved by adjusting ratio of the current injected into the front electrode $I_1$ to the total current $I_t$. The lasing wavelength shifts continuously over 2.4 nm (290 GHz) when changing the current ratio from 10 to 60%. In this constant total current condition (110 mA), output power also changes between 1.5 and 7 mW.

length is effective for linewidth reduction, because of small cavity loss. Introduction of multiple quantum well structure in the active layer is also effective because of its small linewidth enhancement factor. Using these two methods, linewidth as narrow as 350 kHz was realized [56].

Large coupling coefficient $\kappa$ in the grating also reduces linewidth because of reduction of cavity loss. However, in large-$\kappa$ devices spatial hole burning effects may limit linewidth reduction, because of strong nonuniformity in optical field associated with large $\kappa$. Multielectrode structure can be used to compensate for spatial hole burning [57].

Figure 4.13 shows linewidth of a long-cavity multielectrode DFB laser as a function of inverse output power [57]. The device is a 1.2-mm-long two-electrode DFB laser with both facets cleaved. The coupling coefficient $\kappa L$ is 1 and the active layer thickness $d_a$ is 0.08 $\mu$m. Two 600-$\mu$m-long

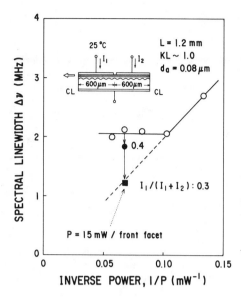

**Figure 4.13.** Observed linewidth as a function of inverse output power in a 1.2-mm-long two-electrode DFB laser with both facets cleaved. Open circles show the linewidth in uniform excitation ($I_1 = I_2$). Filled circle shows the linewidth for current ratio of 0.4, and filled square shows linewidth for current ratio of 0.3. Dashed line is extrapolated from the linewidth to inverse power line below 10 mW [57].

electrodes are formed on top of the device. Currents injected to the front electrode $I_1$ and the rear electrode $I_2$ are adjusted. With uniform excitation ($I_1 = I_2$), the linewidth decreases linearly to 2 MHz with output power increase up to 10 mW. The linewidth decrease, however, is saturated for over 10 mW and the linewidth is constant at 2 MHz for the output power between 10 and 20 mW. The authors assumed that the spatial hole burning causes the linewidth floor, and tried to compensate for the hole-burning effect by adjusting the current ratio. When decreasing the current to front electrode, the linewidth decreases and is minimized to 1.2 MHz at a current ratio of 0.3. The minimized linewidth is just on the line extrapolated from the linewidth to the inverse power line below 10 mW.

## 4.3. FREQUENCY MODULATION CHARACTERISTICS IN WAVELENGTH-TUNABLE LASERS

### 4.3.1. Direct Frequency Modulation in Conventional Semiconductor Lasers

Direct frequency modulation (FM) of semiconductor lasers is an important option in coherent communication systems. The direct FM requires a large frequency shift by a small modulation current and also a flat response for modulation frequency. In general, however, frequency dependence of direct FM response in conventional lasers is poor because the dominant

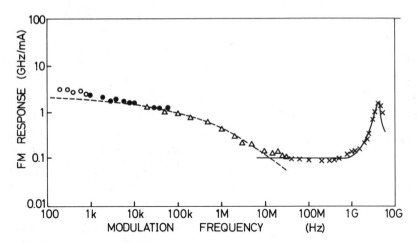

**Figure 4.14.** Typical modulation frequency dependence of FM response in semiconductor lasers [58]. Different symbols represent different measurement methods. See Ref. [58] for details.

mechanism for FM changes from a thermal effect at low frequency to carrier effect at high frequency [58].

Figure 4.14 shows typical modulation frequency dependence of FM response in semiconductor lasers [58]. FM response decreases with modulation frequency in the low modulation frequency range. In this frequency region, FM is dominated by thermal modulation. Since lasing frequency $\nu$ is determined by $2nL = mc/\nu$ with refractive index $n$ and cavity length $L$, frequency deviation $\Delta\nu$ due to temperature change $\Delta T$ is expressed by

$$\frac{\Delta\nu}{\nu} = -(\alpha_L + \alpha_n)\,\Delta T \qquad (4.24)$$

where $\alpha_L = (1/L)\,dL/dT$ is a linear thermal expansion coefficient and $\alpha_n = (1/n)\,dn/dT$ is a thermal refractive index coefficient. Assuming that temperature is uniform in a laser chip, the temperature change caused by current modulation $\Delta I$ can be expressed by

$$\frac{C_T d\,\Delta T}{dt} = -\left(\frac{1}{R_T}\right)\Delta T + r\,\Delta I \qquad (4.25)$$

where $C_T$ is thermal capacity of the laser chip, $R_T$ is heat resistance between the laser chip and a heat sink, and $r$ is the amount of heat created by unit current. Therefore, thermal FM given by Eqs. (4.24) and (4.25) shows rapid rolloff for modulation frequency with cutoff frequency of $1/2\pi C_T R_T$. Since both $\alpha_L$ and $\alpha_n$ are positive, thermal FM exhibits a redshift, that is, shift to lower frequency for current increase. FM response $\Delta\nu/\Delta I$ is given by $-(\alpha_L + \alpha_n)R_T r$ near DC. As heating by currents consists of Joule heat and loss in recombination energy, $r$ is given by $r = 2I_0 R_s + (V_j - h\nu/e\eta_{ex})$ with DC bias current $I_0$, series resistance $R_s$, junction voltage $V_j$, photon energy in electron volts $h\nu/e$, and differential external quantum efficiency $\eta_{ex}$ and is approximated by $r = 2I_0 R_s + h\nu/e(1 - \eta_{ex})$. Using typical values for 1.55-$\mu$m lasers: $\partial\nu/\partial T = -10$ GHz °C$^{-1}$($+0.8$ Å °C$^{-1}$), $R_t = 50$ °C W$^{-1}$, $I_0 = 160$ mA, $R_s = 3$ $\Omega$, $h\nu/e = 0.8$ eV, and $\eta_{ex} = 0.3$, FM response $\Delta\nu/\Delta I$ is approximated to be $\sim -450$ MHz mA$^{-1}$ near DC.

Above the thermal cutoff frequency, FM response is relatively flat but has strong enhancement at the resonance frequency as shown in Figure 4.14. In this frequency region, FM is dominated by carrier density modulation. Carrier-density modulation causes FM through several distinct mechanisms: relaxation oscillation, spectral hole burning, and spatial hole burning. First we ignore the distribution of carriers along the laser cavity, that is, ignore spatial hole burning. Then the rate equations for carrier density $N$ and photon density $P$ are expressed as

$$\frac{dN}{dt} = \frac{I}{eV} - \frac{N}{\tau_s} - G(N, P)P \qquad (4.26)$$

$$\frac{dP}{dt} = -\frac{P}{\tau_p} + G(N, P)P + \beta\frac{N}{\tau_s} \qquad (4.27)$$

where $I$ is the injection current, $e$ is the electron charge, $V$ is the cavity volume, $\tau_s$ is the carrier lifetime, $G(N, P)$ is the temporal gain for the lasing mode, $\tau_p$ is the photon lifetime, and $\beta$ is the spontaneous emission factor. Considering small signal sinusoidal modulation, $P$ and $N$ are expressed by $P = P_0 + pe^{j2\pi ft}$ and $N = N_0 + ne^{j2\pi ft}$ with modulation frequency $f$. Using threshold condition $-1/\tau_p + G(N_0, P_0) = -\beta N_0/(\tau_s P_0)$ and neglecting terms of $e^{j4\pi ft}$ and the small spontaneous emission term $\beta n/\tau_s$, then the expression for $n$ can be given by

$$\left(\frac{\partial G}{\partial N}\right)n = \left(\frac{j2\pi ft}{P_0} + \frac{\beta N_0}{P_0^2\tau_s} - \frac{\partial G}{\partial P}\right)p \qquad (4.28)$$

Since frequency deviation $\Delta\nu$ for the refractive index change $\Delta n_r$ is given by $\Delta\nu/\nu = -\Delta n_r/n_r$, Eq. (4.28) yields the expression for $\Delta\nu$ as

$$\Delta\nu = \frac{\alpha}{4\pi}\left(\frac{j2\pi ft}{P_0} + \frac{\beta N_0}{P_0^2\tau_s} - \frac{\partial G}{\partial P}\right)p \qquad (4.29)$$

Here we use definition of linewidth enhancement factor $\alpha$ expressed by

$$\alpha = \frac{\partial n_r / \partial N}{-(\partial G / \partial N)/2} \cdot \frac{2\pi\nu}{n_r} \tag{4.30}$$

The first term on the right-hand side of Eq. (4.29), $jf(\alpha/2)(p/P_0)$, represents the contribution of the relaxation oscillation to FM. As intensity modulation (IM) $p$ is constant for modulation frequencies below the resonance frequency [59]; FM response due to this contribution is proportional to modulation frequency $f$. Since $p$ is in-phase with modulation current $i$, the phase of FM is $\pi/2$ ahead of that of the modulation current. This term dominates FM only in the vicinity of the resonance frequency because of strong modulation frequency dependence. As photon density is proportional to the output power, $p/P_0$ is given by the IM modulation index, which is approximated as $\Delta I/(I_0 - I_{th})P_0$ well below the resonance frequency with the threshold current $I_{th}$. Using typical values, $I_{th} = 30$ mA, $\alpha = 4$, and those listed earlier, FM response $\Delta f/\Delta I$ is approximated to be $\Delta\nu/\Delta I = 0.06\ f$ and 60 MHz mA$^{-1}$ at modulation frequency of 1 GHz.

The second term in Eq. (4.29), $(\beta N_0/\tau_s P_0)(\alpha/4\pi)(p/P_0)$, represents the contribution of the spontaneous emission. Equation (4.26) produces a steady-state solution for the gain as $G(N, P) = 1/\tau_p - \beta N/\tau_s P$. Therefore $G(N, P)$ is slightly lower than the cavity loss $1/\tau_p$ because of the contribution of the spontaneous emission. With increase in the photon density, this contribution $\beta N/\tau_s P$ decreases and $G(N, P)$ increases up to $1/\tau_p$. It causes increase in carrier density and consequently causes a blue-shift FM. The FM response is proportional to the IM response $p$, which is independent of modulation frequency below the resonance frequency, but is very small. The FM response is evaluated from the steady-state solution of Eqs. (4.26, 4.27). Using approximated solutions; $N_0/\tau_s = I_{th}/eV$ and $P_0 = \tau_p(I - I_{th})/eV$, the FM response $\Delta\nu/\Delta I$ is given by $\Delta\nu/\Delta I = (\beta/2\pi\tau_p)I_{th}/(I - I_{th})(\alpha/2)(p/P_0)$. Using $\tau_p = 1$ ps, $\beta = 10^{-5}$, $I = 2I_{th}$, and other parameters listed earlier, the FM response is approximated to be 0.1 MHz mA$^{-1}$.

The third term in Eq. (4.29), $(\alpha/4\pi)(\partial G/\partial P)p$, represents the contribution of the nonlinear gain due to the spectral hole-burning effect [60]. The spectral hole burning modifies carrier distribution in the energy levels depending on the photon density $P$. It reduces carrier number resonant with lasing frequency. Thus, the gain $G$ decreases with increase in photon density $P$ even if total carrier density is constant. As spectral hole burning is a rapid process, $\partial G/\partial P$ is independent of frequency. Therefore, FM response due to this contribution has the same frequency dependence as IM. Since both $\alpha$ and $\partial G/\partial P$ are negative, FM due to this contribution shows blue-shift character, that is, shift to higher frequency for light power increase and current increase. Although the value of nonlinear gain $\partial G/\partial P$ is not certain to date, there are some reports on it [61, 62]. It was empirically evaluated by frequency dependence of IM response [62] and $\partial G/\partial P_{out}$ of $3.0$–$6.9 \times 10^9$ s$^{-1}$ mW$^{-1}$ is reported, where $\partial G/\partial P_{out}$ is gain saturation coefficient for output power $P_{out}$. Using $\partial G/\partial P_{out} = 5 \times 10^9$s$^{-1}$ mW$^{-1}$ and the others listed earlier, the FM response is approximated to be 100 MHz mA$^{-1}$.

The carrier density distribution in a laser cavity produces a spatial hole-burning effect. Here we consider the spatial hole-burning formed by optical field distribution along the laser cavity. Defining a $z$ axis along with laser cavity, field intensity $I(z)$ for the lasing mode may be written by

$$I(z) = I[1 + f(z)] \tag{4.31}$$

where $f(z)$ satisfies $\int f(z)\,dz = 0$. Then, carrier density can be approximated by

$$N(z,t) = N_a + N_d f(z) + [n_a + n_d f(z)]e^{j2ft} \tag{4.32}$$

Ignoring the spectral hole burning and the spontaneous emission for simplicity, we obtain expressions for $n_a$ and $\Delta\nu$ similar to Eqs. (4.28) and (4.29):

$$\left(\frac{\partial G}{\partial N}\right)n_a = \frac{j2\pi f}{P_0}p - \frac{\partial G}{\partial N_d}n_d \tag{4.33}$$

$$\Delta\nu = \frac{jf\alpha}{2P_0}p + \left(\frac{\partial\nu}{\partial N_d} - \frac{\alpha}{4\pi}\frac{\partial G}{\partial N_d}\right)n_d \tag{4.34}$$

the second term in Eq. (4.34) represents the frequency deviation due to the spatial hole burning. It should be noted that the spatial hole burning does not always cause frequency deviation. In an FP laser with cavity length $L$, the modal gain $G$ only depends on the averaged local gain $g(z)$ and lasing frequency only depends on the averaged refractive index $\bar{n}_r = \int n_r(z)\,dz/L$. Therefore, $\partial G/\partial N_d$ is given by $(c/\bar{n}_r)\int(\partial g/\partial N)f(z)\,dz/L$ and $\partial\nu/\partial N_d$ is given by $(\nu/\bar{n}_r)$ $\int(\partial n_r/\partial N)f(z)\,dz/L = (c/\bar{n}_r)\int(\alpha/4\pi)(\partial g/\partial N)f(z)\,dz/L$. Therefore, if $\alpha$ is constant throughout the cavity $\partial\nu/\partial N_d = (\alpha/4\pi)\partial G/\partial N_d$, and consequently the bracketed term in Eq. (4.34) is 0. However, if $\alpha$ varies along the cavity, the spatial hole burning produces frequency deviation even in FP lasers [63]. In DFB lasers the situation is different. As discussed in Section 4.2.5, lasing wavelength in DFB lasers is strongly affected by the carrier density nonuniformity and is not determined only by averaged refractive index. Furthermore the carrier density nonuniformity also changes cavity loss, as shown previously in Figure 4.10. It introduces a new term, $(\partial\tau_p/\partial N_1)n_1$, in Eqs. (4.33) and (4.34). Therefore, the spatial hole burning produces frequency shift in DFB lasers, even if $\alpha$ is constant throughout the cavity [64].

These mechanisms are competing in FM response of semiconductor lasers depending on the modulation frequency. In general, FM response is dominated by the thermal effect in the low-frequency region. Therefore, FM response in low-frequency is red-shift and exhibits rapid rolloff in frequency. With increase in frequency, the thermal FM vanishes and FM response is dominated by the spectral or spatial hole-burning effects. In this frequency range FM response is blue-shift and exhibits flat frequency dependence. To date it is not clear which one dominates the FM in this mid-frequency range because important parameters, $\partial G/\partial P$ and $\alpha(N)$, are not clear. In higher frequency ranges, the FM response shows significant increase due to resonance frequency. Around the resonance frequency, the relaxation oscillation dominates the FM response.

This competition between mechanisms makes the frequency dependence of the FM complicated in terms of both the FM response and the phase difference. These poor FM characteristics cause significant degradation in coherent transmission with direct FM [65].

### 4.3.2.  Direct Frequency Modulation Using Wavelength-Tunable Lasers

Wavelength tunability in tunable lasers provides another option for direct frequency modulation. As wavelength tuning, discussed in Section 4.2, is based on carrier-density effects except for external-cavity lasers, tuning speed is as fast as carrier lifetime, that is, 1 ns or less. Therefore, wavelength tuning in these devices is fast enough to be used in direct FM for at least several hundreds of megahertz. Figure 4.15 explains direct FM, using wavelength tunability of a 1.5-$\mu$m 3-electrode DFB laser [47]. Figure 4.15a shows lasing wavelength of the laser as a function of the center electrode current $I_c$, where total current is kept constant at 100 mA to eliminate temperature change. Lasing wavelength shifts continuously, maintaining single-longitudinal-mode operation, except for a longitudinal-mode change around the center current of 55 mA. An increase in the center current causes the lasing frequency to shift to a lower or higher frequency depending on the current level with maximum tuning range of 30 GHz (2.2 Å). If the device is biased at the current in which the frequency is sensitive to the current, modulation in the current should cause a large frequency modulation. Moreover, frequency shift direction can be controlled by choosing the proper bias point, depending on the slope in the frequency–current curve.

Figure 4.15b shows the frequency modulation response of the device at several bias levels, where 100-MHz sinusoidal current is superimposed on

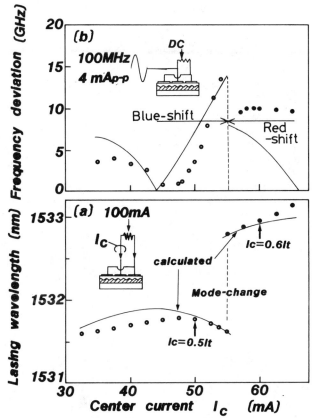

**Figure 4.15.**  Direct FM using frequency-tunability of a three-electrode DFB laser [47]. (*a*) Observe lasing wavelength versus center electrode current $I_c$. Total current is held at 100 mA. FM characteristics at bias point indicated by arrows ($I_c = 0.5I_t$ and $0.6I_t$) are shown in Figures 4.15(*c*) and 4.15(*d*). (*b*) Width of frequency deviation under 100-MHz sinusoidal modulation versus DC bias $I_c$. Modulation current of 4 mA$_{p-p}$ is applied to center electrode with DC bias $I_c$. (*c*) Phase delay of the frequency shift to the modulation current as a function of the modulation frequency. Open circles show the phase delay for the red-shift bias point; triangles show the phase delay for blue-shift bias point in Figure 4.15*a*. Dots show the phase delay in a conventional DFB laser. (*d*) Modulation frequency dependence of FM response. Symbols are the same as in Figure 4.15*c*.

DC bias to the center electrode. Even though the modulation current is kept constant at 4 mA$_{p-p}$ (peak-to-peak amperage), the frequency modulation response changes with the bias condition. Solid curves in the figure show calculated modulation response, which is obtained from the slope on the static response of frequency for DC current in Figure 4.15*a*. As shown, FM response varies with slope of the frequency shift in DC measurements. The maximum FM response is over 3 GHz mA$^{-1}$, which is much larger than FM responses due to any mechanisms discussed in the previous

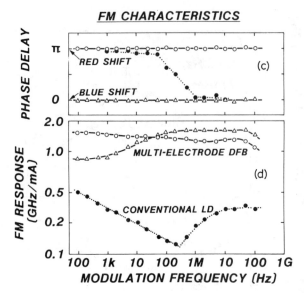

**Figure 4.15.** *(Continued)*

section. Thus strong carrier density modulation dominates FM at all frequencies and eliminates strong frequency dependence.

Figure 4.15c shows dependence of FM efficiency on modulation frequency. Phase delay of the frequency shift with the modulation current is shown in Figure 4.15d. The device is biased at two different points in Figure 4.15a. Open circles represent the red-shift bias point ($I_c$ = 60 mA) and open triangles, the blue-shift bias point ($I_c$ = 50 mA). For comparison, the result for a conventional DFB laser is shown by dots. There is a clear dip in the frequency-dependence curve of the modulation efficiency for the conventional DFB laser (around 100 kHz in Figure 4.15a). The FM efficiency decreases to 120 MHz mA$^{-1}$ from 500 MHz mA$^{-1}$ in the low-frequency region. According to the change in FM efficiency, phase delay changes completely from $\pi$ (red-shift) to 0 (blue-shift). In contrast with the conventional laser diode (LD), the FM efficiency of the multielectrode DFB laser exhibits flat response to modulation frequency up to a 100 MHz for both red-shift and blue-shift cases. The FM response is two to three times larger than the conventional LD. The phase delay also remains constant at 0 for blue-shift and at $\pi$ for red-shift.

A simultaneous modulation of two electrodes produces another interesting application of wavelength-tunable laser in FM. Figure 4.16 shows temporal lasing frequency shift when modulation current is supplied to different electrodes [47]. The device is biased at red-shift region ($I_c$ = 60 mA) in previous figures. A 100-MHz sinusoidal current is applied to the center (circles in Figure 4.16) or the front (triangle) electrodes. As shown

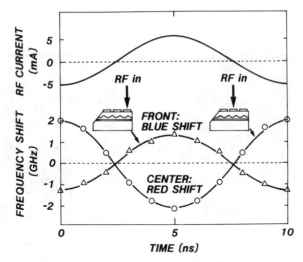

**Figure 4.16.** Temporal lasing frequency shift during 100-MHz sinusoidal modulation for two distinct modulation schemes in three-electrode DFB lasers [47]. Circles show frequency shift when the center electrode is supplied with the modulation current. Triangles show frequency shift when the front electrode is supplied with the modulation current. DC bias currents are set at the red-shift point in Figure 4.15a.

in Figure 4.16, when the modulation current applied to the center elec-trode frequency shift is red-shift, while it is blue-shift when the modula-tion current is applied to the side electrode. This is because by changing the position of electrode-applied modulation current, carrier distribution created by the modulation current is inverted spatially. On the other hand, amplitude modulation is in-phase with the current as expected. If these red- and blue-shift frequency modulations are combined, amplitude and frequency can be modulated independently.

Results of the independent modulation in amplitude and frequency are shown in Figure 4.17, with two extreme cases [47]. Figure 4.17, left panels, shows amplitude modulation with FM suppression where modulation currents applied to the electrodes are nearly in-phase. The amplitudes of the modulation currents are adjusted to cancel the FM (30 mA$_{p-p}$ for the front and 20 mA$_{p-p}$ for the center). The upper panels show intensity, and the lower panels show spectrum under the modulation. Even with 50% amplitude modulation, the lasing spectrum remains sharp with a 2-GHz linewidth. This width under 1 GHz modulation is 20% that of conven-tional DFB. Figure 4.17, right panels, shows pure FM modulation without spurious amplitude modulation. Modulation currents are applied to the electrode out-of-phase, and amplitudes are adjusted to cancel amplitude modulation (10 mA$_{p-p}$ for the front and 20 mA$_{p-p}$ for the center). Amplitude modulation in the upper panels is completely suppressed. Even

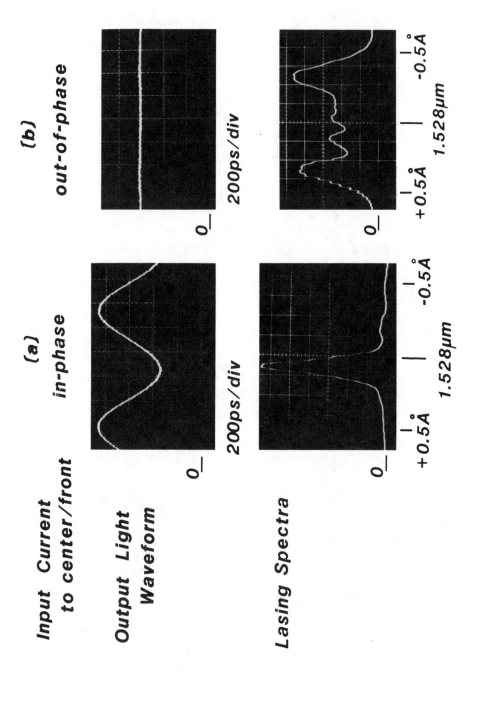

Input Current to center/front

(a) in-phase

(b) out-of-phase

Output Light Waveform

0 — 200ps/div (a)

0 — 200ps/div (b)

Lasing Spectra

+0.5Å | 1.528µm | -0.5Å

0 —

+0.5Å | 1.528µm | -0.5Å

0 —

130

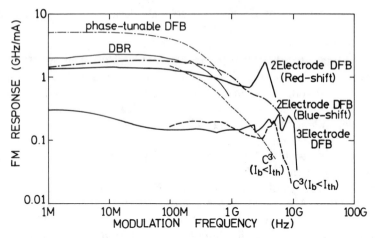

**Figure 4.18.** Modulation frequency dependence of FM response in a phase-tunable DFB laser [38], a DBR laser [22], a two-electrode DFB laser [68], a three-electrode DFB laser [56], and a C$^3$ laser [31].

in this constant-output condition, frequency modulation spread over the lasing spectrum. The frequency modulation derviation reaches 12 GHz (0.9 Å).

Recently efforts have been made to extend strong FM response in wavelength-tunable lasers to the multigigahertz region. Figure 4.18 shows the frequency dependence of the FM response in several wavelength-tunable lasers. The phase-tunable DFB [35] and DBR [66] lasers have FM responses larger than 1 GHz mA$^{-1}$ at low frequencies, but show rapid rolloff with cutoff frequencies of a few hundred megahertz. As the modulation current is supplied to the passive tuning region in these devices, carrier density in the modulated section is affected only by carrier injection and spontaneous emission. Therefore, frequency response is simply determined by spontaneous carrier lifetime $\tau_s$ with cutoff frequency $1/2\pi\tau_s$.

---

**Figure 4.17.** Independent modulation of amplitude and frequency observed in a three-electrode DFB laser [47]. Both center and front electrodes are supplied with 1-GHz modulation currents. Phases and amplitudes of the modulation currents are adjusted to control AM and FM independently. (*a*) In-phase modulation; currents in the front and center electrodes are in-phase. Amplitudes of the modulation currents are adjusted to cancel FM. Top figure shows output lightwave form. Bottom figure shows lasing spectrum under modulation. In spite of large AM in the top figure, lasing spectrum in the bottom figure is not widened by FM. (*b*) Out-of-phase modulation; currents in the front and center electrodes are out-of-phase. Amplitudes of the modulation currents are adjusted to cancel AM. Top figures shows output lightwave form. Bottom figure shows lasing spectrum under modulation. Although AM is completely suppressed in the top figure, lasing spectrum in the bottom figure is widened by FM.

The FM response of $C^3$ lasers is strongly dependent on the DC bias current, as discussed in Section 4.2. When the modulated section is biased below the threshold, the FM response of $C^3$ lasers is similar to that of the DBR and phase-tunable DFB lasers. The LD biased below threshold acts as a tunable etalon, so that it has large FM response at low frequencies but with rapid rolloff. When the modulated section is biased above the threshold, low-frequency FM response significantly decreases and frequency dependence becomes flat because of the carrier density clamping in the LD. The FM response decreases slightly above 1 GHz but increases again as a result of the enhancement at the resonance frequency. The 3-dB cutoff frequency reaches 6 GHz.

The FM response of multielectrode DFB lasers also has wide modulation bandwidth. A modulation bandwidth over 10 GHz is shown in Figure 4.18 with a three-electrode MQW DFB laser. Although the FM response is small due to the long cavity length (1.2 mm), frequency dependence is flat from 100 kHz to 10 GHz. Frequency dependence of FM response in multielectrode DFB lasers also varies with DC bias current. In Figure 4.18 frequency dependence of FM response in a two-electrode DFB laser is shown for two distinct DC biases: the red-shift and the blue shift bias points [67]. For the red-shift bias, the resonance enhancement is more clear and bandwidth is larger than for the blue-shift bias. This difference in frequency response is theoretically explained by small signal analysis [68].

## 4.4.  CONCLUSION

In concluding this chapter, we shall refer again to the experimental results summarized in Figures 4.18 and 4.19. Figure 4.19 shows experimentally observed linewidth and wavelength tuning ranges reported to date. The external-cavity laser shows the best value among all devices in linewidth (2 kHz) and tuning range (154 nm) as well as the continuous tuning range (15 nm). In spite of all these attractive features, problems in long-term reliability and mechanical stability limit application mainly to laboratory use. Relatively poor direct FM characteristics may also limit their use in optical communication systems.

Cleaved coupled cavity lasers have the widest discrete tuning range among monolithic devices and a flat FM response up to 6 GHz. The critical bias dependencies, however, of the lasing characteristics make its practical application difficult, because of their delicate operation mechanism based on the interference.

Distributed Bragg reflector lasers have the widest continuous tuning range among the monolithic devices. The greatest potential use of the DBR devices is in WDM–FDM applications. Relatively poor frequency

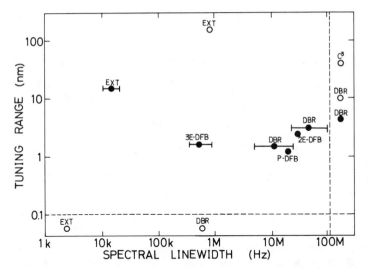

**Figure 4.19.** Tuning range and the spectral linewidth of EXT—external-cavity lasers [10, 12, 13], DBR lasers [19, 21, 35], C³ lasers [28], P-DFB—phase-tunable DFB lasers [38], 2E-DFB—two-electrode DFB lasers [44], and 3E-DFB—three-electrode DFB lasers [56]. Filed circles represent continuous tuning range. Open circles represent total tuning range. Tuning range is not reported for data under dashed line. Spectral linewidth is not reported for data in right hand-side of dashed line.

dependence of FM, due to rapid rolloff limited by spontaneous lifetime, may limit its applications.

The phase-tunable DFB lasers for the first time demonstrated that wavelength tuning possibility in DFB lasers. Research interest in this device is shifting to the multielectrode DFB, which has simpler structure and is much easier to fabricate.

In addition to its simple structure, the multielectrode DFB lasers have many attractive features. Recent reports on narrow-linewidth and high-frequency FM response has made the future of multielectrode DFB lasers appear promising.

## REFERENCES

1. K. Utaka, S. Akiba, K. Sakai, and Y. Matsushima, *Electron. Lett.* **17**, 961–963 (1981).

2. T. Matsuoka, H. Nagai, Y. Itaya, Y. Noguchi, Y. Suzuki, and T. Ikegami, *Electron. Lett.* **18**, 27–28 (1981).

3. Y. Suematsu, S. Arai, and K. Kishino, *IEEE J. Lightwave Technol* **LT-1**, 161–176 (1983).

4. K. Emura, M. Shikada, S. Fujita, I. Mito, H. Honmou, and K. Minemura, *Electron. Lett.* **20**, 1022–1023 (1984).

5. See, for example, F. Kazovsky, "Coherent optical communication systems," *Proc. Optical Fiber Commun. Tutorial*, San Francisco, 1990, p. 63.

6. H. Yamazaki, T. Sasaki, N. Kido, M. Kitamura, and I. Mito, "250 kHz linewidth operation in long cavity 1.5 μm multiple quantum well DFB-LDs with reduced linewidth enhancement factor," *Postdeadline Papers Optical Fiber Commun. Conf.*, San Francisco, PD33-1–PD33-4, 1990.

7. M. W. Fleming and A. Mooradian, *IEEE J. Quantum Electron.* **QE-17**, 44–59 (1981).

8. R. Wyatt and W. J. Devlin, *Electron. Lett.* **19**, 110–112 (1983).

9. C. H. Henry, *IEEE J. Lightwave Technol.* **LT-4**, 258 (1986).

10. R. A. Linke and K. J. Pollock, "Linewidth vs. length Dependence for an external cavity laser," *Proc. 10th IEEE International Semiconductor Laser Conf.*, Kanazawa, I-1, 1986.

11. T. Okoshi, K. Kikuchi, and A. Nakayama, *Electron. Lett.* **16**, 630–631 (1980).

12. M. Notomi, O. Mitomi, Y. Yoshikuni, F. Kano, and Y. Tohmori, *Photon. Technol. Lett.* **2**, 85–87 (1990).

13. F. Favre, D. Le Guen, J. C. Simon, and B. Landousies, *Electron. Lett.* **22**, 795–796 (1986).

14. J. Mellis, S. A. Al-chalabi, K. H. Cameron, R. Wyatt, J. C. Regnault, W. J. Devlin, and M. C. Brain, *Electron. Lett.* **24**, 988–989 (1988).

15. S. Saito, S. Nilsson, and Y. Yamamoto, *IEEE J. Quantum Electron.* 961–970 (1982).

16. E. Brinkmeyer, W. Brennecke, M. Zürn, and R. Ulrich, *Electron. Lett.* **22**, 134–135 (1986).

17. Y. Tohmori, Y. Suematsu, H. Tsushima, and S. Arai, *Electron. Lett.* **19**, 656–657 (1983).

18. Y. Tohmori, K. Komori, S. Arai, Y. Suematsu, and H. Oohashi, *Trans. IECE Jpn.* **68**, 84–86 (1985).

19. S. Murata, I. Mito, and K. Kobayashi, *Electron. Lett.* **23**, 403–405 (1987).

20. S. L. McCall and P. M. Platzman, *IEEE J. Quantum Electron.* **QE-21**, 1899–1918 (1985).

21. S. Murata, I. Mito, and K. Kobayashi, *Electron. Lett.* **24**, 577–579 (1988).

22. S. Murata, I. Mito, and K. Kobabyashi, *IEEE J. Quantum Electron.* **QE-23**, 835–838 (1987).

23. S. Murata, I. Mito, K. Kobayashi, *Proc. Opt. Fiber Commun.* **WD3**, 154 (1987).

24. K. Kobayashi and I. Mito, *IEEE J. Lightwave Technol.* **LT-6**, 1623–1633 (1988).

25. T. L. Koch, U. Koren, R. P. Gnall, C. A. Burrus, and B. I. Miller, *Electron. Lett.* **24**, 1431–1433 (1988).

26. F. Kano, Y. Tohmori, Y. Kondo, M. Nakano, M. Fukuda, and K. Oe, *Electron. Lett.* **25**, 709–710 (1989).

27. W. T. Tsang, N. A. Olson, and J. A. Ditzenberger, *Appl. Phys. Lett.* **43**, 1003–1005 (1983).

28. W. T. Tsang, N. A. Olsson, R. A. Linke, and R. A. Logan, *Electron. Lett.* **19**, 415–416 (1983).

29. W. T. Tsang and N. A. Olsson, *Appl. Phys. Lett.* **43**, 527–529 (1983).

30. W. T. Tsang, N. A. Olsson, and R. A. Logan, *Electron. Lett.* **19**, 488–490 (1983).

31. J. E. Bowers, R. S. Tucker, and C. A. Burrus, *IEEE J. Quantum Electron.* **QE-20**, 1230–1232 (1984).

32. T. Matsuoka, Y. Yoshikuni, and H. Nagai, *IEEE J. Quantum Electron.* **QE-21**, 1880–1886 (1985).

33. W. Streifer, R. D. Burnham, and D. R. Scifres, *IEEE J. Quantum Electron.* **QE-11**, 154–161 (1975).

34. M. Kitamura, M. Yamaguchi, K. Emura, I. Mito, and K. Kobayashi, *IEEE J. Quantum Electron.* **QE-21**, 415–417 (1985).

35. S. Yamazaki, K. Emura, M. Shikada, M. Yamaguchi, and I. Mito, *Electron. Lett.* **21**, 283–285 (1985).

36. K. Kaede, I. Mito, M. Yamaguchi, M. Kitamura, R. Ishikawa, R. Lang, and K. Kobayashi, "Spectral chirping suppression by compensation current in modified DFB-DC-PBH LD," *Proc. 8th Conf. Optical Fiber Commun.*, San Diego, 1985, pp. 104–105.

37. L. D. Westbrook, A. W. Nelson, P. J. Fiddyment, and J. V. Collins, *Electron. Lett.* **20**, 957–959 (1984).

38. S. Murata, I. Mito, and K. Kobayashi, *Electron. Lett.* **23**, 12–14 (1987).

39. S. Murata, S. Yamazaki, I. Mito, and K. Kobayashi, *Electron. Lett.* **22**, 1197–1198 (1986).

40. T. P. Lee, S. G. Menocal, S. Sakano, V. Valster, and S. Tsuji, "Linewidth and FM characteristics of a distributed feedback laser monolithically integrated with a tunable external cavity," *Electron. Lett.* **23**, 153–154 (1987).

41. S. Murata, I. Mito, M. Shikada, and K. Kobayashi, "Narrow spectral linewidth DFB lasers with a monolithically integrated optical waveguide," IOOC-ECOC '85, Venice, *Technical Digest* I, pp. 299–302.

42. H. Soda, Y. Kotaki, H. Sudo, H. Ishikawa, S. Yamakoshi, H. Imai, *IEEE J. Quantum Electron.* **QE-23**, 804–814 (1987).

43. N. K. Dutta, A. B. Piccirilli, T. Cella, and N. R. L. Brown, *Appl. Phys. Lett.* **48**, 1501–1503 (1986).

44. Y. Yoshikuni, K. Oe, G. Motosugi, and T. Matsuoka, *Electron. Lett.* **22**, 1153–1154 (1986).

45. K.-Y. Liou, C. A. Burrus, U. Koren, and T. L. Koch, *Appl. Phys. Lett.* **52**, 1899–1904 (1985).

46. Y. Yoshikuni and G. Motosugi, "Independent modulation in amplitude and frequency regimes by a multi-electrode distributed feedback laser," *Proc. 9th Conf. Optical Fiber Commun.*, Atlanta, 1986, pp. 32–33.

47. Y. Yoshikuni and G. Motosugi, *J. Lightwave Technol.* **LT-5**, 516–522 (1987).

48. G. Björk and O. Nilsson, *J. Lightwave Technol.* **LT-5**, 140–146 (1987).

49. M. Yamada and K. Sakuda, *Appl. Opt.* **26**, 3474–3478 (1987).

50. H. Kogelnik and C. V. Shank, *J. Appl. Phys.* **42**, 2327–2335 (1972).

51. C. H. Henry, *IEEE J. Quantum Electron.* **QE-18**, 259–264 (1982).

52. I. D. Henning and J. V. Collins, *Electron. Lett.* **19**, 927–929 (1983).

53. L. D. Westbrook, *Electron. Lett.* **21**, 1018–1019 (1985).

54. L. D. Westbrook, *Proc. IEE*, **133**, Part J, 223–225, 1985.

55. K. Kojima, K. Kyuma, *Electron. Lett.* **20**, 869–871 (1984).

56. Y. Kotaki, T. Fujii, S. Ogita, M. Matsuda, and H. Ishikawa, "Narrow linewidth and wavelength tunable multiple quantum well $\lambda/4$ shifted distributed feedback laser," *Proc. Optical Fiber Commun.*, San Francisco, THE3, 1990.

57. H. Yasaka, M. Fukuda, and T. Ikegami, *Electron. Lett.* **24**, 760–762 (1988).

58. S. Kobayashi, Y. Yamamoto, M. Ito, and T. Kimura, *IEEE J. Quantum Electron.* **QE-18**, 582–595 (1982).

59. T. Ikegami and Y. Suematsu, *IEEE J. Quantum Electron.* **QE-4**, 148–151 (1968).

60. M. Yamada and Y. Suematsu, *J. Appl. Phys.* **52**, 2563–2664 (1981).

61. C. B. Su, V. Lanzisera, and R. Olshansky, *Electron. Lett.* **21**, 893–894 (1985).

62. R. Olshansky, P. Hill, V. Lanzisera, and W. Powazinik, *IEEE J. Quantum Electron.* **QE-23**, 1410–1418 (1987).

63. O. Nilsson and Y. Yamamoto, *Appl. Phys. Lett.* **46**, 223–225 (1985).

64. P. Vankwikekberge, F. Buytaert, A. Franchois, R. Baets, P. I. Kuindersma, and C. W. Fredriksz, *J. Quantum Electron.* **QE-25**, 2239–2254 (1989).

65. K. Iwashita, N. Takachio, Y. Nakano, and N. Tsuzuki, *Electron. Lett.* **23**, 1022–1023 (1987).

66. S. Murata, I. Mito, and K. Kobayashi, *IEEE J. Quantum Electron.* **QE-23**, 835–838 (1987).

67. A. E. Willner, M. Kuznetsov, I. P. Kaminow, J. Stone, L. W. Stulz, and C. A. Burrus, "FM/FSK characterization of tunable two-electrode DFB lasers and their performance with noncoherent detection," *Proc. Optical Fiber Commun.*, San Francisco, THI6, 1990.

68. N. Kuznetsov, A. E. Willer, and I. P. Kaminow, *Appl. Phys. Lett.* **55**, 1826–1828 (1989).

# 5

# Spectroscopy by Semiconductor Lasers

M. Ohtsu and K. Nakagawa

*Graduate School at Nagatsuta, Tokyo Institute of Technology, Midori-ku, Yokohama, Kanagawa, Japan*

## 5.1. INTRODUCTION

Stable gas, dye, and solid-state lasers have been employed as coherent light sources for high resolution, nonlinear laser spectroscopy. Recent progresses in semiconductor devices have made it possible to achieve high output power, low frequency fluctuations, and wideband tunability in room-temperature and continuous-wave (CW) operated semiconductor lasers. It can be claimed that the performance of some of these lasers has already surpassed that of other types of lasers. This progress has made it possible to use semiconductor lasers for high-resolution laser spectroscopy. Furthermore, it is interesting to note that these semiconductor lasers oscillate at the wavelength range of between 0.7 and 1.6 $\mu$m, which has not been obtained by other types of lasers. Because of this wavelength range, a new field of spectroscopy, for example, spectroscopy of overtones or combination tones in organic molecular vapors, can be exploited. Furthermore, low price, small volume, and low power consumption in semiconductor lasers are quite attractive for spectroscopists who have been troubled with expensive maintenance charges and huge laser systems.

Considering these facts pointed out above, semiconductor lasers could play an essential role in several fields of high-resolution spectroscopy. This chapter will demonstrate such techniques. Prior to this demonstration, FM noise characteristics of semiconductor lasers and their control techniques will be reviewed in Sections 5.2 and 5.3. Very high quality and low noise semiconductor lasers are required for high-resolution laser spectrometer systems, which is quite different from other application systems. A part of these sections may overlap with those of other chapters in this book.

*Coherence, Amplification, and Quantum Effects in Semiconductor Lasers*, Edited by Yoshihisa Yamamoto.
ISBN  0-471-512494  © 1991 John Wiley & Sons, Inc.

However, most of the parts are discussed from a viewpoint of its applications to spectroscopy. Several topics in the laser spectroscopy and their applications will be described in Sections 5.4 and 5.5. A summary is given in 5.6.

## 5.2. CHARACTERISTICS OF SEMICONDUCTOR LASERS TO BE CONSIDERED TO REALIZE SINGLE-MODE OSCILLATION

A semiconductor laser exhibits essentially multilongitudinal-mode oscillation because the linewidth of its gain spectrum is about 100 times broader than separation between the adjacent longitudinal modes. Such the laser has the following specific oscillation characteristics:

1. *Mode-Hopping.* Semiconductor lasers usually shows multilongitudinal-mode oscillation if they employ a conventional Fabry-Perot-type cavity that is fabricated by cleaving or chemical etching of the crystal facets. In this case, power switching between the longitudinal modes occurs.

2. *Mode Partition.* Even though the power of the specific longitudinal mode is larger than those of others, that is, in the case of nearly single-longitudinal-mode oscillation, transient decrease of the main mode power and increase of the other submode power occurs. This decrease in the main mode power is called *power dropout*.

Both phenomena have been interpreted as quantum-mechanical phenomena, which are triggered by mutually uncorrelated spontaneous emission fluctuations introduced into each mode. Power fluctuation characteristics due to these phenomena are governed by cross-saturation characteristics of the gains in these modes, for which intraband relaxation of the carrier in the conduction band plays an essential role. Mode power fluctuations stemmed from these phenomena follow the statistics of a Poisson process, and unified theoretical models have been presented by introducing an analogy with phase transition of the first order in classical thermodynamics [1]. From this theoretical work, it was pointed out that the frequency of occurrences of these power fluctuations decrease exponentially with the increase of the bias level. It can also be decreased by enhancing the cross-gain saturation. For this enhancement, the tellurium (Te) was doped into the clad layer so that the DX center formed by this dopant worked as a saturable absorber [2, 3].

The 1.3- and 1.5-$\mu$m-wavelength InGaAsP lasers have more advanced cavity structures: DFB (distributed feedback) or DBR (distributed Bragg reflector)-type laser cavities, to realize a more reliable nearly single-longitudinal-mode oscillation for applications in optical communication sys-

tems. This DFB structure has recently been employed successfully for
0.8-$\mu$m-wavelength AlGaAs lasers [4]. Even for these longitudinal-mode
controlled lasers, the mode partition phenomenon cannot be neglected if
the power ratio between the main and sublongitudinal modes are less than
40 dB, which could induce errors in spectral assignments in high resolution
and highly sensitive laser spectrometers [5]. To overcome this difficulty,
electrodes for current injection into the laser were divided into two or
three parts to suppress the spatial hole burning of the carrier so that the
power of the sublongitudinal modes could be reduced [6].

Even when the nearly single-longitudinal-mode oscillation can be realized by solving these problems, deterministic instability can be induced if
the emitted light from the laser is injected back into the laser itself from
the external reflecting surface [7, 8]. To avoid this instability, a high-quality
optical isolator with the isolation of higher than 60 dB is required.

For the discussions given in Section 5.3, it is assumed that each problem
presented above has been solved to realize a reliable nearly single-longitudinal-mode oscillation.

## 5.3. REALIZATION OF A WIDEBAND HYPERCOHERENT OPTICAL SWEEP GENERATOR FOR SPECTROSCOPY

A highly coherent and widely tunable light source is required for spectrometer systems. It is demonstrated in this section that such a light source
can be realized by utilizing efficiently frequency-controllable semiconductor lasers.

### 5.3.1. Frequency Fluctuations of a Single-Longitudinal-Mode Semiconductor Laser

The primary measures that have been used for representing the magnitude
of optical frequency fluctuations (FM noise) are power spectral density
and the Allan variances [9, 10]. They are measures in the Fourier frequency and the time domain, respectively. In the case of white FM noise,
the field spectral profile of laser oscillation takes a Lorentzian shape and
its half-linewidth could be used as a convenient measure for representing
the FM noise magnitude because the linewidth is proportional to the
magnitude of the white noise power spectral density.

Figure 5.1 shows a power spectral density of the intrinsic FM noise.
Most fundamental noise source is the spontaneous emission fluctuations
(curve $A$). It gives a quantum noise limit of the free-running laser, which
corresponds to the FM noise of the coherent state or is called the
*Schawlow–Townes limit* of the field spectral linewidth [11]. Probability of

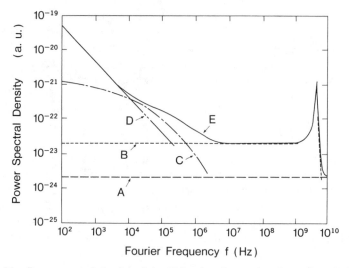

**Figure 5.1.** Power spectral density of the FM noise of a semiconductor laser induced by intrinsic noise sources. Curve $A$, spontaneous emission fluctuations; $B$, carrier-density fluctuations; $C$, temperature fluctuations; $D$, $1/f$ fluctuations. Curve $E$ represents the total intrinsic FM noise by superposing the values of curves $A$–$D$.

occurrences of spontaneous emission events is proportional to the number of cavity modes, and the magnitude of FM noise due to spontaneous emission are proportional to cavity loss.

Carrier density could also be varied by the spontaneous emission, which could induce the fluctuations of refractive index of the cavity and give additional FM noise (curve $B$ in Figure 5.1). Since the carrier density fluctuations are very fast, the adiabatic approximation of this fluctuation cannot be made. Frequency modulation noise due to the carrier density fluctuations have the second-order lag characteristics due to the interband relaxation time and photon lifetime, which exhibits the low-pass characteristics with the cutoff determined by the relaxation oscillation frequency of several gigahertz. The power spectral density of this FM noise exhibits a resonant peak at this frequency. Since center frequencies of the real and imaginary parts of the complex gain spectrum of the laser medium differ, the FM noise could be induced by the carrier density fluctuations, which simultaneously induce the power fluctuations (IM noise) also. The ratio between the FM noise magnitudes due to spontaneous emission and carrier density fluctuations can be expressed by $\alpha^2$. The quantity $\alpha$ has been called a *linewidth enhancement factor* or "$\alpha$-parameter," which takes the values of 2–9 [12].

The third source of the FM noise is the temperature fluctuation (curve $C$ in Figure 5.1) induced by the carrier density fluctuation. There is also a low-pass characteristic with a cutoff frequency determined by the thermal response time constant of the laser device. These three kinds of FM noise give the intrinsic quantum noise originated from the laser.

In addition to these factors, $1/f$ noise could be observed at a low Fourier frequency range (curve $D$ in Figure 5.1). It could give the power-independent width of the field spectrum [13]. Although the origin of this fluctuation has not yet been identified, possible origins are fluctuations of carrier mobility [14], fluctuations of state occupation probability of active carriers [15], and so on. Furthermore, current fluctuations from the current source and ambient-temperature fluctuations could contribute as external noise sources to generate additional FM noise. All of these noise sources contribute to the FM noise of a free-running laser [16].

### 5.3.2. Frequency Modulation Characteristics

Direct frequency modulation (FM) capability by modulating the injection current plays an essential role in reducing FM noise. Therefore, it could be important to investigate the direct FM response characteristics because they could limit the performance of negative electrical feedback and optical feedback systems designed for FM noise reduction. Figure 5.2 shows those of a 1.5-$\mu$m InGaAsP laser and a 0.8-$\mu$m-wavelength AlGaAs laser (CSP, or channeled substrate planar, type) [17]. Mechanisms of the direct FM are attributed to thermal effect and carrier density effect at lower and higher modulation frequency ranges, respectively. They have a low-pass characteristic with the cutoff determined by a thermal response time constant of the device and relaxation oscillation frequency, respectively. Since the direct FM response characteristics are a superposition of these two effects, modulation efficiency shown by Figure 5.2$a$ is, in general, inhomogeneous for a wide modulation frequency range. These characteristics depend on the transverse spatial distributions of the carrier density and $\alpha$ parameter [18]. In particular, the phase lag characteristics shown by Figure 5.2$b$ limit the frequency control bandwidth. Homogeneous modulation characteristics have been obtained recently by controlling the spatial carrier density distribution by segmenting electrodes for current injection [19]. Curve $C$ in Figure 5.2 represents the result.

### 5.3.3. Principle and Method of Frequency Control

The magnitude of laser FM noise $\delta\nu(t)$ can be expressed as [20, 21]

$$\delta\nu(t) = \kappa_{cl} \cdot (1 + \alpha^2) \cdot \Gamma_s(t) + \Gamma_{ex}(t)$$

$$- \int_0^\infty h(\tau) \cdot \{\delta\nu(t - \tau) + \Gamma_n(t - \tau)\} \cdot d\tau \qquad (5.1)$$

where $\kappa_{cl}$ is the cavity loss, $\alpha$ is the $\alpha$ parameter, $\Gamma_s(t)$ is the spontaneous

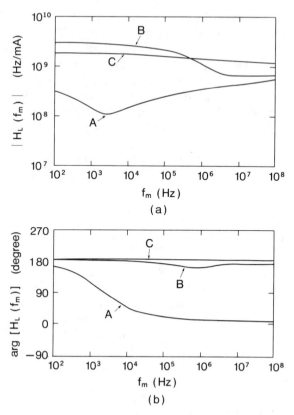

**Figure 5.2.** Complex transfer function $H_L(f_m)$ representing the direct FM responses of a 1.5-$\mu$m InGaAsP laser (curve $A$) and a 0.8-$\mu$m CSP-type AlGaAs laser (curve $B$) [17, © 1988 IEEE]. Curve $C$ represents a 1.5-$\mu$m segmented electrode InGaAsP laser [19, © 1987 IEEE]. Panels ($a$) and ($b$) illustrate for the absolute value and argument of $H_L(f_m)$, respectively.

emission fluctuation, and $\Gamma_{ex}(t)$ is the fluctuation due to external noise sources. These terms represent the magnitude of the FM noise of the free-running laser. An convolution integral in the right-hand side of this equation represents the effect of negative electrical feedback described below. In this integral, $h(\tau)$ is the impulse response function of the feedback loop, $\delta\nu(t-\tau)$ is the magnitude of the FM noise detected by the feedback loop, and $\Gamma_n(t-\tau)$ is the noise magnitude originating from the frequency discriminator element of the feedback loop. On the basis of this equation, at least four methods can be found to reduce the FM noise: (1) negative electrical feedback [21], (2) optical feedback [22], (3) improvements in performance of laser devices, and (4) suppressing the spontaneous emission by the method of cavity quantum electrodynamics (cavity QED). These methods are described and compared with each other in the following.

**5.3.3.1. Negative Electrical Feedback.** As is represented by the convolution integral in Eq. (5.1), an operating parameter (e.g., the injection current) is controlled to vary the laser frequency to compensate for the detected FM noise. Features of this method are:

1. A laser cavity structure does not have to be modified.
2. The feedback loop has a high gain and narrow-to-medium bandwidth.
3. Selective feedback is possible within a specific range of Fourier frequency. In this case, quite a high gain of the feedback loop can be realized, which will provide a promising system for FM laser spectroscopy, a very long baseline laser interferometer for a resonant-type antenna of gravitational wave detection, and so on.
4. The magnitude of the FM noise can be reduced below the spontaneous emission level (coherent state) of the free-running laser by using a high-gain and low-noise feedback loop. That is, Fourier transform of Eq. (5.2) gives

$$F(f) = \frac{\kappa_{cl} \cdot (1 + \alpha^2)}{1 + H(f)} \cdot \Pi_s(f) + \frac{1}{1 + H(f)} \cdot \Pi_{ex}(f)$$
$$- \frac{H(f)}{1 + H(f)} \cdot \Pi_n(f) \tag{5.2}$$

where $f$ is a Fourier frequency and $F$, $\Pi_s$, $\Pi_{ex}$, $\Pi_n$, and $H$ represent the Fourier transforms of $\delta\nu$, $\Gamma_s$, $\Gamma_{ex}$, $\Gamma_n$; and $h$ of Eq. (5.1), respectively. This equation means that the first and second terms of Eq. (5.2), that is, the contributions from the quantum and external noise sources, can be suppressed by the infinite gain of the feedback loop ($|H| \rightarrow \infty$). The magnitude of the last term converges to $|\Pi_n|$ by the infinite gain; thus, the FM noise reduction limit is given by the magnitude of the shot noise from the photodetector used for FM noise detection. In other words, the shot noise corresponds to the quantum noise limit of the laser under negative electrical feedback condition. The magnitude of FM noise at the shot noise limit can be lower than that of the spontaneous emission limit (coherent state) of the free-running laser. Therefore, such a state of low FM noise by negative electrical feedback can be called the *hypercoherent state* [23].
5. Stability of the system is high because the feedback is negative.
6. The feedback loop can be designed optimally by computer simulation of the feedback loop through analogy with the design criteria of the conventional analog electrical feedback circuit. By this optimum design technique, highly reproducible experimental results can be obtained.

**5.3.3.2. Optical Feedback.** The convolution integral in Eq. (5.1) does not appear in this method. Instead, the cavity loss $\kappa_{cl}$ is reduced by modifying cavity structures. Characteristics of this method are:

1. Feedback loop has a medium to high gain and medium to wide bandwidth that depends on the inverse of the photon lifetime in the modified cavity.

2. Modified cavity structure is simple. For this reason, this method has been used for more than 15 years [22].

3. Deterministic instability, such as coherent collapse [24], can be induced. Effective optical coupling and moderate optical feedback should be required to avoid this instability by fixing thin films on a laser cavity facet for an antireflection coating. Furthermore, feedback control of the external cavity length is also required by using, for example, a piezoelectric transducer. Therefore, this method can be called an *optomechanical feedback*. The hypercoherent state is not generated because of the absences of a FM noise detection system and an external feedback loop.

4. Mode-hopping between adjacent longitudinal modes of the external cavity can be induced. A method of monitoring and suppressing this mode-hopping has been proposed [25, 26].

5. Efficiency of direct frequency modulation is drastically reduced.

By injecting a reflected light from an external reflector, the field spectral linewidth of 2 kHz has been realized [27]. Laser frequency can be swept by using a diffraction grating as the external reflector. However, continuous sweep is difficult because of the hopping between the external longitudinal modes, as well as that between the internal longitudinal modes of the laser itself. In spite of these mode-hoppings, a total sweep range as wide as 100 nm has been demonstrated by fabricating the laser device by introducing a quantum well structure [28].

To keep the optical feedback system stable enough by controlling the phase of the reflected light, the optical path length between the external reflector and the laser facet should be controlled by using, a device such as a piezoelectric transducer (PZT). In order to avoid this control system, a self-pumping-type phase conjugation mirror has been used as an external reflector [29]. However, the stability characteristics of this optical feedback system have been limited because of the slow response of the phase conjugation medium.

As a more stable version of optical feedback, the method of using a confocal Fabry–Perot interferometer as an external reflector has been proposed [30], and a field spectral linewidth of about 10 kHz has been realized. The experimental setup of this system is shown in Figure 5.3. In

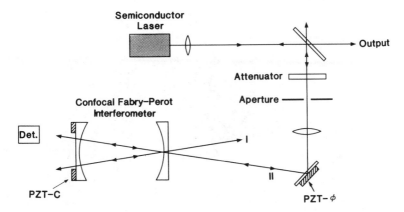

**Figure 5.3.** Experimental setup of FM noise reduction by optical feedback from a confocal Fabry–Perot interferometer [30].

this system, transmitted light, after resonating in the Fabry–Perot interferometer, is injected into the laser. Since the Fabry–Perot interferometer works as a frequency-dependent reflector, the laser frequency is pulled into the resonance frequency of this interferometer, and the laser FM noise is reduced. Control bandwidth is given by the half-width at the half-maximum of the resonance curve of the Fabry–Perot interferometer, which corresponds to the inverse of the photon lifetime of the interferometer. As is the case of using a single reflecting plate described above, the optical pass length between the laser and the Fabry–Perot interferometer should be controlled to maintain the stable optical feedback system. However, the precision of its control may be lower than that of the previous system [31]. In addition to this fact, this system has two more advantages. First, strong optical feedback from the Fabry–Perot interferometer is not required, by which intrinsic laser cavity structure is maintained. This laser system shows that the external mode suppression ratio is much higher than that of the conventional external-cavity lasers. Second, direct frequency modulation is still possible at the modulation frequency which is related by rational fractions to the free-spectral range of the Fabry–Perot interferometer [31, 32].

A compact and frequency-modulatable laser module has been fabricated by employing the techniques of microoptics and fiber optics to make a compact Fabry–Perot interferometer [33–35]. This module is demonstrated by Figure 5.4. This technique has also been applied for a 0.67-$\mu$m visible AlGaInP laser to reduce its FM noise. As is shown in Figure 5.5, the half-linewidth of its field spectrum has been reduced to 50 kHz [36, 37]. This measured value was limited by a resolution of the delayed self-homodyne system of the linewidth measurements. It was confirmed that this value was $2 \times 10^{-4}$ times that of the free-running laser. By

**Figure 5.4.** A compact and frequency-modulatable AlGaAs laser module [33–35, © 1990 IEEE]. The Fabry–Perot interferometer for optical feedback was made by employing the technique of microoptics.

employing simultaneously the optical feedback from a diffraction grating, the wavelength of this visible laser could be swept for over 5 nm.

**5.3.3.3. Improvements in Performances of the Laser Devices.** External optical and electrical elements should be connected to the laser devices for application of two methods described in Sections 5.3.3.1 and 5.3.3.2. However, it could sometimes be inconvenient for some practical industrial applications to use additional elements because of the increasing cost and

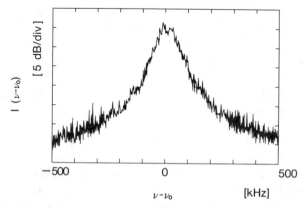

**Figure 5.5.** Field spectral profile $I(\nu - \nu_0)$ of a 0.67-$\mu$m AlGaInP laser whose FM noise was reduced by the optical feedback of Figure 5.3 [36, 37] ($\nu_0$ is the center frequency of the spectrum). Its half-linewidth was 50 kHz, which was estimated by using a delayed self-homodyne technique with an optical fiber about 2 km long.

system volume. To overcome these difficulties, performance of laser device itself has been improved to realize a stable laser, which can be an interesting approach for laser device designers. This method provides a low to medium feedback gain, and the bandwidth is medium to wide because the photon lifetime in the laser cavity is shorter than that of the method described in Section 5.3.3.2. Even for this method, the negative electrical feedback loop should usually be added to improve the stability and reliability of these improved laser devices, and to increase the feedback gain. This method can be divided into two categories:

1. *Reductions of the Cavity Loss and α Parameter.* Cavity loss of the laser device can be reduced by increasing the cavity length. Furthermore, segmented electrodes are fixed to the laser in order to reduce the spatial hole burning by controlling the currents injected into the segmented electrodes. As a more advanced method, a quantum well structure is introduced to adjust the center frequency of the real and imaginary parts of the complex gain spectrum to reduce the value of the $\alpha$ parameter [38].

2. *Integration of the System Described in Section 5.3.3.1 or 5.3.3.2.* An external reflector and an optical waveguide for the optical feedback have been integrated with laser devices [39]. To avoid occurrences of deterministic instability, a phase controller was also integrated between the gain part for the laser oscillation and an external distributed Bragg reflector. As a result of this integration, a field spectral linewidth of about 100 kHz has been realized [40, 41]. Furthermore, corresponding to the optical feedback system illustrated in Figure 5.3 two distributed Bragg reflectors and an optical waveguide were integrated to reduce the field spectral linewidth of a 1.5-$\mu$m InGaAsP laser to 150 kHz [42]. The future problem of this system is the integration of these external components with the laser. For an negative electrical feedback system, the principal factor for limiting the feedback bandwidth is the length of the feedback loop. If frequency demodulators, photodetectors, and amplifiers can be integrated with the laser devices, its bandwidth can be expanded. From this point of view, electrical negative feedback is compatible with the technique of optoelectronics integration.

**5.3.3.4. Using the Method of Cavity Quantum Electrodynamics.** The magnitude of the spontaneous emission fluctuations, specifically, the value of $\Gamma_s$ of Eq. (5.1), can be reduced by the method of cavity QED [43]. In this method, spontaneous emission rate is reduced by reducing the number of cavity modes by using a microcavity. Since this method is still at the early stage of development, several problems remain to be solved; for instance, the spontaneous emission rate is enhanced at the resonance frequency of the cavity. It should be inhibited to realize a low-noise laser.

Several experiments have already been carried out at the microwave frequency region. Similar experiments have recently been carried out at the optical frequency region to control the spontaneous emission rate [44], and fabrication techniques of optical microcavity have been developed [45]. As another approach to control the spontaneous emission rate, realization of band structure of photon energy has been proposed by localizing the photon in the medium with the three-dimensionally periodical distribution of the refractive index [46, 47]. By utilizing these new quantum optical approaches, realization of novel laser devices with low quantum noise is expected in the future.

### 5.3.4. Example of a Hypercoherent Optical Sweep Generator by Negative Electrical Feedback

To realize a stable optical sweep generator for high-resolution laser spectroscopy, a reliable method is the negative electrical feedback. Although the feedback bandwidth is about 100 MHz, it is wide enough for the spectroscopy because most of the spectrometer system have a bandwidth narrower than 100 MHz. The bandwidth of the negative electrical feedback can be expanded by employing the optoelectrical integration technique in the future.

At least five subjects should be considered to realize a reliable optical sweep generator: (1) stabilization of the center frequency of the field spectrum, (2) improvement of the accuracy of the stabilized center frequency, (3) linewidth reduction of the field spectrum, (4) frequency tracking to the other coherent laser, and (5) accurate and wideband frequency sweep. Experimental results obtained by the authors' group and future outlook of the improvement of the system performances are introduced in the following.

Figure 5.6 represents the synthesized system of negative electrical feedback to solve simultaneously the five problems presented above [21]. It appears rather complicated; however, the total volume of the system can be maintained far smaller than those by using other kinds of lasers. Further reductions of the total volume can be expected if several optical or electrical components will be integrated with the lasers in the future.

### 5.3.4.1. Center Frequency Stabilization of the Field Spectrum. The center frequency of the field spectrum can be stabilized if a slow FM noise components (e.g., $f < 1$ Hz) are reduced by using a DC servocontrol loop. Since these slow FM noise components are usually caused by temperature fluctuations and $1/f$ fluctuations, a low drift and stable frequency demodulator, composed of frequency reference and discriminator, is indispensable for this feedback system. Resonance spectral lines in atomic or molecular vapors can be used for such the frequency demodulators.

Absorption spectral lines due to higher harmonics or combination tones of vibration–rotation transitions in organic molecular vapors (e.g., $NH_3$ [48], $H_2O$ [48, 49], $C_2H_2$ [50], HCN [51]) can be used conveniently because

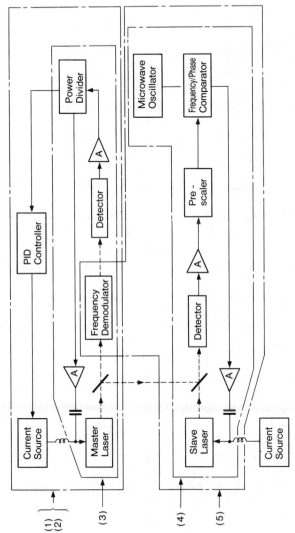

**Figure 5.6.** The synthesized system of negative electrical feedback to solve the five problems simultaneously [21, © 1988 IEEE]. The five blocks in this figure represent (1) stabilization of the center frequency of the field spectrum, (2) improvement of the accuracy of the stabilized laser frequency, (3) linewidth reduction of the field spectrum, (4) frequency tracking to the other coherent laser, and (5) accurate and wideband frequency sweep.

a great number of these spectral lines are distributed around the wave-length region of 0.7–1.6 $\mu$m. The problems are their broad spectral linewidths, weak absorption coefficients, and difficulties in spectral assign-ments. To overcome these difficulties, strong spectral lines at the 0.8-$\mu$m-wavelength region due to electronic transitions in atomic vapor (e.g., Rb [52], Cs [53]) have been used because their spectral assignments have been completed. They have further advantages in realizing a Doppler-free, narrow-linewidth saturation absorption spectral lines, reducing the volume of the gas cells, and so on. Since optogalvanic spectral lines in rare gases exhibit a large signal-to-noise ratio, it can be considered as another candidate [54]. Although it is rather difficult to realize a high-frequency stability because the reference frequency would be shifted as a result of plasma instability and the Stark effect in the discharged gases, a compact discharge lamp has been employed to realize a low-price frequency de-modulator for 1.5-$\mu$m InGaAsP lasers in order to realize a practical light source for optical communication systems [55]. As a novel method, the second harmonic frequency, which is generated from the active layer of the InGaAsP laser, has been stabilized to the Rb linear absorption spectral line in order to stabilize the 1.56-$\mu$m InGaAsP laser frequency [56]. Although this second harmonic power is low, detection sensitivity of absorption spectral line can be increased by using a heterodyning tech-nique with a 0.78-$\mu$m AlGaAs laser, and furthermore, a technique of optical–optical double resonance could make it possible to detect Doppler-free absorption spectral lines in Rb by using a 0.78-$\mu$m AlGaAs laser as a pumping light source [57]. From these facts, it can be claimed that low power in the internally generated second harmonics does not present any essential problems. Using a nonlinear organic waveguide, investigators have recently attempted to increase the efficiency of second harmonic generation [57]. For a stable frequency demodulator for a 0.67-$\mu$m AlGaInP laser, several atomic vapors (e.g., Li) can be used.

Although a Fabry–Perot interferometer does not provide any absolute reference frequencies, it has been conveniently used as a frequency demodulator because of its simple structure. Since the value of its finesse has recently been increased up to several ten thousands, it has been demonstrated that high short-term stability (at the Fourier frequency range of $f > 0.1$ mHz or the integration time range of $\tau < 1 \times 10^4$ s) can be realized by using this supercavity, which has been higher than by using atomic or molecular spectral lines [58].

By using the frequency demodulators described above, a frequency stability as high as $2 \times 10^{-12}$ has been obtained at the integration time of 100 s [59]. Although this value has not yet reached the shot noise limit generated from the photodetector in the feedback loop, theoretical analy-sis has estimated that the stability as high as $1 \times 10^{-15}$ can be obtained if this shot noise limit is realized [21]. The most accurate frequency stability measurement is made by measuring the residual FM noise of the beat frequency between the independently stabilized two lasers. For this purpose, an Allan variance real-time processing system has been developed for the accurate evaluation of frequency stability [60, 61].

### 5.3.4.2. Improvement in the Accuracy of the Stabilized Center Frequency.

In high-resolution, high-precision laser spectroscopy system, the stabilized laser frequency should be accurately calibrated for precise spectral assignments and determining structural constants of atoms and molecules. From these critical requirements, accuracy of the stabilized laser frequency has to be improved. One of the principal phenomenon of limiting the accuracy of semiconductor laser frequency is a blue-shift (about 20 MHz h$^{-1}$) due to the drift of the thermal resistance of the free-running laser device [62]. Furthermore, a drift of the reference frequency also limits the accuracy of the stabilized frequency. For example, a saturated absorption spectral line in Rb exhibits a frequency shift induced by the changes in incident laser power and ambient temperature [63]. Typical values of this frequency shift have been evaluated as $-5$ MHz/(mW cm$^{-2}$) and $-0.8$ MHz K$^{-1}$, respectively. Those of H$_2$O vapor are about 10 times larger [64]. Furthermore, those of optogalvanic spectral lines in rare gases could be much larger because of the plasma instability and the Stark effect in the discharged gas, which can be concluded from a popular historical story that a Lamb-dip stabilized He–Ne laser has been replaced by a CH$_4$-stabilized He–Ne laser to avoid the frequency shift of the Lamb-dip of the He–Ne discharged gas [65]. From the comparison between the results reported so far, it can be concluded that spectral lines due to electronic transitions in atomic vapors are used as more accurate frequency demodulators than those of the higher harmonics or combination tones of vibration–rotation transitions in organic molecular vapors.

To calibrate accurately the stabilized laser frequency, absolute measurements of laser frequency are required. Harmonic mixers using metal–insulator–metal (MIM) point-contact diodes or Josephson devices have been conventionally used for this purpose [66]. However, since the MIM diodes have low sensitivities and the response of Josephson devices is slow [67], it is not practical to use them for the semiconductor lasers with wavelengths shorter than 1.5 $\mu$m. Instead of measuring frequency, wavelength measurements have been employed by a scanning Michelson interferometer-type wavemeter for convenient and rough evaluations of the accuracy with the error of $1 \times 10^{-7}$–$10^{-8}$ [48, 49].

Several trials have recently been reported to realize a sensitive and fast photodetector by using a tunnel junction between thin films of a high-$T_c$ superconductor and a metal film [68], and by using a virtual charge-induced optical nonlinearity in a quantum well structure semiconductor device [69]. If these devices could be used as reliable harmonic mixers, practical systems of absolute frequency measurement may be realized to improve the accuracy as high as $1 \times 10^{-10}$.

**5.3.4.3. Linewidth Reduction of the Field Spectrum.** The half-linewidth of the field spectrum can be reduced by negative electrical feedback if its bandwidth $B$ is wider than the half-linewidth of the free-running laser $\Delta\nu_{FR}$ (i.e., $B > \Delta\nu_{FR}$). A sensitive and wideband frequency demodulator is required to realize a high gain and wideband feedback loop. For this purpose, a high-finesse Fabry–Perot interferometer described in Section 5.3.4.1, especially its reflection mode, has been used [70, 71]. Since the reflection mode works as an optical frequency differentiator, it exhibits a wider bandwidth of frequency demodulation than does the transmission mode [70–72]. Furthermore, to obtain a dispersive frequency demodulation output signal without modulating the laser frequency, several types of frequency demodulators have been proposed, such as (1) installing a Fabry–Perot interferometer in a Mach–Zehnder interferometer [73], (2) installing a polarizer in a Fabry–Perot interferometer [74], and so on. Figure 5.7 shows such a dispersive profile of the output signal from the frequency demodulator (installation method 1). It has also been confirmed that the gain of this demodulator is 10 dB larger and the bandwidth is equal to those of the reflection mode of the Fabry–Perot interferometer [73].

Optimum design of a phase compensation circuit in the feedback loop can be carried out in order to realize the largest gain and widest bandwidth of the feedback loop. For this design, computer-aided analog network design criteria can be employed based on the frequency response characteristics of each feedback element, such as a laser (see Figure 5.2), a frequency demodulator, and so on. Curve $B$ of Figure 5.8 shows an experimental result for an 0.8-$\mu$m AlGaAs laser (CSP type) by using such the optimized slow and fast feedback loops simultaneously [70, 71]. It represents the power spectral density of the residual FM noise. Curve $A$ is the value of the free-running laser. It can be seen by comparing these curves that the feedback bandwidth is 40 MHz and the FM noise was reduced to 60–70 dB within the Fourier frequency range of 100 Hz–1 kHz. It means that the gain–bandwidth product of this feedback loop was as

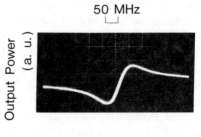

**Figure 5.7.** Relation between the laser frequency and the output power from the Mach–Zehnder interferometer, in which a Fabry–Perot interferometer is installed [73]. The sharp slope is used as a frequency demodulator.

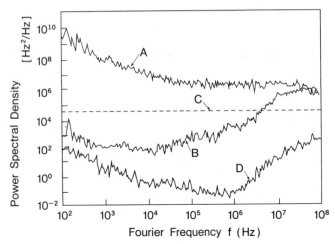

**Figure 5.8.** Power spectral density of the FM noise of a CSP-type AlGaAs laser [70, 71, © 1990 IEEE]. Curve *A*, free-running laser; *B*, under condition of negative electrical feedback; *C*, magnitude of the spontaneous emission noise, that is, FM noise in the coherent state of the free-running laser or the so-called Schawlow–Townes limit; *D*, limit of FM noise detection of the present experimental setup.

large as 50 THz (terahertz). For most laser spectrometers, the bandwidth obtained here could be sufficiently wide. Curve *C* represents the magnitude of the spontaneous emission noise, that is, the FM noise in the coherent state of the free-running laser, which was estimated by assuming the value of the $\alpha$ parameter as 9 [12]. It can be confirmed that the value of curve *B* is lower than that of curve *C* at the Fourier frequency range below 4.4 MHz; thus, the hypercoherent state was realized in this range.

Curve *D* in Figure 5.8 represents the sensitivity of the frequency demodulator employed here, that is, the limit of FM noise reduction, which was determined by intrinsic IM noise of the laser. An optical balanced detector [75–77] could be effective to eliminate the contributions from this IM noise in order to realize the shot-noise limit of the feedback loop. It has been confirmed that the contribution from the IM noise can be reduced to 20–30 dB within the Fourier frequency range below 20 MHz [78]. Curves *A* and *B* of Figure 5.9 represent the field spectral profiles derived by applying the computer program of the fast Fourier transform to curves *A* and *B* of Figure 5.8, respectively [70, 71]. Curve *A* represents that the half-linewidth of the free-running laser is 4.5 MHz, while it can be seen from by comparing curves *A* and *B* that the FM noise within the ±40-MHz region around the center has been drastically reduced by the feedback. From the magnified profile given by Figure 5.9*b*, it is confirmed that the half-linewidth was reduced to 560 Hz, which is, to the authors' knowledge, the narrowest linewidth among the formally reported values.

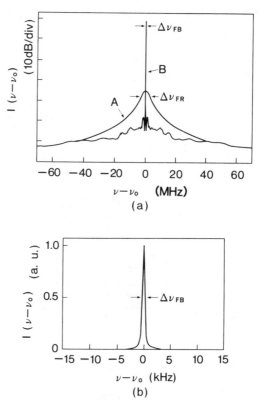

**Figure 5.9.** Field spectral profile $I(\nu - \nu_0)$ of a CSP-type AlGaAs laser [70, 71, © 1990 IEEE], where $\nu_0$ is the center frequency of the spectrum. (a) Curves A and B correspond to curves A and B in Figure 5.8, respectively. (b) Magnified profile of curve B in Figure 5.9a. The values of the half-linewidths for curves A and B are $\Delta\nu_{FR} = 4.5$ MHz and $\Delta\nu_{FB} = 560$ Hz, respectively.

This linewidth has been narrow enough even for ultra-high-resolution laser spectroscopy such as the subnatural linewidth spectroscopy, laser cooling, and so on. The half-linewidth estimated from curve D of Figure 5.8 is about 1 Hz, which corresponds to the linewidth reduction limit of the present experimental setup. It was estimated that the shot-noise limit of the present experimental setup could realize the half-linewidth of about 50 mHz [21].

**5.3.4.4. Frequency Tracking to the Other Coherent Laser.** If a stable laser can be realized by solving the problems mentioned in Sections 5.3.4.1–5.3.4.3, frequency tracking of the second laser (a "slave" laser) to this coherent laser (a "master" laser) is also an essential technique for a heterodyne-type or homodyne-type detection system for laser spectroscopy. Block 4 in Figure 5.6 represents such a frequency tracking loop, that is, a frequency locked loop. To realize a heterodyne-type frequency locked loop, injection current of the slave laser is controlled so that the beat frequency between the two lasers is locked to the microwave local oscillator frequency, which has been called the *frequency offset locking* [79]. By this technique, the residual FM noise of the beat signal has been reduced to 0.2 Hz at the integration time of 100 s, which corresponds to the square root of the Allan variance of $5 \times 10^{-16}$, where the magnitude of this FM noise was normalized to the optical frequency [80]. This value is much smaller than that of the residual FM noise of the master laser, which means that the slave laser frequency tracks very accurately to the stable master laser. The capture and locking ranges of this loop were as wide as 2 GHz, which were limited by the bandwidth of the photodetector used as a heterodyne receiver [80].

It has been estimated that this FM noise can be reduced to about $1 \times 10^{-18}$ by improving the performance of the feedback elements such as a frequency–phase comparator [33, 80]. By these improvements, phase fluctuations (PM noise) of the beat signal can also be reduced to lower than 1 radian; in other words, a heterodyne-type phase locked loop can be realized [81]. Figure 5.10 shows experimental results to demonstrate the low residual PM noise of the beat signal between two semiconductor lasers obtained by the heterodyne optical phase-locked loop [33]. Curves $A$ and $B$ of Figure 5.10$a$ represent the signal waveforms of the local microwave oscillator and of the beat signal, respectively. By the magnitude of the timing jitter on curve $B$, the root mean-square value of the residual PM noise has been estimated as 0.6 radian [33, 82]. Figure 5.10$b$ shows the field spectrum of the beat signal under optical phase locking, of which the control bandwidth was about 1 MHz. The beat signal was also observed at 2-kHz span, where its half-linewidth was 60 Hz, which was limited by the resolution bandwidth of the RF spectrum analyzer. A homodyne-type phase locked loop can be also realized by fixing the frequency of the local microwave oscillator to zero, where the photodetector becomes an optical phase comparator. Figure 5.11 shows the experimental result of the power spectral density of the PM noise of the homodyne phase-locked loop [83]. Within the Fourier frequency range below 12 MHz, drastic reduction of the PM noise can be seen. The bandwidth of this phase-locked loop, 12 MHz, could be large enough for the most of the homodyne phase-locked system for high-resolution laser spectroscopy. To evaluate quantitatively the magnitude of these residual FM or PM noise, the measurement system with about 100 times more accurate [82] than those of conventional ones [60, 61] has been developed.

For FM laser spectroscopy [84], the beat signal between the two lasers has to be measured. In such a case, measurement accuracy could be limited by the magnitude of the FM noise of the beat signal, which depends on the uncorrelated spontaneous emissions introduced into each laser mode. To improve this accuracy, a method of making a correlation

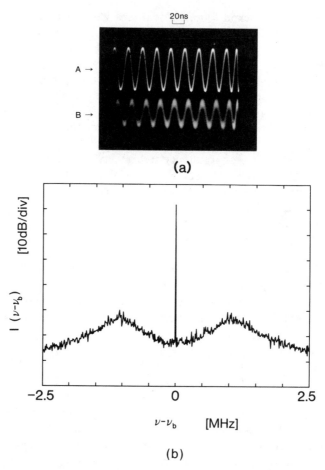

**Figure 5.10.** (*a*) A waveform of the beat signal obtained by the heterodyne-type phase-locked loop [33]. Curves *A* and *B* are the waveforms from the microwave local oscillator and the beat signal, respectively. Their frequencies are 50 MHz. (*b*) A field spectrum $I(\nu - \nu_b)$ of the beat signal under optical phase locking ($\nu_b$ is the center frequency of the spectrum). The resolution bandwidth and sweep time were 1 kHz and 30 s, respectively.

between these spontaneous emission fluctuations has been proposed [85]. Since the spontaneous emission corresponds to the stimulated emission driven by the zero-point fluctuations in vacuum, any correlations would not exist between those for different modes. However, if a number of upper energy levels are engaged in laser transitions, correlation can be generated between these spontaneous emissions by making a quantum correlation between these upper levels by applying a resonant external modulation. The correlated spontaneous emission has already been observed for the external-cavity semiconductor laser [86]. However, one

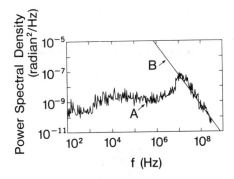

**Figure 5.11.** Power spectral density of the PM noise of the homodyne phase-locked loop [83]. Curve *A* shows experimental results for locked condition. The locking bandwidth was 12 MHz. The solid line *B* represents the unlocked condition estimated by extrapolating an unlocked part of curve *A*.

should not confuse this with the active mode locking phenomenon because some of the characteristics in correlated spontaneous emission are similar to those of the active mode locking. A method of highly precise evaluation has recently been proposed by measuring the Allan variance of short-term PM noise in the beat signal [87].

**5.3.4.5. Accurate and Wideband Frequency Sweep.** In the heterodyne-type frequency locked loop described in Section 5.3.4.4, the slave laser frequency can be swept by sweeping the frequency of the microwave local oscillator while maintaining the stability of the slave laser frequency very high. For a conventional AlGaAs laser with a Fabry–Perot cavity, a continuous tunable range of 64 GHz has been already realized [80], which is large enough for high-resolution laser spectroscopy. By using a longitudinal-mode controlled laser, such as a DFB or DBR laser, tunable range as wide as 1 THz can be realized [88].

Furthermore, as is shown by Figure 5.12, a novel system of realizing a petahertz (PHz; $10^{15}$ Hz) class coherent optical sweep generator has been proposed [89, 90]. In this system, four commercially available semiconductor lasers, with wavelengths of 1.56, 1.34, 0.78, and 0.67 $\mu$m, are employed as primary light sources. Negative electrical feedback loops as described above are applied to these lasers to reduce their FM noises. As additional passive elements, nonlinear optical waveguides using organic materials such as MNA and DAN, are used for second harmonic generation, parametric frequency conversion, sum, and difference frequency generations. In this system, frequency tracking between the 1.56-$\mu$m laser and the 0.78-$\mu$m laser has been realized by locking simultaneously the frequencies of the 0.78-$\mu$m laser and the second harmonics of the 1.56-$\mu$m laser to the center frequency of the pump–probe spectral shape in Rb atomic vapor (see Figure 5.16 later in this chapter). Similar frequency tracking between the 1.34-$\mu$m laser and the 0.67-$\mu$m laser can be realized by using the Li atomic vapor. By incidenting these stabilized lasers into organic nonlinear waveguides used for parametric frequency converters, fre-

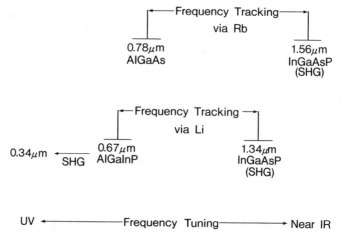

**Figure 5.12.** Schematic explanation of the principle for a petahertz-class hypercoherent optical sweep generator [89, 90]. Four coherent semiconductor lasers are used as primary light sources. Frequency links between these lasers are realized by using absorption lines of Rb and Li vapors as frequency references for negative electrical feedback. Nonlinear organic waveguides are used for generations of the second harmonic wave for this link. For wideband frequency sweep, other organic nonlinear waveguides are used to realize parametric frequency conversion and sum or difference frequency generation. Another inorganic oxide nonlinear optical crystal is used to generate the second harmonic wave from the 0.67-$\mu$m laser. Maximum of the wavelength tuning range expected by this system is 1.22 $\mu$m (i.e., from 1.56 $\mu$m to 0.34 $\mu$m), which corresponds to 700-THz frequency range.

quency-converted coherent lights can be generated, which can be frequency-tunable within the wavelength range of 1.56–0.67 $\mu$m. Inorganic oxide nonlinear optical crystals should be used for efficient second harmonic generation from the 0.67-$\mu$m laser. By summarizing these techniques, rough frequency tuning between the wavelength range of 1.56–0.34 $\mu$m can be expected, which corresponds to the frequency-tunable range of 700 THz, or approximately 1 PHz. Such a wideband frequency tuning, to the authors' knowledge, has never been realized by other kind of laser systems. For practical applications to high-resolution laser spectroscopy, expensive dye laser systems may be replaced by this inexpensive and compact system. Fine and accurate tuning can be carried out by employing simultaneously the technique of frequency offset locking. Essential problems to be solved to realize the petahertz frequency sweep range could be crystal growing and fabrication of efficient organic nonlinear optical waveguides. Theoretical analysis, crystal growing, and fabrication of the DFB-type channel waveguides are now in progress [91].

Table 5.1 summarizes the present status and future outlook of performance of the optical sweep generator that could be realized by considering five subjects described above [90]. It can be claimed that a highly precise

**Table 5.1 Present Status and Future Outlook of Optical Sweep Generator Performance [90]**

| | Present Performance | Future (Predicted) Performance |
|---|---|---|
| Center frequency stabilization | $2 \times 10^{-12\,a}$ | $1 \times 10^{-15\,a}$ |
| Reproducibility and accuracy | $1 \times 10^{-8}$ | $1 \times 10^{-10}$ |
| Linewidth reduction | 560 Hz | 50 mHz |
| Frequency/phase tracking | $5.5 \times 10^{-16\,a}$ | $< 1 \times 10^{-18\,a}$ |
| | 0.6 radian[b] | $< 0.1$ radian[b] |
| Accurate and continuous frequency sweep | 1 THz | 700 THz |

[a] The value at the integration time of 100 s.
[b] Root mean square.

optical sweep generator has been realized, and further improvements can be expected in the near future.

## 5.4. TOPICS OF SPECTROSCOPY

Spectroscopic data of atoms and molecules at the near-infrared region of about 0.7–1.6-$\mu$m wavelength have not been documented sufficiently because of the lack of the reliable coherent light sources. However, the improvements of semiconductor lasers have made it possible to carry out highly sensitive and high-resolution spectroscopy at this wavelength region. Several examples in experimental results of spectroscopy, obtained by frequency controlled semiconductor lasers, are demonstrated in the following subsections.

### 5.4.1. Linear and Nonlinear Laser Spectroscopy

A great number of Doppler-broadened spectral lines due to higher harmonics and combination tones of vibration–rotation transitions in organic molecular vapors can be observed within the wavelength range of 0.7–1.6 $\mu$m. Although their absorptions are weak, spectral lines due to $2\nu_1$, $2\nu_3$, or $\nu_1 + \nu_3$ vibration transitions in $NH_3$ and $2\nu_2 + \nu_3$ vibration transition in $H_2O$ have been observed at the 1.5-$\mu$m wavelength with the detection sensitivity of $2.3 \times 10^{-3}$ torr per a meter optical pass length [48]. Furthermore, those due to the $2\nu_1 + \nu_2 + \nu_3$ vibration transition in $H_2O$ have been measured at the wavelength range of 0.8 $\mu$m [49]. In this case, fairly strong absorptions were observed because the $2\nu_1 + \nu_2 + \nu_3$ band is coupled with the $\nu_2 + \nu_3$ band by Darling–Dennison resonance [92]. The wavelengths of these absorption lines have been calibrated within the error of $1 \times 10^{-7}$ to $1 \times 10^{-8}$. After these pioneering works, spectroscopic data have been accumulated for several organic molecular vapors. For example, spectral lines in $C_2H_2$ [50] and HCN [51] have been measured by sweeping a DFB laser frequency for over 1.5 THz. A highly sensitive spectrometer for studying the wavelength regions of 0.75–0.88 $\mu$m, 1.3 $\mu$m, and 1.5 $\mu$m has recently been constructed [93], which was sensitive to the absorption of $2 \times 10^{-6}$. By using this spectrometer, the systematic study on the characteristics of spectral measurements of the vibration transitions in $NH_3$ has been carried out. As a result of this study, the transfer from the normal mode vibration to local mode vibration was found with increase of the vibration quantum number. These experimental data were confirmed with theoretical analysis. Further improvements in sensitivity of these spectrometers can be expected by employing the technique of frequency modulation (FM) spectroscopy [84], which is compatible to semiconductor lasers because of their high efficiency of direct frequency modulation. Several experimental results of FM spectroscopy have already been demonstrated [94]. A problem to be solved in FM spectroscopy by semiconductor lasers is that IM is induced simultaneously by FM, which limits the measurement sensitivity. However, the contribution from IM could be reduced by employing a two-tone FM technique. As a result of it, the sensitivity was improved by 10 times that of the conventional one [95].

Spectral lines due to electronic transitions in atomic vapors have been measured at the 0.8-$\mu$m wavelength region. Systematic studies on Rb [52] and Cs [53] have been reported. For rare gases, the technique of optogalvanic spectroscopy has been employed because a great number of optogalvanic spectral lines are distributed between the near-infrared and the visible spectra [96]. Although sufficiently reproducible frequency measurements could be rather difficult because of the frequency shift due to plasma instability and the Stark effect, fairly high sensitivity in measurements have been obtained [54].

Since most spectral lines due to electronic transitions in atomic vapors are distributed within the visible region, short wavelength lasers could be required for these measurements. For this purpose, a second harmonic wave generated from the active layer of an AlGaAs laser has been used to measure the strong spectral lines in K (transitions $5p^2P_{1/2}-4s^2S_{1/2}$, and $5p^2P_{3/2}-4s^2S_{1/2}$) and Al (transition $3p^2P_{3/2}-4s^2S_{1/2}$) vapors at the 0.4-$\mu$m wavelength range [97]. Similar measurements of Rb lines by using the second harmonic wave generated internally from a 1.56-$\mu$m InGaAsP laser have also been carried out [56].

Doppler-free spectroscopy of atomic vapors has been carried out by using a technique of atomic beam or nonlinear spectroscopy. Atomic beam spectroscopy has been carried out, for Cs, Rb, and so on. Examples of Doppler-free fluorescence spectroscopy experimental and calculated re-

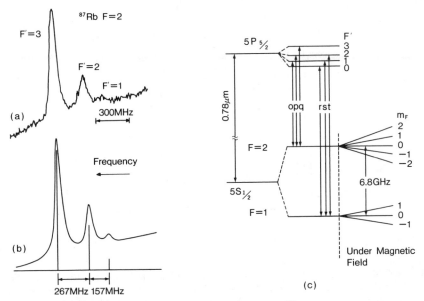

**Figure 5.13.** Doppler-free fluorescence spectral shape of $^{87}$Rb atomic beam [98]. (*a*) Experimental results; (*b*) calculated results; (*c*) relevant energy-level diagram of the Rb atom.

sults are shown by Figures 5.13*a* and 5.13*b*, respectively [98]. It was obtained by using a 0.78-$\mu$m AlGaAs laser, and the half-linewidths of the fluorescence spectral profiles due to the transitions from $F' = 1$, 2, and 3 levels of the excited state to the $F = 2$ level of the ground state was as narrow as 70 MHz. This linewidth was determined by the residual Doppler broadening due to the 5° divergence of the atomic beam [98]. However, it is narrow enough that each spectral component represented by Figure 5.13*c* is clearly resolved.

As the first example of nonlinear spectroscopy, a saturated absorption technique has been employed for Cs [53], Rb [52, 99], and other elements. Figure 5.14*a* shows the derivatives of saturated absorption lineshapes in the $^{87}$Rb $D_2$ line obtained by using a 0.78-$\mu$m AlGaAs laser, in which 11 narrow spectral lines, including cross-resonance lines, are resolved [99]. They have linewidths as narrow as 40 MHz, which is limited by the lifetimes of relevant energy levels. Their spectral strengths and positions on the abscissa agree well with those estimated by theoretical analyses (see Figure 5.14*b*).

As the second example, optical–optical double resonance has been carried out for the $D_1$ [100] and $D_2$ lines in Rb by using a 0.79-$\mu$m and 0.78-$\mu$m AlGaAs lasers, respectively. Figure 5.15 shows the double reso-

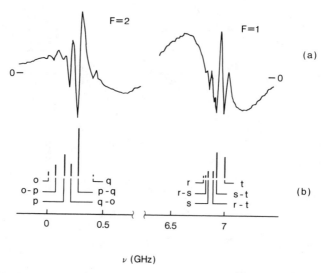

ν (GHz)

**Figure 5.14.** Derivative of saturated absorption spectral lineshapes in $^{87}$Rb [99, © 1985 IEEE]. (*a*) Experimental results. (*b*) Assigned spectral lines. For the notations of this spectral lines, see the relevant energy-level diagram presented in Figure 5.13*c*. For the transition from the $F = 1$ level of the ground state, two saturated absorption lines (*s* and *t*) and three cross-resonance lines (*r–s*, *r–t*, and *s–t*) were observed. For the transition from the $F = 2$ level, three saturated absorption lines (*o*, *p*, and *q*) and three cross-resonance lines (*o–p*, *q–o*, and *p–q*) were observed.

nance spectral shapes for the $D_2$ component of an $^{87}$Rb atomic beam [98], which exhibit clear Doppler-free spectral shapes. Because the pump laser frequency was fixed to the transition frequency between the $F = 2$ and $F' = 3$ levels, the strength of the spectral component $F' = 3$ is weaker than that of the other two components as the result of saturation. Figure 5.16*a* shows a novel experimental setup for a nonlinear pump–probe

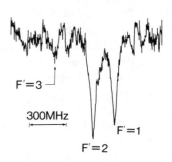

**Figure 5.15.** Optical–optical double-resonance spectral shapes of $^{87}$Rb atomic beam [98]. The pump laser frequency was fixed to the transition frequency from $F = 2$ to $F' = 3$.

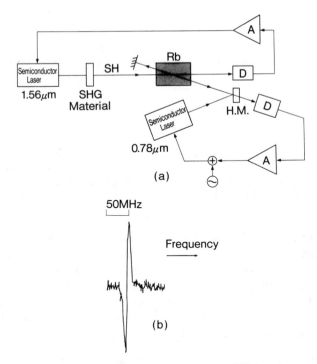

**Figure 5.16.** Novel Doppler-free pump–probe spectroscopy of Rb by using an AlGaAs laser and the second harmonics of the InGaAsP laser as the pumping and probe light sources, respectively [90]. (*a*) Experimental setup; (*b*) an example of the first derivative of the spectral profile.

spectroscopy [90]. Doppler-free spectral profiles, which are due to the saturation by the pumping from a 0.78-$\mu$m AlGaAs laser, can be probed by using a second harmonic wave generated from a 1.56-$\mu$m InGaAsP laser. Figure 5.16*b* shows an experimental result of the first derivative of the Doppler-free spectral shape. This spectral shape can be used as a frequency reference to stabilize the frequency of the pump and probe lasers simultaneously, which can realize a stabilized frequency link between the 0.78-$\mu$m AlGaAs and 1.56-$\mu$m InGaAsP lasers, and will be used as stable master lasers for wideband optical sweep generator represented by Figure 5.12. Similar results can be expected for Li atoms by using 0.67-$\mu$m AlGaInP and 1.34-$\mu$m InGaAsP lasers.

The third example is the Doppler-free two-photon spectroscopy for Li atomic vapor, which has been carried out by using a dye laser and an AlGaAs laser [101]. By using a thermoionic heat pipe diode as a sensitive detector for spectral measurements, clear spectral profiles, shown by Figure 5.17, have been obtained. It can be seen from this figure that the hyperfine components in $^6$Li and $^7$Li have been clearly resolved.

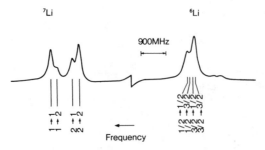

**Figure 5.17.** Spectral profile due to the Doppler-free two photon transitions $2S-2P-3S$ in the $^6$Li and $^7$Li atoms [101]. The numbers below the spectra denote the quantum numbers $F$ and $F'$ of the initial and final states, respectively.

As one of the other applications of semiconductor lasers to alkali vapor spectroscopy, Rydberg states of alkali atoms can be produced by stepwise excitation, which is shown in Figure 5.18 for Cs and Rb [102]. By using this spectroscopic method, information about highly excited states can be obtained [103]. The Rydberg atoms prepared by this method have also been used for one atom maser in the microwave region, which is one branch of the cavity QED described in Section 5.4.4.

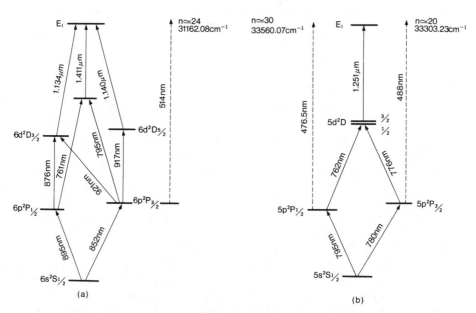

**Figure 5.18.** Stepwise excitation schemes to obtain the Rydberg states of Cs ($a$) and Rb ($b$); $E_I$ is the ionization level [102]. Most of the transitions indicated by solid arrows in this figure can be driven by the semiconductor lasers. The broken arrows indicate transitions driven by argon ion lasers.

### 5.4.2. Test of Parity Nonconservation

The existence of weak interaction between elemental particles in atoms has been predicted by the standard electroweak theory developed by Glashow, Weinberg, and Salam. Experimental tests for this theory have been carried out in the field of high-energy physics using a huge accelerator. However, it has been found that a method of precision laser spectroscopy could be used as a more convenient and less expensive tool for this test [104, 105]. Since the parity conservation of the wave function of the atom could be violated by the weak interaction, a weak optical transition due to this violation can be detected by a carefully designed sensitive laser spectrometer. Figure 5.19 shows such an experimental setup for detecting the change of polarization of the transmitted light through the Pb atomic vapor, in which a 1.3-$\mu$m InGaAsP laser is used as a coherent light source [106]. The optical transition due to magnetic dipole interaction in Pb atoms was monitored to detect the change of polarization induced by the parity nonconservation. This sensitive spectrometer, employing the techniques of polarization modulation and phase-sensitive detection, was sensitive to the polarization rotation of 0.1 $\mu$radian. Stark coefficients in Cs atoms have also been accurately measured by using a carefully frequency controlled AlGaAs laser, by which several structural constants of Cs have been estimated to test the parity nonconservation [107].

Semiconductor lasers can be also used for accurate measurement of the Lamb shift of muonic atoms [108]. A frequency controlled 0.73 $\mu$m AlGaAs laser is used as the master laser for the injection locking of a high-power solid-state Alexandrite laser to reduce its FM noise. The third harmonic radiation from the locked Alexandrite laser is generated by using nonlinear optical media. This ultraviolet light can be used as a pumping source of the muonic atoms.

### 5.4.3. Manipulations of Atoms and Ions

Experimental studies of deceleration of atomic motion in vacuum (i.e., laser cooling [109]) and confinement of atoms in a limited volume by laser beams have been recently progressed rapidly with the aid of improvements in the laser frequency control technique. Deceleration of atomic motion is possible by the light pressure imposed by the laser beam. The lowest equivalent temperature of the cooled atom is lower than the value determined by the Doppler effect [$< 40 \ \mu$K (microkelvins)] [110, 111]. Furthermore, a recent experiment has revealed a temperature lower than 2 $\mu$K, which corresponded to the thermal velocity of several centimeters per second [112]. This is lower than the value determined by the recoiling between the atom and photon.

It has become possible to confine a number of atoms (e.g., $1 \times 10^7$ atoms) in a limited volume in vacuum for more than several minutes with

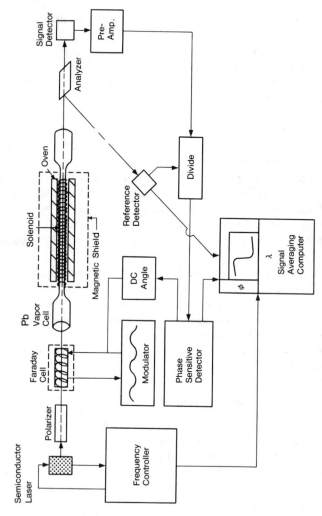

**Figure 5.19.** Experimental setup for testing the parity nonconservation in Pb atoms [106]. A 1.3-μm InGaAsP laser beam was incident into the Pb atomic vapor. Its frequency was tuned to the transition frequency due to the magnetic dipole moment interaction in Pb. Change in the polarization of the transmitted light was precisely measured by employing the techniques of the polarization modulation and the phase-sensitive detection.

atomic density as high as $1 \times 10^{11}$ cm$^{-3}$ [113]. In such high-density situation, the confined atomic mass exhibits a specific characteristics similar to those of viscous fluids. For this reason, this confined atomic mass has been called "optical molasses."

For these experiments, it is advantageous to utilize a high efficiency of frequency modulation and controllability in semiconductor lasers. Figure 5.20 shows such an experimental setup using 0.85-$\mu$m AlGaAs lasers for optical pumping, laser cooling, and producing optical molasses of Cs atoms [114]. Laser coolings of Rb atoms [115] and rare gases [116] have also been carried out by using AlGaAs lasers. These experiments have been carried out by using expensive and complicated atomic beam apparatus. However, a simple experimental configuration using an inexpensive Cs vapor cell has recently been employed to produce the optical molasses [117].

The experimental study of an ion trap, that is, confining a single ion in a limited volume of an electromagnetic potential such as a Penning trap or Paul trap, has been remarkable progress by simultaneously utilizing the technique of laser cooling. Several specific characteristics, such as quantum jump [118], have been observed for a cooled and confined single ion [119]. By further improvements of performances in confinement technique, squeezing between uncertainties of the momentum and the position of the ion have recently been observed [120]. Furthermore, by confining and cooling high-density atoms, it was observed that their characteristics were similar to those of ionic crystal and solid-state plasma [121].

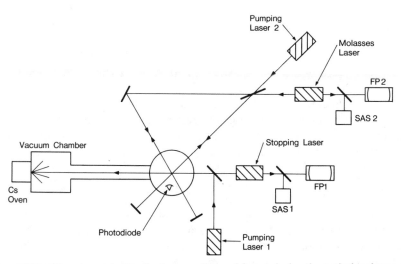

**Figure 5.20.** Experimental setup for laser cooling and for producing the optical molasses of Cs atoms [114]. AlGaAs lasers were used for optical pumping, laser cooling, and producing optical molasses.

Since most of the ions have their resonant frequencies in the visible–ultraviolet region, several expensive dye lasers and their second harmonic or sum-frequency radiations have been used. However, the efficiency of these wavelength conversion is rather low in these short-wavelength regions. To overcome these difficulties, the possibility of using easily controllable semiconductor lasers have been examined [120]. For this purpose, it is expected that a hypercoherent optical sweep generator such as that described in Section 5.3.4 can be used as a reliable coherent light source.

### 5.4.4. Cavity Quantum Electrodynamics

As was described in Section 5.3.3 [4], enhancement or inhibition of spontaneous emission from atoms can be realized by using a microcavity with the dimension of a wavelength order. At the initial stage of the cavity QED study, experiments were carried out in the microwave frequency region. Controlled rate of the spontaneous emission and specific Rabi oscillation driven by vacuum fluctuations were observed by using a microcavity and single-atom maser [122]. To prepare the atoms of a large transition dipole moment, Cs atoms were excited to the Rydberg state by using AlGaAs lasers as pumping sources.

Because of the technical difficulty of fabricating a microcavity in the optical frequency region, experiments of optical cavity QED have been carried out by installing atoms in a high-finesse and mode-degenerated confocal Fabry–Perot cavity. Enhancement and inhibition of the spontaneous emission from Yb atoms have been observed at the resonant and nonresonant frequencies of the confocal Fabry–Perot cavity, respectively [123]. Experimental study of the single-atom laser is also in progress. As is shown by Figure 5-21, FM sidebands on the spontaneous emission spectral line from Na atoms in the high-finesse Fabry–Perot cavity have recently been observed, which were attributed to vacuum Rabi oscillation [124]. The techniques of cavity QED could be useful to realize a novel light-emitting semiconductor device, such as a low-threshold semiconductor laser. For this purpose, the Fabry–Perot microcavity with the dimensions of optical wavelength has been fabricated, as is shown by Figure 5.22 [45]. Enhancement and inhibition of spontaneous emission from a semiconductor quantum well have also been demonstrated by using a multilayer distributed Bragg reflector used as a microcavity [125]. Since the study of optical cavity QED is still at the early stage, several problems remain to be solved. For example, conventional studies on cavity QED have demonstrated that spontaneous emission has been enhanced at the cavity resonance frequency. However, for low-noise laser devices, it should be inhibited to reduce the quantum noise. It is expected that these problems will be solved in the near future, and that novel and stable semiconductor

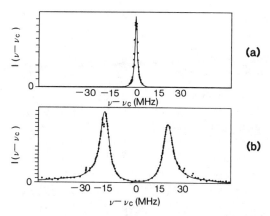

**Figure 5.21.** Experimental results for measuring the vacuum Rabi oscillation in Na atoms [124]. (a) Spectral profile $I(\nu - \nu_c)$ of the empty Fabry–Perot cavity. (b) Spectral profile $I(\nu - \nu_c)$ of the cavity with Na atoms. This profile has two peaks representing the FM sidebands due to the vacuum Rabi oscillation ($\nu_c$ is the cavity resonance frequency).

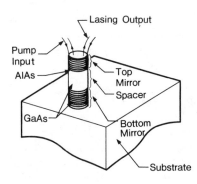

**Figure 5.22.** Schematic diagram of the Fabry–Perot microcavity created by employing a semiconductor device fabrication technique [45]. Its diameter and lengths are about 1.5 and 4.0 $\mu$m, respectively. Laser operation has been observed by this cavity.

lasers will be realized by utilizing the cavity QED technique. Furthermore, such advanced lasers could be used as powerful light sources for further improvements in the study of laser spectroscopy.

## 5.5. APPLICATIONS OF SPECTROSCOPY

Results of atomic and molecular spectroscopy have been used for a great number of applications. In the case of spectroscopy by semiconductor lasers, wider fields of practical applications can be opened because of low price, small size, and low power consumption of semiconductor lasers. These include several specific applications that cannot be essentially realized without using semiconductor lasers. These examples are described in the following subsections.

### 5.5.1. Analytical Spectroscopy

**5.5.1.1. Isotope Separation.** Several kinds of lasers have been used for laser isotope separation of U for nuclear fusion and Li, Ba, and Ga for tracers in medical diagnoses, and so on. Frequency-stable, frequency-tunable, and high-power lasers should be used for efficient selective excitation of a specific isotope by utilizing isotope shifts of the center frequencies of resonant spectral lines between each species of isotopes. For this purpose, several high-power lasers, such as dye lasers pumped by copper-vapor lasers, have been conventionally employed. However, some of the dye lasers have recently been replaced by frequency controlled semiconductor lasers to reduce the cost of the laser systems.

One successful example is the isotope separation of Rb and Cs by the light-induced drift method [126]. Although conventional systems have used dye lasers, the efficiency of isotope separation has not been sufficiently high because of the deexcitation occurred at the process of selective optical pumping. To improve this efficiency, dye lasers have been replaced by AlGaAs lasers. Since the direct FM response characteristics of these semiconductor lasers exhibit the resonant peak at the relaxation oscillation, a series of FM sidebands appear in the field spectrum of these lasers. By careful adjustment of the bias level, the frequency separation between the optical carrier and the FM sideband of the second order can be coincided with the frequency separation between the two hyperfine energy levels of the ground state (e.g., $F = 1$ and $F = 2$ levels in Rb). By pumping Rb atoms by an optical carrier and this FM sideband simultaneously, cyclic excitation can be realized. By this cyclic excitation, most of $^{87}$Rb atoms are maintained at the excitation states so as to realize a large cross section of the collision with buffer gases. As a result, drift velocity of $^{87}$Rb in buffer gases could become different from that of $^{85}$Rb, by which separation efficiency has been improved. Frequency stability of the laser can be maintained high enough by locking the optical carrier frequency to the Rb resonant frequency by following the technique of negative electrical feedback. As is represented by Figure 5.23 [126], separation efficiency as high as the one obtained by using a conventionally used dye laser (150-mW power) has been realized by the AlGaAs laser with the power of only 3.3 mW. This result has also been confirmed by the theoretical analysis based on the density matrix formulation for Rb atoms [127].

For isotope separations of Li, Ba, Ga, and so on, by using the Doppler-free two-photon transition process, one of the dye lasers has been replaced by a semiconductor laser, and as is shown by Figure 5.17, clearly resolved Doppler-free spectral shapes of $^7$Li and $^6$Li have been obtained and the magnitude of the isotope shift has been evaluated with the

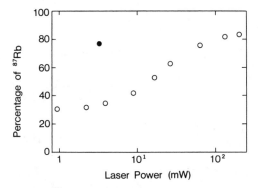

**Figure 5.23.** Percentage of $^{87}$Rb in a natural Rb atomic vapors obtained as the result of laser-induced drift [126]. Closed circle represents the result using the AlGaAs laser with relaxation oscillation sidebands at the power of 3.3 mW. Open circles represent results using a dye laser.

accuracy of about $1 \times 10^{-7}$ [101]. For the isotope separation of $^{235}$U, there are two possibilities [127]:

1. High power AlGaAs lasers can be used as one of the pumping sources for two-step excitation because the $^{235}$U atoms have fairly large excitation cross sections at the wavelength region of about 0.8–0.7 $\mu$m.

2. A stable and coherent visible AlGaInP laser can be used as the master laser for injection locking of a pulsed high-power dye laser.

With all of these possibilities, low-cost light sources for practical isotope separation systems can possibly be realized in the near future.

As has been widely known, several atomic isotopes, such as Li, Ga, and Ba, as described above, can be used as tracers for medical diagnoses. Molecular isotopes can also be used for this medical application. Since stable isotopes contained in a human respiration by doping several molecules, such as $^{13}$CO, can be sensitively detected by using a semiconductor laser spectrometer, the result of this detection can also be used as an useful information for medical diagnoses [128].

**5.5.1.2. Analysis of SiH$_x$ Radicals.** Fabrication of amorphous silicon film by chemical vapor deposition has been a key technology for the semiconductor device industry. For accurate control of deposited film thickness, quantitative analysis and assignment of species of SiH$_x$ are required. For example, it has been confirmed by UV photolysis that the SiH$_2$ radical could emit fluorescence due to transitions between the vibration levels. It can be expected that visible AlGaInP lasers will be used as a coherent

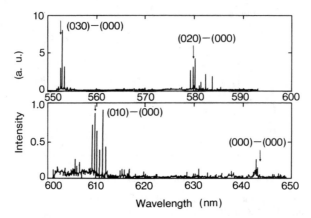

**Figure 5.24.** Laser excitation fluorescence spectra of $SiH_2$ radical [129]. The assignment of the vibrational progression and the band origin are indicated by vibration quantum numbers and arrows, respectively.

light source for efficient and practical spectral analysis in UV photolysis systems because the fluorescence spectra emitted due to the excitation by a pulsed ArF excimer laser are distributed at the wavelength region of 0.6 $\mu$m [127, 129] (see Figure 5.24).

**5.5.1.3. Laser Radar.** Although the power of a semiconductor laser usually is lower than that of $CO_2$, Ar, and YAG lasers, the small size and low power consumption of the semiconductor laser are advantageous features for use as a practical airborne or automobile-borne laser radar (lidar) system. Furthermore, by utilizing a high efficiency of direct modulation of a semiconductor laser, the pseudorandom modulation technique can be employed to realize high sensitivity and high spatial resolution in lidar operation. Figure 5.25 shows an example of the measurement of aerosols and clouds in the sky [130]. It was confirmed from the results of these field tests that this pseudorandom modulation CW lidar had the maximum measurable distance of 1 km for aerosol, and 3–5 km for cloud or dust with the spatial resolution of 9 m.

A FM–CW lidar has been proposed, which requires a < 100-kHz-linewidth AlGaAs laser for a range finding with the target range of 10–100 m, spatial resolution of 10 $\mu$m, and data rate of 10–100 pixels per second [131]. To prepare the coherent semiconductor laser for this system, techniques of negative electrical feedback and optical feedback described in Section 5.3.3 have been utilized.

**5.5.2. Optical Pumping of Atomic Clocks**

Frequency-stable microwave oscillators have been used as key devices for microwave communication, broadcasting, navigation, earthquake predic-

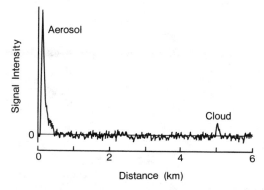

**Figure 5.25.** An example of the measurement of aerosols and clouds in the sky by using a semiconductor laser radar (lidar) by employing the technique of pseudorandom modulation [130].

tion (seismology), astronomy, and so on. For these systems, the most reliable oscillators are atomic clocks. Drastic improvements in the performances in these atomic clocks have been attained by using semiconductor lasers as coherent optical pumping source. Two examples of these improvements are described in the following.

**5.5.2.1. Cesium (Cs) Atomic Clock at 9.2 GHz.** A Cs atomic clock is a microwave oscillator, for which the transition frequency (9.2 GHz) between two hyperfine levels ($F = 4$, $m_F = 0$ and $F = 3$, $m_F = 0$; see Figure 5.26a) in the ground state of Cs has been used as the frequency reference to control the frequency of a voltage-controlled crystal oscillator [102]. Since the frequency accuracy of this clock is very high, it has been used as a primary standard of time. This clock has been studied and developed in a number of national research institutes in the world and used as a time-keeping clock. A compact and portable Cs atomic clock has been developed for a satellite-borne system.

In the conventional Cs atomic clock, deflection of the Cs atomic beam by DC magnetic fields has been used to select the atoms at the energy levels of $F = 4$, $m_F = 0$ and $F = 3$, $m_F = 0$. In this state selection scheme, atoms at $m_F \neq 0$ levels among the $2F + 1$ magnetic sublevels cannot be used, which limited the efficiency of the state selection. Furthermore, this clock has had several error sources limiting frequency accuracy, such as the frequency shift induced by the magnetic fields, which were used to deflect the atomic beam. To overcome these difficulties, an optical pumping scheme has been proposed [132]. Figure 5.26b shows a typical experimental setup for an optically pumped Cs atomic clock. Two 0.85-$\mu$m AlGaAs laser frequencies, $\nu_1$ and $\nu_2$, are tuned to the transition frequencies from $F = 4$ to $F' = 4$, and from $F = 3$ to $F' = 4$, respectively, where

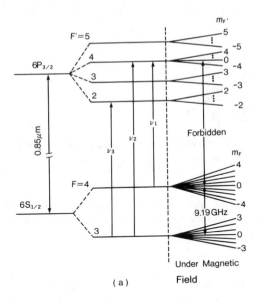

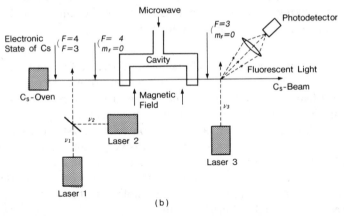

**Figure 5.26.** (*a*) Energy levels of Cs atoms. (*b*) Experimental setup for optical pumping of a Cs atomic clock [102].

$F'$ represents the quantum number of the excited state. After the cyclic transitions between the ground and the excited states due to the simultaneous optical pumping by the two lasers, all the atoms are transferred to the $F = 4$, $m_F = 0$ level of the ground state. Therefore, efficiency of state selection is increased eight times that of the conventional magnetic deflection scheme. After this state selection, atoms pass through the spatially isolated two microwave cavities, by which a fringe-shaped microwave Ramsey spectral profile can be observed. Figure 5.27 represents a typical

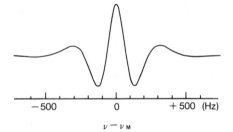

−500        0        +500  (Hz)

$\nu - \nu_M$

**Figure 5.27.** An example of fringe-shaped microwave Ramsey spectral profile obtained by an optically pumped Cs atomic clock [133, © 1988 IEEE]. The half-linewidth of the center part of this fringe was 200 Hz.

fringe-shaped Ramsey spectral profile. The half-linewidth of the center part of this fringe was as narrow as 200 Hz [133], which is determined by the inverse of the separation between the two microwave cavities. This narrow spectral line is used as a sensitive and stable frequency demodulator to control the microwave frequency. The third AlGaAs laser are used to measure this Ramsey spectral shape. Its frequency $\nu_3$ is tuned to the transition frequency between $F = 3$ and $F' = 2$ to excite the atoms of the $F = 3$ level that have been deexcited from the $F = 4$ level due to the microwave transition. Ramsey spectral shape can be measured by detecting the fluorescence from the $F' = 2$ level, which can be emitted by the optical pumping from $F = 3$ to $F' = 2$ by this laser.

Although the optical pumping scheme has been proposed for more than 10 years ago [132], experiments with this scheme have progressed at a slow speed because of the lack of frequency controlled AlGaAs lasers. Rapid progress in developing frequency controlled AlGaAs lasers, as was described in Section 5.3.3, made remarkable progress in this scheme. By this optical pumping, frequency accuracy higher than $5 \times 10^{-14}$ can perhaps be realized in the near future. For further improvements of this optically pumped Cs atomic clock, employments of the laser cooling technique have been proposed [134].

**5.5.2.2.  Rubidium (Rb) Atomic Clock at 6.8 GHz.** Although the frequency accuracy of a Rb atomic clock is lower than that of the Cs atomic clock, it has been popularly used as a compact and low-price microwave oscillator because the short-term frequency stability was rather higher than that of Cs atomic clocks. A block diagram of the Rb atomic clock is shown by Figure 5.28. As is shown by Figure 5.13c, transition frequency (6.8 GHz) between the two hyperfine levels in the ground state ($F = 2$, $m_F = 0$ and $F = 1$, $m_F = 0$) is used as the frequency demodulator to control the voltage controlled crystal oscillator.

To detect the spectral profile of this frequency demodulator, a technique of microwave–optical double resonance has been employed, for which optical pumping from the ground state to the excited state is required by using a light source of 0.78-$\mu$m wavelength. The double-resonance spectral profile can be measured by detecting the light power

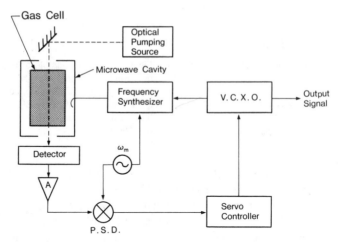

**Figure 5.28.** Experimental setup of a Rb atomic clock. VCXO and PSD represent a voltage-controlled crystal oscillator and a phase-sensitive detector, respectively.

transmitted through the Rb vapor cell. Buffer gases such as Ne, and Ar have been filled in the Rb vapor cell to reduce the double-resonance spectral width by atomic collisions so as to increase the sensitivity of frequency demodulation. By this collisional narrowing, termed *Dicke narrowing*, a spectral width as narrow as 100–500 Hz has been obtained. Microwave frequency is modulated and a phase-sensitive detector is used to measure the derivative of double-resonance spectral shape in order to use it as a frequency demodulator. Conventional Rb atomic clocks have used an incoherent discharge lamp for optical pumping. However, center frequency of the double-resonance spectral profile could be shifted as a result of variations of lamp power. This shift, called "light shift," is induced by the AC Stark effect by the optical field, which could not be avoided as long as the incoherent lamp has been used. This light shift has limited the long-term frequency stability. Furthermore, the frequency accuracy has been limited by the frequency shift induced by the collision between the Rb and buffer gas atoms. For these reasons, the Rb atomic clock has been considered as the secondary standard of time.

Improvements in short- and long-term frequency stabilities, and frequency accuracy also, can be expected by using an AlGaAs laser as a coherent optical pumping source. In this pumping scheme, the effect of microwave frequency modulation could be transferred to the coherent optical field of the laser in the Rb vapor cell. This modulation transfer occurs as a result of the nonlinear susceptibility of the Rb atoms that interact with the optical and microwave fields simultaneously. Since the double-resonance spectral signal is detected by using this modulated laser light, the effect of this modulation transfer induces a kind of interference

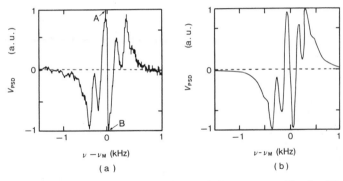

**Figure 5.29.**  Derivative of double-resonance spectral shape measured by the AlGaAs laser pumped Rb atomic clock [135, © 1987 IEEE]. (*a*) Experimental result. The linewidth of the center of this fringe-shaped spectral profile (i.e., the separation between points *A* and *B*) is about $\frac{1}{20}$ times narrower than that for the conventional lamp-pumped Rb atomic clock. (*b*) Calculated result [136].

fringe on the double-resonance spectral profile, as is shown by Figure 5.29*a* [135]. The width of the center part of this fringe-shaped spectral profile was about $\frac{1}{20}$ times narrower than that for the conventional lamp-pumped Rb atomic clock. The principle of obtaining such a fringe-shaped spectral profile is equivalent to that of FM laser spectroscopy [136]. As is shown by Figure 5.29*b*, such a specific spectral profile could be accurately reproduced by using the theoretical model based on the density matrix for the three-level atoms. If has also been confirmed that the fringe-shaped spectral profile of Figure 5.29 was equivalent to the fringe-shaped Ramsey spectral profile in the Cs atomic clock (see Figure 5.27). In the case of the Cs atomic clock, the perturbations from the microwave fields were applied to the Cs atoms at the two separate positions along the spatial axis of Cs atomic beam trajectory. On the other hand, in the Rb atomic clock, the perturbations from the optical fields were applied to the Rb atoms at separate positions along the optical frequency axis, which is a result of modulation transfer. The effect of this separated perturbation induced the fringe on the double-resonance spectral shape. By using this narrow-linewidth spectral shape as a frequency demodulator, it was confirmed that the short-term frequency stability was improved 40 times that of the conventional Rb atomic clocks [136].

Further improvements of short-term frequency stability was confirmed by reducing the FM noise of the laser at the Fourier frequency of the microwave modulation frequency because the noise contained in the double-resonance signal was mainly originated from the laser FM noise [137]. It has been demonstrated by Figure 5.8 that the negative electrical feedback can reduce the laser FM noise as large as 60 dB at around 1-kHz Fourier frequency range, in which region the microwave is frequency-mod-

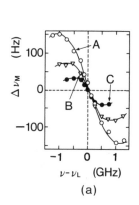

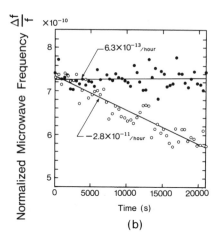

**Figure 5.30.** (*a*) Measured relation between the light shift and laser frequency detuning [138, © 1990 IEEE]. The laser power densities were 1008, 360, and 144 $\mu$W/cm$^2$ for curves *A*, *B*, and *C*, respectively. (*b*) Drift of microwave frequency [138, © 1990 IEEE]. The closed and open circles represent the results obtained with and without using the self-tuning technique, respectively.

ulated. This reduction is large enough to reach the shot-noise level, which is determined by the photodetector for the detection of the double-resonance signal. At the shot-noise limit, it has been estimated that the short-term frequency stability of the Rb atomic clock can be improved to as high as $5 \times 10^{-15} \, \tau^{-1/2}$, where $\tau$ is the integration time [136]. This stability is 1000 times higher than that of the presently developed laser-pumped Rb atomic clock.

The magnitude of the light shift induced by the AC Stark effect due to the optical field can be quantitatively evaluated by varying the laser power and frequency. The result is shown by Figure 5.30*a* [138]. Negative electrical feedback system has been developed to control the laser frequency in order to suppress the light shift of the microwave frequency. By this self-tuned system, drift of the microwave frequency, which originated from the drifts of the optical pumping power and frequency, has been reduced. The result is shown by Figure 5.30*b* [138], from which it can be confirmed that the long-term frequency stability was improved 40 times that of the conventional ones.

It has been also confirmed that the performances of the long-term frequency stability is higher than those of the portable Cs atomic clocks [137]. The replacement of the Rb vapor cell by the Rb atomic beam has been proposed to improve the frequency accuracy by eliminating the atomic collision with the buffer atoms [139].

By the intensive studies described above, it is expected that a novel, high-performance, compact, and low-price Rb atomic clock will be realized in the near future, and will be, thus, employed to the key applications such as the global positioning system (GPS), a novel primary standard of time, and so on.

### 5.5.3. Optical Pumping of Solid-State Lasers

Since a wavelength of an AlGaAs laser coincides with that of the absorption spectral band of a Nd:YAG crystal and the AlGaAs laser power has been increased up to, for example, 76 W, by employing an array structure [140], arc lamps used for optical pumping sources for CW Nd:YAG lasers have been replaced by these efficient AlGaAs lasers. The IM noise of the high-power, multi-longitudinal-mode oscillation of AlGaAs lasers, due to its mode-hopping or mode-partition phenomenon, limits the power and frequency stabilities of the Nd:YAG laser. To solve this problem, a careful design of the AlGaAs laser as a stable optical pumping source has been carried out by using a rate equation formalism [141]. Figure 5.31 shows

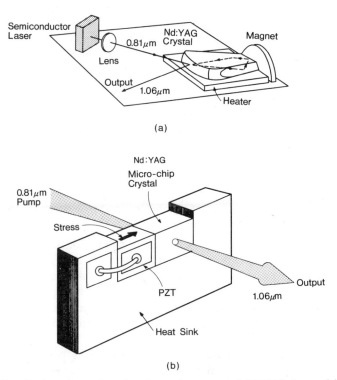

(a)

(b)

**Figure 5.31.** Configurations of semiconductor laser-pumped Nd:YAG lasers. (*a*) NPRO (nonplanar ring oscillator) or MISER (monolithic isolated single-mode end-pumping ring)-type laser [142, © 1989 IEEE]. (*b*) A Nd:YAG laser by using a microchip YAG crystal [145, reprinted with permission of Lincoln Laboratory, Massachusetts Institute of Technology, Lexington, Massachusetts].

typical structures of such the all-solid-state Nd:YAG lasers. Figure 5.31*a* shows the laser structure called *nonplanar ring oscillator* (NPRO) or (*monolithic isolated single-mode end-pumping ring* (MISER) type, where the laser crystal itself forms a ring cavity and the end-pumping scheme from the AlGaAs laser is employed [142]. A DC magnetic field is applied to the crystal to control the polarization of the laser beam by utilizing the Faraday effect of the crystal. By this control, a unidirectional ring cavity configuration with only a clockwise running wave could be realized. Wavelength tuning is also possible by varying the crystal temperature. Stability of the laser oscillation frequency is high because the cavity was monolithically integrated with the laser crystal. The field spectral linewidth under the free-running condition was less than 10 kHz. A stress applied to the crystal by a piezoelectric transducer (PZT) can modulate the laser frequency with the FM efficiency of about 500 kHz/V. By utilizing this FM response characteristics, the laser frequency was stabilized to a high-finesse ($> 20,000$) Fabry–Perot cavity by negative electrical feedback, which realized a linewidth below 30 Hz [143]. Recent studies of this system revealed the linewidth to be as narrow as 3 Hz [144]. Since the Schawlow–Townes limit of the free-running laser was estimated as 1 Hz at 1 MW of output power, this value of 3 Hz has almost reached this limit. It can be expected that the hypercoherent state is realized by improving the performance of the feedback loop in the near future.

Figure 5.31*b* shows the laser using a microchip of the YAG crystal ($0.7 \times 1.0 \times 2.0$ mm in size) to realize a single-longitudinal-mode oscillation by end-pumping scheme [145]. The laser frequency can be modulated by applying a stress to the crystal by a PZT. The FM efficiency was about 0.3 MHz/V up to the modulation frequency of 80 kHz. The maximum modulatable frequency was about 25 MHz, which was limited by the acoustic resonances of the Nd:YAG crystal.

In addition to the Nd:YAG lasers, a great number of solid-state lasers can be optically pumped by semiconductor lasers. For example, a Ho:YAG laser oscillates at 2.1-$\mu$m wavelength [146], which can be applied to the laser radar system because it guarantees safety for the eye and low propagation loss in the atmosphere. Furthermore, if a Ti:sapphire laser can be pumped by an AlGaAs laser-pumped Nd:YAG laser, a low noise and all-solid-state laser with an oscillation wavelength of 0.67–1.1 $\mu$m could be realized, which could, thus, replace the dye lasers and semiconductor lasers for various applications, such as high-resolution spectrometer for atoms and molecules [147].

These low-noise solid-state lasers can be used as powerful coherent light sources for basic studies of science. For example, a low-noise Nd:YAG laser has been designed for a very long-baseline interferometer for gravitational wave detection [148]. Typical performance required for this laser is frequency stability higher than $1 \times 10^{-15}$, output power higher than 100 W, and so on. To realize these performances, negative electrical

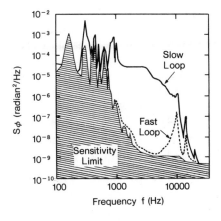

**Figure 5.32.** Power spectral density of the PM noise representing the phase fidelity of the injection-locked high-power Nd:YAG laser to the master laser [150].

feedback loop, composed of an electrooptical modulator and high-finesse Fabry–Perot interferometer, has been designed to obtain a very high gain with a bandwidth of about 50 kHz [149]. Furthermore, to obtain high power, the injection locking of a 13-W continuous-wave Nd:YAG laser has been carried out by using a semiconductor laser-pumped Nd:YAG laser with 40 mW of output power as the master laser [150]. Figure 5.32 shows the phase fidelity of the injection-locked laser obtained by using a slow and fast feedback loop, which is similar to the case for the frequency control of a semiconductor laser (see Figure 5.8 [70, 71]). It can be seen from this figure that phase noise reduction is about 40 dB for the Fourier frequency range of 1–20 kHz, from which the total root-mean-square phase noise can be estimated as about 0.3 radian. Further increases in injection-locked laser power up to 100 W can be expected while maintaining the frequency stability as low as that of the master laser.

Performance of these lasers can be improved to as high as that required for the gravitational wave detection system in the near future. The problems remained to be solved include (1) using a phase-squeezed light to save the laser power in order to avoid optical damages to the high-quality interferometer mirrors and (2) realizing the quantum nondemolition measurement system to measure the tiny displacement of the interferometer mirror by the gravitational wave [151].

## 5.6. SUMMARY

In this chapter, the topics of laser spectroscopy were reviewed by following the performance of frequency controlled semiconductor lasers and a petahertz-class hypercoherent optical sweep generator. These findings are based on at least two factors: (1) high-resolution spectroscopy cannot be realized without using these stable light sources, and (2) novel results of spectroscopy can be used to improve the performance of these lasers; for example, very narrow spectral lines in atoms or molecules can be used as a sensitive frequency demodulator for the laser frequency control. These facts indicate that progress in spectroscopy and progress in laser technology are mutually positively fed back and thus are interrelated. It can be concluded from this discussion that further development of semiconductor laser spectroscopy is essential for progress in quantum electronics, quantum optics, atomic physics, and a wide variety of practical industrial application systems. Further details of the characteristics of semiconductor laser operations and possible applications can be found in ref. 102.

## ACKNOWLEDGMENTS

The authors would like to thank Mr. C.-H. Shin of their research group for his critical reading and comments on the manuscript, and to graduate students for their active cooperation in the authors' work. Figure 5.31($b$) was reprinted with permisssion by the Lincoln Laboratory, Massachusetts Institute of Technology, Lexington, Massachusetts.

## REFERENCES

1. M. Ohtsu and Y. Teramachi, *IEEE J. Quantum Electron.* **25**, 31–38 (1989).
2. N. Chinone, T. Kuroda, T. Otoshi, T. Takahashi and T. Kajimura, *IEEE J. Quantum Electron.* **QE-21**, 1264–1269 (1985).
3. M. Ohtsu, Y. Teramachi, and T. Miyazaki, *Opt. Commun.* **61**, 203–207 (1987).
4. K. Kojima, S. Noda, K. Mitsunaga, K. Fujiwara, and K. Kyuma, "Low threshold current AlGaAs/GaAs DFB lasers grown by MBE," in *IEICE Technical Report on Optical and Quantum Electronics*, June 1985 (IEICE, Tokyo, 1985), pp. 93–98 (in Japanese).
5. K.-Y. Liou, M. Ohtsu, C. A. Burrus, Jr., U. Koren, and T. L. Koch, *IEEE J. Lightwave Technol.* **7**, 632–639 (1989).
6. M. Usami, S. Akiba, and Y. Matsushima, "Mode characteristics of $\lambda/4$-shifted DFB lasers with the distributed current injection along the cavity," *Extended Abstracts of the 49th Autumn Meeting of Jpn. Soc. Appl. Phys., Tokyo 1988* (Jpn. Soc. Appl. Phys., Tokyo, 1988), 6p-ZC-16 (in Japanese).
7. R. Lang and K. Kobayashi, *IEEE J. Quantum Electron.* **QE-16**, 347–355 (1980).
8. M. Ohtsu, "Noises in lasers," *Physics Monthly* **6**, 297–303 (1985) (in Japanese).
9. P. Kartaschoff, *Frequency and Time*, Academic Press, London, 1978.
10. M. Ohtsu, *Lasers and Atomic Clocks*, Ohm-sha, Tokyo, 1986 (in Japanese).
11. A. Blaquiere, *Compt. Rend.* **26**, 2929–2931 (1962).
12. M. Oshinski and J. Buus, *IEEE J. Quantum Electron.* **QE-23**, 9–29 (1987).

13. D. Welford and A. Mooradian, *Appl. Phys. Lett.* **40**, 560–562 (1982).

14. M. Ohtsu and S. Kotajima, *Jpn. J. Appl. Phys.* **23**, 760–764 (1984).

15. K. Vahala and A. Yariv, *Appl. Phys. Lett.* **43** 140–142 (1983).

16. M. Ohtsu, H. Fukada, T. Tako, and H. Tsuchida, *Jpn. J. Appl. Phys.* **22**, 1157–1166 (1983).

17. M. Ohtsu and M. Tabuchi, *IEEE J. Lightwave Technol.* **6**, 357–369 (1988).

18. O. Nilsson and Y. Yamamoto, *Appl. Phys. Lett.* **46**, 223–225 (1985).

19. Y. Yoshikuni and G. Motosugi, *IEEE J. Lightwave Technol.* **LT-5**, 516–522 (1987).

20. Y. Yamamoto, O. Nilsson, and S. Saito, *IEEE J. Quantum Electron.* **QE-21**, 1919–1928 (1985).

21. M. Ohtsu, *IEEE J. Lightwave Technol.* **6**, 245–256 (1988).

22. A. Mooradian, "High resolution tunable infrared lasers," in *Laser Spectroscopy*, R. G. Brewer and A. Mooradian, eds., Plenum Press, New York, 1973, pp. 223–236.

23. M. Ohtsu, "Realization of hyper-coherent light," *Science* (Japanese version of *Scientific American*), March 1989, pp. 64–73 (in Japanese).

24. D. Lenstra, B. H. Verbeek, and A. J. den Boef, *IEEE J. Quantum Electron.* **QE-21**, 674–679 (1985).

25. M. Ohtsu, K.-Y. Liou, E. C. Burrows, C. A. Burrus, and G. Eisenstein, *Electron. Lett.* **23**, 1111–1113 (1987).

26. M. Ohtsu, K.-Y. Liou, E. C. Burrows, C. A. Burrus, and G. Eisenstein, *IEEE J. Lightwave Technol.* **7**, 68–76 (1989).

27. N. A. Olsson and J. P. Van der Ziel, *IEEE J. Lightwave Technol.* **LT-5**, 510–515 (1987).

28. D. Mehuys, M. Mittelstein, and A. Yariv, "(GaAl)As quantum well semiconductor lasers tunable over 105 nm with an external grating," *Proc. Conf. Lasers and Electro-Optics* (CLEO '89), Baltimore, 1989 (LEOS/IEEE and OSA, Washington, D.C., 1989), FL4.

29. K. Vahala, K. Kyuma, A. Yariv, S. K. Kwong, M. Cronin-Golomb, and K. Y. Kau, *Appl. Phys. Lett.* **49**, 1563–1565 (1986).

30. B. Dahmani, L. Hollberg, and R. Drullinger, *Opt. Lett.* **12**, 876–878 (1987).

31. M. Ohtsu, "A frequency modulatable narrow-linewidth semiconductor laser," *IEICE Technical Report on Optical and Quantum Electronics* (1987), pp. 35–40 (IEICE, Tokyo, 1987) (in Japanese).

32. L. Hollberg and M. Ohtsu, *Appl. Phys. Lett.* **53**, 944–946 (1988).

33. C.-H. Shin, M. Teshima, M. Ohtsu, T. Imai, J. Yoshida, and K. Nishide, "Modulatable, high coherent and compact semiconductor laser modules," *Proc. 7th International Conf. Integrated Optics and Optical Fiber Communications* (IOOC '89), July 1989, Kobe, Japan (IEICE, Tokyo, 1989) 21D4-5, pp. 116–117.

34. J. Yoshida, T. Imai, K. Nishide, C.-H. Shin, M. Teshima, and M. Ohtsu, "Modulatable high coherent laser for fiber optic gyroscope," *Reports on Topical Meeting of the Laser Society of Japan, Tokyo, 1989* (Laser Society of Japan, Osaka, 1989), pp. 33–38 (in Japanese).

35. C.-H. Shin, M. Teshima, M. Ohtsu, T. Imai, J. Yoshida, and K. Nishide, *IEEE Photonics Technol. Lett.* **2**, 167–169 (1990).

36. H. Suzuki, I. Koshiishi, K. Nakagawa, and M. Ohtsu, "Frequency control of visible semiconductor laser," *Extended Abstracts of the 59th Autumn Meeting of Jpn. Soc. Appl. Phys., 1989* (Jpn. Soc. Appl. Phys., Tokyo, 1989), 29p-ZL-7 (in Japanese).

37. M. Ohtsu, H. Suzuki, K. Nemoto, and Y. Teramachi, *Jpn. J. Appl. Phys. Part 2*, **29**, L1463–L1465 (1990).

38. Y. Arakawa and A. Yariv, *IEEE J. Quantum Electron.* **QE-22**, 1887–1899 (1986).

39. T. Fujita, J. Ohya, K. Matsuda, M. Ishino, H. Sato, and H. Serizawa, *Electron. Lett.* **21**, 374–376 (1985).

40. I. Mito and K. Kitamura, "GaInAs/GaInAsP multiquantum well DFB lasers," *Proc. Conf. on Lasers and Electro-Optics* (CLEO '89), Baltimore, 1989 (LEOS/IEEE, OSA, Washington, DC, 1989), TUD5.

41. H. Imai, "Optical devices for high speed transmission including widely tunable DFB/DBR lasers," *Proc. Conf. on Lasers and Electro-Optics* (CLEO '89), Baltimore, 1989 (LEOS/IEEE, OSA, Washington, DC, 1989), FB4.

42. N. A. Olsson, C. H. Henry, R. F. Kazarinov, H. J. Lee, B. H. Johnson, and K. J. Orlowsky, *Appl. Phys. Lett.* **51**, 1141–1142 (1987).

43. S. Haroche and D. Kleppner, "Cavity quantum electrodynamics," *Physics Today*, January 1989, pp. 24–30.

44. A. Anderson, S. Haroche, E. A. Hinds, W. Jhe, D. Mechede and L. Moi, "Atomic physics experiments in micronized cavities: suppressing spontaneous decay at optical frequency," *Proc. 15th International Quantum Electronics Conference* (IQEC '87), Baltimore, 1987 (OSA, Washington, DC, 1987), THAA1.

45. J. L. Jewell, S. L. McCall, Y. H. Lee, A. Scherer, A. C. Gossard, and J. H. English, *Appl. Phys. Lett.* **54**, 1400–1402 (1989).

46. E. Yablonovitch, "Inhibited spontaneous emission in solid-state electronics," *Proc. 15th International Quantum Electronics Conference* (IQEC '87), Baltimore, 1987 (OSA, Washington, DC 1987), THAA4.

47. E. Yablonovitch, "Photonic band structure: Observation of an energy gap for light in 3-D periodic dielectric structures," *Proc. Conf. Quantum Electronics and Laser Science* (QELS '89), Baltimore, 1989 (OSA, Washington, DC, 1989), TUKK6.

48. M. Ohtsu, H. Kotani and H. Tagawa, *Jpn. J. Appl. Phys.* **22**, 1553–1557 (1983).

49. K. Fukuoka, M. Ohtsu and T. Tako, *Jpn. J. Appl. Phys.* **23**, L117–L120 (1984).

50. S. Kinugawa, H. Sasada, and K. Uehara, "Detection of $C_2H_2$ absorption lines with 1.5 $\mu$m DFB lasers," *Extended Abstracts of the 49th Autumn Meeting of Jpn. Soc. Appl. Phys., 1988* (Jpn. Soc. Appl. Phys., Tokyo, 1988), 6p-Q-13 (in Japanese).

51. H. Sasada, *J. Chem. Phys.* **88**, 767–777 (1988).

52. H. Tsuchida, M. Ohtsu, T. Tako, N. Kuramochi, and N. Oura, *Jpn. J. Appl. Phys.* **21**, L561–L563 (1982).

53. T. Yabuzaki, A. Ibaraki, H. Hori, M. Kitano, and T. Ogawa, *Jpn. J. Appl. Phys.* **20**, L451–L454 (1981).

54. S. Yamaguchi and M. Suzuki, *Appl. Phys. Lett.* **41**, 597–598 (1982).

55. Y. C. Chung, *Photon. Technol. Lett.* **1**, 135–136 (1989).

56. M. Ohtsu and E. Ikegami, *Electron. Lett.* **25**, 22–23 (1989).

57. E. Ikegami, H. Kusuzawa, K. Nakagawa, and M. Ohtsu, "Nonlinear organic waveguides for a LD-based optical sweep generator I," *Extended Abstracts of the 50th Autumn Meeting of the Jpn. Soc. Appl. Phys., 1989* (Jpn. Soc. Appl. Phys., Tokyo, 1989), 28p-ZP-16 (in Japanese).

58. Ch. Salomon, D. Hills, and J. L. Hall, *J. Opt. Soc. Am. B* **5**, 1576–1587 (1988).

59. M. Ohtsu, *Opt. Quantum Electron.* **20**, 283–300 (1988).

60. T. Kato, K. Kuboki, and M. Ohtsu, "Evaluation of performance of frequency offset locking system," *Extended Abstracts of the 48th Autumn Meeting of Jpn. Soc. Appl. Phys., 1987* (Jpn. Soc. Appl. Phys., Tokyo, 1987), 20p-ZQ-4 (in Japanese).

61. K. Kuboki and M. Ohtsu, *IEEE Trans. Instrum. Meas.* **39**, (in press). August issue, 1990.

62. M. Ohtsu, M. Hashimoto, and H. Ozawa, "A highly stabilized semiconductor laser and its application to optically pumped Rb atomic clock," *Proc. 39th Annual Frequency Control Symp. Philadelphia, 1985* (IEEE, Piscaway, NJ, 1985), pp. 43–53.

63. H. Furuta and M. Ohtsu, *Appl. Opt.* **28**, 3737–3743 (1989).

64. V. Pevtshin and S. Ezekiel, *Opt. Lett.* **12**, 172–174 (1987).

65. M. Ohi, *Jpn. J. Appl. Phys.* **12**, 1377–1381 (1973).

66. D. A. Jennings, K. M. Evenson, and D. J. E. Knight, *Proc. IEEE* **74**, 168–179 (1986).

67. Y. Miki, "Josephson devices for optical frequency measurements," *Proc. Symp. on Quantum Measurements, Tokyo, 1987* (Society of Promotion for Industry and Technology in Japan, Tokyo, 1987), pp. 19–28 (in Japanese).

68. Y. Enomoto and T. Murakami, "Optical detector using oxide superconductor," *Extended Abstracts of the 36th Spring Meeting of Jpn. Soc. of Appl. Phys. and Related Soc., Tokyo, 1989* (Jpn. Soc. Appl. Phys., Tokyo, 1989), 2p-ZD-4, p. 1112 (in Japanese).

69. M. Yamanishi, "Ultrafast optical processes through virtual charge polarization in quantum well structures," *Oyo Buturi* **58**, 1696–1707 (1989) (in Japanese).

70. M. Ohtsu, M. Murata, and M. Kourogi, "Subkilohertz linewidth of a semiconductor laser by electrical feedback and its network analysis," *Proc. Conf. on Lasers and Electro-Optics* (CLEO '89), Baltimore, 1989 (LEOS/IEEE and OSA, Washington, DC, 1989), THK30.

71. M. Ohtsu, M. Murata, and M. Kourogi, *IEEE J. Quantum Electron.* **25**, 231–241 (1990).

72. M. Murata and M. Ohtsu, "FM noise reduction of a semiconductor laser by reflection mode of high finesse Fabry–Perot resonator," *Extended Abstracts of the 35th Spring Meeting of the Jpn. Soc. Appl. Phys. and Related Soc., 1988* (Jpn. Soc. Appl. Phys., Tokyo, 1988), 29p-ZP-9 (in Japanese).

73. M. Kourogi, A. Kiyohara, and M. Ohtsu, "Improvement of the electrical negative feedback system for FM noise reduction of a semiconductor laser," *IEICE Technical Report on Optical and Quantum Electronics* (1989), pp. 19–24 (IEICE, Japan, Tokyo, 1989) (in Japanese).

74. T. W. Hänsch and B. Coullaud, *Opt. Commun.* **35**, 441–444 (1980).

75. H. P. Yuen and V. W. S. Chan, *Opt. Lett.* **8**, 177–179 (1989).

76. G. Abbas, V. W. S. Chan, and T. K. Lee, *IEEE J. Lightwave Technol.* **LT-3**, 1110–1122 (1985).

77. S. B. Alexander, *IEEE J. Lightwave Technol.* **LT-5**, 523–537 (1987).

78. M. Teshima, S.-H. Shin, and M. Ohtsu, "FM characteristics of optical feedback LD for homodyne OPLL," *Extended Abstracts of the 50th Autumn Meeting of the Jpn. Soc. Appl. Phys., 1989* (Jpn. Soc. Appl. Phys., Tokyo, 1989), 29p-ZL-3 (in Japanese).

79. M. Ohtsu, S. Katsuragi, and T. Tako, *IEEE J. Quantum Electron.* **QE-17**, 1100–1106 (1981).

80. K. Kuboki and M. Ohtsu, *IEEE J. Quantum Electron.* **25**, 2084–2090 (1989).

81. C.-H. Shin, K. Kuboki, and M. Ohtsu, "Simulation for the heterodyne type optical phase-locked loop by semiconductor lasers," *Trans. IEE Japan*, **108-C**, 678–684 (1988) (in Japanese).

82. S.-H. Shin, M. Teshima, and M. Ohtsu, "Heterodyne optical phase-locked loop by semiconductor lasers," *Extended Abstracts of the 50th Autumn Meeting of the Jpn. Soc. Appl. Phys., 1989* (Jpn. Soc. Appl. Phys., Tokyo, 1989), 29p-ZL-4 (in Japanese).

83. M. Kourogi, A. Kiyohara, and M. Ohtsu, "Homodyne OPLL of semiconductor lasers by all electrical negative feedback," *Extended Abstracts of the 37th Spring Meeting of the Jpn. Soc. Appl. Phys., 1990*) Jpn. Soc. Appl. Phys., Tokyo, 1990), 31a-G-4 (in Japanese).

84. G. C. Bjorklund, *Opt. Lett.* **5**, 15–17 (1980).

85. M. O. Scully, *Phys. Rev. Lett.* **55**, 2802–2805 (1985).

86. M. Ohtsu and K.-Y. Liou, *Appl. Phys. Lett.* **52**, 10–12 (1988).

87. M. Winters, Joint Institute for Laboratory Astrophysics, University of Colorado, (1989), private communication.

88. Y. Tohmori, K. Komori, S. Arai, and Y. Suematsu, *Electron. Lett.* **21**, 743–745 (1985).

89. H. Kusuzawa, E. Ikegami, K. Nakagawa, and M. Ohtsu, "Nonlinear organic waveguides for a LD-based optical sweep generator II (Frequency sweep)," *Extended Abstracts of the 50th Autumn Meeting of the Jpn. Soc. Appl. Phys., 1989* (Jpn. Soc. Appl. Phys., Tokyo, 1989), 28p-ZP-17 (in Japanese).

90. M. Ohtsu, "Progress toward highly coherent semiconductor lasers," *Proc. 7th International Conf. on Integrated Opt. and Optical Fiber Communications*, (IOOC '89), July 1989, Kobe, Japan, (IEICE of Japan, Tokyo, 1989) 19A3-1.

91. H. Kusuzawa, E. Ikegami, K. Nakagawa, and M. Ohtsu, "Basic study on a LD-based Peta-Hertz class ultrahigh coherent optical sweep generator," *Proc. 4th Meeting on Lightwave Sensing Technology, 1989, Osaka* (Lightwave Sensing Technol. Group, Jpn. Soc. Appl. Phys., Tokyo, 1989), LST4-8 (in Japanese).

92. B. T. Darling and D. M. Dennison, *Phys. Rev.* **57**, 128–139 (1940).

93. K. Nakagawa, "Local mode vibrations in overtone vibration of $NH_3$," Ph.D. thesis, University of Tokyo, 1989.

94. W. Lenth, *Opt. Lett.* **8**, 575–577 (1983).

95. L.-G. Wang, H. Riris, C. B. Carlisle, and T. F. Gallangher, *Appl. Opt.* 2071–2077 (1988).

96. F. M. Phelps III, *M.I.T. Wavelength Tables*, Vol. 2, *Wavelengths by Element*, The MIT Press, Cambridge, MA, 1982.

97. K. Sakurai and N. Yamada, *Opt. Lett.* **14**, 233–235 (1989).

98. H. Furuta, M. Hashimoto, H. Suzuki, K. Nakagawa, and M. Ohtsu, "Optical-microwave spectroscopy for rubidium beam atomic clocks," *Proc. Electronic Circuit Technical Meeting, IEE Jpn., 1989* (IEE Jpn., Tokyo, 1989), ECT-89-12 (in Japanese).

99. M. Ohtsu, M. Hashimoto, and H. Ozawa, "A highly stabilized semiconductor laser and its application to optically pumped Rb atomic clock," *Proc. 39th Annual Symp. Frequency Control, Philadelphia, 1985* (IEEE, Piscaway, NJ, 1985), pp. 43–53.

100. M. Suzuki and S. Yamaguchi, *IEEE J. Quantum Electron.* **24**, 2392–2399 (1988).

101. C. Vadla, A. Obrebski, and K. Niemax, *Opt. Commun.* **63**, 288–292 (1987).

102. M. Ohtsu and T. Tako, "Coherence in semiconductor lasers," in *Progress in Optics XXV*, E. Wolf, ed., Elsevier, Amsterdam, 1988, pp. 191–278.

103. H. Rinnenberg, "Ultrahigh *n* Rydberg atoms," *Technical Digest of 14th International Conf. on Quantum Electronics, San Francisco, 1986* (LEOS of IEEE, OSA, Washington, DC 1986), THKK1.

104. S. L. Gilbert and C. E. Wieman, *Phys. Rev. A* **34**, 792–803 (1986).

105. M. C. Noecker, B. P. Masterson, and C. E. Wieman, *Phys. Rev. Lett.* **61**, 310–313 (1988).

106. T. P. Emmons, J. M. Reeves, and E. N. Forston, *Phys. Rev. Lett.* **51**, 2089–2092 (1983).

107. C. E. Tanner and C. E. Wieman, *Phys. Rev. A* **38**, 162–165 (1988).

108. S. Chu (AT & T Bell Laboratories), private communication, 1986.

109. T. W. Hänsch and A. L. Schawlow, *Opt. Commun.* **13**, 68–69 (1975).

110. P. D. Lett, R. N. Watts, C. I. Westbrook, W. D. Phillips, P. L. Gould, and H. J. Metcaff, *Phys. Rev. Lett.* **61**, 169–172 (1988).

111. R. N. Watts, P. D. Lett, C. I. Westbrook, S. L. Rolston, C. E. Tanner, W. D. Phillips, P. L. Gould, and H. J. Metcalf, "Laser cooling below the Doppler limit," *Proc. Conf. on Quantum Electronics and Laser Science* (QELS '89), Baltimore, 1989 (OSA, Washington, DC, 1989), THLL1.

112. A. Aspect, E. Arimond, R. Kaiser, N. Vansteenkiste, and C. Cohen-Tannoudji, *Phys. Rev. Lett.* **61**, 826–829 (1988).

113. E. L. Raab, M. G. Prentiss, A. E. Cable, S. Chu, and D. E. Pritchard, *Phys. Rev. Lett.* **59**, 2631–2634 (1987).

114. D. Sesko, C. G. Fan, and C. E. Wieman, *J. Opt. Soc. Am. B* **5**, 1225–1227, (1988).

115. R. N. Watts, D. H. Yang, B. Sheehy, and H. Metcalf, "Deceleration and cooling of a thermal beam of rubidium," *Proc. 15th International Quantum Electronics Conference* (IQEC '87), Baltimore, 1987 (OSA, Washington, DC, 1987), TUGG37.

116. H. Katori, K. Yamashita, and F. Shimizu, "Laser cooling of rare gas atoms by semiconductor lasers," *Extended Abstracts of the Autumn Meeting of the Jpn. Soc. of Phys., 1989* (Jpn. Soc. Phys., Tokyo, 1989), 5a-ZF-11 (in Japanese).

117. C. E. Wieman, Joint Institute for Laboratory Astrophysics, University of Colorado, 1989, private communication.

118. E. Schrödinger, *Br. J. Phil. Sci.* **3**, 109–123 (1952).

119. J. C. Bergquist, R. G. Hulet, W. M. Itano, and D. J. Wineland, *Phys. Rev. Lett.* **57**, 1699–1702 (1986).

120. J. Bergquist, National Institute for Standards and Technology, 1989, private communication.

121. S. L. Gilbert, J. J. Bollinger, and D. Wineland, *Phys. Rev. Lett.* **60**, 2022–2025 (1988).

122. D. Meschede, H. Walther, and G. Muller, *Phys. Rev. Lett.* **54**, 551–554 (1985).

123. D. J. Heinzen, J. J. Childs, J. F. Thomas, and M. S. Feld, *Phys. Rev. Lett.* **58**, 1320–1323 (1987).

124. M. G. Raizen, R. J. Thompson, R. J. Brecha, H. J. Kimble, and H. J. Carmichael, *Phys. Rev. Lett.* **63**, 240–243 (1989).

125. K. Igeta, Y. Yamamoto, and S. Machida, "Control of spontaneous emission from a GaAs quantum well," *Extended Abstracts of the Autumn Meeting of the Jpn. Soc. Phys., 1989* (Jpn. Soc. Phys., Tokyo, 1989) 4a-ZF-6 (in Japanese).

126. A. D. Streatere, J. Mooibroek, and J. P. Woerdman, *Appl. Phys. Lett.* **52**, 602–604 (1988).

127. M. Ohtsu, "Investigation on applicability of visible semiconductor lasers to elementary analysis," *Proc. 1st Meeting on Lightwave Sensing Technology, Tokyo, 1988* (Lightwave Sensing Technol. Research Group, Jpn. Soc. Appl. Phys., Tokyo, 1988), LSTI-4 (in Japanese).

128. U. Lachish, S. Rotter, E. Adler, and U. El-Hanany, *Rev. Sci. Instrum.* **58**, 923–927 (1987).

129. G. Inoue and M. Suzuki, *Chem. Phys. Lett.* **105**, 641–644 (1984).

130. N. Takeuchi, *Development of Pseudo-Random Modulation CW Lidar and Its Application to Field Measurement* (Research Report from the National Institute for Environmental Studies, Japan, No. 122, 1989) (in Japanese).

131. R. G. Beausoleil, J. A. McGarvey, R. L. Hagman, and C. S. Hong, "Narrow-linewidth semiconductor laser for frequency modulated continuous-wave ladar," *Proc. Conf. on Lasers and Electro-Optics* (CLEO '89) Baltimore, 1989 (LEOS/IEEE, OSA, Washington, DC, 1989), WN3.

132. J. L. Picque, *Metrologia* **13**, 115–119 (1977).

133. S. Ohshima, Y. Nakadan, and Y. Koga, *IEEE Trans. Instrum. Meas.* **37**, 409–413 (1988).

134. D. W. Sesko and C. E. Wieman, *Opt. Lett.* **14**, 269–271 (1989).

135. M. Hashimoto and M. Ohtsu, *IEEE J. Quantum Electron.* **QE-23**, 446–451 (1987).

136. M. Hashimoto and M. Ohtsu, *J. Opt. Soc. Am. B* **6**, 1777–1789 (1989).

137. M. Ohtsu, H. Hashimoto, and H. Suzuki, "Performance evaluation of a semiconductor laser pumped Rb atomic clock," *Extended Abstracts of the 50th Autumn Meeting of the Jpn. Soc. Appl. Phys., 1989* (Jpn. Soc. Appl. Phys., Tokyo, 1989), 29a-X-5 (in Japanese).

138. M. Hashimoto and M. Ohtsu, *IEEE Trans. Instrum. Meas.* **39**, 458–462 (1990).

139. H. Furuta, K. Nakagawa, and M. Ohtsu, "Rubidium-atomic-beam clock pumped by a semiconductor laser I," *Extended Abstracts of the 50th Autumn Meeting of the Jpn. Soc. Appl. Phys., 1989* (Jpn. Soc. Appl. Phys., Tokyo, 1989), 29a-X-6 (in Japanese).

140. M. Sakamoto, D. F. Welch, J. G. Endriz, D. R. Scifres, and W. Streifer, *Appl. Phys. Lett.* **54**, 2299–2300 (1989).

141. T. M. Baer, "Diode-pumped solid state lasers," *Proc. Conf. on Lasers and Electro-Optics* (CLEO '89), Baltimore, 1989 (LEOS/IEEE and OSA, Washington, DC, 1989), FJ1.

142. A. C. Nilsson, E. K. Gustafson, and R. L. Byer, *IEEE J. Quantum Electron.* **25**, 767–790 (1989).

143. T. Day, A. C. Nilsson, M. M. Fejer, A. D. Farinas, E. K. Gustafson, C. D. Nabors, and R. L. Byer, *Electron. Lett.* **25**, 810–812 (1989).

144. T. Day, E. K. Gustafson, and R. L. Byer, *Opt. Lett.* **15**, 2221–2223 (1990).

145. J. J. Zayhowski and A. Mooradian, *Opt. Lett.* **14**, 618–620 (1989).

146. T. Y. Fan, G. Huber, R. L. Byer, and P. Mitzscherlich, *Opt. Lett.* **12**, 678–680 (1987).

147. K. Ishikawa and S. Imai, "Tunable solid state lasers," Optronics, No. 78 (1988) (Optronics Corp., Tokyo, 1988), pp. 81–86 (in Japanese).

148. R. W. P. Drever, J. Hough, A. J. Munley, S.-A. Lee, R. Spero, S. E. Whitcomb, H. Ward, G. M. Ford, M. Hereld, N. A. Robertson, I. Kerr, J. R. Pugh, G. P. Newton, B. Meers, E. D. Brooks III, and Y. Gursel, "Gravitational wave detectors using laser interferometer and optical cavities: Ideas, principles and prospects," in *Quantum Optics, Experimental Gravitation, and Measurement Theory*, P. Meystre and M. O. Scully, eds., Plenum Press, New York, 1981, pp. 503–514.

149. D. Shoemaker, A. Brillet, C. Nary Man, O. Crequt, and G. Kerr, *Opt. Lett.* **14**, 609–611 (1989).

150. C. D. Nabors, A. D. Farinas, T. Day, S. T. Yang, E. K. Gustafson, and R. L. Byer, *Opt. Lett.* **14**, 1189–1191 (1989).

151. M. Hillery and M. O. Scully, "Quantum noise and QND measurements," in *Quantum Optics, Experimental Gravitation, and Measurement Theory*, P. Meystre and M. O. Scully, eds., Plenum Press, New York, 1981, pp. 661–674.

# 6

# Coherent Detection Using Semiconductor Lasers: System Design Concepts and Experiments

T. G. HODGKINSON
*British Telecom Research Laboratories, Martlesham Heath, Ipswich, Suffolk, England*

## 6.1. INTRODUCTION

When this chapter was written, all operational optical fibre transmission systems worldwide were based on directly detecting an intensity-modulated waveform. However, it had been widely appreciated for some time that coherent optical detection techniques could significantly improve on the performance given by these systems. The reason for this is that coherent detection can in principle achieve shot-noise-limited performance whereas conventional direct detection gives a performance which is limited by receiver thermal noise. The improved performance potential offered by coherent detection, which could be as large as 20 dB, offers increased repeater separation, would allow the data rate (capacity) of long-haul systems to be upgraded without reducing operating margins, and for optically multiplexed distribution networks would increase the available power budget. In addition to the performance advantage, the high selectivity of coherent detection should allow the vast optical bandwidth available in the low-loss transmission window of conventional single-mode fibre (1.3–1.6 $\mu$m) to be efficiently utilised.

Although the principles of coherent detection are not new (they are well established at radio frequencies), it was not until the development of the gas laser in the early 1960s that it became possible to apply and study these principles in the optical domain. It was during the decade following the development of the gas laser that the experiments reported by Enloe

*Coherence, Amplification, and Quantum Effects in Semiconductor Laser*, Edited by Yoshihisa Yamamoto.
ISBN 0-471-512494 © 1991 John Wiley & Sons, Inc.

and Rodda [1] and DeLange and Dietrich [2] first demonstrated that coherent optical detection was, in principle, possible. However, because of the research and technological developments needed in the fields of optical waveguides, semiconductor lasers, and optical components, the use of coherent detection in optical transmission systems was not given serious consideration again until the early 1980s. At about this time, a few research laboratories began to reassess the feasibility and potential of coherent optical transmission systems, and since this time many laboratories worldwide have become actively involved in coherent optical fibre transmission systems research.

Since 1980 a large amount of theoretical and experimental research has been carried out, and the aim of this chapter is to indicate how this work has evolved up to the present time (mid-1989) and led to demonstrations of experimental digital coherent optical fibre transmission systems that have used semiconductor lasers and advanced design concepts. In order to present this as clearly as possible, this chapter is divided into four main sections. Section 6.2 is an introduction to the principles of coherent detection and Section 6.3 considers the system design options available in terms of the different modulation formats, the two types of coherent detection (homodyne and heterodyne), and the requirements that must be satisfied to achieve shot-noise-limited performance. Section 6.4 considers the advanced system design concepts that have been proposed for overcoming the two most significant coherent detection problems (its sensitivity to polarisation and laser phase noise), and Section 6.5 overviews system experiments using 1.5-$\mu$m semiconductor laser sources. However, any significant experimental results reported for experiments using a single semiconductor laser and/or a different wavelength than 1.5 $\mu$m have also been included.

## 6.2. BASIC COHERENT SYSTEM CONCEPTS

### 6.2.1. Coherent System Configuration

Figure 6.1 shows a simplified block diagram of an ideal digital coherent optical fibre transmission system using a dual photodiode receiver design. The transmitter comprises a single-longitudinal-mode laser source followed by an external optical modulator, the output of which is connected to the single-mode transmission fibre. For some modulation formats the laser can be directly current modulated, thus removing the need for an external modulator. Unlike direct detection, which can only extract information from the amplitude of the received optical carrier, coherent detection can also extract information from either its phase or frequency; therefore, the output from the transmitter laser can be either amplitude, frequency, or phase modulated. For digital systems these three options are

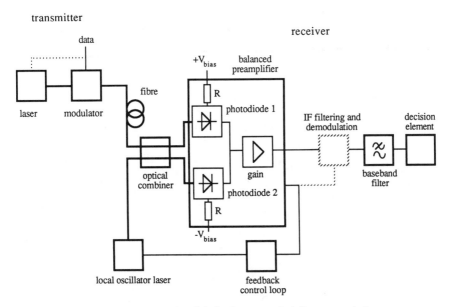

**Figure 6.1.** Block diagram of a digital coherent optical fibre transmission system.

commonly referred to, respectively, as *amplitude, frequency,* and *phase shift-keying* (ASK, FSK, and PSK). When ASK modulation is used the modulation signal switches the optical carrier on and off, for FSK it steps the carrier frequency between two predetermined values that are usually separated by a value in excess of the data rate (wide deviation FSK), and for PSK it steps the carrier phase between two values separated ideally by 180°.

At the receiver a fibre directional coupler would normally, although not necessarily, be used to combine the low-level received signal with the much larger signal from the single-longitudinal-mode local oscillator laser, the resulting output from the coupler being detected using a single (unbalanced) or dual (balanced) photodiode optoelectronic preamplifier. From this point on the receiver takes one of two forms depending on whether homodyne or heterodyne detection is used, the difference between these being that for the former the signal and local oscillator field frequencies are equal, whereas for the latter they are different. In the homodyne receiver the transmitted data is available directly at the output from the preamplifier, so there is only need to follow this with a baseband noise filter and decision-making element. To ensure stable homodyne detection, optical phase-locked-loop feedback control is needed to lock the phase of the local oscillator to that of the received signal. In the heterodyne receiver the output from the preamplifier is centred about an intermediate frequency (IF), so IF demodulation preceded by bandpass filtering is required between the preamplifier and the baseband noise filter. To stabilise the IF, automatic frequency control (AFC) feedback is needed to lock the frequency of the local oscillator to that of the received signal.

### 6.2.2. Coherent Detection Principles

**6.2.2.1. Optical Mixing.** When the modulated signal and local oscillator fields are combined (added) and photodetected at the input of a coherent receiver, optical mixing occurs as a result of the photocurrent being proportional to the square of the resultant field (i.e., its power). To achieve ideal mixing the two fields should be combined in a waveguide device such as a fibre coupler, as this avoids the problems associated with spatial misalignment and wavefront matching. Because the optical mixing produces a photocurrent component which is proportional to the product of the two input fields, the received modulation spectrum is down-converted and centred on an intermediate frequency equal to the frequency difference between the signal and local oscillator fields. Consequently, if these two frequencies are chosen such that the IF falls within the operating frequency range of conventional electronics, the original baseband signal can be recovered using standard radio communication demodulation techniques. For multichannel communication system applications this gives a level of selectivity that is far superior to that offered by present optical filters; consequently, multichannel coherent systems can be operated with much narrower channel separations than those using direct detection.

Apart from its high selectivity, optical mixing also gives a performance advantage because it can overcome the thermal noise performance limitation usually associated with directly detecting the intensity of the received signal. The reason for this is that the amplitude of the intermediate frequency is proportional to the amplitude of the local oscillator field, and as this is increased the received signal-to-noise ratio improves until the local oscillator shot noise dominates over the receiver thermal noise. Once this point is reached, the receiver performance becomes shot-noise-limited and independent of any further increase in local oscillator power.

**6.2.2.2. System Performance Analysis.** The performance of a digital system is usually measured in terms of its error probability at the input to the decision element, and this convention will be adhered to here. For analysis purposes the receiver model given in Figure 6.2 will be used. In this model current source $i_{P1}$ represents the photocurrent generated by photodiode 1, and the mean-square noise currents represent shot noise ($\langle i_{IS}^2 \rangle$), local oscillator excess intensity noise ($\langle i_{I_X}^2 \rangle$), and preamplifier thermal noise ($\langle i_{IE}^2 \rangle$). The current sources and the photodiode enclosed by the dashed box show the model modifications needed when a balanced receiver design

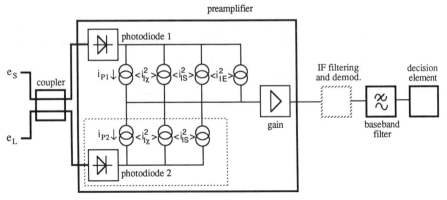

**Figure 6.2.** Coherent receiver model used for system performance analysis purposes. After Hodgkinson [12], © 1987 IEEE.

is to be analysed. The receiver noise sources are assumed to be white, Gaussian, and, apart from the excess intensity noise sources, uncorrelated. Although the shot-noise statistics are Poisson, the Gaussian approximation introduces very little error when the mean number of photons per bit period is large; this condition will always be satisfied provided sufficient local oscillator power is available. This model has ignored photodiode dark current, but this can be easily included in the analysis if required.

To avoid unnecessary complication at this stage, the error probability will be derived assuming rectangular pulse binary modulation, an unbalanced preamplifier design, a receiver transfer characteristic matched to the received rectangular pulse shape, and synchronous IF demodulation for the case of the heterodyne receiver. For the time being it will also be assumed that the local oscillator shot noise is the dominant noise term.

*6.2.2.2.1. Signal at the Input to the Decision Element.* When an unmodulated received optical field ($e_S$) is combined with the local oscillator field ($e_L$) in a coupler with power coupling ratio $\alpha$, the photocurrent ($i_{P1}$) produced by the power incident on photodiode 1, assuming plane waves, is proportional to

$$i_{P1} \propto \frac{\eta e \lambda}{hc} \int_t^{t+\Delta t} \left( \sqrt{\alpha}\, e_S + \sqrt{1-\alpha}\, e_L \right)^2 dt \qquad (6.1)$$

$$e_S = E_S \cos(\omega_S t + \phi_S) \qquad (6.2)$$

$$e_L = E_L \cos(\omega_L t + \phi_L) \qquad (6.3)$$

where $\eta$ is the photodiode quantum efficiency, $h$ is Plank's constant, $e$ is electronic charge, $\lambda$ is optical wavelength, $c$ is the velocity of light, $\omega_S$ is the signal angular frequency, $\omega_L$ is the local oscillator angular frequency, and $\phi_S$ and $\phi_L$ are the field phases at $t = 0$. The choice of integration time $\Delta t$ is to some extent arbitrary, but the value used must exceed the period of the optical fields; in practice it will almost certainly be determined by the bandwidth of the photodiode.

If the signal and local oscillator field polarisations are matched, the optical mixing is ideal and the resulting photocurrent, which is derived by substituting Eqs. (6.2) and (6.3) in Eq. (6.1) and integrating, is equal to

$$i_{P1} = \frac{\eta e \lambda}{hc} \{ \alpha P_L + (1 - \alpha) P_S$$

$$+ 2\sqrt{\alpha P_L (1 - \alpha) P_S} \cos[(\omega_L - \omega_S)t + \phi] \} \quad (6.4)$$

where $P_S$ is the received signal power in the absence of modulation, $P_L$ is the local oscillator power, and $\phi$ is the phase difference $\phi_L - \phi_S$.

For ideal coherent detection the local oscillator power must be much larger than the signal power, and when this is the case, the signal photocurrent can be approximated to

$$i_S = \frac{\eta e \lambda}{hc} 2\sqrt{\alpha P_L (1 - \alpha) P_S} \cos(\omega_{IF} t + \phi) \quad (6.5)$$

where $\omega_{IF}$ is the intermediate frequency and equal to the value given by $\omega_L - \omega_S$.

If the received signal is a digitally modulated waveform, Eq. (6.5) can take one of several slightly different forms depending on the modulation format used, but for analytic purposes, knowing that with the appropriate receiver design FSK can be treated as a superposition of two ASK waveforms, the modulated photocurrent ($i_m$) can be expressed generally as

$$i_m = \frac{\eta e \lambda}{hc} 2 f_{(t)} \sqrt{\alpha P_L (1 - \alpha) P_S} \cos(\omega_{IF} t + \phi) \quad (6.6)$$

where $P_S$ now represents the peak signal power and $f_{(t)}$ represents the two binary modulation states: for ASK, $f_{(t)} = 0$ or 1; for FSK, $f_{(t)} = 0$ or $\sqrt{2}$; and for PSK, $f_{(t)} = -1$ or 1.

If this expression is transferred through the receiver, and it is assumed that the appropriate phase-locked-loop is ideal (i.e., unaffected by noise) and maintains the condition $\phi = 0$, the signal amplitude ($V_D$) at the input to the decision element at a decision instant will be found to be

$$V_D = \frac{\eta e \lambda}{hc} 2 K_H f_{(t)} \sqrt{\alpha P_L (1 - \alpha) P_S} \quad (6.7)$$

**Table 6.1   $K_{DM}$ Values for the Various Modulation and Detection Combinations**

|  | $K_{DM}$ | |
|---|---|---|
| MOD | HOM | HET |
| ASK | 8 | 4 |
| FSK | 8 | 4 |
| PSK | 16 | 8 |

where $K_H$ indicates the type of decision used: for homodyne detection $K_H = 1$, for heterodyne detection $K_H = 1/\sqrt{2}$.

If the various $K_H$ and $f_{(t)}$ combinations are substituted into Eq. (6.7), it will be found that the peak-to-peak signal amplitude ($V_{p-p}$) at the input to the decision element can be expressed generally as

$$V_{p-p} = \frac{\eta e \lambda}{hc} \sqrt{K_{DM} \alpha P_L (1 - \alpha) \langle P_S \rangle} \qquad (6.8)$$

where $\langle P_S \rangle$ is the mean received signal power in the presence of modulation and $K_{DM}$ is a parameter whose value depends on the particular modulation and detection combination being used (see Table 6.1). It is clear from this expression that the signal at the input to the decision element is proportional to $\sqrt{\langle P_S \rangle}$, not $\langle P_S \rangle$, as is the case for direct detection, and it also shows that the signal gain produced by optical mixing is proportional to $\sqrt{P_L}$.

**6.2.2.2.2.  Noise Power at the Input to the Decision Element.** Ideally, $\alpha P_L$ would be made to be the dominant DC optical power term in Eq. (6.4), and when this is the case the double-sided shot-noise spectral density ($S_{IS}$), derived using the usual shot-noise equation, approximates to

$$S_{IS} = e\langle i_{P1} \rangle = \frac{\eta e^2 \lambda}{hc} \alpha P_L = \alpha S'_{IS} \qquad (6.9)$$

where $\langle i_{P1} \rangle$ is the mean photocurrent and $S'_{IS}$ is the shot-noise spectral density that would be measured at the output of the local oscillator laser.

Because all noise sources other than shot noise have been ignored, the mean-square noise voltage $\langle v_n^2 \rangle$ at the input to the decision element is

$$\langle v_n^2 \rangle = \frac{\eta e^2 \lambda}{hc\tau} \alpha P_L \qquad (6.10)$$

where $1/\tau$ is the double-sided noise bandwidth for a matched filter receiver, $\tau$ being the modulation bit period.

**6.2.2.2.3. Receiver Sensitivity.** For binary modulation with equiprobable data states and Gaussian noise statistics, the error probability $(P_e)$ is

$$P_e = \frac{1}{2}\text{erfc}\left(\frac{Q}{\sqrt{2}}\right) \tag{6.11}$$

where $Q$ is the normalised Gaussian variable used by Personick in his analysis of direct detection systems [3], and for a decision threshold set at half the $V_{p\text{-}p}$ value, it is equal to

$$Q = \frac{V_{p\text{-}p}}{2\sqrt{\langle v_n^2 \rangle}} \tag{6.12}$$

Substituting Eqs. (6.8) and (6.10) into Eq. (6.12) and rearranging, the receiver sensitivity, which is usually defined to be the mean received signal power needed to achieve a $10^{-9}$ error probability $(Q = 6)$, can be shown to be

$$\langle P_S \rangle = \frac{4hcQ^2}{K_{DM}\eta\lambda\tau(1 - \alpha)} \tag{6.13}$$

**6.2.2.2.4. Comparison of Unbalanced and Balanced Receiver Sensitivities.** To derive the receiver sensitivity for the balanced receiver design the extra photodiode and shot-noise source enclosed by the dashed box in Figure 6.2 need to be introduced into the earlier analysis. However, as a full analysis would show, the resulting signal and noise expressions at the input to the decision element are given by doubling Eqs. (6.8) and (6.10) after substituting $\alpha = 0.5$. Substituting these new expressions into Eq. (6.12), it can be shown that the receiver sensitivity for the balanced receiver, assuming the coupler has zero excess loss, is given by

$$\langle P_S \rangle = \frac{4hcQ^2}{K_{DM}\eta\lambda\tau} \tag{6.14}$$

Comparing this with the equivalent result for the unbalanced receiver, it is easily seen that the receiver sensitivity can be expressed generally as

$$\langle P_S \rangle = \frac{4hcQ^2}{K_{DM}\eta\lambda\tau K_B(1 - \alpha)} \tag{6.15}$$

for the unbalanced receiver $K_B = 1$, for the balanced design $K_B = 2$, and $\alpha = 0.5$.

*6.2.2.2.5. Shot-Noise-Limited Receiver Sensitivity.* If the receiver is balanced and the photodiode is perfect ($\eta = 1$), the resulting receiver sensitivity is the ideal shot-noise-limited value for the particular modulation and detection combination being used, but for practical purposes the shot-noise-limited performance ($\langle P_{\mathrm{SL}} \rangle$) should be taken to be given by

$$\langle P_{\mathrm{SL}} \rangle = \frac{4hcQ^2}{K_{\mathrm{DM}}\eta\lambda\tau} \tag{6.16}$$

The sensitivity of a coherent receiver is sometimes expressed in terms of the mean number of photons ($N_{\mathrm{p}}$) received per data bit as this has the advantage of being independent of both the data rate and the wavelength, and its relationship with a particular error probability is given by

$$N_{\mathrm{P}} = \frac{4Q^2}{\eta K_{\mathrm{DM}}} \tag{6.17}$$

## 6.3. SYSTEM DESIGN CONSIDERATIONS

### 6.3.1. Modulation Formats

The receiver sensitivities for the various modulation and detection combinations have been derived from Eq. (6.16), and the results, referenced to the ASK heterodyne receiver sensitivity, are given in Table 6.2. The performance given by PSK is 3 dB better than that given by ASK because optical power is not wasted in a continuous-wave carrier component. A further advantage, which is not obvious from Table 6.2, is that for peak

**Table 6.2   Receiver Sensitivities for the Various Modulation and Detection Combinations Referenced to the ASK Heterodyne Receiver Sensitivity**

| Detection | Modulation | Realative Improvement (dB) |
|-----------|------------|----------------------------|
| Homodyne | PSK | 6 |
|  | FSK | 3 |
|  | ASK | 3 |
| Heterodyne | PSK | 3 |
|  | FSK | 0 |
|  | ASK | reference |

power-limited optical sources FSK and PSK both give 3 dB more transmitter power than when ASK is used. The performance figures given in Table 6.2 assume synchronous demodulation, but for heterodyne detection there is the option of using nonsynchronous IF demodulation, but the outcome of this is that the resulting receiver sensitivity is usually slightly worse than for the synchronous case (of the order of 1 dB).

Other modulation schemes that may be of advantage when using heterodyne detection are DPSK, QPSK, multifrequency FSK, and continuous-phase FSK (CPFSK). DPSK is standard PSK with the difference that the use of differential encoding results in the phase of the optical carrier being changed only whenever a binary 0 is to be transmitted, when a binary 1 is to be transmitted the phase is left in its previous state. The advantage of using DPSK is that a performance similar to that given by a PSK system can be achieved without the need for synchronous IF demodulation at the receiver. QPSK is a phase modulation format that uses four quadrature phase states to replace the four combinations that exist when a binary modulation signal is encoded by processing 2 data bits at a time. The outcome of this is that the receiver bandwidth is halved for a given data rate, but the sensitivity to laser phase noise is increased. Multifrequency FSK is a modulation technique that uses a different optical frequency to represent each possible combination of a predetermined number of modulation data bits. The advantage of using this modulation format is that it gives improved receiver sensitivity, but unfortunately the modulation spectral width is many times wider than the data rate. Although this is not a problem when using a fibre transmission medium, it does require a receiver bandwidth that is much wider than usual. CPFSK modulation can be considered as being narrow deviation FSK because the frequency separation between the two modulation frequencies is chosen to be small enough to produce a specific phase shift during a modulation bit period. The reason for this is that when the appropriate IF demodulation is used at the receiver, this phase information can be utilised to improve the system performance, and because of the small frequency separations used the receiver bandwidth requirement is less than that for wide-deviation FSK. A disadvantage of CPFSK systems is that they are more sensitive to laser phase noise than are wide-deviation FSK systems. The ideal CPFSK system would use a frequency separation equal to half the data rate in hertz [this is known as *minimal shift-keying* (MSK)], because this gives the most compact IF spectrum and a performance similar to that given by a PSK system can be achieved.

Digitally modulating the polarisation of the optical carrier between orthogonal states is another type of modulation format that could be used [4], but this offers no particular advantage over the more conventional schemes already discussed. However, the use of polarisation modulation in conjunction with the other schemes is an option that may be of use for achieving polarisation-insensitive coherent detection (see Section 6.4.1.3).

Of all the modulation options available, PSK, DPSK, CPFSK, and FSK were, at time of writing, emerging as the ones most likely to be used. The advantage of the FSK schemes is that the desired modulation can be produced by directly frequency modulating the laser, and the advantage of the PSK schemes is that, ideally, they give the best performance.

### 6.3.2. Homodyne Receivers

Comparison of the performance improvements in Table 6.2 shows that for a given modulation format a homodyne receiver is always 3 dB more sensitive than the equivalent heterodyne receiver. The reason for this is that the baseband signal power at the input to the decision element of a heterodyne receiver is half that of a homodyne receiver, but, as has been shown using the concept of image band noise [5], the shot-noise spectral density is the same in both cases. Apart from improved performance, homodyne detection has the additional advantage of using receivers with baseband bandwidths, but it has the disadvantage of requiring an optical phase-locked-loop and hence low phase noise (narrow linewidth) optical sources (see Section 6.3.5).

**6.3.2.1. Pilot Carrier Optical Phase-Locked-Loop.** A balanced pilot carrier optical phase-locked-loop receiver takes the form shown in Figure 6.3. The received signal and local oscillator fields are combined in a 1 : 1 fibre

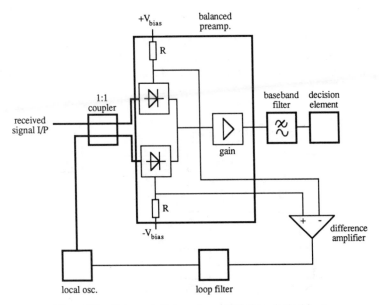

**Figure 6.3.** Balanced pilot carrier optical phase-locked-loop.

directional coupler the two outputs of which are connected to a dual photodiode preamplifier. The advantage of the balanced design is that it overcomes the problems associated with DC drift and any low frequency variations in local oscillator power that may exist (excess intensity noise). The phase-locked-loop maintains a constant 90° phase difference between the received signal and local oscillator fields, which is the desired condition for demodulating antipodal (180° modulation depth) PSK modulation. Whenever the phase difference between the local oscillator and received signal fields deviates from 90°, the difference amplifier produces a control signal that after being filtered is used to adjust the frequency of the local oscillator laser to bring it back into phase quadrature with the received signal. Unfortunately, with antipodal PSK modulation there is no carrier component present in the received modulation spectrum, so to be able to use this type of phase-locked-loop the modulation depth must be reduced to generate a pilot carrier component. However, the process of generating the carrier introduces a performance degradation (see Section 6.3.4.4).

The limitations of the pilot carrier phase-locked-loop are that it must be DC-coupled, a pilot carrier must be available for phase tracking purposes, and there is a possibility of tracking out the low-frequency content of the modulation spectrum. These limitations are overcome by the Costas phase-locked-loop design.

**6.3.2.2. Costas Optical Phase-Locked-Loop.** The balanced Costas phase-locked-loop takes the form shown in Figure 6.4. The local oscillator and the received signal fields are combined in a balanced 90° optical hybrid (techniques for realising this hybrid are discussed in Section 6.4.2.2), and as a consequence of this the signals in the I (in-phase) and Q (quadrature) arms have a quadrature phase relationship. Therefore, by multiplying the I and Q arm signals, an error signal is produced whenever the phase difference between the received signal and local oscillator fields

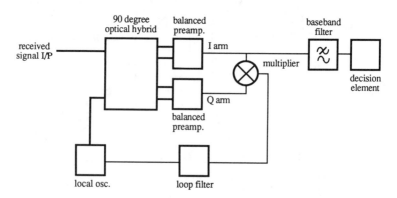

**Figure 6.4.** Balanced Costas optical phase-locked-loop.

deviates from 90°. This error signal is filtered and then used to adjust the frequency of the local oscillator laser in the same way as in the pilot carrier phase-locked-loop. When phase-locked conditions exist the base-band data waveform, which is obtained from the I arm, takes its maximum possible value.

Usually, the I and Q arm signal-to-noise ratios are independent of the signal split used, so there is no performance disadvantage associated with dividing the signal power equally between the two arms; this is the split that maximises the feedback control signal. However, for homodyne detection the dominant noise terms are situated within the phase-locked-loop, and the outcome of this is that the receiver performance is degraded by the split ratio (see Section 6.3.4.4). To minimise this degradation an unequal split in favour of the I arm should be used, but it can never be reduced to zero because the required phase tracking can be achieved only if a fraction of the received signal power is sacrificed to the Q arm. To achieve an unequal power split between the two arms of the receiver, it is necessary to use the polarisation hybrid design, which produces quadrature phase signals by using the properties of polarised light.

### 6.3.3. Heterodyne Receivers

**6.3.3.1. Phase-Locked-Loop IF Demodulation.** When heterodyne detection is used, the IF can be demodulated using a standard electrical pilot carrier phase-locked-loop design and the operating principles are similar to those for the homodyne case. The basic difference between this and the homodyne case is that an electrical voltage controlled oscillator (VCO) is phase-locked to the IF instead of the local oscillator being phase-locked to the received optical signal. The pilot carrier phase-locked-loop design limitations discussed earlier still apply and as before they can be overcome by using a Costas phase-locked-loop, but they can also be overcome by the squaring and reverse modulation control loops. However, these three control loop options are essentially different practical implementations of the same function, the squaring loop being the design usually used.

The squaring phase-locked-loop heterodyne receiver takes the form shown in Figure 6.5, but if the preamplifier bandwidth is narrow enough to prevent noise saturation of the demodulator, the IF bandpass filter can usually be omitted. The phase-locked-loop IF demodulator basically consists of a signal path and a control path, the IF signal being divided equally between them. To be able to phase-lock the VCO to the control path signal the phase modulation must be removed, and this is achieved by squaring as this produces an unmodulated carrier component, but at twice the original IF frequency. This is then mixed with the output from the electrical VCO, and whenever the phase difference between the control signal and the VCO output deviates from 90°, the error signal generated adjusts the frequency of the VCO to bring it back into phase quadrature

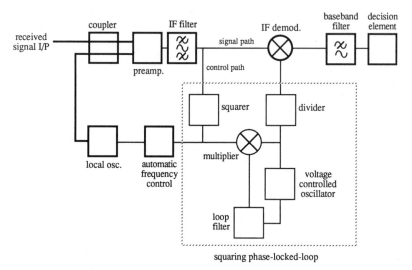

**Figure 6.5.** Heterodyne receiver with squaring phase-locked-loop IF demodulation.

with the control signal. To demodulate the signal path waveform the VCO frequency is divided by two and then used as the local oscillator drive to the electrical phase detector situated within the signal path.

Irrespective of whether the Costas, squaring or reverse modulation phase-locked-loop design is used, none of the received signal is sacrificed to achieve phase tracking as is the case for the homodyne Costas phase-locked-loop. The reason for this is that the dominant noise sources are external to the phase-locked-loop.

**6.3.3.2. Nonsynchronous IF Demodulation.** As mentioned earlier, with heterodyne detection there is the option of using nonsynchronous IF demodulation, of which there are two categories: envelope (linear or square-law) and delay demodulation. The advantages of these two types of demodulation are that there is no need for phase-locking, a performance close to that given by synchronous demodulation is in principle possible and they are usually much simpler to implement as they use less component parts. As will become clear later, they also have the advantage of being more tolerant of phase noise, which becomes an important consideration when laser phase noise is taken into account.

For standard nonsynchronous coherent systems, any value IF can usually be used provided it is high enough for the data rate being used. However, DPSK systems are an exception to this because the desired phase relationship between the two signals input to the delay IF demodulator can be achieved only if the IF is a multiple of half the data rate in hertz. The reason for this is that the demodulation delay must equal the modulation bit time, so the phase relationship between the demodulator input signals can be optimised only by adjusting the IF. An outcome of this is that the IF centre frequency stability must be of the order of 1% of the data rate in hertz if degradations in excess of 1 dB are to be avoided [6].

### 6.3.4. Performance Degradations Due to Nonideal Operating Conditions

One of the main reasons for using coherent detection is to achieve a performance that is as close to the shot-noise limit as possible, but this may be difficult to achieve in practice unless certain conditions are satisfied. The factors that are most likely to affect system performance are polarisation dispersion, chromatic dispersion, coupling ratio of the coupler, preamplifier thermal noise, local oscillator excess intensity noise, polarisation mismatch between the received signal and local oscillator fields, laser phase noise, and for synchronous detection, the phase-locked-loop design.

**6.3.4.1. Polarisation Dispersion and Chromatic Dispersion.** Although the transmission distance limit imposed by both polarisation dispersion [7] and chromatic dispersion [8] have been studied for various modulation formats and data-rates, the use of different parameters makes a direct comparison of the results difficult. However, assuming the relative difference between the polarisation and chromatic dispersion transmission distance limits are independent of both modulation format and data-rate, it can be concluded from reported FSK results [9] that the transmission distance limit for 1.5-$\mu$m systems will be determined by linear chromatic dispersion. However, if the launch polarisation matches a fibre eigenmode, or if dispersion shifted fibre or dispersion compensating receiver designs [10] are used, it is possible that the limit will then be set by polarisation dispersion.

**6.3.4.2. Polarisation Mismatch.** If the received signal and local oscillator fields have different polarisation states the optical mixing is nonideal, and the outcome of this is that the reduced amplitude of the resulting homodyne or heterodyne signal, as appropriate, degrades the receiver sensitivity. To include the effect of polarisation mismatch in the receiver sensitivity expression derived earlier it is necessary to redefine Eq. (6.1).

When the signal and local oscillator fields are arbitrarily polarised, it is necessary to resolve them into orthogonal components with respect to the same $X, Y$ reference axes. If the signal field components are denoted $e_{SX}, e_{SY}$ with phase relationship $\delta_S$ and the local oscillator components are expressed in the same way but using an L subscript in place of the S, the power incident on the photodiode is given by the sum of the $X$ and $Y$ plane powers (these two powers can be summed because they are produced by orthogonal fields). Therefore, Eq. (6.1) is replaced by

$$i_{P1} \propto \frac{\eta e\lambda}{hc} \int_t^{t+\Delta t} \left\{ \left(\sqrt{\alpha}\, e_{SX} + \sqrt{1-\alpha}\, e_{LX}\right)^2 + \left(\sqrt{\alpha}\, e_{SY} + \sqrt{1-\alpha}\, e_{LY}\right)^2 \right\} dt$$

(6.18)

$$e_{SX} = \sqrt{K_{SX}}\, E_S \cos(\omega_S t + \phi_S)$$  (6.19)

$$e_{SY} = \sqrt{K_{SY}}\, E_S \cos(\omega_S t + \delta_S + \phi_S)$$  (6.20)

$$e_{LX} = \sqrt{K_{LX}}\, E_L \cos(\omega_L t + \phi_L)$$  (6.21)

$$e_{LY} = \sqrt{K_{LY}}\, E_L \cos(\omega_L t + \delta_L + \phi_L)$$  (6.22)

$K_{SX}$, $K_{SY}$ and $K_{LX}$, $K_{LY}$ represent the fraction of $P_S$ and $P_L$ in the $X, Y$ planes, respectively.

Substituting these expressions in the earlier analysis, it can be shown that the modualted photocurrent is now equal to

$$i_{m} = \frac{\eta e\lambda}{hc} 2 f_{(t)} \sqrt{\alpha P_L(1-\alpha) P_S K_P}\, \cos(\omega_{IF} t + \phi + \phi_P)$$  (6.23)

$$K_P = K_{SX} K_{LX} + K_{SY} K_{LY} + 2\sqrt{K_{SX} K_{LX} K_{SY} K_{LY}}\, \cos(\delta_L - \delta_S)$$  (6.24)

$$\phi_P = \tan^{-1}\left\{ \frac{\sqrt{K_{SY} K_{LY}}\, \sin(\delta_L - \delta_S)}{\sqrt{K_{SX} K_{LX}} + \sqrt{K_{SY} K_{LY}}\, \cos(\delta_L - \delta_S)} \right\}$$  (6.25)

If the optical or IF phase-locked loop, as appropriate, is ideal and maintains the condition $\phi + \phi_P = 0$, or if the IF demodulator is phase insensitive, the performance degradation ($PD_{KP}$) associated with polarisation mismatch is simply

$$PD_{K_P} = -10 \log_{10}(K_P) \qquad 0 \le K_P \le 1$$  (6.26)

To illustrate how $K_P$ varies with polarisation mismatch, it has been plotted in Figure 6.6 as a function of amplitude ($\Delta_K$) and phase ($\Delta_P$) mismatch parameters for the specific case of a right-hand circularly polarised local oscillator. When the local oscillator polarisation is right-hand circular $K_{LX} = K_{LY} = 0.5$, $\delta_L = 90°$, and the mismatch parameters are given by $\Delta_K = K_{SX} - 0.5$ and $\Delta_P = 90° - \delta_S$. It is clear from Figure 6.6 that polarisation mismatch can cause large performance degradations, as there is a complete system performance fade ($K_P = 0$) whenever the signal and local oscillator are orthogonally polarised (i.e., $K_{LX} = K_{SY}$, $K_{LY} = K_{SX}$, and $\delta_L - \delta_S = 180°$).

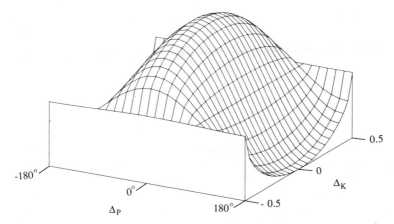

**Figure 6.6.** Effect of polarisation misalignment on the value of polarisation parameter $K_P$, and hence receiver performance, when using a standard coherent receiver. After Hodgkinson et al. [47], © 1988 IEE.

Because of this sensitivity to polarisation misalignment and the fact that the polarisation at the output of conventional single-mode fibre varies slowly and randomly with time, practical coherent systems will have to incorporate some means for achieving polarisation-insensitive detection (see Section 6.4.1).

### 6.3.4.3. Thermal Noise and Local Oscillator Excess Intensity Noise

*6.3.4.3.1. Unbalanced Receiver.* If the earlier analysis is used to derive the noise power at the input to the decision element when conditions are such that thermal noise and local oscillator excess intensity noise cannot be ignored, it will be found that

$$\langle v_n^2 \rangle = \frac{\alpha S_{IS}'}{\tau} \left\{ 1 + \frac{S_{IE}}{\alpha S_{IS}'} + \frac{S_{I\chi}}{\alpha S_{IS}'} \right\} \qquad (6.27)$$

where $S_{IE}$ is the double-sided power spectral density for the preamplifier thermal noise and $S_{I\chi}$ is the double-sided power spectral density for the local oscillator excess intensity noise. When $S_{I\chi}$ is expressed in terms of the excess noise factor, $\chi$, it is equal to [11]

$$S_{I\chi} = \alpha^2 \eta \chi S_{IS}' \qquad (6.28)$$

Should either the thermal noise or excess intensity noise prove to be frequency dependent, Eq. (6.27) is still valid provided the noise spectral density terms are taken to represent their equivalent frequency-independent value.

Substituting Eq. (6.27) for Eq. (6.10) in the earlier analysis and comparing the resulting receiver sensitivity with the shot-noise-limited value, it will be found that the performance degradation ($PD_{K_c}$) caused by power coupling ratio, thermal noise, and local oscillator excess intensity noise is given by

$$PD_{K_c} = 10 \log_{10} \left\{ \frac{1 + \alpha\eta\chi + (S_{IE}/\alpha S_{IS}')}{1 - \alpha} \right\} \qquad (6.29)$$

Because coupling ratio $\alpha$ appears in both the numerator and denominator of Eq. (6.29) there is an optimum value that minimises the performance degradation provided the other variables are specified. The reason for the existence of an optimum coupling ratio is that the signal path loss decreases with increasing coupling ratio, whereas the local oscillator path loss increases, thus reducing the effective local oscillator power (e.g., with a coupling ratio of $1:10$ the signal path loss is 0.5 dB and the local oscillator path loss is 10 dB [12]). Using differentiation techniques, it can be shown that the optimum coupling ratio ($\alpha_{opt}$) is equal to

$$\alpha_{opt} = \frac{1}{(1 + \eta\chi)} \left( \sqrt{\frac{S_{IE}}{S_{IS}'}\left(1 + \eta\chi + \frac{S_{IE}}{S_{IS}'}\right)} - \frac{S_{IE}}{S_{IS}'} \right) \qquad (6.30)$$

and that the associated minimum performance degradation ($PD_{K_{c,min}}$) is

$$PD_{K_{c,min}} = -10 \log_{10}\{1 - 2\alpha_{opt}\} \qquad (6.31)$$

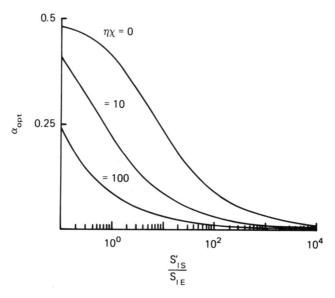

**Figure 6.7.** Optimum coupling ratio dependence on thermal noise and local oscillator excess intensity noise. After Hodgkinson [12], © 1987 IEEE.

The optimum coupling ratio lies in the range $0 < \alpha_{\text{opt}} \leq 0.5$ and decreases for increasing values of $\eta\chi$ and/or increasing shot noise : thermal noise ratio $(S'_{\text{IS}}/S_{\text{IE}})$, (Figure 6.7). Provided the effective local oscillator power is sufficiently large to dominate all other noise sources, the coupling ratio becomes progressively smaller because in this regime the performance degradation is caused purely by the signal path loss, which is given by $-10\log_{10}(1 - \alpha)$. In the absence of excess intensity noise the minimum performance degradation is $\leq 0.5$ dB provided the shot noise : thermal noise power ratio is $\leq 300$ (Figure 6.8). When excess noise is present, even larger noise power ratios than this are needed to keep the performance degradation below 0.5 dB.

It is not obvious from the results plotted in Figure 6.8 but for ASK homodyne the performance degradation does not tend towards infinity as the shot noise : thermal noise power ratio reduces because in the limit of thermal noise dominating it becomes an intensity modulated direct detection system.

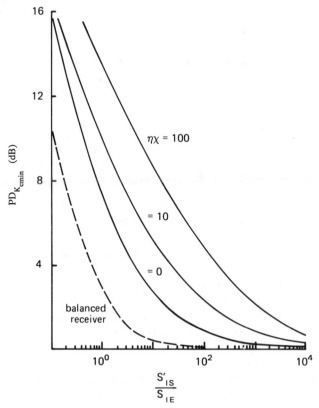

**Figure 6.8.** Performance degradation caused by thermal noise and local oscillator excess intensity noise. After Hodgkinson [12], © 1987 IEEE.

*6.3.4.3.2.  Balanced Receiver.* Using the balanced receiver model given in Figure 6.2, and knowing that the two excess intensity noise sources are correlated and cancel each other in an ideal balanced receiver [13], it can be shown that the mean-square noise voltage at the input to the decision circuit is

$$\langle v_n^2 \rangle = \frac{S'_{IS}}{\tau} \left\{ 1 + \frac{S_{IE}}{S'_{IS}} \right\} \qquad (6.32)$$

and that for the balanced receiver

$$PD_{K_{c,min}} = 10 \log_{10} \left\{ 1 + \frac{S_{IE}}{S'_{IS}} \right\} \qquad (6.33)$$

The dashed curve in Figure 6.8 shows that $PD_{K_{c,min}}$ is $\leq 0.5$ dB when the local oscillator shot noise power is 10 times the thermal noise power. Also, comparing this degradation curve with the other plots shows that the balanced receiver design makes most efficient use of the available local oscillator power as well as overcoming excess intensity noise and signal path loss caused by the optical coupling ratio.

*6.3.4.3.3.  Tuned Heterodyne Receiver.* The thermal noise characteristics of optoelectronic receiver designs are such that as the bandwidth is increased to accommodate a higher intermediate frequency, the thermal noise spectral density usually increases. Consequently, to achieve shot-noise-limited performance an associated increase in local oscillator power is required. However, if there is a constraint placed on the maximum available local oscillator power (by safety considerations, device limitations, etc.), as the IF is increased a point will be reached where the shot noise is no longer dominant and the system performance will begin to degrade. To avoid this problem it has been proposed that an optoelectronic preamplifier tuned to the IF should be used [14–16] as it should in principle be possible to design the receiver to operate at any desired IF without having to tolerate an increase in thermal noise.

**6.3.4.4.  Phase-Locked-Loop Receivers (PSK-Modulated Systems).** When either a pilot carrier or a Costas homodyne phase-locked-loop is used, the receiver performance is degraded because some of the received signal power is sacrificed for phase tracking purposes. For the former design an unmodulated pilot carrier component is needed and for the latter a fraction of the received signal power is coupled into the Q arm of the receiver. If the power used for phase tracking purposes is expressed as $K_{track}\langle P_{ST} \rangle$, where $\langle P_{ST} \rangle$ is the total mean received signal power, it follows that the useful received signal power is given by $\{1 - K_{track}\}\langle P_{ST} \rangle$

and that the performance degradation $(PD_{K_{track}})$ associated with phase tracking is

$$PD_{K_{track}} = -10 \log_{10}\{1 - K_{track}\} \qquad 0 \leq K_{track} \leq 1 \qquad (6.34)$$

Given that it is specified that the performance degradation associated with generating a reference signal for phase tracking purposes is to be less than 0.5 dB, this can be achieved only if the reference signal is less than 10% of the total received signal power.

**6.3.4.5. Phase Error Variance.** In the earlier analysis it was assumed that the phase-locked-loop demodulation was ideal, but in practice the phase-locked-loop phase error $(\phi_{e(t)})$ is a random variable as a result of both additive Gaussian noise and laser phase noise sources (see Section 6.3.5.2). When this phase error is treated as a random variable, the error probability derivation is usually simplified by taking the phase error to be a constant during a modulation bit period and applying conditional error probability theory [17].

The conditional error probability $(P_{e|\phi e})$, which by definition is the probability of an error occurring for a given $\phi_{e(t)}$ value, which will be denoted $\phi_e$, is given by

$$P_{e|\phi_e} = \frac{1}{2} \text{erfc}\left( \frac{Q \cos(\phi_e)}{\sqrt{2}} \right) \qquad (6.35)$$

which is a modified form of Eq. (6.11). The mean error probability $(\langle P_{e|\phi_e} \rangle)$ expected over many bit periods is then derived by multiplying the conditional error probability by its probability of occurrence (this is equal to the probability of $\phi_{e(t)}$ having a specific value) and then averaging the result over all $\phi_e$ values. Given the phase error is a zero-mean Gaussian random variable with variance $\sigma_\phi^2$, the mean error probability is

$$\langle P_{e|\phi_e} \rangle = \frac{1}{\sqrt{2\pi\sigma_\phi^2}} \int_{-\infty}^{\infty} \exp\left( -\frac{\phi_e^2}{2\sigma_\phi^2} \right) P_{e|\phi_e} \, d\phi_e \qquad (6.36)$$

Using numerical techniques a set of error probability curves can be derived for a range of $\sigma_\phi^2$ values, and the sketched curves in Figure 6.9 show the form they take for increasing $\sigma_\phi^2$. In this figure the horizontal axis is mean received optical power normalised with respect to the value needed to achieve a $10^{-9}$ error probability when $\sigma_\phi^2 = 0$. In other words, it represents the performance degradation $(PD_{K\phi})$ caused by phase error variance.

At low received power levels the error probability is dominated by additive noise terms and if the received power is increased, the error

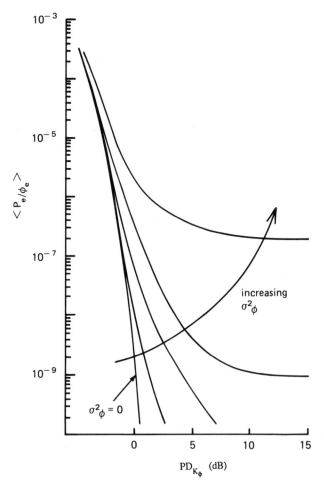

**Figure 6.9.** Influence of phase error variance on the conditional error probability curves. After Hodgkinson [12], © 1987 IEEE.

probability reduces. Normally this improvement would continue as the received power is increased further, but because phase noise effects cannot be overcome by increasing the received power, the error probability gradually makes the transition from being additive-noise-dominated to phase-noise-dominated. If the resulting saturation, which is known as the *irreducible error probability* ($\langle P_{e|\phi e} \rangle_{\mathrm{IR}}$) [17], is expressed in terms of parameters being used in this analysis, it will be found to be equal to [12]

$$\langle P_{e|\phi_e} \rangle = \mathrm{erfc}\left( \frac{\pi}{\sqrt{8\sigma_\phi^2}} \right) \qquad (6.37)$$

It is clear from the plots in Figure 6.9 that for phase-modulated systems the phase error variance must be kept to a minimum if large performance degradations are to be avoided; for example, if the condition $PD_{K\phi} \leq$ 1 dB is specified, this can be achieved only if the condition $\sigma_\phi^2 \leq 0.04$ is satisfied (the relationship between $\sigma_\phi^2$ and laser linewidth is considered in Section 6.3.5).

With the exception of delay IF demodulation, phase error has no meaning for nonsynchronous demodulation and can be ignored for this class of demodulators. However, this does not mean that their performance is unaffected by phase noise (see Section 6.3.5.3.2.).

**6.3.4.6.  Generalised Receiver Sensitivity.** All of the preceding performance degradations are independent of each other and additive; therefore, the total mean received signal power needed to achieve a given error probability can be expressed generally as

$$\langle P_{ST} \rangle = \frac{\langle P_{SL} \rangle}{K_P K_{c,\,min} K_\phi \{1 - K_{track}\}} \qquad (6.38)$$

The $K$ terms are the linear equivalents for the corresponding logarithmic $PD_K$ performance degradation terms.

### 6.3.5.  Laser Linewidth Considerations

**6.3.5.1.  Laser Phase Noise.** For an optical source to be suitable for use as either a coherent system transmitter or a receiver local oscillator it must give very narrow-linewidth single-mode operation, the required linewidth being dependent on the modulation and detection combination used. Ideally, the linewidth would be zero hertz, but in practice the phase of the laser field is perturbed by noise (Figure 6.10). In Figure 6.10 the $E$ vector represents a laser field with angular frequency $\omega$, and the two smaller vectors represent stationary zero-mean Gaussian in-phase ($n_{i(t)}$) and quadrature ($n_{q(t)}$) noise components. As the noise components vary, the tip of the $E$ vector moves around within the limits of the thin line circle boundary, but if the condition $E \gg n_{i(t)}$ exists, the amplitude fluctuations can be ignored and the circle replaced by its horizontal diameter. When this is the case the random phase ($\phi_{(t)}$) of the $E$ field is given by

$$\phi_{(t)} = \tan^{-1}\left\{\frac{n_{q(t)}}{E}\right\} \qquad (6.39)$$

However, because it has been assumed that the condition $E \gg n_{i(t)}$ is satisfied, it is reasonable to assume that $n_{q(t)}/E \ll 1$ is also satisfied, so it follows that $\tan \phi_{(t)} \approx \phi_{(t)}$ and that Eq. (6.39) simplifies to

$$\phi_{(t)} = \frac{n_{q(t)}}{E} \qquad (6.40)$$

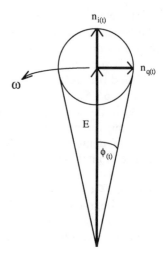

**Figure 6.10.** Phasor representation of laser instability in terms of in-phase and quadrature noise sources. After Hodgkinson [12], © 1987 IEEE.

Provided $E$ is constant, $\phi_{(t)}$ can be treated as a zero-mean Gaussian-random process and when the condition $\phi_{(t)} \approx \phi_{(t+\tau)}$ exists, it can also be assumed to be quasistationary.

The in-phase and quadrature noise terms represent the combined effect of many noise sources [18], but in the limit they represent quantum noise alone, and then the output from the laser will have a Lorentzian line-shape. The relationship between the full-width half-maximum (FWHM) Lorentzian linewidth ($\Delta f$) and the double-sided noise spectral density ($S_{\phi(f)}$) for the field phase instability is [19]

$$S_{\phi(f)} = \frac{\Delta f}{2\pi f^2} \tag{6.41}$$

It is clear from this expression that the phase noise spectral density increases with laser linewidth, and, as will now be shown, the outcome of this is that the laser sources must satisfy certain linewidth requirements for the various coherent system configurations.

**6.3.5.2. Synchronous Receivers.** As mentioned earlier, when laser phase noise and additive Gaussian noise are taken into account, the phase-locked-loop phase error ($\phi_{e(t)}$) becomes a random variable and the result-ing phase error variance ($\sigma_\phi^2$) produces a system performance degradation. It follows from this that the ideal phase-locked-loop would reduce the phase error variance to zero and give no performance degradation, but the nature of phase and additive noise makes this impossible. The reason for this is that the loop bandwidth needs to be as wide as possible to overcome phase noise effects, but as narrow as possible to overcome additive noise effects. The outcome of this is that for given noise conditions there is a particular phase-locked-loop bandwidth that minimises the phase error variance and hence the performance degradation. In practice, this degradation will almost certainly be a quoted design specification, in which case the optimum phase-locked-loop design can be taken to be that which effectively maximises the tolerable laser phase noise (linewidth). This is now considered in greater detail.

### 6.3.5.2.1. Optimum Phase-Locked-Loop Design.

Irrespective of whether homodyne or heterodyne detection is used, the various phase-locked-loop designs can all be analysed using an equivalent linearised second order model of the form shown in Figure 6.11 [20, 21]. In this model $K_D$ is the phase detector gain parameter whose value is determined by which type of loop is being analysed, and $F_{(S)}$ and $K_0/S$ are, respectively, the transfer functions for the loop filter and the optical or electrical voltage controlled oscillator as appropriate. $n'_{(t)}$ is the shot noise generated at the output of the local oscillator laser (this assumes that a shot-noise-limited balanced preamplifier is being used), $\phi_{B(t)}$ is the phase noise generated by the combined effect of the transmitter and local oscillator laser linewidths (this assumes that the heterodyne receiver phase-locked loop VCO introduces negligible phase noise), and $\phi_{e(t)}$ is the random variable loop phase error.

Using usual phase-locked-loop analysis techniques [22], it can be shown that the phase error variance ($\sigma_\phi^2$) is equal to

$$\sigma_\phi^2 = \int_{-\infty}^{\infty} S_{\phi B(f)} |1 - H_{(S)}|^2 \, df + \frac{S'_{IS}}{K_D^2} \int_{-\infty}^{\infty} |H_{(S)}|^2 \, df \qquad (6.42)$$

and that the general $K_D$ expression, for a phase modulation receiver, is

$$K_D = \frac{\eta e \lambda}{hc} \sqrt{\frac{P_L \langle P_{SL} \rangle K_{DM} K_{PLL}}{K_\phi \{1 - K_{track}\}}} \qquad (6.43)$$

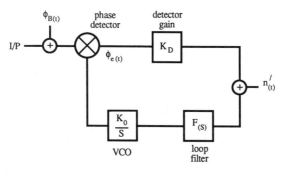

**Figure 6.11.** Linearised second-order phase-locked-loop model. After Hodgkinson [12], © 1987 IEEE.

where $S_{\phi B(f)}$ is the sum of the transmitter and local oscillator phase noise spectral densities $S_{\phi T(f)}$ and $S_{\phi L(f)}$, respectively; $S'_{IS}$, $K_H$, $K_\phi$, $K_{track}$, and $\langle P_{SL} \rangle$ are as defined earlier; and $K_{PLL}$ is a parameter whose value is determined by the type of receiver used: for pilot carrier and homodyne Costas phase-locked-loop receivers $K_{PLL} = K_{track}$; for the heterodyne Costas loop and its equivalent variants $K_{PLL} = 1$. $H_{(S)}$ is the Laplace ($S$) domain closed-loop transfer function, which for a second order loop is equal to

$$H_{(S)} = \frac{2\zeta\omega_n S + \omega_n^2}{S^2 + 2\zeta\omega_n S + \omega_n^2} \tag{6.44}$$

Substituting for $S_{\phi B(f)}$ and solving the integrals [12], Eq. (6.42) reduces to

$$\sigma_\phi^2 = \frac{\pi \Delta f_B}{2\zeta\omega_n} + \frac{S'_{IS}}{K_D^2}\left\{\zeta + \frac{1}{4\zeta}\right\}\omega_n \tag{6.45}$$

where $\Delta f_B$ is the sum of the transmitter and local oscillator FWHM linewidths (this is commonly referred to as the *beat linewidth*), $\omega_n$ is the natural loop angular frequency, and $\zeta$ is the loop damping factor.

In practice, $K_{PLL}$, $K_{track}$, and $K_\phi$ (note that the latter effectively specifies $\sigma_\phi^2$) will almost certainly be system design parameters; therefore, for system design purposes the expression that relates these parameters to the beat linewidth will probably be most useful. Using substitutions for $K_D$, $\langle P_{SL} \rangle$, and $S'_{IS}$ in Eq. (6.45) and rearranging, it can be shown that for PSK systems the tolerable normalised beat linewidth ($\Delta f'_B$) is equal to

$$\Delta f'_B = \frac{2\zeta\sigma_\phi^2}{\pi}\omega'_n - \frac{(4\zeta^2 + 1)}{2\pi Q^2}\frac{K_\phi(1 - K_{track})}{K_{PLL}}\omega'^2_n \tag{6.46}$$

$\Delta f'_B$ and $\omega'_n$ are $\Delta f_B$ and $\omega_n$ normalised with respect to the data rate in hertz. (*Note*: Negative solutions from this equation indicate that the shot noise has become dominant as a result of the loop bandwidth being too wide.)

The form of Eq. (6.46) is such that as the phase-locked-loop bandwidth is increased it passes through an optimum value ($\omega'_{opt}$), this being the bandwidth that maximises the tolerable beat linewidth ($\Delta f'_{max}$). The analytic expressions for these optima are

$$\omega'_{opt} = \frac{2\zeta Q^2 \sigma_\phi^2}{(4\zeta^2 + 1)}\frac{K_{PLL}}{K_\phi(1 - K_{track})} \tag{6.47}$$

$$\Delta f'_{max} = \frac{2\zeta^2 Q^2 \sigma_\phi^4}{\pi(4\zeta^2 + 1)}\frac{K_{PLL}}{K_\phi(1 - K_{track})} \tag{6.48}$$

*6.3.5.2.2. Maximum Tolerable Beat Linewidth.* It is clear from Eq. (6.48) that the maximum tolerable beat linewidth increases as $PD_{K\phi}$ and/or $PD_{K,\text{track}}$ are made larger, therefore, a trade-off exists between tolerable beat linewidth and system performance. In practice, it would not be expected that the performance degradation due to these two parameters would be allowed to be much in excess of 1 dB for most applications, so if the conditions $PD_{K\phi} = 1$ dB ($\sigma_\phi^2 \leq 0.04$), $\zeta = 0.7071$, $Q = 6$ ($\langle P_{e|\phi e} \rangle = 10^{-9}$), $K_{\text{PLL}} = 1$ and $K_{\text{track}} = 0$ are assumed to be reasonable parameter values, the maximum tolerable beat linewidth for the heterodyne Costas loop and its variants is 0.77% of the data rate in hertz. For the homodyne and heterodyne pilot carrier phase-locked-loop receivers and the Costas loop homodyne receiver operating with a similar total performance degradation, and with this divided equally between $PD_{K\phi}$ (0.5 dB, $\sigma_\phi^2 \leq 0.028$) and $PD_{K,\text{track}}$ (0.5 dB, $K_{\text{track}} = 0.11$), the maximum tolerable beat linewidth reduces to 0.04% of the data rate. The associated optimum loop bandwidths are 86 and 7% of the data rate in hertz, respectively. In practice, achieving the optimum phase-locked-loop design may be impossible because the loop bandwidth requirement is greater than present technology can provide; this is especially so for the heterodyne Costas loop. So, if for the sake of argument it is assumed that for all of the phase-locked-loop designs it will be possible to achieve a bandwidth of the order of 7% of the data rate for data rates up to a few gigabits per second, the heterodyne Costas loop would now be able to tolerate a beat linewidth only in the region of 0.08% of the data rate, which is an order of magnitude smaller than the optimum value and now only twice the optimum value for the other loops. If practical constraints prevent loop bandwidths equal to 7% of the data rate from being achieved, it is reasonable to assume that the beat linewidth requirements could easily be reduced by an additional order of magnitude.

**6.3.5.3. Nonsynchronous Receivers.** The effect of laser phase noise on nonsynchronous heterodyne receivers is not easily analysed as a result of the nonlinear nature of the IF demodulation, and this nonlinearity has been studied both recently [23–30] and in the past in conjunction with radio communications [31]. The basic analysis difficulty is that the demodulation nonlinearity produces Rician noise statistics, and this gives nonanalytic error probability expressions at the output of the baseband filter. However, a significant understanding of phase noise effects can be gained simply by considering the phase error variance for delay IF demodulation (DPSK and CPFSK receivers), or by comparing IF and baseband signal-to-noise ratios for envelope demodulation (ASK and FSK receivers). It must be stressed that these two approaches are only being used for illustrative purposes and although they give results similar to those obtained from a full analysis, it is not intended that they should be considered to be alternative analysis techniques.

*6.3.5.3.1. Delay Demodulation.* For a receiver using a delay IF demodulator, conditional error probability theory is still used, so the error probability curves will still be of the form given in Figure 6.9. Therefore, for similar operating conditions the performance degradation associated with a particular value of phase error variance will be similar for both PSK and DPSK receivers. For the PSK receiver the phase error variance is determined by the phase-locked-loop design, but for the DPSK receiver it is determined by the IF demodulation delay time (this equals the modulation bit-period $\tau$), and its relationship with this delay is given by [19]

$$\sigma_\phi^2 = 2\pi\,\Delta f_B \tau \qquad (6.49)$$

Assuming that a phase noise performance degradation $\leq 1$ dB is a reasonable practical value, the phase error variance must be $\leq 0.04$, so it follows from Eq. (6.49) that the maximum beat linewidth that can be tolerated is in the region of 0.65% of the data rate in hertz. (It is worth noting that this result is similar to that for the optimum Costas phase-locked-loop heterodyne receiver.)

*6.3.5.3.2. Envelope Demodulation.* Because of the mathematical similarities between linear and square-law envelope demodulation there is little to choose between them in practice, so for discussion purposes the latter will be assumed. When an IF waveform is square-law-demodulated, the amplitude of the demodulated signal is dependent on how much of the IF power spectrum lies within the passband of the IF filter. Therefore, it follows that provided the filter bandwidth is wide enough to accommodate any spectral spreading caused by laser phase noise, the receiver performance ought to be phase-noise-independent. However, this is not the case because the noise-times-noise term generated by square-law demodulation results in the noise spectral density at the output of the baseband filter being IF bandwidth dependent. Consequently, to maintain a constant baseband signal-to-noise ratio, and hence error probability, the IF signal-to-noise ratio must be increased as the IF bandwidth is widened. The increase in received signal power needed to achieve the desired IF signal-to-noise ratio represents the performance degradation caused by phase noise.

An indication of the likely magnitude of this performance degradation can be easily obtained by analysing the demodulation of an unmodulated carrier using the procedure given in Ref. 32. The resulting expression for the shot-noise-limited signal-to-noise ratio ($\gamma_{BB}$) at the output of a baseband filter with rectangular bandwidth $1/2\tau$ hertz, can be shown to be proportional to

$$\gamma_{BB} \propto \frac{K_m \gamma_{IF}}{1 + \{(4m - 1)/8K_m \gamma_{IF}\}} \qquad (6.50)$$

where $m$ is the rectangular bandwidth of the IF filter normalised to the data rate ($m \geq 1$), $\gamma_{IF}$ is the carrier-to-noise ratio at the output of the IF filter when $m = 1$, and $K_m$ is the increase in signal power needed to maintain $\gamma_{BB}$ constant as $m$ is increased.

The term enclosed by brackets in the denominator of Eq. (6.50) represents the effect of the noise-times-noise term, and for zero phase noise conditions ($K_m = 1$ and $m = 1$) and typical error probability values ($\gamma_{IF} \approx 20$ times) it is negligibly small. When the noise-times-noise term can be ignored $\gamma_{BB}$ is directly proportional to $\gamma_{IF}$, thus the square-law demodulator approximates a linear synchronous demodulator. If the IF bandwidth ($m$) is increased to accommodate the spectral broadening associated with phase noise, the noise-times-noise term increases and the received signal must be increased to maintain a constant baseband signal-to-noise ratio. This increase, which has been defined to be $K_m$, represents the associated performance degradation ($PD_{K_m}$), and this can be shown to be equal to

$$PD_{K_m} = 10 \log_{10}\left( \frac{1}{2}\left\{ 1 + \sqrt{1 + \frac{4m - 1}{2\gamma_{IF}}} \right\} \right) \qquad (6.51)$$

when it is assumed that for zero phase noise conditions Eq. (6.50) can be replaced by $\gamma_{BB} \propto \gamma_{IF}$.

Using this result, it will be found that for the phase noise performance degradation to be $\leq 1$ dB, the IF bandwidth must not exceed 14 times the data rate in hertz. Therefore, given that 95% of the power contained in a Lorentzian lineshape would be passed by an IF bandwidth equal to 12 times the FWHM linewidth, beat linewidths of the order of the data rate in hertz can be tolerated when the condition $PD_{K\phi} \leq 1$ dB is to be satisfied. This is a significant relaxation on the linewidth requirements for synchronous and DPSK receivers.

## 6.4.   ADVANCED COHERENT SYSTEM DESIGN CONCEPTS

### 6.4.1.   Polarisation-Insensitive Detection

For operational systems, the transmission fibre will almost certainly be subjected to temperature and environmental (mechanical, acoustic) variations, and for conventional single-mode fibre the resulting stress induced birefringence variation is known to cause the polarisation at the output of the fibre to slowly vary in an unpredictable manner [33, 34]. It is also now known that these variations exhibit a low-pass characteristic with a corner frequency that at worst is only of the order of a few tens of hertz [34–36]. Therefore, if practical coherent optical fibre transmission systems are to be operated over conventional fibre, the baseband signal output from the receiver must be made insensitive to these polarisation changes if large performance degradations are to be avoided (see Section 6.3.4.2). Several techniques can be used to achieve this, and they fall into one of three categories: polarisation tracking receivers, polarisation diversity receivers, and schemes using orthogonally polarised optical fields. In principle these techniques are completely general and applicable for either homodyne or heterodyne detection, but for descriptive/explanatory purposes heterodyne detection will be assumed.

**6.4.1.1. Polarisation Tracking Receivers.** Once it was known that the polarisation at the output of conventional single-mode fibre changed only slowly with time, it appeared feasible that these changes could be tracked out at the receiver by inserting a suitable number of polarisation controllers in either the local oscillator or the received signal path and incorporating them in a feedback control loop. The polarisation controllers that have emerged as being most suited to automatic control of the polarisation state in a fibre have been based on either piezoelectric fibre squeezers, Faraday rotators, electrooptic waveguide devices, or mechanical fibre cranks.

*6.4.1.1.1. Polarisation Controllers*

(*a*) *Piezoelectric Fibre Squeezers.* These devices compress the fibre as the applied drive voltage is increased, and this induces linear birefringence into the fibre, the resulting birefringent axes being parallel and perpendicular to the applied stress axis. If two squeezers are aligned such that their stress axes are at 45° with respect to each other, it is possible with suitable drive voltages to convert from any input polarisation state to any desired output state. An advantage of the fibre squeezer is that it has low insertion loss, but its speed of operation is limited to kilohertz rates and the continual stressing may cause fibre fatigue.

(*b*) *Faraday Rotators.* These devices apply an axial magnetic field to the fibre that induces circular birefringence into the fibre with a value determined by the Verdet constant and the magnitude of the drive current applied to the electromagnet. Because the Verdet constant for fibre is very small, these devices are practicable only if many turns of fibre are wound through the centre of the wire coils forming the electromagnet. Two Faraday rotators interconnected by a length of linearly birefringent fibre that introduces a 90° phase shift between the two orthogonal field components can convert any input state of polarisation to any linear output state. The advantages of the Faraday rotator are low insertion loss and no moving parts; hence no fibre fatigue. However, the response time is limited by how rapidly the magnetic field can be changed, and the high drive currents may restrict their range of application.

(c) *Electrooptic Waveguide Devices.* Various lithium niobate (LiNbO$_3$) devices exist that can convert an arbitrary polarisation state to any desired state and these have the advantages of having no moving parts and high-speed operation. However, the lowest drive voltage devices are highly wavelength-selective and at the time of writing they also had a significant insertion loss.

(d) *Mechanical Fibre Cranks.* These are fibre loops or U-shaped cranks of such a diameter that the stress-induced linear birefringence gives fibre half- and quarter-wave plate equivalents. Polarisation control is achieved by using a suitable motor drive arrangement to rotate the fibre crank, but this has the disadvantages of having moving parts and a very slow response time.

### 6.4.1.1.2. Achieving Continuous Polarisation Tracking.

Apart from the rotating fibre crank, none of the controllers mentioned above are capable of directly giving continuous tracking when incorporated in a control feedback loop. The reason for this is that they are effectively finite-range controllers, so if the received state of polarisation drifts outside the range of the control loop, the feedback signal limits and prevents further control. However, by using extra controllers in the feedback loop it is possible to achieve continuous tracking by having a control algorithm whereby controllers that have reached their limit are progressively reset [37].

If both the signal and local oscillator polarisation states are time-varying and endless tracking is required from finite-range controlling elements, the minimum number of controllers needed is four [37], but under some conditions this can be reduced further. Three can be used if the local oscillator polarisation is both stable and an eigenmode of the second controller [38], or two can be used if some passive polarisation selective optics are incorporated into the receiver design along with a second optoelectronic receiver and associated IF processing electronics [39].

### 6.4.1.2. Polarisation Diversity Receivers.

To avoid using polarisation tracking receivers, but still retain polarisation-independent performance, the alternative approach of using polarisation diversity detection can be considered. Both unbalanced and balanced receiver designs are possible; however, the balanced one has all the advantages discussed earlier. The principle of operation, assuming a balanced design (Figure 6.12), is that the received signal and local oscillator fields are combined in a standard 1 : 1 fibre directional coupler, each output of which is connected to some form of polarisation-selective coupler. (Preferably this would be a fibre device.) The outputs from these couplers are then paired according to their state of polarisation, and each pair is connected to a standard balanced optoelectronic preamplifier and associated IF demodulation stage. The baseband signal is then obtained by summing the demodulator outputs.

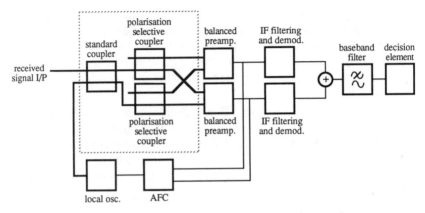

**Figure 6.12.** Balanced polarisation diversity heterodyne receiver.

Using the polarisation notation defined earlier, the shot-noise-limited baseband signal-to-noise ratio for an ideal polarisation diversity receiver ($\gamma_D$), expressed in terms of the equivalent signal-to-noise ratio for standard polarisation aligned heterodyne detection ($\gamma_S$), can be shown to be given by

$$\gamma_D = \gamma_S \frac{(K_{SX}K_{LX} + K_{SY}K_{LY})^2}{(K_{SX}K_{LX}^2 + K_{SY}K_{LY}^2)} \qquad (6.52)$$

from which it can be concluded that polarisation-insensitive performance is achieved only when the local oscillator polarisation is set such that its output power is divided equally between the two receiver arms ($K_{LX} = K_{LY}$ and by definition $K_{SX} + K_{SY} = 1$). In practice, it may be necessary to use polarisation-maintaining fibre in the local oscillator path to ensure this operating condition remains stable once set.

With the appropriate local oscillator polarisation it follows from Eq. (6.52) that an ideal polarisation diversity receiver has a performance potential equal to that of a polarisation tracking receiver ($\gamma_D = \gamma_S$), but practical limitations may prevent this from being achieved. When synchronous IF demodulation is used, Eq. (6.52) is valid only if the gain of each arm of the polarisation diversity receiver is continuously adjusted (weighted) in proportion to the modulus of the signal in that particular arm [40]; otherwise the performance will fluctuate by 3 dB. This weighting could be implemented by squaring the synchronously demodulated signal, but practical limitations will probably make gain control the better choice. When square-law or delay IF demodulation is used, this weighting occurs naturally, but because of the limitations of practical demodulation devices, the presence of noise-times-noise terms and the dual-arm nature of the receiver, it is expected that DPSK [41] systems will suffer a 0.4-dB performance degradation. For similar reasons this is also expected to be the case for FSK systems, whereas for ASK systems the degradation could be as large as 2 dB [42].

Although polarisation diversity detection removes the need for polarisation controlling elements and associated processing electronics, it has the disadvantage of requiring two optoelectronic preamplifiers and associated IF processing electronics.

**6.4.1.3. Schemes Using Orthogonally Polarised Optical Fields.** It is not immediately obvious, but by considering Poincare's sphere it becomes clear that although orthogonally polarised fields may undergo absolute polarisation changes when propagating through a single-mode fibre transmission medium, their orthogonality is unaffected. Therefore, if it is arranged for the same modulation information to be transmitted in the two orthogonal polarisation states, a suitably designed receiver can extract and process this information to give polarisation-insensitive detection. In practice, there will eventually be some deterioration of orthogonality between the signals at the receiver if the transmission medium exhibits either polarisation selective loss or polarisation dispersion [43], but this is not expected to be a problem for typical transmission distances. However, should this prove not to be the case, polarisation-insensitive detection can still be achieved using conventionally modulated signals provided the local oscillator power is distributed between two orthogonal polarisation states.

*6.4.1.3.1. Dual-Frequency Orthogonally Polarised Sources.* Given that a continuous-wave optical signal constituting two different-frequency orthogonally polarised optical fields exists, polarisation-insensitive detection can be achieved by using this as either the transmitter source or local oscillator source as required. To generate a dual-frequency orthogonally polarised optical field, the outputs from two separate sources could be combined, or the output from a single source could be divided into two, with one half then being frequency-shifted prior to being recombined with the other half. If the dual-frequency source is used at the transmitter and it is externally modulated using conventional modulators, the waveforms at the input to the receiver for ASK, FSK, and DPSK/PSK modulations, for the case of linear received polarisation states and a 1–0 data sequence, are as given in Figure 6.13 [44].

Irrespective of where the dual-frequency source is located, the optical mixing process at the receiver now generates two intermediate frequencies (for some polarisation combinations only one of these will exist), and these are processed using a standard dual-filter FSK receiver (Figure 6.14), the filters now being used to separate out the information contained in the two orthogonal polarisation states. After demodulating the two IF waveforms, the resulting baseband signals are summed. Analysing this detec-

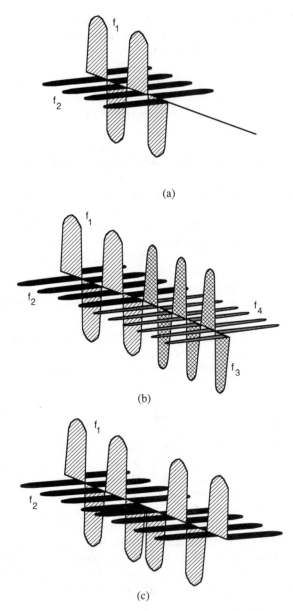

(a)

(b)

(c)

**Figure 6.13.** Modulated waveforms associated with using a dual-frequency orthogonally polarised transmitter source: (*a*) ASK; (*b*) FSK; (*c*) PSK. After Hodgkinson and Cook [44], © 1988 John Wiley & Sons, Inc.

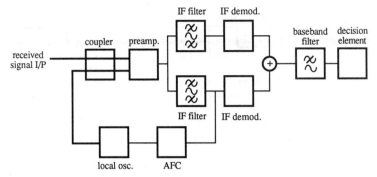

**Figure 6.14.** Dual-filter FSK heterodyne receiver.

tion scheme in detail using the polarisation notation defined earlier, it can be shown that the resulting baseband signal-to-noise power ratio ($\gamma_0$), expressed in terms of that given by the equivalent ideal standard heterodyne receiver ($\gamma_S$), is given by

$$\gamma_0 = \gamma_S \frac{(K_{P1} + K_{P2})}{2} \tag{6.53}$$

where

$$K_{P1} = K_{SX}K_{LX} + K_{SY}K_{LY} + 2\sqrt{K_{SX}K_{LX}K_{SY}K_{LY}} \cos(\delta_L - \delta_S) \tag{6.54}$$

$$K_{P2} = K_{SY}K_{LX} + K_{SX}K_{LY} - 2\sqrt{K_{SX}K_{LX}K_{SY}K_{LY}} \cos(\delta_L - \delta_S) \tag{6.55}$$

The derivation of Eq. (6.53) ignores the noise-times-noise components generated by nonsynchronous IF demodulation, and for the case of synchronous IF demodulation it assumes that the gain of each receiver arm is continuously adjusted in proportion to the modulus of the signal in that particular arm.

Summing Eqs. (6.54) and (6.55) shows that for all modulation formats, the baseband signal-to-noise ratio is polarisation-independent ($K_{P1} + K_{P2} = 1$) but 3 dB worse than for ideal, standard polarisation aligned heterodyne detection.

The advantage of using a dual-frequency orthogonally polarised source to achieve polarisation-insensitive detection is that it removes the need for the complete extra receiver needed when using a polarisation diversity receiver. Furthermore, if the dual-frequency source is used at the transmitter, the receiver complexity is kept to a minimum. For the special case of ASK and single filter detection FSK systems, a standard square-law IF demodulation heterodyne receiver can be used (Figure 6.15) provided the two orthogonally polarised optical carriers are separated in frequency by

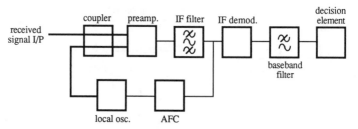

**Figure 6.15.** Single-filter FSK and ASK heterodyne receiver.

at least three times the data rate in hertz. The disadvantages of using a dual-frequency source are that the receiver performance is 3 dB worse than that given by a polarisation tracking receiver and wider IF bandwidths are needed; typically this is twice the data rate plus the frequency separation between the two orthogonal optical carriers. However, for some applications (e.g., medium-data-rate, local loop networks) this wider bandwidth requirement may not be of concern.

***6.4.1.3.2. Orthogonally Polarised FSK.*** If the orthogonal optical fields output from a dual-frequency orthogonally polarised transmitter source are separately ASK-modulated using antiphase modulation waveforms, the resulting waveform is identical to that given when the two tones of conventional FSK modulation are transmitted in orthogonal polarisation states (Figure 6.16). It follows then that orthogonally polarised FSK modulation is a special case of using a dual-frequency orthogonally polarised source and that polarisation-insensitive detection can be achieved using the receiver given in Figure 6.14.

When this type of modulation is used, there is only one IF generated at the receiver and depending on the received state of polarisation, it lies somewhere between the limits of being either an ASK or an FSK modulated waveform. As a result of this it is expected that with a fixed decision threshold there will be a small polarisation dependent penalty when using

**Figure 6.16.** Modulated waveform associated with orthogonally polarised FSK modulation. After Hodgkinson and Cook [44], © 1988 John Wiley & Sons, Inc.

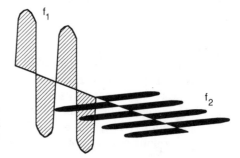

nonsynchronous IF demodulation because of the different optimum decision thresholds for these two types of modulation. However, automatic decision threshold tracking should remove this additional penalty.

*6.4.1.3.3. Polarisation Scrambling.* This is a technique whereby polarisation-insensitive detection can be achieved using a standard heterodyne receiver design given that the increased IF bandwidth requirement for scrambled systems is achievable. The principle of operation is that the transmitter or local oscillator polarisation is varied (scrambled) cyclically between orthogonal states at a rate that would ideally equal the data rate, but in practice it may be necessary to use higher rates to overcome practical device limitations. The effect of the scrambling is that irrespective of any applied modulation, both the amplitude and the phase of the IF generated within the receiver are caused to vary in such a way that after demodulation and averaging over a modulation bit period (this is automatically provided by the baseband filter), a constant-amplitude signal is obtained irrespective of the received state of polarisation. However, although the performance is polarisation-insensitive, it is 3 dB worse than that given by the equivalent polarisation tracking receiver.

In principle any waveshape can be used to scramble the polarisation of a modulated field, but there is an optimum shape that minimises the spectral spread of the IF waveform. The optimum waveform to use is the one that causes the polarisation to traverse around a major circle on Poincare's sphere at a constant angular velocity as this generates only one frequency-shifted replica of the original modulated field, in contrast to the multiplicity generated by square-wave scrambling. Although not immediately obvious, the cyclic polarisation change associated with this type of motion around Poincare's sphere is in fact identical to the change produced when two different-frequency orthogonally polarised fields are combined, the rate at which the circle is traversed being determined by their frequency difference. It follows then that ideal polarisation scrambling is equivalent to using a dual-frequency orthogonally polarised source. Therefore, depending on the modulation format used, an optical field that has been scrambled using the ideal waveshape should resolve into two different frequency orthogonal polarisation components, and give the waveforms in Figure 6.13.

### 6.4.2. Phase Diversity Detection

A significant advantage of homodyne detection over heterodyne detection is that it uses baseband receiver bandwidths (a standard untuned heterodyne receiver requires at the very least three times this value), but it has the major disadvantage of needing an optical phase-locked-loop to make the local oscillator laser track any phase instability on the received optical carrier. However, at the time of writing there were significant practical problems to be overcome before the optical phase-locked loop would be practicable, and for this reason phase diversity detection [which is also known as *multiport detection* and *in-phase* and *quadrature* (I & Q) detection] appeared to be an attractive alternative because it removed the need for the optical phase-locked-loop without sacrificing the reduced bandwidth advantage.

**6.4.2.1. Achieving Phase Diversity.** The phase diversity receiver is effectively a dual-arm heterodyne receiver that uses a local oscillator which is operated at nominally the same frequency as the received optical field; in other words, a zero-hertz IF is used. At the receiver a 90° optical hybrid is used for optical mixing purposes as this generates two equal-magnitude quadrature-phase photocurrents. The instantaneous values for these I and Q arm currents are proportional to $\cos(\phi_{(t)})$ and $\sin(\phi_{(t)})$, respectively, where $\phi_{(t)}$ is the phase instability between the received signal and local oscillator fields. It follows that by squaring (i.e., using the appropriate IF demodulator) and summing these two waveforms, a baseband signal that is independent of phase instability $\phi_{(t)}$ will be produced, as is the case for an optical phase-locked-loop homodyne receiver. Provided the appropriate IF demodulator is used, phase diversity detection can be applied to ASK, DPSK, and FSK systems [45]. The disadvantages of phase diversity detection when compared with homodyne detection are that it requires two complete optoelectronic preamplifiers and associated IF processing electronics, and as a result of using quadrature phase signals it gives a performance that is at least 3 dB less sensitive. However, because a zero-hertz IF is used the receiver bandwidth remains the same as for a homodyne receiver.

At first sight it appears that phase diversity detection ought to be insensitive to laser phase noise, but this is not so. The reason for this is that any effects that are caused by phase instability between the received signal and local oscillator fields are removed after IF demodulation, whereas the sensitivity to laser phase noise is determined at the input to the IF demodulator. In other words, although the phase diversity receiver operates without the need for an optical phase-locked-loop, its sensitivity to laser phase noise is expected to be similar to that for the equivalent standard heterodyne receiver.

**6.4.2.2. Optical Hybrid Designs.** The 90° optical hybrid can be realised using an ideal 4 × 4 fibre directional coupler or by using a combination of four 2 × 2 fibre directional couplers. (Both of these are multiport coupler hybrid designs.) However, the former is difficult to fabricate and the latter must be either a very stable construction or use active feedback control to stabilise relative optical path delays on a scale much shorter than the wavelength of the optical field [45]. To overcome these limitations a 120° optical hybrid (3 × 3 coupler) can be used, but there is a loss of perfor-

mance and three optoelectronic preamplifiers and associated IF processing electronics are needed. An alternative approach, which does not
introduce any performance degradation and has the advantage of being a
dual-arm design, is to use the hybrid design, which generates quadrature
phase signals by utilising the properties of polarised light.

The phase diversity receiver design which uses a balanced optical
polarisation hybrid, is similar to the polarisation diversity receiver shown
in Figure 6.12; the dashed box represents the components used to construct the polarisation hybrid. The reason for the lack of component
differences within the receiver, apart from the bandwidth of the IF filter
being different, is that the type of diversity achieved is determined purely
by the received signal and local oscillator polarisation settings. For phase
diversity a circularly polarised local oscillator is used with a linearly
polarised signal oriented such that the received power is divided equally
between the two arms of the receiver, whereas for polarisation diversity it
is necessary only for the local oscillator polarisation to be set such that its
output power is divided equally between the two arms of the receiver.

An alternative polarisation hybrid design is to replace the polarisation
couplers with linear, but orthogonally aligned, polarisers and use unbalanced receivers to detect the resulting optical signals [46]. However,
because the linear polarisers block half of the received signal power, there
is a 3-dB performance degradation when compared with the previous
design.

**6.4.2.3. Sensitivity to Polarisation Mismatch.** The sensitivity of phase
diversity detection to polarisation mismatch is dependent on the type of
hybrid used. For receivers using the multiport coupler hybrid, the performance degradation caused by polarisation mismatch is identical to that for
both standard homodyne and heterodyne detection, so Eq. (6.26) and the
contours plotted in Figure 6.6 are still valid. For receivers using
the polarisation hybrid with a stable circularly polarised local oscillator,
the performance degradation $(PD'_{K_P})$ can be shown to be given by [47]

$$PD'_{K_P} = -10\log_{10}\left\{\tfrac{1}{2}\left(1 + \sqrt{1 - 4\Delta_K^2}\,\cos(\Delta_P)\right)\right\} \qquad (6.56)$$

which is plotted in Figure 6.17 (the polarisation offset parameters $\Delta_K$ and
$\Delta_P$ are as defined earlier in Section 6.3.4.2). The difference between these
contours and those in Figure 6.6 is such that a polarisation hybrid is
approximately an order of magnitude more sensitive to polarisation mismatch than the multiport coupler hybrid.

Assuming the availability of polarisation-maintaining fibre will enable a
stable local oscillator polarisation to be produced, it follows that for
practical phase diversity receivers it will be necessary to automatically
control the received state of polarisation to maintain the desired linear

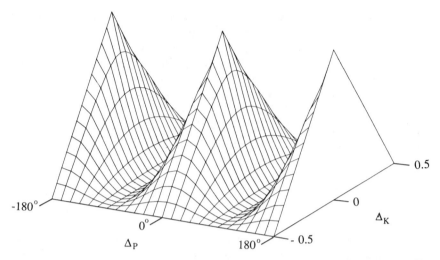

**Figure 6.17.** Effect of polarisation misalignment on the value of polarisation parameter $K_P$, and hence receiver performance, when using a 90° polarisation hybrid phase diversity receiver. After Hodgkinson et al. [47], © 1988 IEE.

state. For receiver designs using the polarisation hybrid, the polarisation control will have to be to a high degree of accuracy in both phase and amplitude. However, one advantage of this type of receiver over standard and multiport coupler designs is that it is more suited to polarisation control because when the local oscillator polarisation is stable and known, the I and Q arm signals define the input polarisation exactly. This makes detecting polarisation changes a simpler process than for the other types of receiver as it is simply a case of monitoring the I and Q arm signals to detect any amplitude difference (subtract I and Q arm signals) and/or deviation from phase quadrature (multiply I and Q arm signals) the resulting signals being used to drive polarisation controllers [47–49]. Alternatively, to avoid the need for polarisation controllers, polarisation diversity detection can be achieved by splitting the received signal into its orthogonal polarisation components and then using a phase diversity receiver to detect each of these: Figure 6.18 shows such a receiver design based on unbalanced 90° polarisation hybrids.

The main disadvantage of phase diversity detection is that two or more complete heterodyne receivers are needed depending on which implementation is used, and this may inhibit their use where cost is of prime importance. In principle, the need for extra receivers can be overcome by modifying a standard heterodyne receiver to operate with a zero-hertz IF and using this in conjunction with cyclically varying the phase of either the transmitter or local oscillator laser at a rate that exceeds the data rate [50]. (This is the phase domain equivalent of polarisation scrambling.) However,

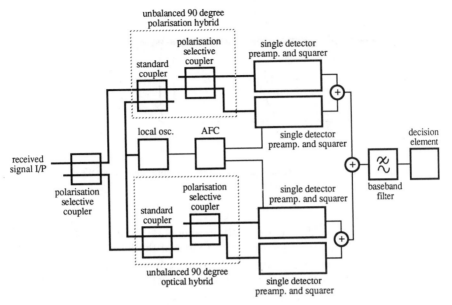

**Figure 6.18.**   Heterodyne receiver giving both phase and polarisation diversity detection.

it is expected that some reduction in potential performance may have to be tolerated and that for some modulation schemes the tolerable laser linewidths may be narrower than usual. Also, it is expected that the receiver bandwidth will have to be increased, so this technique may not offer much advantage, if any, over standard heterodyne detection.

## 6.5.   EXPERIMENTAL COHERENT OPTICAL FIBRE TRANSMISSION SYSTEM STUDIES

### 6.5.1.   Semiconductor Lasers Suitable for Practical Coherent Detection

Achieving the desired source spectral purity was a major factor in determining whether semiconductor lasers would be suitable for use in coherent optical fibre transmission systems, but as device research and development has progressed, two laser structures have emerged as being particularly suitable for use in such systems.

**6.5.1.1.   External-Cavity Semiconductor Laser.** The use of frequency-selective optical feedback to achieve single-mode narrow-linewidth operation from conventional multi-longitudinal-mode 1.5-$\mu$m semiconductor laser structures, has been achieved using the external-cavity configuration shown in Figure 6.19 [51]. The external-cavity is formed between the

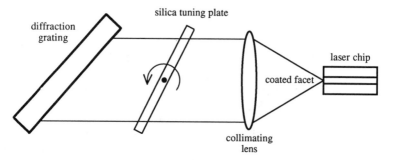

**Figure 6.19.** External-cavity semiconductor laser configuration. After Wyatt et al. [51], © 1985 British Telecommunications plc.

frequency selective diffraction grating, which is set at a high angle of incidence, and the antireflection coated facet of the laser chip, the lens in the cavity being used to collimate the output from the laser. The wavelength selectivity of the grating, which gives a reflection bandpass characteristic of the order of 10 GHz, allows the desired lasing wavelength to be selected by rotating the grating about its vertical axis, and operation well away from the gain peak of the active material is possible. Another feature of the external-cavity laser is that the frequency can be tuned by changing the length of the optical cavity, hence the provision of the silica tuning plate; changing the angle of the plate gives fine-tuning of the lasing frequency, and for heterodyne receivers this gives an optical source that can be easily incorporated in an automatic frequency tracking control AFC loop. With output powers and cavity lengths of the order of a few milliwatts and 10 cm, respectively, the external-cavity semiconductor laser structure has produced kilohertz-order linewidths, making this a suitable source for both synchronous and nonsynchronous coherent systems. This is especially so now that a miniaturised packaged version has been reported [52].

**6.5.1.2. Distributed Feedback Lasers.** Early distributed feedback (DFB) lasers were not suitable coherent optical system sources because of their poor linewidth, but they have now been developed to the stage where values of the order of a few tens of megahertz are available, and submegahertz values have been observed from research devices using the separate-confinement heterostructure quantum well design [53]. A major advantage associated with the DFB laser is that it can be directly FSK-modulated, which removes the need for an external modulator, but when direct modulation is used the refractive index change caused by both temperature and carrier density changes produces a nonlinear frequency modulation characteristic. Another disadvantage of the DFB laser is that it is very sensitive to optical reflection back into the lasing cavity. However, it is now well known that the nonlinear modulation characteristic can be overcome by using either passive equalisation of the modulation signal .[54], Manchester [55], bipolar or differential encoding techniques [56–58], multicontact DFB laser structures [also known as *distributed Bragg reflector* (DBR) lasers] [59] or feedback equalisation [60, 61]. It is also known that the sensitivity to reflections is reduced to an acceptable level if the optical isolation between the laser and the transmission fibre is of the order of 50 dB. The DFB laser was, at the time of writing, best suited to nonsynchronous FSK coherent detection systems, but DFB lasers with external cavities had been used in DPSK experiments. Consequently, it is expected that further device improvements will result in the non-external-cavity DFB laser becoming a suitable source for high-data-rate DPSK coherent systems.

### 6.5.2. Homodyne Systems

**6.5.2.1. Optical Phase-Locked-Loop System.** At the time of writing this section, homodyne system performance measurements had not been reported for systems using only semiconductor lasers, those that had been reported were for systems using gas lasers. However, a semiconductor laser pilot carrier optical phase-locked-loop suitable for use as a homodyne receiver had been demonstrated using the experimental arrangement in Figure 6.20 [62].

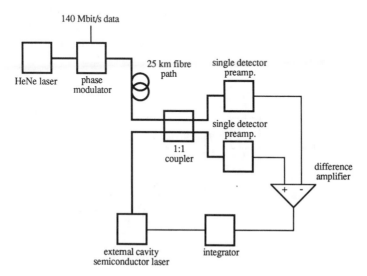

**Figure 6.20.** Optical phase-locked-loop using an external-cavity semiconductor laser. After Malyon et al. [62], © 1986 IEE.

The output from the 1.5-$\mu$m HeNe laser was phase modulated at 140 Mbit/s using an external LiNbO$_3$ modulator, and a 144° modulation depth was used so that a carrier component was present. After transmission through 25 km of fibre, the modulated signal was detected by a balanced homodyne receiver constructed from two single-detector preamplifiers. To control the local oscillator frequency the outputs from these two preamplifiers were combined in a difference amplifier, and after being integrated the resulting output was used to drive a phase modulator situated in the external-cavity of the local oscillator laser.

This experiment demonstrated that a semiconductor laser homodyne phase-locked-loop is capable of tracking the phase of the received signal at input powers typical of the values required by Gbit/s systems. Also, it highlighted that local oscillator power reflected back down the transmission medium can result in Rayleigh backscatter being the dominant cause of loop phase error, rather than the Lorentzian linewidth of the semiconductor laser.

Although the interest in homodyne detection was still relatively low-key at the time of writing, mainly as a result of the significant implementation problems associated with practical optical phase-locked-loops, an optical receiver design suitable for homodyne detection had been demonstrated to be capable of successfully phase-locking two external-cavity semiconductor lasers [63].

**6.5.2.2. Brillouin Amplifier System.** Homodyne detection without the need for an optical phase-locked-loop can in principle be achieved by selectively amplifying a residual carrier component, and then recombining this with the modulation sidebands prior to photodetection [64] (Figure 6.21). To achieve the selective gain, Brillouin amplification has been proposed, and it has been shown that a gain of 26 dB can be achieved [65].

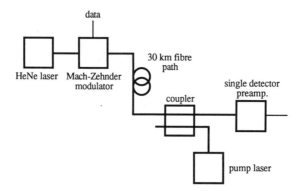

**Figure 6.21.** Homodyne receiver using Brillouin amplification of the pilot carrier. After Smith [64], © 1987 IEEE.

However, at the time of writing the bit-error-rate performance given by a homodyne receiver using a Brillouin amplified carrier had not been assessed.

### 6.5.3. Synchronous Heterodyne Systems

Because of their stringent laser phase noise requirement, synchronous heterodyne systems were not considered to be viable in the shorter term, and this resulted in there being very little experimental work prior to mid-1989. In fact, at the time of writing the PSK heterodyne system experiments reported in Refs. 66–68 were the only experiments to have incorporated a 1.5-$\mu$m semiconductor laser.

For the 140-Mbit/s experiment reported in Ref. 66, the output from a HeNe laser was PSK-modulated using a LiNbO$_3$ modulator. The resulting signal was transmitted over 109 km of conventional single-mode fibre prior to being detected by an external-cavity semiconductor laser local oscillator heterodyne receiver of the form shown in Figure 6.5. The measured receiver performance was 4 dB worse than the ideal value, but at least 2 dB of this was caused by receiver thermal noise. Despite this 4-dB degradation, the measured performance was 14 dB better than could be achieved by the best direct detection systems at that time.

The experiment reported in Ref. 67, was similar to the one just described except that it was operated back to back, the data rate was 565 Mbit/s, the receiver was a polarisation tracking balanced preamplifier design and the IF demodulator was a Costas phase-locked-loop. The measured receiver performance was 8 dB worse than the ideal value, but at least 3 dB of this was caused by receiver thermal noise.

The technique of squaring the IF, dividing the result by 2, and using this to directly demodulate the IF has recently been studied in a 560-Mbit/s experiment using external-fibre-cavity DFB lasers [68]. With an IF beat linewidth of 2 MHz, the measured performance was approximately 7 dB worse than the shot-noise-limited value, but 1 dB better than the performance obtained with the receiver configured to detect DPSK modulation. Whether this technique has the potential of giving a performance similar to that for a phase-locked-loop receiver is not clear, but it is expected that there will be some degradation due to the carrier-to-noise ratio at the output of the dividing circuit not being as good as that at the output of a phase-locked-loop.

### 6.5.4. Nonsynchronous Heterodyne Systems

#### 6.5.4.1. Standard System Experiments

*6.5.4.1.1. DPSK Systems [6, 66, 69–75].* The component parts used for these DPSK experiments have all been similar to those used in the synchronous PSK experiments except that an electrical mixer delay IF demodulator was used as opposed to a phase-locked-loop. At the time of writing the highest data rate reported was 2 Gbit/s [70], and the longest unrepeated transmission distance was 260 km for a system operating at 400 Mbit/s [74]. In both of these experiments the transmitter and local oscillator sources were external-cavity semiconductor lasers, but results have been reported for 1.2 Gbit/s systems using monolithic external-cavity DFB lasers [71] and short-fibre external-cavity DFB lasers [72]. A 400-Mbit/s system using external optical feedback DBR lasers has also been reported [69].

An important aspect which has emerged from these experiments is that in all cases the performance with respect to the shot-noise limit has degraded with increasing data rate, with values of the order of 10 dB having been measured for the Gbit/s rates. This dependence on data rate has been attributed to nonuniform frequency response and bandwidth limitations introduced by the phase modulator, its associated drive electronics and the receiver. Despite these degradations, it has been shown that for data rates $\leq 2$ Gbit/s, long transmission path lengths can be used without incurring any additional penalty, and they show that laser phase noise effects are negligible when external-cavity semiconductor lasers are used (beat linewidth $\approx 40$ kHz). This was not the case for the 1.2-Gbit/s experiment using monolithic external-cavity DFB lasers, because the increased beat linewidth ($\approx 3.5$ MHz) degraded performance by approximately 1 dB.

A detailed study of the effect of laser phase noise on the performance of a 400-Mbit/s DPSK receiver has shown that practical receivers require beat linewidths $\leq 0.3\%$ for the associated performance degradation to be $\leq 1$ dB [75]. This result is slightly worse than theory predicts, and it is thought that this is the outcome of practical design limitations [29]. Other work carried out at 147 Mbit/s has shown that the parameters most likely to degrade system performance are the IF not being a multiple of half the data rate in hertz, IF centre frequency instability, reduced modulation depth, and IF filter design [6].

***6.5.4.1.2. Wide Deviation FSK Systems [55, 57, 59–61, 76–80].*** To assess the performance of wide deviation FSK systems using DFB transmitter and local oscillator lasers, the receiver configurations given in Figures 6.14 and 6.15 have been used with square-law IF demodulation. The dual filter design has been used when the preamplifier bandwidth has been wide enough to pass both modulation tones, and the single-filter design has been used when it has only been wide enough to pass one of the two modulation tones. Irrespective of which design was used, the FSK modulation was produced by directly current-modulating the DFB transmitter laser. To compensate for the nonideal frequency modulation characteristic of the laser, the various techniques referred to earlier have been used.

At the time of writing, the highest data rate reported was 600 Mbit/s (single-filter receiver) [80], and the longest unrepeated transmission distance was 301 km for a system operating at 34 Mbit/s with a tone separation of 500 MHz (dual-filter receiver) [77]. System measurements have shown that all of these experiments have given a performance 4–10 dB away from the shot-noise limit; the reasons for this are attributed to receiver thermal noise, nonideal modulation characteristic, practical receiver design limitations and for those experiments using unbalanced receiver designs, local oscillator excess intensity noise. The main outcome from these experiments is that they have shown direct FSK modulation of the laser to be practicable and that linewidths of the order of the data rate can be tolerated (the beat linewidth for the 34-Mbit/s experiment was 20 MHz). They have also shown that long transmission path lengths have negligible effect on the performance of medium-data-rate systems.

Because wide-deviation FSK systems require wide-bandwidth receivers (the minimum value that can be used is equal to tone separation plus the data rate in hertz), high-data-rate systems will almost certainly be restricted to using single-filter designs. However, because this type of receiver effectively wastes half of the received signal power, a better solution for overcoming limited receiver bandwidth appears to be the use of CPFSK modulation.

***6.5.4.1.3. CPFSK Systems [78, 81–87].*** The reported CPFSK experiments have all used a delay IF demodulator receiver similar to the type used for DPSK modulation; the main difference is the value of the delay used in the demodulator. They have also used both DFB and external-cavity semiconductor lasers as transmitter and local oscillator sources. For typical DFB laser linewidths it has been found that a tone separation of the order of 0.7 times the data rate is needed if a reasonable performance is to be achieved as well as a fairly compact IF spectrum.

At the time of writing, the highest data rate reported was 4 Gbit/s [87], and the longest unrepeated transmission distance was 290 km for a 400-Mbit/s system [83]. For similar reasons as given for the wide-deviation FSK systems, the measured performances have been in the range 5–11 dB worse than the shot-noise-limited value. The main points to emerge from these experiments are that a receiver bandwidth of the order of twice the data rate in hertz can be used, and that the beat linewidth must be < 0.5% of the data rate in hertz for the associated performance degradation to be < 1 dB (as might be expected this is similar to the DPSK requirement). An additional important aspect to emerge is that the effects of chromatic dispersion cannot be ignored for systems operating at several gigabits per second over long transmission distances. This was highlighted by the 1.8-dB performance degradation caused by chromatic dispersion in the 4-Gbit/s experiment over 202 km of fibre. This particular experiment has also shown that delay equalisation of the IF waveform can be used to remove this degradation.

*6.5.4.1.4. ASK Systems [88, 89].* Once FSK/CPFSK-type systems using a directly modulated DFB laser at the transmitter had been shown to be feasible, there was little further interest in conventional ASK systems, the main reasons for this being that they offer no receiver sensitivity advantage and they have the significant disadvantage of requiring an external amplitude modulator. However, a few ASK experiments using both external-cavity semiconductor lasers and DFB lasers have been reported operating at data rates of 150 Mbit/s [88] and 400 Mbit/s [89]. For these experiments the transmitter laser was externally modulated by a $LiNbO_3$ directional coupler type of modulator, otherwise the experimental arrangements were identical to those used for single-filter detection of wide-deviation FSK. From these experiments it was concluded that the external-cavity semiconductor lasers gave the best performance, but it is now known that it was not really justifiable to compare the performance results because there was insufficient IF bandwidth available to be able to achieve the best possible performance when using the broad-linewidth DFB lasers ($\approx 50$ MHz each).

*6.5.4.1.5. Systems Using Optical Amplifiers [90–93].* Now that optical amplification has become a viable means for increasing optical transmission path length, there has been growing interest in what performance degradations, if any, are likely to be introduced by either semiconductor or rare-earth-doped fibre optical amplifiers. In a 565-Mbit/s DPSK heterodyne system experiment using five travelling-wave semiconductor optical amplifiers, interconnected via optical attenuators to simulate fibre loss, it has been shown that in the absence of optical filtering and optical isolation between the amplifiers, a 6.5-dB performance degradation occurs [90]. The source of this degradation has been attributed to spontaneous noise and amplifier backward gain. An earlier 400-Mbit/s FSK experiment using four semiconductor optical amplifier repeaters, but with optical isolation between all the amplifiers and a 1.5-nm-bandwidth optical filter between the last two amplifiers, was operated over 370 km of fibre and the performance was degraded by only 1.5 dB [91]. A 565-Mbit/s DPSK experiment has also shown that a diode-pumped erbium-doped fibre amplifier is suitable for use as an optical repeater and that the associated performance degradation is in the region of 0.5 dB for a single repeater [92].

The possibility of using a semiconductor optical amplifier as a preamplifier at the input to a heterodyne receiver has been studied in a 45-Mbit/s FSK experiment [93]. The outcome was that for receivers operating close to the shot-noise limit, the spontaneous noise power output from the optical amplifier will always degrade the system performance. For this particular experiment the measured degradation was approximately 10 dB.

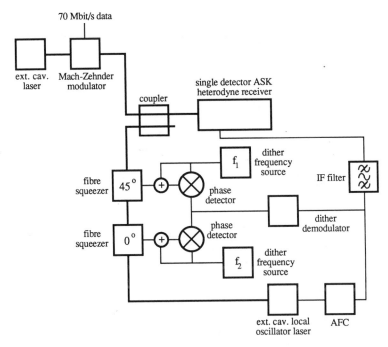

**Figure 6.22.** Polarisation tracking receiver using dither feedback control. After Harmon et al. [96], © 1987 SPIE.

### 6.5.4.2.   Polarisation-Insensitive System Experiments

*6.5.4.2.1. Polarisation Tracking Receivers [67, 94–96].* Experimental systems using automatic polarisation control have been demonstrated at data rates of up to 565 Mbit/s, and three types of feedback control loop have emerged as being suited for use in polarisation tracking receivers, the difference between them being the technique used to derive the feedback signal. A microprocessor has been used to make small step changes to the drive voltage applied to the polarisation controllers in order to "peak-search" the IF amplitude [94], a fraction of the received optical power has been tapped off and analysed using bulk optic polarisation sensitive elements [95] and the local oscillator polarisation has been dithered to produce polarisation-offset-dependent, low-level IF amplitude modulation [67, 96].

The experimental system used to demonstrate the dither feedback control technique is shown in Figure 6.22. At the transmitter the output from an external-cavity semiconductor laser was ASK-modulated at 70 Mbit/s by an external LiNbO$_3$ modulator. At the receiver the polarisation of the external-cavity semiconductor laser local oscillator was square-

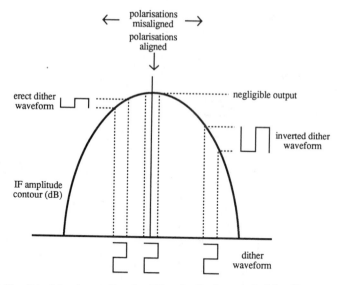

**Figure 6.23.** Principle of operation for dither feedback control. After Harmon et al. [96], © 1987 SPIE.

wave-modulated (dithered) at 125 and 300 Hz by two piezoelectric fibre squeezers oriented at 45° with respect to each other. The outcome of dithering the local oscillator polarisation is that as the polarisations become misaligned an increasing amount of low-level amplitude modulation is superimposed on the IF (Figure 6.23). Therefore, by demodulating the dither information, error signals are generated that can then be used to adjust the fibre squeezers until the polarisations are realigned. This experiment demonstrated the feasibility of using dither polarisation control, but to achieve continuous polarisation tracking, additional controllers would be required [67].

***6.5.4.2.2. Polarisation Diversity [41, 42, 97–101].*** At the time of writing, polarisation diversity systems experiments had been operated at data rates of up to 1.2 Gbit/s using all the modulation formats except PSK, and a 560 Mbit/s experiment had been operated over 150 km of conventional single-mode fibre [98]. Both external-cavity and DFB lasers have been used, and the measured performance results have all been of the order of 10 dB worse than the shot-noise-limited value. However, comparing these measurements with those for the equivalent standard polarisation aligned receiver shows that the diversity receivers gave performances in the range 0.3–2 dB worse than the standard receiver and, apart from the 600-Mbit/s single-filter FSK experiment [41], they all exhibited a polarisation-dependent performance variation of the order of 1 dB. The explanation for this

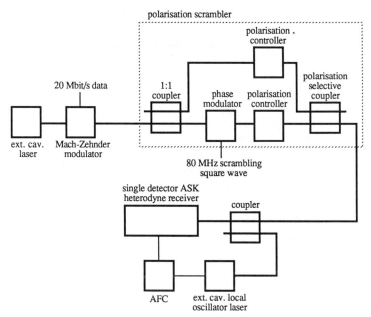

**Figure 6.24.** Heterodyne system configuration using polarisation scrambling at the transmitter to achieve polarisation-insensitive receiver performance. After Hodgkinson et al. [102], © 1987 IEE.

and the poor performance with respect to the shot-noise limit has been attributed to receiver thermal noise, imbalance between the two receiver arms, the extra noise associated with the dual-arm receiver design, and nonideal IF demodulation.

***6.5.4.2.3. Polarisation Scrambling [102].*** The use of polarisation scrambling to achieve polarisation-insensitive coherent detection was first demonstrated in a 20-Mbit/s ASK system experiment (Figure 6.24). The output from the external-cavity transmitter laser was amplitude-modulated by a Mach–Zehnder modulator and the polarisation of the modulated optical carrier was square-wave-modulated at four times the data rate by the polarisation scrambler. To achieve modulation between orthogonal polarisation states, the drive applied to the phase modulator in the scrambler was set to give a 180° phase shift. The receiver used was a standard ASK square-law IF demodulator heterodyne receiver using an external-cavity semiconductor laser local oscillator. By manually varying the signal polarisation the receiver performance was found to vary by approximately 1.0 dB and this is thought to have been caused by nonideal square-law IF demodulation.

*6.5.4.2.4. Systems Using Dual-Frequency Orthogonally Polarised Optical Fields [44, 103–105].* The feasibility of this type of polarisation-insensitive detection has been demonstrated in an orthogonally polarised FSK modulation experiment [104], and in a single-filter FSK experiment using a dual-frequency orthogonally polarised local oscillator constructed from two separate DFB lasers [105]. For both of these experiments the performance dependence on polarisation was similar to that for the other polarisation-insensitive schemes ($\approx 1$ dB), and it occurred for similar reasons.

The orthogonally polarised 500-Mbit/s FSK experiment was basically a standard FSK system with the addition of a passive birefringent element at the output of the DFB laser transmitter. The birefringent element, which was a 175-m length of highly birefringent fibre, is needed to convert standard FSK, produced by directly modulating the transmitter laser, into orthogonally polarised FSK. The conversion principle utilises the ability of birefringence to change polarisation and the fact that the change is frequency-dependent. Therefore, with the appropriate frequency separation between the two tones of a conventional FSK waveform, their respective polarisation states can be made to be orthogonal: for this experiment the tone separation needed was 1.3 GHz. The transmission path used in this experiment was 153 km of conventional single-mode fibre, and the local oscillator source was an external-cavity laser. The measured performance had a 2-dB sensitivity to polarisation variation and, as would be expected, the best performance was 3 dB worse than that given by the equivalent polarisation aligned receiver. The performance dependence on polarisation was attributed to the decision threshold not being at the optimum setting (see Section 6.4.1.3.2).

The main points to emerge from this experiment were that there was no significant degradation of orthogonality between the two FSK tones after propagation through 153 km of fibre, and that the dispersion of the birefringent element used at the transmitter places an upper limit on the data rate that can be used. The implication of the latter is that once the data rate has been specified this will determine the maximum tolerable birefringence, which, in turn, will limit the minimum tone separation that can be used.

### 6.5.4.3. Phase Diversity System Experiments [45–47, 101, 106–113].

Phase diversity detection experiments using 120° and 90° optical hybrid receiver designs have been studied in detail, and both have been shown to have the same sensitivity to laser linewidth and IF centre frequency stability as the equivalent standard heterodyne receiver. These experiments have demonstrated the use of ASK, FSK, and DPSK modulation formats in conjunction with external-cavity and DFB lasers. At the time of writing, the highest data rate reported was 5 Gbit/s [113], the longest unrepeated transmission distance was 100 km for a 4-Gbit/s system [113], and the reported performance measurements were all in the range 7–14

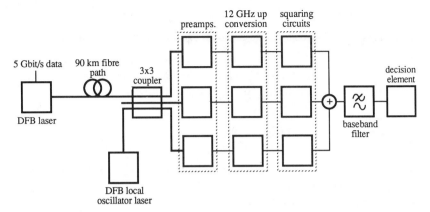

**Figure 6.25.** A 5-Gbit/s phase diversity detection experiment using a 120° optical hybrid receiver. After Emura et al. [113], © 1989 IEE.

dB away from the shot-noise-limited value. These performance degradations were caused mainly by receiver thermal noise, polarisation fluctuation, local oscillator intensity noise, and gain and delay mismatch between the receiver arms.

The 5-Gbit/s experiment used the experimental arrangement shown in Figure 6.25. The transmitter and local oscillator sources were both DFB lasers, with the transmitter being directly FSK-modulated. The transmission path length was 90 km and the receiver was a single-filter design based on the 120° optical hybrid (3 × 3 multiport). To overcome the practical limitations associated with using electrical mixers for squaring baseband signals, the signal of each arm was unconverted by 12 GHz prior to squaring. The measured performance was in the region of 14 dB worse than the shot-noise-limited value, the reasons for this being receiver thermal noise, nonideal electronic processing and chromatic dispersion.

**6.5.4.4.  Combined Polarisation and Phase Diversity Experiment.** Using a receiver configuration similar to that shown in Figure 6.18, a 200-Mbit/s DPSK experiment has demonstrated that simultaneous polarisation and phase diversity detection is feasible [114]. Using unbalanced 90° polarisation hybrids and 1.3-μm DFB optical sources, a performance approximately 10 dB worse than the shot-noise-limited value was measured; the reasons for this were attributed to the factors that have affected the performance of all the other polarisation and phase diversity experiments.

**6.5.4.5.  Heterodyne Detection Using a Low-Value IF [50, 115].** A 1-Gbit/s DPSK waveform has been detected using a receiver operating with a 15-MHz IF. This was made possible by using a 1-GHz sinusoidal waveform to vary the phase of the transmitted signal through a peak-to-peak change of 140°. However, this experiment, which gave a performance 18 dB away from the shot-noise limit, gives no indication of whether it will be possible to use bandwidths as narrow as those used for homodyne detection. The reason for this is that a 1.7-GHz IF bandwidth was used, whereas this should have been in the region of 700 MHz to be similar to the bandwidth of a 1-Gbit/s homodyne receiver.

### 6.5.5.  FSK and DPSK Field Trials

By mid-1989, coherent system and optical component developments had reached the point where laboratory experiments were beginning to be replaced by field-trial-type demonstrations, and by this time two such systems had been reported. The first reported an undersea 560-Mbit/s CPFSK system operating over a transmission path length of 90 km and using external grating DFB lasers and a polarisation diversity receiver [116]. The measured performance was of the order of 16 dB worse than the shot-noise-limited value, with almost half of this being caused by degraded IF stability outside the laboratory environment. The second field trial, which used miniature packaged external-cavity semiconductor lasers and a continuous polarisation tracking balanced receiver [117], reported a land 565-Mbit/s DPSK system operating over a transmission path length of 176 km. The polarisation control was achieved by using four polarisation-maintaining-fibre coils wound on piezoelectric cylinders situated in the local oscillator path. To achieve operation over 176 km of fibre, the modulated optical signal was amplified by a travelling-wave semiconductor optical amplifier prior to transmission. This amplification introduced no additional performance degradation, but the measured performance was 10 dB worse than the shot-noise-limited value due to receiver thermal noise, practical filter design, and residual amplitude modulation introduced by the phase modulator. The entire transmission system was assembled in modular form and inserted in a standard 19-in. rack.

### 6.5.6.  Multichannel Systems

The successful demonstration of coherent optical detection techniques has led to considering their application potential for future wideband communication networks [118–120]. At the time of writing, various new network structures were being studied, and common aspects being addressed were laser frequency stabilisation [121], producing optical frequency reference combs [122–130], interchannel interference [118, 131–133], and crosstalk caused by optically generated nonlinearities [118, 134, 135].

#### 6.5.6.1.  Optical Frequency Reference Combs.  Numerous schemes have been proposed for producing optical frequency reference combs, but despite the number reported only three basic techniques appear to be

involved. One technique is to modulate a single-laser source to produce a multiplicity of sidebands; this can be achieved by using either a nonlinear modulation scheme (e.g., phase modulation) or a harmonically rich modulation waveform [129, 130]. The other two techniques, which can be used individually or combined, use heterodyne offset frequency-locking loops [122, 123] or Fabry–Perot interferometer resonance peaks [125–128] to lock several laser sources to a single master laser. If absolute stability is required the master laser can be locked to an atomic frequency standard (absorption line) using previously reported techniques [121].

**6.5.6.2. Channel Spacing Considerations.** When using coherent transmission techniques it should in principle be possible to use channel separations equal to the modulation bandwidth; for heterodyne systems this would require the use of an optical image band rejection filter [136, 137]. To achieve such a narrow channel spacing in practice would almost certainly require the use of very-narrow-linewidth sources and pretransmission filtering to minimise interchannel interference [132]. Although the use of channel separations as small as the ideal value had not been shown to be feasible at the time of writing, an experimental heterodyne system had been successfully operated with a channel separation equal to four times the data rate [132].

**6.5.6.3. Optical Spread Spectrum Multiple Access Network.** It has been proposed that for some applications where narrow channel spacings are required, an optical spread spectrum system may be worth considering [138]. This technique minimises the interchannel interference, allows a broad-linewidth laser source to be used, removes the need for a reference frequency comb, uses simple terminal equipment, and because self-heterodyne detection is used a high degree of sensitivity is achieved. It is estimated that without the use of optical amplifiers the capacity of such a system (Figure 6.26) would be loss-limited to eight 140-Mbit/s channels.

The operating principle is that the output from a frequency-ramped laser is divided equally between each transmission channel and a reference channel. Prior to transmission each channel is delayed by a unique amount to create channel orthogonality, and the reference signal is frequency shifted (alternatively the frequency shift could be applied to each transmission channel). At the distribution point the individual channels and the reference are distributed equally between all of the receiving terminals that are designed to use the reference as a local oscillator for self-heterodyne detection. The desired channel is selected by matching the receiver and transmitter delays. The feasibility of such a system has been verified in a two-channel, 2-Mbit/s experiment using a DFB laser [138], but for such systems to achieve their full potential they will need to incorporate optical amplifiers.

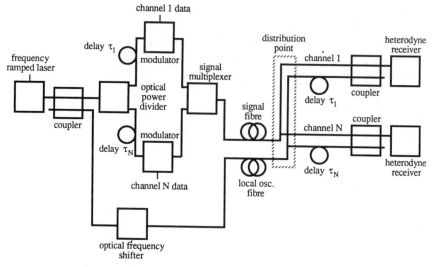

**Figure 6.26.** Optical spread spectrum multiple-access network configuration. After Healey et al. [138], © 1988 IEE.

### 6.5.6.4. Crosstalk Due to Nonlinear Optical Effects.

When large multiplexes are assembled, the power build up within the fibre may reach levels such that stimulated Brillouin scattering, stimulated Raman scattering, parametric four-wave mixing, and Kerr nonlinearities begin to generate intermodulation products [134, 135, 139–144]. The resulting crosstalk will place an upper limit on either the number of channels or the maximum power per channel, or both, but at the time of writing it was not clear what the actual limits would be [74, 118].

Stimulated Raman gain increases linearly with channel separation, so it is expected that Raman crosstalk effects will be minimised by keeping channel separations to a minimum [74]. Theoretical [139] and experimental [140] studies have shown Raman crosstalk can be significant at power levels that are a few orders of magnitude smaller than the single-channel power threshold ($\approx 1$ W).

Stimulated Brillouin gain is a narrower-bandwidth, higher-gain effect, but by avoiding either modulation formats with a residual carrier component, channel separations equal to the Brillouin shift (11 GHz) or bidirectional working, its effect on system performance should be minimised. Experimental work has shown that when two signals are separated by the Brillouin shift, the gain coefficient reduces by 80% if one of these is PSK-modulated at a rate $> 100$ Mbit/s [141].

Basically, the Kerr effect is the refractive index variation that occurs within a fibre when the optical power changes, and under certain operating conditions [134] it can cause amplitude–phase conversion, nonlinear pulse broadening, or modulation instability (an interplay between the Kerr nonlinearity and group-velocity dispersion). To avoid Kerr crosstalk, it is clear that the total multiplex power should be time-invariant, but in practice this may be difficult to achieve. The reasons for this are that power variations occur if the multiplex either constitutes amplitude-modulated carriers (ASK) or if there are residual amplitude effects as a result of nonideal angle modulation (FSK, PSK) [142, 143]. Also, if conditions are such that the individual channel frequencies are highly stable with similar polarisation states and the appropriate phase relationships, inter-channel beat powers will be established. For synchronous coherent systems it is thought that the amplitude–phase conversion effect will be the dominant one, because it is expected to have a lower threshold power than the other Kerr generated crosstalk terms.

Parametric four-wave mixing, which is closely related to the Kerr modulation instability, could give rise to significant amounts of crosstalk because its gain coefficients can be as great as those for stimulated Raman scattering. Crosstalk generated by this effect can be minimised by using wide channel separations and restricting the power per channel, with the limiting values being dependent on the transmission distance and the fibre chromatic dispersion [144].

Four-wave mixing-type interactions can also occur when an optical amplifier is used to amplify an optical multiplex because any optical input power variation causes a carrier density change, which in turn causes a refractive index change. This effect in semiconductor optical amplifiers has received considerable theoretical attention recently [145–147] and it has been shown that the magnitude of the effect peaks when the channel spacing (in radians) is approximately equal to the reciprocal of the spontaneous carrier lifetime [146, 147]. Early experimental studies of the effect of four-wave mixing on the different types of coherent systems have now been reported. One experiment has measured a 5-dB performance degradation when using a semiconductor laser amplifier to amplify an optical multiplex comprising four 45-Mbit/s FSK channels separated by 200 MHz [148]. This experiment also showed that increased channel separation significantly reduced the observed degradation. A three-channel 560-Mbit/s CPFSK experiment using a 2-GHz channel separation frequency has also reported the observation of a performance degradation caused by four-wave mixing [149].

### 6.5.7. Future Possibilities

From a communications system point of view, first-generation, medium-data-rate, engineered heterodyne systems are now seen as being a viable prospect, especially as field trial demonstrations are beginning to be reported. The poor performance given by the high-data-rate experiments is probably a transitionary problem related to the thermal noise characteristics of practical wideband receiver designs, which tuned receiver developments should help overcome. It is also envisaged that homodyne detection may become practicable if optical integration developments offer solutions to the present problems associated with practical optical phase-locked-loop design. Coherent detection techniques are also beginning to be considered for use in frequency-switching and multiple-access applications.

At the time of writing, rare-earth-doped fibre amplifiers were just beginning to emerge as alternatives to semiconductor optical amplifiers, and because of their longer spontaneous carrier lifetime their linearity was expected to be better than could be achieved by the semiconductor amplifiers. However, this and their suitability for use in coherent transmission systems was only just beginning to be addressed.

The possibility of achieving transmission distances in excess of several thousand kilometers [150] by using soliton propagation and Raman gain to overcome fibre loss was also being studied at the time of writing, but the indications were that much research would have to be carried out before soliton-based coherent communication systems would become practicable. Another potential means of increasing transmission distance being studied at the time of writing, was the possibility of achieving a detection sensitivity below the quantum noise limit using squeezed light [151, 152], but as with soliton propagation, much research still appeared to be needed before squeezed-light coherent communication systems would be considered feasible.

As advanced coherent communication techniques and optical nonlinearities begin to be better understood, it is envisaged that they may eventually be utilised for providing optical signal processing and/or optical routing and control functions in future advanced optical distribution networks.

## REFERENCES

1. L. H. Enloe and J. L. Rodda, *Proc. IEEE* **53**, 165–166 (1965).

2. O. E. Delange and A. F. Dietrich, *Bell Syst. Tech. J.* **47**, 161–178 (1968).

3. S. D. Personic, *Bell Syst. Tech. J.* **52**, 843–886 (1973).

4. E. Dietrich, B. Enning, R. Gross, and H. Knupke, *Electron. Lett.* **23**, 421–422 (1987).

5. S. D. Personic, *Bell Syst. Tech. J.* **59**, 213–216 (1971).

6. J. M. P. Delavaux, L. D. Tzeng, C. Y. Kuo, R. E. Tench, and M. Dixon, *IEEE J. Lightwave Technol.* **7**, 138–150 (1989).

7. R. E. Wagner and A. F. Elrefaie, "Polarisation-dispersion limitations in lightwave systems," *Conf. Proc. OFC '88*, Paper TU16, pp. 37 (1988).

8. A. F. Elrefaie, R. E. Wagner, D. A. Atlas, and D. G. Daut, *IEEE J. Lightwave Technol.* **6**, 704–709 (1988).

9. M. Tsubokawa and Y. Sasaki, *Electron Lett.* **24**, 350–352 (1988).

10. J. H. Winters, *IEEE J. Lightwave Technol.* **7**, 813–815 (1989).

11. Y. Yamamoto, *IEEE J. Quantum Electron.* **QE-19**, 34–58 (1983).

12. T. G. Hodgkinson, "Receiver analysis for synchronous coherent optical fibre transmission systems," *IEEE J. Lightwave Technol.* **LT-5**, 573–586 (1987).

13. G. L. Abbas, V. W. S. Chan, and T. K. Yee, *IEEE J. Lightwave Technol.* **LT-3**, 1110–1122 (1985).

14. K. Kikuchi, T. Okoshi, and K. Emura, "Achievement of nearly shot-noise-limited operation in a heterodyne-type PCM-ASK optical communication system," *Conf. Proc. ECOC '82*, Paper AX11-6, pp. 419–424 (1982).

15. T. G. Hodgkinson, R. Wyatt, D. W. Smith, D. J. Malyon, and R. A. Harmon, "Studies of 1.5 $\mu$m coherent transmission systems operating over installed cable links," *Conf. Proc. IEEE Globecom '83*, pp. 21.3.1–21.3.5 (1983).

16. G. Jacobsen, J. X. Kan, and I. Garrett, *IEEE J. Lightwave Technol.* **LT-7**, 105–114 (1989).

17. J. J. Stiffler, *Theory of Synchronous Communications*, Prentice-Hall, Englewood Cliffs, NJ, 1971.

18. L. G. Kazovsky, *IEEE J. Lightwave Technol.* **LT-4**, 182–195 (1986).

19. J. B. Armor, Jr., "Phase-lock control considerations for multiple, coherently combined lasers," Master's thesis, Air Force Institute of Technology, Wright-Patterson Air Force Base, OH, Report GEO/EE/77D-2 (1977).

20. T. G. Hodgkinson, *Electron. Lett.* **21**, 1202–1203 (1985).

21. T. G. Hodgkinson, *Electron. Lett.* **22**, 394–396 (1986).

22. F. M. Gardner, *Phaselock Techniques*, 2nd ed., Wiley, New York, 1979.

23. G. Nicholson, *Opt. Quantum Electron.* **17**, 399–410 (1985).

24. J. Franz, C. Rapp, and G. Soder, *J. Opt. Commun.* **7**, 15–20 (1986).

25. I. Garrett, and G. Jacobsen, *IEEE J. Lightwave Technol.* **LT-5**, 551–560 (1987).

26. G. Jacobsen and I. Garrett, *IEEE J. Lightwave Technol.* **LT-5**, 478–484 (1987).

27. G. J. Foschini, L. J. Greenstein, and G. Vannucci, *IEEE Trans. Commun.* **36**, 306–314 (1988).

28. I. Garrett and G. Jacobsen, *IEEE J. Lightwave Technol.* **6**, 1415–1423 (1988).

29. E. Patzak and P. Meissner, *IEE Proc. J.* **135**, 355–357 (1988).

30. J. Franz and H. N. Schaller, *J. Opt. Commun.* **10**, 28–32 (1989).

31. W. W. Harman, *Principles of Statistical Theory of Communication*, McGraw-Hill, New York, 1963.

32. H. Taub and D. L. Schilling, *Principles of Communications Systems*, McGraw-Hill, Kogakusha, 1971.

33. D. W. Smith, T. G. Hodgkinson, and R. A. Harmon, *Br. Telecom Technol. J.* **1**, 12–16 (1983).

34. L. Giehmann, and M. Rocks, *Opt. Quantum Electron.* **19**, 109–113 (1987).

35. Y. Namihira, S. Ryu, M. Kuwazuru, K. Mochizuki, and Y. Iwamoto, *Electron. Lett.* **23**, 343–344 (1987).

36. C. D. Poole, N. S. Bergano, H. J. Schulte, R. E. Wagner, V. P. Nathu, J. M. Amon, and R. L. Rosenberg, *Electron. Lett.* **23**, 1113–1115 (1987).

37. N. G. Walker and G. R. Walker, *Br. Telecom. Technol. J.* **5**, 63–76 (1987).

38. R. Noe, H. Heidrich, and D. Hoffmann, *IEEE J. Lightwave Technol.* **6**, 1199–1208 (1988).

39. C. J. Mahon and G. D. Khoe, *Electron. Lett.* **23**, 1234–1235 (1987).

40. T. Imai, *Electron. Lett.* **24**, 979–980 (1988).

41. B. Glance, *IEEE J. Lightwave Technol.* **LT-5**, 274–276 (1987).

42. T. G. Hodgkinson, R. A. Harmon, and D. W. Smith, *Electron. Lett.* **24**, 58–59 (1988).

43. L. J. Cimini, Jr., I. M. I. Habbab, R. K. John, and A. A. M. Saleh, *Electron. Lett.* **23**, 1365–1366 (1987).

44. T. G. Hodgkinson and A. R. J. Cook, "Polarisation insensitive coherent detection using orthogonally polarised optical fields," *Microwave Opt. Technol. Lett.* **1**, 246–249 (1988).

45. A. W. Davis, M. J. Pettitt, J. P. King, and S. Wright, *IEEE J. Lightwave Technol.* **LT-5**, 561–572 (1987).

46. L. G. Kazovsky, L. Curtis, W. C. Young, and N. K. Cheung, *Appl. Opt.* **26**, 437–439 (1987).

47. T. G. Hodgkinson, R. A. Harmon, D. W. Smith, and P. J. Chidgey, "In-phase and quadrature detection using a 90° optical hybrid receiver: Experiments and design considerations," *IEE Proc. J.* **135**, 260–267 (1988).

48. O. Strobel, H. Schmuck, and H. Krimmel, *J. Opt. Commun.* **9**, 128–132 (1988).

49. V. Napasab and T. Okoshi, *J. Opt. Commun.* **9**, 102–107 (1988).

50. I. M. I. Habbab, J. M. Kahn, and L. J. Greenstein, *Electron. Lett.* **24**, 974–976 (1988).

51. R. Wyatt, K. H. Cameron, and M. R. Matthews, "Tunable narrow line external-cavity lasers for coherent optical systems," *Br. Telecom. Technol. J.* **3**, 5–12 (1985).

52. J. Mellis, S. Al-Chalabi, K. H. Cameron, R. Wyatt, J. C. Regnault, W. J. Devlin, and M. C. Brain, "Miniature packaged external-cavity semiconductor lasers for coherent communications," *Conf. Proc. ECOC '88*, **1**, pp. 219–222 (1988).

53. S. Takano, T. Sasaki, H. Yamada, M. Kitamura, and I. Mito, *Electron. Lett.* **25**, 356–357 (1989).

54. S. B. Alexander, D. Welford, and D. L. Marquis, *IEEE J. Lightwave Technol.* **7**, 11–23 (1989).

55. K. Emura, M. Shikada, S. Fujita, I. Mitu, H. Homnou, and K. Minemura, *Electron. Lett.* **20**, 1022–1023 (1984).

56. R. S. Vodhanel and B. Enning, *Electron. Lett.* **24**, 163–165 (1988).

57. R. C. Steele and M. Creaner, *Electron. Lett.* **25**, 732–734 (1989).

58. R. S. Vodhanel, B. Enning, and A. F. Elrefaie, *IEEE J. Lightwave Technol.* **6**, 1549–1553 (1988).

59. S. Yamazaki, K. Emura, M. Shikada, M. Yamaguchi, and I. Mito, *Electron. Lett.* **21**, 283–285 (1985).

60. M. Ohtsu, *IEEE J. Lightwave Technol.* **6**, 245–256 (1988).

61. B. Enning and R. S. Vodhanel, "Adaptive quantised feedback equalisation for FSK heterodyne transmission at 150 Mbit/s and 1 Gbit/s," *Post-Deadline Conf. Proc. OFC '88*, pp. PD231/1–PD23/4 (1988).

62. D. J. Malyon, D. W. Smith, and R. Wyatt, "Semiconductor laser homodyne optical phase-locked-loop," *Electron. Lett.* **22**, 421–422 (1986).

63. J. M. Kahn, B. L. Kasper, and K. J. Pollock, *Electron. Lett.* **25**, 626–628 (1989).

64. D. W. Smith, "Techniques for multigigabit coherent optical transmission," *IEEE J. Lightwave Technol.* **LT-5**, 1466–1478 (1987).

65. C. J. Atkins, D. Cotter, D. W. Smith, and R. Wyatt, *Electron. Lett.* **22**, 556 (1987).

66. R. Wyatt, T. G. Hodgkinson, and D. W. Smith, *Electron. Lett.* **19**, 550–552 (1983).

67. M. J. Creaner, R. C. Steele, G. R. Walker, and N. G. Walker, *Electron. Lett.* **24**, 270–271 (1988).

68. S. Watanabe, T. Chikama, T. Naito, and H. Kuwahara, *Electron. Lett.* **25**, 588–590 (1989).

69. K. Emura, M. Shikada, S. Yamazaki, K. Komatsu, I. Mito, and K. Minemura, *Electron. Lett.* **21**, 1121–1122 (1985).

70. A. H. Gnauck, R. A. Linke, B. L. Kasper, K. J. Pollock, K. C. Reichmann, R. Valenzuela, and R. C. Alferness, *Electron. Lett.* **23**, 286–287 (1987).

71. S. Yamazaki, S. Murata, K. Komatsu, Y. Koizumi, S. Fujita, and K. Emura, *Electron. Lett.* **23**, 860–862 (1987).

72. H. Kuwahara, T. Chikama, H. Onaka, M. Seino, and T. Kiyonaga, "Stable DPSK lightwave transmission at 1.2 Gbit/s using compact fibre external cavity lasers," *Conf. Proc. OFC '88*, Paper WC2, pp. 52 (1988).

73. J. M. P. Delavaux, L. D. Tzeng, and M. Dixon, *Electron. Lett.* **24**, 941–942 (1988).

74. R. A. Linke and A. H. Gnauck, *IEEE J. Lightwave Technol.* **6**, 1750–1769 (1988).

75. C. Y. Kuo, J. M. P. Delavaux, D. A. Ackerman, and L. D. Tzeng, *IEEE J. Lightwave Technol.* **7**, 520–524 (1989).

76. S. Yamazaki, K. Emura, M. Shikada, M. Yamaguchi, I. Mito, and K. Minemura, *Electron. Lett.* **22**, 5–7 (1986).

77. K. Emura, S. Yamazaki, S. Fujita, M. Shikada, I. Mito, and K. Minemura, *Electron. Lett.* **22**, 1096–1097 (1986).

78. R. S. Vodhanel, J. L. Gimlett, N. K. Cheung, and S. Tsuji, *IEEE J. Lightwave Technol.* **LT-5**, 461–468 (1987).

79. K. Emura, S. Yamazaki, M. Shikada, S. Fujita, M. Yamaguchi, I. Mito and K. Minemura, *IEEE J. Lightwave Technol.* **LT-5**, 469–477 (1987).

80. T. Chikama, H. Onaka, T. Kiyonaga, M. Suyama, and H. Kuwahara, "An

optical transmission experiment using solitary DFB lasers for broadband distribution networks," *Conf. Proc. ECOC '87*, **1**, pp. 349–352 (1987).

81. T. G. Hodgkinson, D. W. Smith, R. Wyatt, and D. J. Malyon, *Br. Telecom. Technol. J.* **3**, 5–18 (1985).

82. K. Iwashita, T. Imai, and T. Matsumoto, *Electron. Lett.* **22**, 164–165 (1986).

83. K. Iwashita and T. Matsumoto, *Electron. Lett.* **22**, 791–792 (1986).

84. K. Iwashita and N. Takachio, *Electron. Lett.* **23**, 1022–1023 (1987).

85. K. Manome, K. Emura, S. Yamazaki, S. Fujita, S. Takano, M. Shikada, and K. Minemura, "A 1.2 Gbit/s CPFSK heterodyne detection transmission experiment with optimum system configuration for solitary laser diodes," *Conf. Proc. ECOC '87*, **1**, pp. 333–336 (1987).

86. J. L. Gimlett, R. S. Vodhanel, M. M. Choy, A. F. Elrefaie, N. K. Cheung, and R. E. Wagner, *IEEE J. Lightwave Technol.* **LT-5**, 1315–1324 (1987).

87. K. Iwashita and N. Takachio, *Electron. Lett.* **24**, 759–760 (1988).

88. Y. K. Park, J. M. P. Delavaux, N. A. Olsson, T. V. Nguyen, D. E. Tamburino, R. W. Smith, S. K. Korotky, and M. Dixon, *Electron. Lett.* **22**, 283–284 (1986).

89. R. A. Linke, B. L. Kasper, R. C. Alferness, A. R. McCormick, N. A. Olsson, L. L. Buhl, T. L. Koch, S. K. Korotky, and J. J. Veselka, "Coherent lightwave transmission experiments using amplitude and phase modulation at 400 Mbit/s and 1 Gbit/s data rates," *Conf. Proc. OFC '86*, Paper WE1, pp. 86–87 (1986).

90. D. J. Malyon, R. C. Steele, M. J. Creaner, M. C. Brain, and W. A. Stallard, *Electron. Lett.* **25**, 354–356 (1989).

91. N. A. Olsson, M. G. Oberg, L. A. Koszi, and G. Przybylek, *Electron. Lett.* **24**, 36–38 (1988).

92. T. J. Whitley, M. J. Creaner, R. C. Steele, M. C. Brain, and C. A. Millar, *IEEE Photon. Technol. Lett.* **1**, 425–427 (1989).

93. B. Glance, G. Eisenstein, P. J. Fitzgerald, K. J. Pollock, and G. Raybon, *Electron. Lett.* **24**, 1229–1230 (1988).

94. T. Okoshi, Y. H. Cheng, and K. Kikuchi, *Electron. Lett.* **21**, 787–788 (1985).

95. H. Honmou, S. Yamazaki, K. Emura, R. Ishikawa, I. Mito, M. Shikada, and K. Minemura, *Electron. Lett.* **22**, 1181–1182 (1986).

96. R. A. Harmon, G. R. Walker, and T. K. White, "Polarisation control in a coherent optical fibre system using a dither technique," *Proc. SPIE Fibre Optics '87: Fifth International Conference on Fibre Optics and Opto-electronics*, **734**, pp. 63–67 (1987).

97. T. E. Darcie, B. Glance, K. Gayliard, J. R. Talman, B. L. Kasper, and C. A. Burrus, *Electron. Lett.* **23**, 1369–1370 (1987).

98. S. Ryu, S. Yamamoto, and K. Mochizuki, *Electron. Lett.* **23**, 1382–1384 (1987).

99. L. D. Tzeng, T. W. Cline, and A. A. M. Saleh, *Electron. Lett.* **24**, 330–332 (1988).

100. S. Watanabe, T. Naito, T. Chikama, T. Kiyonaga, Y. Onoda, and H. Kuwahara, *Electron. Lett.* **25**, 383–384 (1989).

101. L. G. Kazovsky, *IEEE J. Lightwave Technol.* **7**, 279–292 (1989).

102. T. G. Hodgkinson, R. A. Harmon, and D. W. Smith, "Polarisation-insensitive heterodyne detection using polarisation scrambling," *Electron. Lett.* **23**, 513–514 (1987).

103. I. M. I. Habbab and L. J. Cimini, Jr., *IEEE J. Lightwave Technol.* **6**, 1537–1548 (1988).

104. R. Noe, J. L. Gimlett, and R. S. Vodhanel, *Electron. Lett.* **25**, 4–5 (1989).

105. H. Tsushima and S. Sasaki, *Electron. Lett.* **25**, 539–541 (1989).

106. T. G. Hodgkinson, R. A. Harmon, and D. W. Smith, *Electron. Lett.* **21**, 867–868 (1985).

107. A. W. Davis, S. Wright, M. J. Pettitt, J. P. King, and K. Richards, *Electron. Lett.* **22**, 9–11 (1986).

108. M. J. Pettitt, D. Remedios, A. W. Davis, A. Hadjifotiou, and S. Wright, *Electron. Lett.* **23**, 1075–1076 (1987).

109. R. Noe, W. B. Sessa, R. Welter, and L. G. Kazovsky, *Electron. Lett.* **24**, 567–568 (1988).

110. R. Schneider and J. Pietzsch, "Coherent 565 Mbit/s DPSK transmission experiment with a phase diversity receiver," *Post-Deadline Papers Conf. Proc. ECOC '87*, pp. 5–8 (1987).

111. L. G. Kazovsky, R. Welter, A. F. Elrefaie, and W. Sessa, *IEEE J. Lightwave Technol.* **6**, 1527–1536 (1988).

112. G. Nicholson, *J. Opt. Commun.* **9**, 13–16 (1988).

113. K. Emura, R. S. Vodhanel, R. Welter, and W. Sessa, "5 Gbit/s optical phase diversity homodyne detection experiment," *Electron. Lett.* **25**, 400–401 (1989).

114. T. Okoshi and Y. H. Cheng, *Electron. Lett.* **23**, 377–378 (1987).

115. J. M. Kahn, I. M. I. Habbab, and C. R. Giles, *Electron. Lett.* **23**, 1455–1457 (1988).

116. S. Ryu, S. Yamamoto, Y. Namihira, K. Mochizuki, and H. Wakabayashi, *Electron. Lett.* **24**, 399–400 (1988).

117. M. J. Creaner, R. C. Steele, I. Marshall, G. R. Walker, N. G. Walker, J. Mellis, S. A. Al-Chalabi, I. Sturgess, M. Rutherford, J. Davidson, and M. C. Brain, *Electron. Lett.* **24**, 1354–1356 (1988).

118. I. W. Stanley, G. R. Hill, and D. W. Smith, *IEEE J. Lightwave Technol.* **LT-5**, 439–451 (1987).

119. B. S. Glance, J. Stone, K. J. Pollock, P. J. Fitzgerald, C. A. Burrus, Jr., B. L. Kasper, and W. Stulzi, *IEEE J. Lightwave Technol.* **6**, 1770–1781 (1988).

120. E. J. Bachus, R. P. Braun, C. Caspar, H. M. Foisel, E. Grossmann, B. Strebel, and F. J. Westphal, *IEEE J. Lightwave Technol.* **7**, 375–382 (1989).

121. B. Villeneuve, N. Cyr, and M. Tetu, *Electron. Lett.* **24**, 736–737 (1988).

122. E. J. Bachus, R. P. Braun, W. Eutin, E. Grossmann, H. M. Foisel, K. Heimes, and B. Strebel, *Electron. Lett.* **21**, 1203–1205 (1985).

123. W. A. Stallard, D. J. T. Heatley, R. A. Lobbet, A. R. Beaumont, D. J. Hunkin, B. E. Daymond-John, R. C. Booth, and G. R. Hill, *Br. Telecom. Technol. J.* **4**, 16–22 (1986).

124. B. Glance, P. J. Fitzgerald, K. J. Pollack, J. Stone, C. A. Burrus, G. Eisenstein, and L. W. Stulz, *Electron. Lett.* **23**, 750–752 (1987).

125. K. Nosu, T. Hiromu, and I. Katsushi, *IEEE J. Lightwave Technol.* **LT-5**, 1301–1308 (1987).

126. I. Mito, K. Kaede, and M. Sakaguchi, "Frequency tunable laser diodes and their application to coherent system," *Conf. Proc. ECOC '88*, **1**, pp. 74–77 (1988).

127. H. Shimosaka, K. Kaede, and S. Murata, "Frequency locking of FDM optical sources using widely tunable DBR LDs," *Conf. Proc. OFC '88*, Paper THG3 (1988).

128. P. Gambini, M. Puleo, and E. Vezzoni, "Laser frequency stabilisation for multichannel coherent systems," *Conf. Proc. ECOC '88*, **1**, pp. 78–81 (1988).

129. T. G. Hodgkinson and P. Coppin, *Electron. Lett.* **25**, 509–510 (1989).

130. M. W. Maeda, and L. G. Kazovsky, *IEEE Photon. Technol. Lett.* **1**, 455–457 (1989).

131. L. G. Kazovsky, *IEEE J. Lightwave Technol.* **LT-5**, 1095–1102 (1987).

132. Y. K. Park, S. S. Bergstein, R. E. Tench, R. W. Smith, S. K. Korotky, K. J. Burns, and S. W. Granlund, *IEEE J. Lightwave Technol.* **6**, 1312–1320 (1988).

133. L. G. Kazovsky and J. L. Gimlett, *IEEE J. Lightwave Technol.* **6**, 1353–1365 (1988).

134. D. Cotter, *Opt. Quantum Electron.* **19**, 1–17 (1987).

135. W. J. Tomlinson and R. H. Stolen, *IEEE Commun. Magazine* **26**, 36–44 (1988).

136. B. S. Glance, *IEEE J. Lightwave Technol.* **LT-4**, 1722–1725 (1986).

137. T. Naito, T. Chikama, M. Suyama, T. Kiyonaga, and H. Kuwahara, "Crosstalk penalty in a two channel 560 Mbit/s DPSK heterodyne optical communications system using an image rejection receiver," *Conf. Proc. OFC '89*, Paper THC3, p. 141 (1989).

138. D. W. Smith, P. Healey, G. P. Fry, and K. Clayton, "A frequency chirped heterodyne spread spectrum optical fibre multiple-access technique," *Conf. Proc. ECOC '88*, **1**, pp. 82–85 (1988).

139. A. R. Chraplyvy, *Electron. Lett.* **20**, 58–59 (1984).

140. J. Hegarty, N. A. Olsson, and M. McGlashan-Powell, *Electron. Lett.* **21**, 395–397 (1985).

141. A. Bolle, G. Grosso, and B. Daino, *Electron. Lett.* **25**, 2–3 (1989).

142. A. R. Chraplyvy and J. Stone, *Electron. Lett.* **20**, 996–997 (1984).

143. D. Cotter, *Br. Telecom. Technol. J.* **1**, 17–19 (1983).

144. R. G. Waarts and R. P. Braun, *Electron. Lett.* **22**, 873–875 (1986).

145. R. M. Jopson and T. E. Darcie, *Electron. Lett.* **24**, 1372–1374 (1988).

146. T. G. Hodgkinson and R. P. Webb, *Electron. Lett.* **24**, 1550–1552 (1988).

147. T. G. Hodgkinson and R. P. Webb, "Analysis of intermodulation distortion in a travelling wave semiconductor laser amplifier," *Workshop Proc., Tirrenia International Workshop on Digital Communications: Coherent Optical Communications and Photonic Switching*, pp. 171–185 (1989).

148. B. S. Glance, G. Eisenstein, P. J. Fitzgerald, K. J. Pollock, and G. Raybon, *IEEE J. Lightwave Technol.* **7**, 759–765 (1989).

149. S. Ryu, K. Mochizuki, and H. Wakabayashi, "Influence of nondegenerate four-wave mixing on coherent transmission systems using in-line semiconductor laser amplifiers," *Conf. Proc. OFC '89*, Paper THC4, pp. 142 (1989).

150. L. F. Mollenauer and K. Smith, "Ultralong range soliton transmission," *Conf. Proc. OFC '89*, Paper WO 1, 97 (1989).

151. J. Arnaud, *Electron. Lett.* **25**, 1–2 (1989).

152. E. Giacobino, C. Fabre, and G. Leuchs, *Physics World* (February), 31–34 (1989).

# 7

# Traveling-Wave Semiconductor Laser Amplifiers

TADASHI SAITOH AND TAKAAKI MUKAI
*NTT Basic Research Laboratories, Nippon Telegraph and Telephone Corporation, Musashino-shi, Tokyo, Japan*

## 7.1. INTRODUCTION

Laser amplifiers, which can directly amplify optical signals without converting them into electric signals, are key devices for future optical communications systems, because they simplify the systems and facilitate changes in both data rate and data format. Laser amplifiers are based on the same principle of stimulated emission in the population inverted laser medium as laser oscillators. Even though the amplifying mechanism of laser amplifiers is the same as that of laser oscillators, laser amplifiers need to avoid optical feedback and also require a higher pumping rate or longer interaction length than laser oscillators to obtain a reasonable signal gain. Semiconductors have attracted much attention as high-gain materials suitable for laser amplifiers [1], while rare-earth-doped optical fibers satisfy the requirement for long interaction length [2]. Two types of semiconductor laser amplifier (SLA) structures are sketched in Figure 7.1 with their schematic gain spectra. An SLA is easily obtained by setting the bias current below the oscillation threshold. The SLA acts as a resonant amplifier, that is, as a Fabry–Perot amplifier (FPA), in which the incident signal is amplified while bouncing back and forth between the two end-mirrors. Unfortunately, the FPA requires critical frequency matching between the input signal and the FP resonant mode. When the reflectivities of both facets are suppressed, such as by precise antireflection (AR) coating, it works as a single-pass amplifier. This is commonly referred to as a traveling-wave amplifier (TWA) [3–6]. The TWA is superior to the FPA in gain bandwidth, signal gain saturation, and noise figure.

*Coherence, Amplification, and Quantum Effects in Semiconductor Lasers*, Edited by Yoshihisa Yamamoto.
ISBN 0-471-512494 (©) 1991 John Wiley & Sons, Inc.

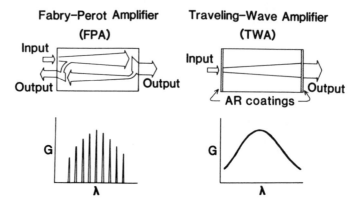

**Figure 7.1.** Classification of semiconductor laser amplifiers.

Research on SLAs based on homostructure devices started [7–10] just after the invention of semiconductor lasers in 1962 [11–13]. Even though these studies were not developed for practical application because the devices could not operate at room temperature, the principle of optical amplification in semiconductor lasers was confirmed. After the achievement of continuous-wave (CW) oscillation [14, 15] at room temperature in GaAs lasers by introducing DH (double-heterostructure) structures, SLAs were reviewed by Zeidler and Schicketanz of Siemens Research Laboratories [16–18] for use in repeaters for optical fiber transmission systems in the 1970s. However, it was difficult to assess the properties of SLAs because stable single-frequency semiconductor lasers and other high-performance optical devices such as optical isolators were not available then. In 1980 both theoretical and experimental SLA research was started by a group in NTT. They demonstrated a stable signal gain measurement [19] and succeeded in reducing the receiving power level in an AlGaAs FPA [20]. They were able to overcome some of the obstacles encountered by the earlier researchers, because single-frequency lasers and low-loss optical isolators were available by then. Fundamental characteristics of SLAs, such as signal gain, gain bandwidth, gain saturation, and noise, were successively reported [21–27].

Precise AR coating on laser diode (LD) facets for TWAs has been studied [28–31] since the first GaAs TWA was reported by Simon in 1982 [25]. However, long-wavelength TWAs were only achieved in 1986 when Saitoh and Mukai made a 1.5-$\mu$m GaInAsP TWA [32]. This achievement stimulated the current research on TWAs and their applications. The fundamental characteristics of long-wavelength TWAs and near-TWAs have been examined [3, 4, 33, 34] for application to high-bit-rate or coherent [35] optical fiber transmission systems. System applications of TWAs reported so far include preamplifiers [36–39] and in-line repeater amplifiers [40–44]. Furthermore, multichannel amplification [45–48] and optical short-pulse amplification [49–52] experiments have made good use of the wide gain bandwidth of TWAs. Recently, TWA structures incorporating angled facets [53–56] and window facets [57, 58] have also been developed as alternatives to precise AR coating.

Besides the use for signal amplification, SLAs have been intensively studied for use as an optical switch [59, 60], an optical modulator [61–63], a frequency converter [64, 65], and a frequency chirper [66] for pulse compression.

An excellent text on SLAs was published in 1985 by Mukai et al. [1]. Although it is only one chapter of a book, it describes the important fundamental characteristics of SLAs in detail, mainly treating AlGaAs FPAs. Some review papers [4, 6, 67] on SLAs are also helpful in grasping the state of the art of SLAs and their applications for communications systems.

This chapter compares long-wavelength TWAs with FPAs and describes the fabrication of TWAs, fundamental device characteristics, multichannel amplification, optical short-pulse amplification, and system applications of SLAs. Section 7.2 describes how cavity resonance is suppressed, that is, how TWAs are obtained. Then, AR coating on LD facets, an angled facet structure, and a window facet structure are introduced as realistic structures for TWAs. Section 7.3 treats fundamental device characteristics of SLAs, specifically signal gain, signal gain saturation, and noise. TWA device characteristics are compared with those of FPAs. Design criteria for improving SLA characteristics are discussed in terms of material gain parameters and amplifier operating conditions. Interchannel crosstalk induced by signal gain saturation and four-wave mixing in multichannel amplification are described in Section 7.4. Nearly degenerate four-wave mixing in a TWA is discussed from the viewpoint of frequency conversion as well as a source of crosstalk in frequency-division-multiplexed (FDM) systems. Section 7.5 discusses characteristics of short optical pulse amplification, including the amplification of isolated pulses and repetitive pulses. The self-phase modulation effect accompanying high-energy pulse amplification is also considered from a pulse compression point of view. Section 7.6 introduces system applications of SLAs, such as preamplifiers, in-line repeaters, and booster amplifiers. Requirements for each application are summarized in terms of amplifier performance. Simple signal-to-noise ratio design is also presented for cascaded linear amplifier chains.

## 7.2. SUPPRESSION OF CAVITY RESONANCE

In fabricating a TWA, Fabry–Perot (FP) cavity resonance in the laser diode (LD) must be suppressed due to its end facet reflectivity. The value

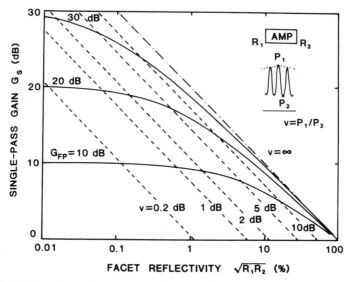

**Figure 7.2.** Fabry–Perot signal gain $G_{FP}$ (solid lines) and signal gain undulation $v$ (dashed lines) as functions of single-pass gain $G_s$ and facet reflectivity $\sqrt{R_1 R_2}$.

of the required facet reflectivity $R(= \sqrt{R_1 R_2})$ depends on the single-pass gain $G_s$ and the allowed signal-gain undulation $v$ due to the residual FP resonance [68]:

$$R = \frac{1}{G_s} \cdot \frac{\sqrt{v} - 1}{\sqrt{v} + 1} \tag{7.1}$$

This relationship is shown in Figure 7.2. The signal gain undulations are shown as functions of the single-pass gain and the facet reflectivity. As shown by the broken lines, the undulation is uniquely determined by the product of the single-pass gain and the facet reflectivity [See Eq. (7.12)]. When $G_s\sqrt{R_1 R_2} < 0.17$, the signal gain undulation becomes less than 3 dB, that is, the TWA condition. AR coating is not the only means of reducing facet reflectivity; angled facet and window facet structures have also been studied. However, even if these structures are used, AR coating is essential for improving the coupling efficiency and for obtaining sufficiently low reflectivity.

### 7.2.1. AR Coating

**7.2.1.1. Theory.** An antireflection coating is often used to eliminate reflection from the surface of optical components such as lenses. The

AR-coating conditions for a plane wave with a wavelength of $\lambda$, which is normally incident from air to a homogeneous isotropic medium with a refractive index of $n_{\text{sub}}$, are well known as $n_f = \sqrt{n_{\text{sub}}}$ and $n_f h_f = \lambda/4$, where $n_f$ and $h_f$ are the refractive index and the thickness of the AR-coating film. However, AR-coating conditions required for guided waves in an LD are different from those for a simple plane wave, because the guided mode is composed of elementary plane waves with different wave vectors.

LD facet reflectivities have been calculated by several methods. Clarke [28] and Eisenstein [29] calculated the facet reflectivity of AR coated LDs using an angular spectrum approach [69, 70] assuming a Gaussian field distribution in a homogeneous active material. They clarified that the optimum values of refractive index and thickness of AR coating film on the LD facet depend on the beam spot size, which is the only parameter describing the Gaussian field distribution and its angular spectrum. Kaplan and Deimel [71] showed that the correction to the residual reflectivity due to the index step $\Delta n$ between active and cladding layers is significant by calculating an infinite perturbation expansion in terms of a suitably chosen integral of $\Delta n$. The dependence of optimum AR coating film parameters on the active-layer thickness is shown in the following analysis, which is based on both the field distribution in the slab waveguide and on an angular-spectrum approach [30].

A symmetric slab waveguide with a single-layer AR coating is shown in Figure 7.3. The active layer of refractive index $n_a$ and thickness $d_a$ is sandwiched by cladding layers of the same refractive index $n_{\text{cl}}$. The angular spectrum of the TE-polarized incident field $F_{\text{inc}}(s)$ is derived from the Fourier transform of the incident field $E_{\text{inc}}(x)$ at the boundary ($z = 0$)

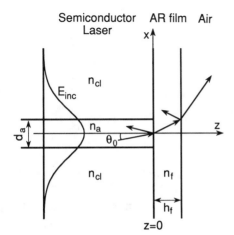

**Figure 7.3.** Symmetric slab waveguide with a single-layer AR coating.

between the LD and the AR coating as

$$
\begin{aligned}
F_{\text{inc}}(s) &= 2\int_0^{d_a/2} E_{\text{inc}}(x)\exp(ik_a sx)\,dx + 2\int_{d_a/2}^{\infty} E_{\text{inc}}(x)\exp(ik_{\text{cl}}sx)\,dx \\
&= A\left[\frac{\sin\{(\kappa + k_a s)d_a/2\}}{\kappa + k_a s} + \frac{\sin\{(\kappa - k_a s)d_a/2\}}{\kappa - k_a s}\right] \\
&\quad + \frac{2A}{\gamma^2 + k_{\text{cl}}^2 s^2}\cos\left(\frac{d_a \kappa}{2}\right)\left[\gamma\cos\left(\frac{d_a k_{\text{cl}} s}{2}\right) - k_{\text{cl}}s\,\sin\left(\frac{d_a k_{\text{cl}} s}{2}\right)\right] \\
&\equiv F_{\text{inc},a}(s) + F_{\text{inc},\text{cl}}(s),
\end{aligned}
\tag{7.2}
$$

where $s = \sin\theta_0$, $\theta_0$ is the incident angle, $x$ is the transverse coordinate, $k_a = 2\pi n_a/\lambda$, $k_{\text{cl}} = 2\pi n_{\text{cl}}/\lambda$, and $A$ is an arbitrary constant. The propagation constants $\kappa$ and $\gamma$ in the $x$ direction are related through the propagation constant $\beta$ in the $z$ direction as

$$
\kappa^2 = k_a^2 - \beta^2
\tag{7.3}
$$

and

$$
\gamma^2 = \beta^2 - k_{\text{cl}}^2
\tag{7.4}
$$

Using the Fresnel reflection coefficients for the field components in the active region $r_a(s)$ and the cladding region $r_{\text{cl}}(s)$, we get the reflected-field angular spectrum $F_{\text{ref}}(s)$:

$$
F_{\text{ref}}(s) = -\left[r_a(s)F_{\text{inc},a}(s) + r_{\text{cl}}(s)F_{\text{inc},\text{cl}}(s)\right]
\tag{7.5}
$$

The reflected field $E_{\text{ref}}(x)$ at $z = 0$ is obtained by the inverse Fourier transform of $F_{\text{ref}}(s)$. The power reflectivity $R_{\text{AR}}$ for AR-coated LD facets is given by the square of the coupling coefficient between $E_{\text{inc}}(x)$ and $E_{\text{ref}}(x)$; that is

$$
R_{\text{AR}} = \frac{\left|\int_{-\infty}^{\infty} E_{\text{inc}}(x)E_{\text{ref}}(x)\,dx\right|^2}{\left|\int_{-\infty}^{\infty} E_{\text{inc}}^2(x)\,dx\right|^2}
\tag{7.6}
$$

The facet reflectivity for the TM mode is obtained in the same way.

The calculated reflectivity of single-layer AR-coated LD facets for the TE mode is shown in Figure 7.4 as a function of film thickness. The parameters used in these calculations are $\lambda = 1.55\ \mu\text{m}$, $n_a = 3.524$, $n_{\text{cl}} = 3.169$, and $d_a = 0.11\ \mu\text{m}$. With the appropriate combination of refractive index and film thickness, reflectivity of less than $10^{-4}$ can be obtained even for LD facets with a single-layer AR coating.

The optimum refractive index $n_{\text{opt}}$ and the optimum normalized film thickness $n_{\text{opt}}h_{\text{opt}}/\lambda$ for the facet of the slab waveguide are shown in

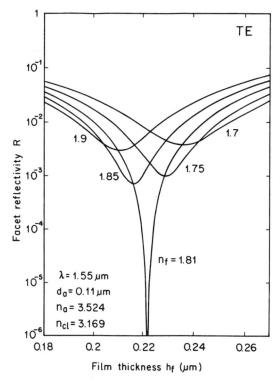

**Figure 7.4.** Calculated reflectivity of a single-layer AR-coated LD facet as a function of film thickness for various refractive indices of the coating film [30].

Figure 7.5 as a function of active-layer thickness $d_a$. The $n_{opt}$ has a peak value larger than $\sqrt{n_a}$ at around $d_a = \lambda/n_a$ and approaches $\sqrt{n_{cl}}$ and $\sqrt{n_a}$ for small and large values of $d_a$, respectively. Around $d_a = \lambda/n_a$, $n_{opt}$ becomes large to reduce the elementary-plane wave components whose reflection coefficients cannot be reduced by adjusting the film thickness because they are totally reflected from the boundary between the AR coating film and the LD. The normalized film thickness is also dependent on $d_a$, and is always larger than 0.25, corresponding to the plane wave case.

Antireflection coating conditions for TE and TM modes are different [72–74] in the same way that the uncoated facet reflectivity depends on the polarization direction [75]. This difference becomes very noticeable around $d_a = \lambda/n_a$. Multilayer AR coatings have also been investigated [73] to satisfy conditions for both polarizations as well as to broaden the low-reflectivity wavelength range.

**7.2.1.2. Fabrication.** To satisfy the AR coating conditions, both the refractive index and the thickness of the coating film must be controlled.

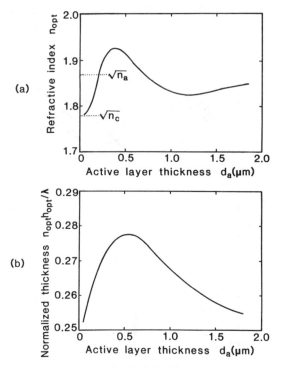

**Figure 7.5.** Optimum AR-coating conditions for an LD facet: (*a*) refractive index; (*b*) normalized film thickness of coating film; $\lambda = 1.55~\mu$m, $n_a = 3.524$, and $n_{cl} = 3.169$ [30].

Many dielectric materials, such as SiO [30, 76, 77], SiN [78–80], $SiO_2$–$Si_3N_4$ [81], and PbO–$SiO_2$ [82], have been used as AR coating films on LD facets. The refractive index is usually controlled by the evaporation or sputtering conditions. For example, the refractive index of thermally evaporated $SiO_x$ film at $\lambda = 1.536~\mu$m is shown in Figure 7.6 as a function of the oxygen partial pressure with evaporation rate as a parameter [30]. The refractive index decreases from 1.9 to 1.5 with an increase in the oxygen partial pressure and a decrease in the evaporation rate.

The wavelength giving the minimum reflectivity for both facets must be adjusted to the operating wavelength of the TWA. Methods of monitoring the output power during evaporation have been adopted to control the minimum reflectivity wavelength. Edmonds et al. [83] have tried to reduce the facet reflectivity by monitoring the spontaneous emission power from an LD biased *below* the original oscillation threshold. However, their method is not suitable for fabricating a TWA because the operating current of the TWA is a few times larger than that of the original LD threshold. Low reflectivity at the operating wavelength has been achieved [84] using a method involving monitoring the output power of LDs biased

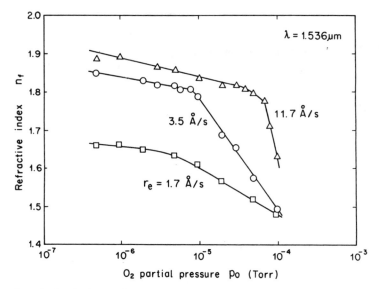

**Figure 7.6.** Refractive index of evaporated $SiO_x$ film at 1.53-$\mu$m wavelength as a function of oxygen partial pressure $p_o$ for various evaporation rates $r_e$.

at a few times the original oscillation threshold during evaporation and stopping the evaporation when the output power reaches the minimum point.

An experimentally obtained monitored output power signal as a function of evaporation time is shown in Figure 7.7 with a schematic graph of the dependence of the facet reflectivity and the gain coefficient on wavelength. Output power, at first, increases due to a decrease in facet reflectivity, that is, an increase in differential quantum efficiency. Then, monitored power rapidly decreases because the threshold current exceeds the prescribed bias current when the film thickness approaches the optimum, which gives the minimum reflectivity. Thus, the lowest facet reflectivity at the gain-peak wavelength can be obtained by stopping the evaporation at the minimum output power.

When biased far above the threshold, the injected carrier density increases as the facet reflectivity decreases and the gain spectrum simultaneously shifts toward the shorter wavelengths due to the band-filling effect. Each numbered gain spectrum corresponds to a numbered point in the monitored output power signal and the reflectivity curve in Figure 7.7. At the minimum output power, point 3, the minimum reflectivity wavelength automatically equals the gain-peak wavelength $\lambda_2$. The AR coating thus obtained from the first-facet evaporation is suitable for applications that only need an AR coating on a single facet. Note that the minimum reflectivity wavelength must be set to a shorter wavelength for TWAs,

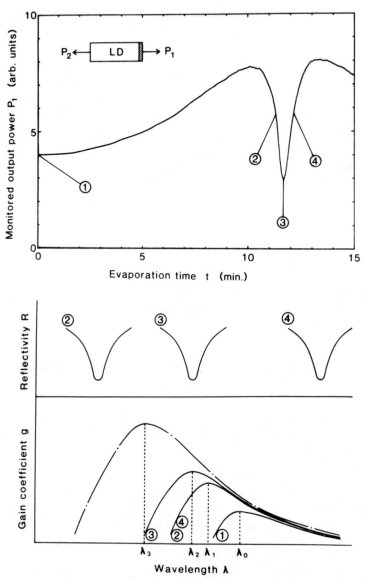

**Figure 7.7.** Monitored output power during evaporation (upper segment), and the dependence of facet reflectivity and gain coefficient on wavelength (lower segment). Numbered curves correspond to each other [67].

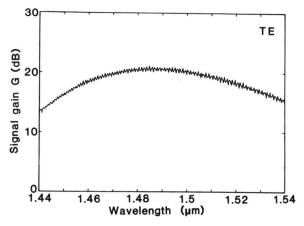

**Figure 7.8.** TE-mode amplified spontaneous emission power spectrum of AR-coated TWA at bias current of 60 mA. The curve is calibrated to amplifier signal gain using the signal gain value at 1.48 $\mu$m measured by external signal light injection [3].

since the gain curve shifts toward shorter wavelengths, for example, $\lambda_3$ in the figure, after the second-facet AR coating. This indicates that the AR coating film of the first facet must be thinned by a few nanometers for TWAs with very low facet reflectivities.

A TE-mode amplified spontaneous emission (ASE) spectrum of a TWA made by precise single-layer AR coating is shown in Figure 7.8 over a wavelength range of 100 nm [3]. The ASE spectrum was calibrated to give the signal gain spectrum. The 3-dB bandwidth for a 21-dB peak signal gain is broadened to about 70 nm without being restricted by the FP gain profile. The geometric average of the two facet reflectivities is estimated as 0.04% by the signal-gain undulation.

### 7.2.2. Angled Facet Structure

By slanting the active region stripe from the cleavage plane as shown in Figure 7.9, the effective reflectivity, that is, the coupling between the guided field and the reflected field from the cleaved facet, can also be reduced. This structure is called an angled facet structure or an angled stripe structure. It has been studied for fabricating superluminescent diodes [85] and external-cavity LDs [86] for mode-locking. Recently, this structure has attracted considerable attention for fabricating TWAs [53–56] and has been analyzed theoretically [87, 88].

The effective reflectivity of an angled facet is obtained by calculating the coupling between the incident field and the reflected field, in a manner similar to the process for calculating the AR coated facet reflectivity. Using the Gaussian beam approximation, we obtain a simple formula for

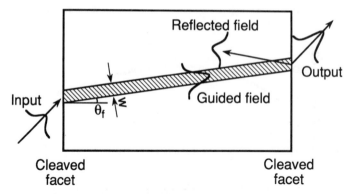

**Figure 7.9.** TWA with angled facets.

the effective reflectivity of an angled facet $R_{ang}(\theta_f)$ [88]:

$$R_{ang}(\theta_f) = R_f(\theta_f)\exp\left[-\left(k_{cl}w_0\theta_f\right)^2\right]$$
$$\equiv R_0 R_{ang, rel}(\theta_f) \tag{7.7}$$

where $R_f(\theta_f)$ and $R_0$ are the Fresnel reflectivities of plane waves that are reflected from the angled and normal facets, respectively, $w_0$ is the Gaussian beam spot size, and $\theta_f$ is the angle between the beam propagation direction and the normal of the end facet (see Figure 7.9). The relative reflectivity $R_{ang, rel}(\theta_f)$ is defined as the effective reflectivity of the angled facet structure having 100% end-facet reflectivity.

The calculated relative reflectivity $R_{ang, rel}(\theta_f)$ is shown in Figure 7.10 as a function of the facet angle with the active region width as a parameter.

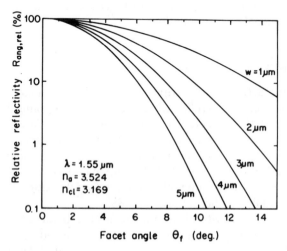

**Figure 7.10.** Calculated relative reflectivity of an angled facet as a function of the facet angle for several waveguide widths.

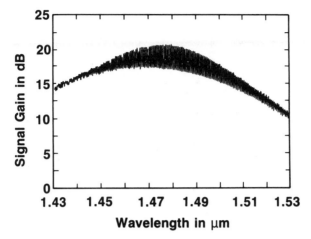

**Figure 7.11.** TE-mode amplified spontaneous emission power spectrum of a TWA with angled facets at a 20-dB peak signal gain [53].

The relative reflectivity decreases with an increase in the facet angle $\theta_f$. However, the coupling efficiency between the TWA and an optical fiber is degraded for large facet angles due to the asymmetry of the far-field pattern. When the active region width increases, the effective reflectivity decreases because the angular spectrum of the reflected wave is narrowed. However, the active region width is limited by the single transverse mode condition. If the active region is broadened only near the end facet, low reflectivity will be obtained, preserving the single transverse mode operation [89].

The TE-mode ASE spectrum of an angled facet TWA *without* AR coating [53] is shown in Figure 7.11. The device is a ridge-waveguide structure with a ridge width of 3.8 $\mu$m and a facet angle of about 7°. Effective reflectivity is estimated to be 0.2% from the signal gain and the undulation of the ASE spectrum. Effective reflectivity as low as 0.01% can be expected from a combination of the angled facet structure and an approximately 1% AR coating.

### 7.2.3. Window Facet Structure

The effective reflectivity can also be reduced by introducing a window facet structure (or a buried facet structure), which has a transparent window region between the end of the active layer and the cleaved end facet. The window facet structure has been used to suppress Fabry–Perot resonances in distributed feedback (DFB) semiconductor lasers [90]. TWAs with polarization-insensitive gain have recently been fabricated [57, 58] using window facet structures with thick active layers for which AR coating conditions are very different between polarizations.

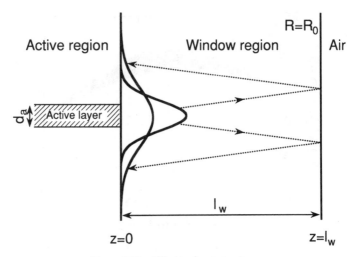

**Figure 7.12.**   Window facet structure.

A schematic diagram of the window facet structure is shown in Figure 7.12. The guided field in the waveguide is emitted into the window region, which is composed of transparent material. The field propagates with some radiation angle due to diffraction, and is reflected at the cleaved facet. The effective reflectivity of the window facet structure is obtained from the coupling between the guided field and the reflected field at the end of the active layer. The effective reflectivity of the window facet structure $R_{\text{win}}(l_w)$ for the slab waveguide is given by a Gaussian beam approximation [91] as

$$R_{\text{win}}(l_w) = R_0 \frac{\sqrt{1 + \left(l_w^2/k_{\text{win}}^2 w_0^4\right)}}{\sqrt{\left[1 + \left(l_w^2/k_{\text{win}}^2 w_0^4\right)\right]^2 + \left(l_w^2/k_{\text{win}}^2 w_0^4\right)}}$$

$$\equiv R_0 R_{\text{win, rel}}(l_w) \tag{7.8}$$

where $R_0$ is the Fresnel reflectivity of the cleaved end facet, $l_w$ is the window region length, $k_{\text{win}} = 2\pi n_{\text{win}}/\lambda$, $n_{\text{win}}$ is the refractive index of the window material, and $w_0$ is the Gaussian beam spot size. The relative reflectivity $R_{\text{win, rel}}(l_w)$ is defined as the effective reflectivity of the window facet structure having 100% end-facet reflectivity. Note that interface reflectivity $R_{\text{int}}$ between the active and window regions limits the total reflectivity of the window facet structure when $R_{\text{win}}$ becomes very low. Its value is about $10^{-4}$ for the structure having a 1.55-$\mu$m wavelength range GaInAsP-active layer and an InP-window region, and it can be reduced by matching the refractive index of the window region to the effective refractive index of the active region.

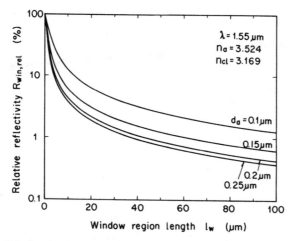

**Figure 7.13.** Calculated relative reflectivity of a window structure as a function of the window region length for several active-layer thicknesses.

Figure 7.13 shows the calculated relative reflectivity $R_{win, rel}(l_w)$ of the slab waveguide for several active-layer thicknesses as a function of the window region length $l_w$. As the window region length $l_w$ and the active-layer thickness $d_a$ increase, the relative reflectivity decreases because the reflected beam spot at $z = 0$ grows in size. However, a long window region degrades the coupling between the TWA and the optical fiber. For rectangular cross-sectional waveguides, the window facet structure more effectively reduces the effective reflectivity. However, if the active-layer thickness is increased far beyond $d_a = \lambda/n_a$, then the relative reflectivity increases because the diffraction angle of the incident beam at $z = 0$ decreases.

Cha et al. [57] have fabricated a TWA with window facet structures (Figure 7.14). Front and rear window region lengths are 35 and 55 $\mu$m, respectively, and the active-layer thickness is 0.26 $\mu$m. Both cleaved facet

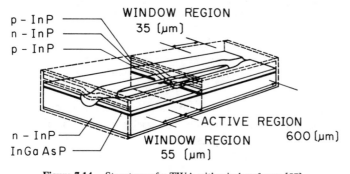

**Figure 7.14.** Structure of a TWA with window facets [57].

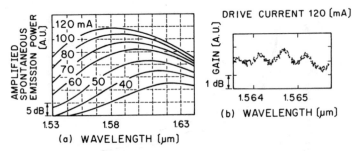

**Figure 7.15.** (*a*) Amplified spontaneous emission power spectra of the TWA with window facets for various bias currents; (*b*) unsaturated signal gain spectrum for TE-polarized input signal [57].

reflectivities are reduced to about 1% by AR coatings. ASE spectra and a signal-gain spectrum for the TE-polarized input signal of the TWA are shown in Figure 7.15. The undulation of the ASE due to Fabry–Perot resonance is almost completely suppressed within a wavelength range of 100 nm. It is difficult to suppress the facet reflectivity over such a wide wavelength range by single-layer AR coatings alone. The average reflectivity of both facets is estimated to be less than 0.03% from the undulation of the signal-gain spectrum.

## 7.3. FUNDAMENTAL DEVICE CHARACTERISTICS

This section compares the fundamental characteristics of TWAs—small-signal gain, signal-gain saturation, and noise—with those of FPAs.

### 7.3.1. Small-Signal Gain

**7.3.1.1. Small-Signal Gain and Its Bandwidth.** Small-signal gain for an SLA is given [21] by the power transmission coefficient of an active Fabry–Perot etalon, which includes a gain medium in the cavity, as a function of input signal frequency $\nu$ in the form of

$$G(\nu) = \frac{(1 - R_1)(1 - R_2)G_s}{\left(1 - \sqrt{R_1 R_2}\, G_s\right)^2 + 4\sqrt{R_1 R_2}\, G_s \sin^2\!\left[\pi(\nu - \nu_0)/\Delta\nu\right]} \quad (7.9)$$

where $R_1$ and $R_2$ are input and output facet reflectivities, $\nu_0$ is the cavity resonant frequency, and $\Delta\nu$ is the free spectral range (FSR) of the SLA. The single-pass gain $G_s$ is expressed as

$$G_s = \exp[(\Gamma g - a)L] \quad (7.10)$$

where $\Gamma$ is the optical mode confinement factor for the active layer, $g$ is the material gain coefficient, $a$ is the absorption coefficient, and $L$ is the amplifier length. For Eq. (7.9), the 3-dB bandwidth $B$ (full-width at half-maximum) of an FPA is expressed as

$$
B = \frac{2\Delta\nu}{\pi} \sin^{-1}\left[ \frac{1 - \sqrt{R_1 R_2}\, G_s}{\left(4G_s\sqrt{R_1 R_2}\right)^{1/2}} \right] \tag{7.11}
$$

On the other hand, the 3-dB bandwidth of a TWA is three orders of magnitude larger than that of the FPA, since the 3-dB bandwidth of the TWA is determined by the full gain width of the amplifier medium itself, without being restricted by the FP gain profile. Gain undulation $\Delta G$ (dB)($= v$), which is defined as the difference between resonant and nonresonant signal gain, is derived from Eq. (7.9) as

$$
\Delta G = \left( \frac{1 + \sqrt{R_1 R_2}\, G_s}{1 - \sqrt{R_1 R_2}\, G_s} \right)^2 \tag{7.12}
$$

It should be less than 3 dB for TWAs over the entire signal-gain spectrum.

The first 1.5-$\mu$m true TWA with a residual facet reflectivity of 0.04% and $\Delta G$ for a 24.5-dB peak signal gain of 1.5 dB was fabricated by Saitoh and Mukai in 1986 [32]. The TWA signal-gain spectra measured for TE- and TM-polarized optical signals are shown in Figure 7.16 with an FPA signal-gain spectrum. Signal gain fluctuates smoothly over the entire

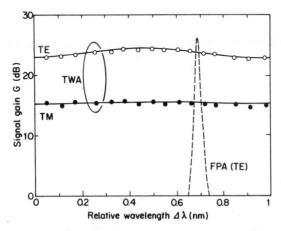

**Figure 7.16.**   Unsaturated signal gain spectra of a TWA and an FPA within one free spectral range. Circles ○ and ● denote the TWA experimental signal gains for TE- and TM-polarized input signals. Solid curves represent theoretical FP resonance curves fitted to the TWA experimental data. The dashed curve represents experimental FPA data [84].

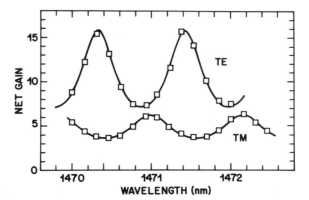

**Figure 7.17.** Signal gain spectra for TE- and TM-polarized input signals. Solid lines are calculated FP resonance curves [92].

FSR of the TWA, in contrast to the FPA, where signal gain is obtained only in the vicinity of the resonant frequencies. Furthermore, the 3-dB bandwidth of the signal gain is broadened to 70 nm, that is, more than 9 THz, without being restricted by the FP mode. For the TWA, the obtained gain difference $\Delta G_{TE/TM}$ between TE and TM polarizations is still large, although it is far smaller than those of FPAs [23]. These polarization-dependent signal-gain characteristics arise mainly from the difference in the optical mode confinement factor. In communications system design, $\Delta G_{TE/TM}$ must be evaluated in the worst case by the difference between the maximum of the TE gain and the minimum of the TM gain.

**7.3.1.2. Reduction of Polarization Sensitivity.** Signal-gain spectra of a near-TWA for both TE and TM signal polarization states [92] are shown in Figure 7.17. The spectra have different amplitudes (dichroism) and different FSRs (birefringence). This is the origin of the reduction in TWA effective bandwidth, which arises from the random change over time in the polarization state at the fiber output. Birefringence is negligible when the signal-gain undulation disappears due to sufficiently low facet reflectivity. Several methods of reducing or compensating for the signal-gain difference between polarizations have been demonstrated. These include a thick narrow active-layer structure, separate confinement heterostructure (SCH), or large optical cavity (LOC) structures, twin-amplifier configurations, and a double-pass configuration.

If we can make the active-layer width the same as the active-layer thickness, then signal gain will become completely insensitive to polarization. Problems exist in maintaining the single transverse mode conditions or the reproducibility in fabricating narrow active-layer structures. The gain difference between polarizations has been reduced to less than 1.3 dB

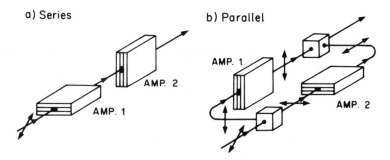

**Figure 7.18.** Twin-amplifier configurations: (*a*) series and (*b*) parallel arrangements [95].

for a 26-dB TE signal gain [57] by making the active layer 0.26 $\mu$m thick and 0.4 $\mu$m wide. However, a TWA with a thick active layer has two disadvantages: low saturation output power if the operating carrier density is low (see Section 7.3.2) or thermally caused gain saturation if the operating current density is high.

Besides the thick narrow active layer structure, the SCH or LOC structures are also useful for reducing polarization sensitivity [93]. A gain difference of less than 1 dB has been obtained using a *modified LOC structure* [94]. The SCH or LOC structures can obtain high saturation output power because they have thin active layers resulting in high operating carrier density.

Twin-amplifier configurations [95] are shown in Figure 7.18. In series, the TE polarization in the first amplifier becomes TM polarization in the second amplifier and vice versa. If both amplifiers exhibit equal gain characteristics, we expect the combined system to exhibit polarization insensitivity. Even though the series configuration is simple, it suffers from the problem of mutual coupling between amplifiers. In the parallel configuration, the two amplifiers each see a TE-polarized signal. Therefore, the signal gain can be adjusted to the optimized TE polarization. Signal-gain spectra for both polarizations are shown in Figure 7.19. The TE and TM gain spectra of a single amplifier are shown in Figure 7.19*a*. The maximum $\Delta G_{\text{TE/TM}}$ of a single amplifier is about 7 dB. Figure 7.19*b* shows the signal-gain spectra of the parallel arrangement. This configuration gives $\Delta G_{\text{TE/TM}}$ of less than 0.6 dB at any wavelength.

A double-pass configuration [96] is schematically shown in Figure 7.20. In this configuration, the signal lights pass through the same amplifier twice, and the polarization is rotated by 90° between passes. Therefore, this configuration provides an intermediate value of TE and TM gains for input signals with any polarization state, similar to the series arrangement in Figure 7.18. In spite of a 6-dB loss due to the 3-dB fiber coupler, this configuration gives relatively high gain as a single amplifier because it operates twice on the same signal. Signal gains for different polarizations

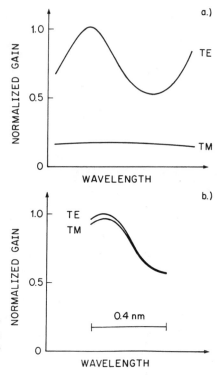

**Figure 7.19.** Normalized signal gain for two orthogonal polarizations of input signal: (*a*) for a single amplifier (TE and TM); (*b*) for amplifiers in parallel configuration [95].

are shown in Figure 7.21 as a function of amplifier bias current. The gain difference $\Delta G_{\mathrm{TE/TM}}$ between polarizations is reduced to about 0.2 dB when a Faraday rotator is inserted (broken lines in Figure 7.21), while $\Delta G_{\mathrm{TE/TM}}$ is about 4 dB without a Faraday rotator. The disadvantage of this configuration is the low saturation output power because the signal passes through the amplifier twice.

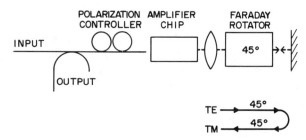

**Figure 7.20.** Double-pass configuration amplifier [96].

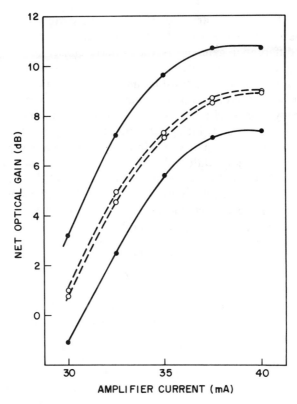

**Figure 7.21.** Measured signal gain of a double-pass configuration amplifier. Solid lines represent signal gain without the Faraday rotator. Dashed lines denote signal gain with the Faraday rotator [96].

### 7.3.2. Signal Gain Saturation

The signal-gain saturation of an SLA is caused by a reduction of the population inversion in the active layer due to an increase in the stimulated emission. The gain saturation characteristics are especially important in optical repeaters and multichannel amplifiers, which require high-power operation.

The injected carrier density can be determined by using the following single-mode rate equations [97] to which a term for the optical injection is added:

$$\frac{dN_e}{dt} = \frac{J}{qd_a} - \frac{N}{\tau_c} - A_g v_g (N_e - N_0) S \qquad (7.13)$$

$$\frac{dS}{dt} = -\frac{S}{\tau_p} + \frac{A_g v_g \Gamma N_e}{V_0} + A_g v_g \Gamma (N_e - N_0) S + S_{\text{in}} \qquad (7.14)$$

where $N_e$ is the injected carrier density, $J$ is the current density, $q$ is the electron charge, $d_a$ is the active-layer thickness, $\tau_c$ is the carrier lifetime, $A_g$ is the differential gain, $v_g$ is the group velocity in the amplifying medium, $N_0$ is the carrier density required for transparency, $S$ is the photon density, $\tau_p$ is the photon lifetime, $\Gamma$ is the optical mode confinement factor, $V_0$ is the optical mode volume, and $S_{in}$ is the injected signal photon density. For precise analysis, we should use multimode rate equations [23] that can properly include the effect of the gain saturation due to total spontaneous emission power. However, the single-mode rate equations are simple and physically sufficient to treat the gain saturation due to amplified signal power. By representing the number of the photons in the active layer by the optical intensity $I$ ($= h\nu v_g S$) and using the linear approximation of the gain coefficient given as $g = A_g(N_e - N_0)$ [see Eq. (7.25)], the following equation is obtained:

$$g = \frac{g_0}{1 + I/I_s} \tag{7.15}$$

where $g_0$ is the unsaturated gain coefficient and $I_s$ is the saturation intensity defined by the optical intensity that reduces $g$ to $g_0/2$; $I_s$ is given by

$$I_s = \frac{h\nu}{A_g \tau_c} \tag{7.16}$$

The growth rate for the signal intensity $I(z)$ along the amplifier thus becomes

$$\frac{dI(z)}{dz} = \frac{\Gamma g_0}{1 + I(z)/I_s} I(z) \tag{7.17}$$

Here, the absorption coefficient $a$ is omitted. The signal gain $G$ is derived by integrating Eq. (7.17) from the input port to the output port of the laser amplifier [98]:

$$G \equiv \frac{I_{out}}{I_{in}} = G_0 \exp\left[-\frac{I_{out} - I_{in}}{I_s}\right]$$

$$= G_0 \exp\left[-\frac{G-1}{G}\frac{I_{out}}{I_s}\right] \tag{7.18}$$

where $I_{in}$ and $I_{out}$ are the input and the output optical intensities and $G_0$ is the unsaturated signal gain in the steady state. Regardless of the unsaturated value, the signal gain is reduced to $1/e$, that is by 4.34 dB when the output intensity $I_{out}$ is equal to the saturation intensity $I_s$. The

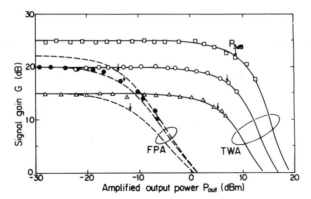

**Figure 7.22.** Theoretical and experimental signal gain of a TWA and an FPA as a function of amplified output power. Solid curves represent calculated results for the TWA, dashed curves for the FPA. Saturation output power $P_{3\,dB}$ is indicated by arrows [99].

saturation intensity $I_s$ is a physical parameter and is related to the saturation output intensity $I_{3\,dB}$, which is by definition the intensity at which the signal gain has fallen by 3 dB from the unsaturated value:

$$I_s = \frac{G_0 - 2}{G_0 \ln 2} I_{3\,dB} \qquad (7.19)$$

Equation (7.19) indicates that $I_s$ is larger than $I_{3\,dB}$ by 1.5–1.6 dB.

Experimental signal gain saturation characteristics of TWAs and FPAs [99] are shown in Figure 7.22 along with theoretical curves. The signal gain clearly decreases with an increase in amplified output power. The saturation characteristics can be evaluated from the saturation output power $P_{3\,dB}$ $(= I_{3\,dB} d_a w / \Gamma)$. The $P_{3\,dB}$ of the TWA ranges from 5 to 9 dBm, which is about 20 dB larger than that of the FPA. The increase in the $P_3$ dB of the TWA is attributed to the suppression of the resonant effect; that is, a slight saturation in gain coefficient results in large signal-gain saturation in FPAs, and to the increase in the saturation intensity $I_s$ of the amplifying medium due to high operating carrier density [3]. Note that in the TWA $P_{3\,dB}$ increases with the signal gain, in contrast to the FPA, in which the slight increase in $I_s$ is screened by the resonant effect.

Figure 7.23 shows the dependence of $I_s$ for 1.55-$\mu$m GaInAsP material on the gain coefficient for different concentrations of the p-type dopant [100]. Compared with the undoped GaAs material indicated by the dash–dotted line, GaInAsP materials have smaller $I_s$ values in the small-g region because the photon energy is halved [see Eq. (7.16)]. The increase in $I_s$ in the region of large $g$ results from $\tau_c$ decreasing with the carrier density. This effect is very strong in 1.55-$\mu$m GaInAsP material whose carrier lifetime is rapidly reduced in this region as a result of the

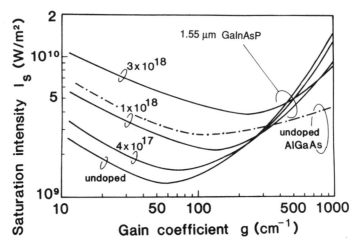

**Figure 7.23.** Calculated saturation intensities of 1.55-$\mu$m GaInAsP (solid lines) and AlGaAs (dash-dotted line) as functions of gain coefficient for several doping concentrations [100].

Auger recombination process [101]. The large value of $I_s$ in the region of small $g$ is due to $A_g$ becoming smaller in the small-gain coefficient region.

If the device length $L$, the optical mode confinement factor $\Gamma$, and the facet reflectivity $R$ are decreased, then the saturation intensity $I_s$ of the TWA increases because operation takes place in the large-$g$ region. Large $I_s$ can also be obtained, if the carrier lifetime is shortened, such as by carrier diffusion from nearby carrier storage regions [102]. If $I_s$ is not degraded, an increase in the mode cross-sectional area $S_e$ $(= d_a w/\Gamma)$ is another effective way to increase the TWA output power.

Henning et al. [103] have shown theoretically that the $P_{3\ dB}$ can be increased by setting the signal wavelength longer than the gain-peak wavelength. Their prediction has been demonstrated experimentally [104] by the wavelength-dependent signal-gain saturation characteristics in a TWA. Figure 7.24 shows the saturation output power and the unsaturated signal gain of the TWA as a function of the input signal wavelength. The $P_{3\ dB}$ monotonically increases with an increase in the signal wavelength. This is because the gain saturation reduces the injected-carrier density and causes the gain peak to shift toward a longer wavelength in accordance with the reverse process of band-filling, which compensates for the decrease in gain at the longer wavelength.

This unique feature in a semiconductor gain medium has clearly been demonstrated in a TWA (Figure 7.25), where the saturated signal-gain spectrum coincides exactly with the unsaturated spectrum under less biased conditions [105]. This experimental result also clarifies that the semiconductor laser gain saturates *homogeneously* over the *entire* gain spectrum, which is especially important for understanding saturation-

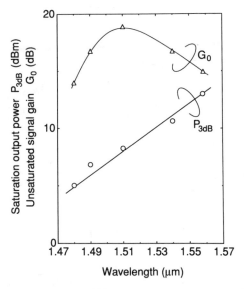

**Figure 7.24.** Saturation output power (○) and unsaturated signal gain (△) of a TWA as a function of input signal wavelength. (Redrawn from Figure 2 of Inoue et al. [104].

induced crosstalk in multichannel amplification. The choice of operating wavelength is important for both high-power and low-noise (see Section 7.3.3.1) operation.

### 7.3.3. Noise

Noise, which cannot be completely eliminated, is one of the most important characteristics of SLAs, especially when they are used as optical in-line repeaters, since it determines the maximum number of repeaters and the spacing of digital regenerative repeaters [26, 106].

According to an analysis based on the photon statistic master equation [107], the photon fluctuation at the amplifier output is expressed [26, 106] by

$$
\begin{aligned}
\sigma_{\text{out}}^2 &\equiv \langle n_{\text{out}}^2 \rangle - \langle n_{\text{out}} \rangle^2 \\
&= G\langle n_{\text{in}} \rangle \\
&\quad + (G - 1)n_{\text{sp}}m_t\Delta f_1 \\
&\quad + 2G(G - 1)n_{\text{sp}}\chi\langle n_{\text{in}} \rangle \\
&\quad + (G - 1)^2 n_{\text{sp}}^2 m_t \Delta f_2 \\
&\quad + G^2\big(\langle n_{\text{in}}^2 \rangle - \langle n_{\text{in}} \rangle^2 - \langle n_{\text{in}} \rangle\big)
\end{aligned}
\tag{7.20}
$$

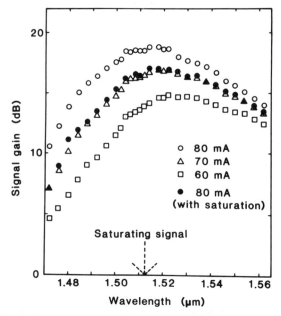

**Figure 7.25.** Signal gain spectra measured for a weak probe beam with and without a strong injected saturating beam. Circle ● denotes the saturated signal gain for a TWA with a bias current of 80 mA. Symbols ○, △, and □ are unsaturated signal gains measured at bias currents of 80, 70, and 60 mA, respectively, with no saturating beam [105].

where $\sigma_{\text{out}}^2$ is the variance in the photon number per second at the amplifier output, $\langle n_{\text{in}} \rangle$ is the mean value of the input photon number per second incident on the amplifier, $n_{\text{sp}}$ is the population inversion parameter of the amplifier medium, $\chi$ is the excess noise coefficient for the signal–spontaneous beat noise, $\Delta f_1$ and $\Delta f_2$ are the equivalent noise bandwidths for the spontaneous emission shot noise and spontaneous–spontaneous beat noise, and $m_t$ is the number of the effective transverse modes. The five terms on the right-hand side of Eq. (7.20) respectively represent the amplified signal shot noise, spontaneous emission shot noise, signal–spontaneous beat noise, spontaneous–spontaneous beat noise, and signal excess noise. The first and third terms denote the noise proportional to the signal photon number, while the second and the fourth terms indicate the noise generated from the broadband ASE that exists independently of the input signal. The variance in the input photons for a completely coherent signal with Poisson distribution and a completely incoherent signal with Bose–Einstein distribution is of the form [106]

$$\sigma_{\text{in}}^2 \equiv \langle n_{\text{in}}^2 \rangle - \langle n_{\text{in}} \rangle^2 = \begin{cases} \langle n_{\text{in}} \rangle & \text{coherent} \\ \langle n_{\text{in}} \rangle^2 + \langle n_{\text{in}} \rangle & \text{incoherent} \end{cases}$$

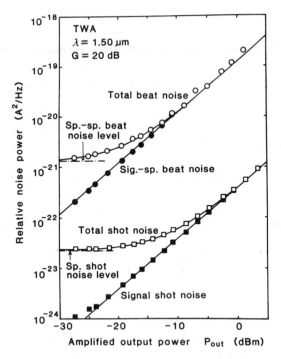

**Figure 7.26.** Relative noise power of TWA output for a 20-dB signal gain at 1.50 $\mu$m as a function of amplified output power. An etalon filter with an FSR of 1.5 THz and a finesse of 11 is used in front of the APD to suppress both the spontaneous–spontaneous beat noise and spontaneous emission shot noise [3].

Therefore, the last term disappears when the input signal is completely coherent. On the other hand, when amplifiers are cascaded in tandem, the input signal into the $i$th amplifier ($i \geq 2$) becomes incoherent because of the ASE power from the preceding amplifiers. Then, input signal excess noise $\sigma_{in}^2 - \langle n_{in} \rangle$ is amplified by a factor of $G^2$ via the fifth term of Eq. (7.20), which is essential in calculating the noise accumulation in the cascaded amplifier system [1, 26]. The two beat noises are the dominant noise source in an SLA because their values are $G$ times larger than the shot noise levels.

Figure 7.26 shows the relative noise power versus the amplified output power of a 1.5-$\mu$m GaInAsP TWA [3] measured at a 20-dB signal gain. These characteristics are measured by an avalanche photodiode (APD) followed by electronic amplifiers and an RF spectrum analyzer. Beat noise is estimated by the RF spectrum analyzer, while shot noise is estimated from the APD photocurrent after the amplifier output is attenuated by a factor of $\eta_{out}$. Since the beat noise decreases with $\eta_{out}^2$, while the shot noise decreases with $\eta_{out}$ [108], the measured noise levels are converted to values at the amplifier output. The noise power at the amplifier output is divided by the values of the load resistance and the electrical bandwidth to express the noise power in the units of noise power spectrum density [square amperes per hertz ($A^2/Hz$)], as shown in Figure 7.26. The frequency characteristics of beat noise are almost flat at low frequencies because their bandwidths are the same as those of the signal gain [108]. Therefore, beat noise powers were measured at 100 MHz in this case.

The beat noise levels are larger than the shot-noise levels by about 20 dB, corresponding to the signal-gain value $G$. This is also understood by Eq. (7.20). The dominant noise in the high-power region, as in repeater application, is the signal–spontaneous beat noise, while the spontaneous–spontaneous beat noise is dominant in the low-power region. The signal–spontaneous beat noise is inherent to SLAs because it arises from the beat between the amplified signal power and the ASE power around the signal frequency. On the other hand, the spontaneous–spontaneous beat noise can be reduced by loading a narrowband optical filter matched to the signal frequency, since this noise arises from the beat between the ASE components over a wide gain spectrum.

**7.3.3.1. Signal–Spontaneous Beat Noise.** First, we focus discussion on the signal–spontaneous beat noise. Experimental signal–spontaneous beat noise for both a TWA [109] and an FPA [100] at 20-dB signal gains is shown in Figure 7.27 and compared with the calculated theoretical values.

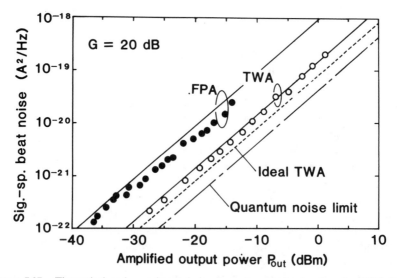

**Figure 7.27.** Theoretical and experimental signal-spontaneous beat noise of a TWA ($\bigcirc$) and an FPA ($\bullet$) for a 20-dB signal gain at 1.5 $\mu$m as a function of amplified output power [67].

Signal–spontaneous beat noise level, as expressed by $4q^2 n_{sp} \chi (G - 1) P_{out}/h\nu$, increases in proportion to signal output power $P_{out}$.

Signal–spontaneous beat noise characteristics can be evaluated by the noise figure $F$, which is defined by the degradation in the signal-to-noise ratio before and after amplification; that is, the noise figure appears as an increment from the quantum noise limit indicated by the dash–dotted line in Figure 7.27. Assuming that the input signal to the amplifier is shot-noise-limited and considering two dominant beat noise components at the amplifier output, the noise figure $F$ is given [110] as

$$F \equiv \frac{(S/N)_{in}}{(S/N)_{out}}$$

$$\simeq 2 n_{sp} \chi + \frac{n_{sp}^2 m_t \Delta f}{\langle n_{in} \rangle} \qquad \text{for} \quad G \gg 1 \qquad (7.21)$$

Here, the population inversion parameter $n_{sp}$, which represents the degree of imperfect population inversion, is given by [100, 111] by

$$n_{sp} \equiv n_{sp1} \cdot n_{sp2}$$

$$= \frac{1}{1 - \exp\left[ (h\nu - \Delta E_f)/k_B T \right]} \cdot \frac{\Gamma g}{\Gamma g - a}$$

$$\simeq \frac{N_e}{N_e - N_0} \cdot \frac{\Gamma g}{\Gamma g - a} \qquad (7.22)$$

where $\Delta E_f$ is the energy separation between quasi-Fermi levels, $k_B$ is the Boltzmann constant, and $T$ is the absolute temperature. The first term $n_{sp1}$ on the right-hand side of Eq. (7.22) represents the ratio of spontaneous to net stimulated transition rate per mode. It expresses the effect of the stimulated absorption loss in the resonant transition process; the second term $n_{sp2}$ represents the additional contribution from the nonresonant losses $a$, such as free carrier loss and scattering loss. The excess noise coefficient $\chi$, which represents signal–spontaneous beat noise enhancement resulting from input mirror reflectivity $R_1$, is expressed [26] as

$$\chi = \frac{(1 + R_1 G_s)(1 - R_2)(G_s - 1)}{(1 - R_1)(1 - R_2)G_s - \left(1 - \sqrt{R_1 R_2} G_s\right)^2}$$

$$\simeq \frac{(1 + R_1 G_s)(G_s - 1)}{(1 - R_1)G_s} \qquad \text{for} \quad G \gg 1 \qquad (7.23)$$

Note that $\chi \simeq 1$ for TWAs $(R \simeq 0)$. The contribution of the spontaneous–spontaneous beat noise [the second term in Eq. (7.21)] can be reduced by using an optical frequency filter ($\Delta f \to 0$) or by operating in the high-power region ($1/\langle n_{\text{in}} \rangle \to 0$). When the signal–spontaneous beat noise exceeds the spontaneous–spontaneous beat noise, the noise figure $F$ can be reduced to

$$F = 2n_{\text{sp}}\chi \qquad (7.24)$$

The factor 2 in Eq. (7.24) indicates that even an ideal TWA has a noise figure of 3 dB, which originates from the uncertainty in quantum statistics and is the general quantum limit for phase-insensitive linear amplifiers [112, 113].

The broken line indicates the signal–spontaneous beat noise for an ideal TWA ($R_1 = R_2 = 0$) with $n_{\text{sp}}\chi = 1$, where $F = 3$ dB. Experimental results have demonstrated a noise figure $F$ of 5.2 dB in the TWA, which is an improvement of 6 dB compared with the FPA. This improvement is attributed to both the small $n_{\text{sp}}$ value resulting from TWA operation at a high carrier density and to the reduced $\chi$ value due to the low input facet reflectivity. As is easily expected from Eq. (7.22), operating at high carrier density or high gain coefficient, accomplished by decreasing $L$, $\Gamma$, and $R$, is a good way to reduce the value of $F$. Öberg and Olsson [34] measured the wavelength dependence of the noise figure of a 1.5-$\mu$m TWA and demonstrated that operating at the longer wavelength side of the amplifier gain spectrum provide smaller $F$ values. They attributed the wavelength dependence of the noise figure to the decrease in the waveguide loss and in the spontaneous emission factor with increasing wavelength. Further improvements in $F$ require small $N_0$ values, which are possible with a highly doped $n$-type active layer [1] or a quantum well structure [114, 115].

**7.3.3.2. Spontaneous–Spontaneous Beat Noise.** The dependence of the spontaneous–spontaneous beat noise of a 1.5-$\mu$m GaInAsP TWA measured without signal injection [109] is shown in Figure 7.28 as a function of the small-signal gain. The spontaneous–spontaneous beat noise increases in proportion to $G^2$ as expressed by the fourth term in Eq. (7.20). This contrasts with the FPA case where the spontaneous–spontaneous beat noise increases with the signal gain by between $G^{1.1}$ [100] and $G^{1.6}$ [108] due to the existence of the active Fabry–Perot noise filter. TWAs have greater spontaneous–spontaneous beat noise than FPAs in the high-$G$ region because of the strong dependence of the spontaneous–spontaneous beat noise on the signal gain. However, this beat noise can be reduced by about 20 dB (broken line in Figure 7.28), if the amplified signal is selectively detected using an optical frequency filter with a bandwidth equal to the FSR of the TWA.

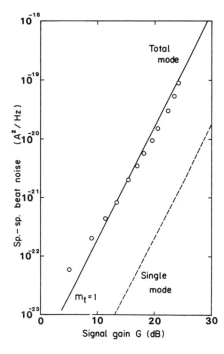

**Figure 7.28.** Theoretical and experimental spontaneous–spontaneous beat noise of TWA as a function of signal gain. Dashed and solid lines respectively represent the theoretical results with and without an optical frequency filter whose bandwidth coincides with one FSR of the TWA [109].

### 7.3.4. Design Criteria for SLAs

The preceding subsections have discussed fundamental device characteristics for SLAs, that is, signal gain, gain saturation and noise. Here, we discuss the design criteria for realizing high-gain, wide-bandwidth, high-power, and low-noise SLAs. Figure 7.29 summarizes the relationships among fundamental device characteristics, physical mechanisms, operating conditions, amplifier material parameters, and structural and device parameters. Upward and downward arrows indicate the direction in which each quantity should be designed to improve device performance.

Maximum signal gain without any optical signal injection is restricted by the gain saturation induced by the ASE power. Therefore, it can be improved by reducing the spontaneous emission factor $\beta$ [116], that is, by increasing the saturation intensity. Since the 3-dB signal gain bandwidth $B$ is restricted by the resonance effect arising from the residual reflectivity at both ends of the amplifier, the cavity resonance must be suppressed, using the methods described in Section 7.2, to produce broadband optical amplifiers.

Polarization sensitivity of the signal gain between TE and TM modes $\Delta G_{TE/TM}$ for TWAs is attributed to the difference in the optical mode confinement factor between TE and TM modes [3]. To reduce $\Delta G_{TE/TM}$, it is important to use waveguide design to make $\Gamma_{TM}/\Gamma_{TE}$ approach 1. One approach is to simply increase the active-layer thickness $d$ up to the

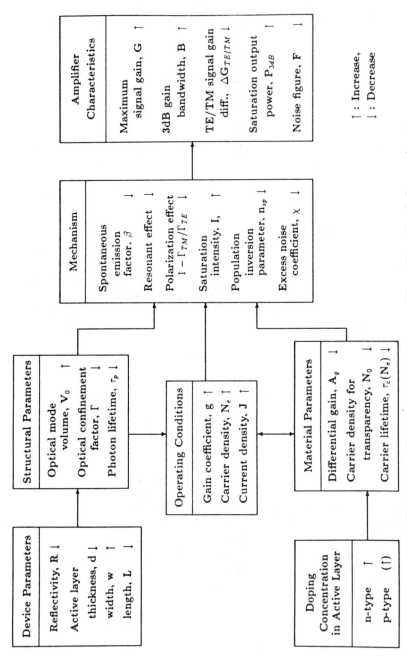

**Figure 7.29.** Design criteria for semiconductor laser amplifiers. Upward and downward arrows indicate the directions in which each quantity should be designed to improve the device performance.

limit of the single transverse mode conditions. This, however, requires reducing the amplifier length $L$ to about 50 $\mu$m to avoid degradations in saturation and noise characteristics [117]. On the other hand, when a waveguide layer is introduced between the active and cladding layers, that is, the SCH structure, $\Delta G_{TE/TM}$ can be reduced to 1.4 dB without degrading either saturation or noise characteristics for a thin active layer of ordinary length $L$ [93].

One important feature of SLAs is that material gain parameters strongly depend on both the dopant type and its concentration and on the injected carrier density. As a result, the saturation intensity and population inversion parameter, which determine the saturation and noise characteristics, respectively, become dependent on the amplifier operating conditions, as discussed next.

Figure 7.30 shows peak gain coefficient $g$ (cm$^{-1}$) versus injected carrier density $N_e$ for the 1.55-$\mu$m GaInAsP materials as a function of background doping level [100]. The dependence of the peak gain coefficient on carrier density was calculated using the density of states with a Kane function interpolated to the Halperin–Lax bandtail and Stern's improved matrix element [118, 119]. From this figure, the peak gain coefficient can

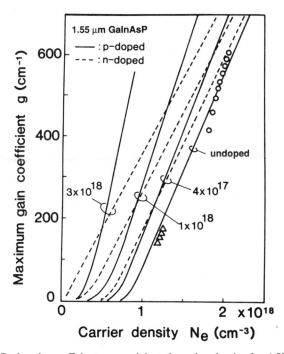

**Figure 7.30.** Peak gain coefficient versus injected carrier density for 1.55-$\mu$m GaInAsP lasers. Solid and dashed lines denote $p$-type and $n$-type active layers, respectively. Numbers indicate background dopant concentration [100].

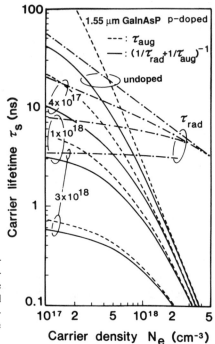

**Figure 7.31.** Carrier lifetime $\tau_c$ versus injected carrier density $N_e$ for $p$-doped 1.55-$\mu$m GaInAsP lasers. The dash-dotted line denotes radiative lifetime, while the dashed line represents Auger recombination lifetime. Total carrier lifetime is given by the solid line [100].

be approximated as a linear function of the carrier density as

$$g = A_g(N_e - N_0) \tag{7.25}$$

where $A_g$ is the differential gain and $N_0$ is the carrier density where the stimulated emission rate exceeds the stimulated absorption rate, that is, transparency. As the doping concentration increases, $A_g$ increases in the $p$-doped active layers, while it decreases in the $n$-doped active layers. The carrier density $N_0$ decreases with an increase in doping concentration irrespective to the dopant type. Note that $N_0 \simeq 0$ is achieved in a heavily doped $n$-type active layer.

Calculated carrier lifetime $\tau_c$ versus injected carrier density $N_e$ for $p$-doped 1.55-$\mu$m GaInAsP materials is shown in Figure 7.31 as a function of background doping level [100]. The dash–dotted lines denote the radiative lifetime $\tau_{rad}$ calculated in the above gain spectrum calculations, while dashed lines represent the Auger recombination lifetime $\tau_{aug}$ [120] which becomes significant in quarternary materials in the long wavelength region. Taking into account the contributions from both $\tau_{rad}$ and $\tau_{aug}$, the carrier lifetime $\tau_c$ is given by the solid lines. The $\tau_c$ rapidly decreases with an increase in $N_e$ due to the dominant contribution of the Auger recombination process. As a result, the saturation intensity $I_s$ given by Eq. (7.16)

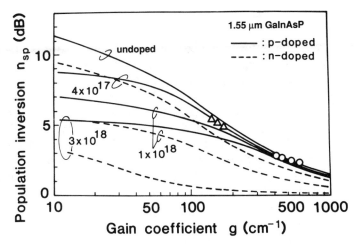

**Figure 7.32.** Population inversion parameter $n_{sp1}$ versus gain coefficient $g$ for 1.55-$\mu$m GaInAsP materials [100].

steeply increases in the high-$g$ region, that is, high-$N_e$ region [100], as shown in Figure 7.23. SLAs operating in such a high-$I_s$ region can provide high saturation output power $P_{3\,dB}$. For a fixed $I_s$ value, an increase in the mode cross-sectional area $S_e$ ($= dw/\Gamma$) is an effective way to increase $P_{3\,dB}$.

Amplifier noise figure $F$ is given by $2n_{sp1}n_{sp2}\chi$ [see Eq. (7.24)]. In TWAs, the dominant contribution comes from $2n_{sp1}$. The population inversion parameter $n_{sp1}$ defined by the first term of Eq. (7.22) represents the effect of the carrier density $N_0$, which corresponds to the stimulated absorption loss in the resonant transition process. The $n_{sp1}$ for 1.55-$\mu$m GaInAsP materials is shown in Figure 7.32 as a function of gain coefficient $g$ [100]. There are two approaches to reducing $n_{sp1}$. The first is to set the amplifier operating point in the high-$g$ region. Here, $n_{sp1}$ is effectively reduced by increasing $N_e$ with respect to a fixed $N_0$ value. The second is to use a gain material whose $N_0$ value is small, ideally $N_0 \simeq 0$. A heavily doped $n$-type active layer offers this.

In summary, Section 7.3.4 has shown two important criteria for improving SLA characteristics:

1. Use a gain medium whose material parameters $A_g$, $N_0$, and $\tau_c$ are small.
2. Choose a device design that sets the amplifier operating point in the high-$g$ region.

A heavily doped $n$-type active layer is promising for the first criterion. Effective ways to meet the second criterion include reducing facet reflectivity $R$, amplifier length $L$, and optical mode confinement factor $\Gamma$ [1, 100].

These criteria for optical amplifiers are completely opposite to those for laser oscillators. In designing LD oscillators, it is essential to make the threshold current as small as possible, in order to achieve operation at a higher pumping rate, which improves AM and FM noise characteristics as well as optical output power [116]. Conversely, for SLAs, which are operated well below the oscillation threshold, both criteria (1 and 2 above) require a higher threshold current and operation in the high-$g$ and high-$N_e$ region. In practice, the SLA characteristics cannot be improved infinitely by designing the structure to increase the threshold current; they are limited by the thermal saturation of gain, since the temperature rise associated with the increased injection current reduces the gain coefficient of the amplifying medium. To minimize this limitation, it is important to improve thermal properties by employing a device structure that ensures carrier confinement, for example, the buried heterostructure, prior to applying the second criterion.

## 7.4.  MULTICHANNEL AMPLIFICATION

A TWA can greatly simplify optical repeaters used in wavelength-division-multiplexed (WDM) systems, since its wide gain bandwidth allows a single TWA to simultaneously amplify signals of different wavelengths (i.e., multichannel amplification) [26]. This section discusses nonlinear phenomena in SLAs that cause additional crosstalk in multichannel amplification. As summarized in Figure 7.33, two significant nonlinear effects are signal gain saturation, occurring in a large input signal (see Section 7.3.2), and nearly degenerate four-wave mixing (NDFWM), occurring in closely spaced multifrequency input signals [121, 136]. The dominant

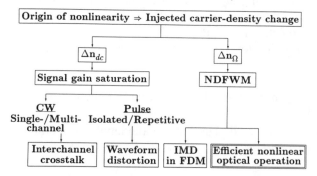

**Figure 7.33.** Nonlinear phenomena in semiconductor laser amplifiers: $\Delta n_{dc}$, DC carrier component; $\Delta n_{\Omega}$, carrier beat component [121].

nonlinearity in SLAs results from changes in the injected-carrier density as a result of signal amplification. Signal-gain saturation arises from the reduction in DC carrier component $\Delta n_{dc}$, while the NDFWM is attributed to the generation of carrier beat component $\Delta n_\Omega$ at the beat frequency $\Omega$. Gain saturation causes interchannel crosstalk in multichannel amplification and waveform distortions in pulse amplification (see Section 7.5). The NDFWM also causes intermodulation distortion (IMD) in closely spaced FDM systems, but it is a highly efficient optical nonlinear operation, suitable for frequency conversion.

### 7.4.1. Saturation-Induced Crosstalk

In multichannel amplification, interchannel crosstalk due to signal-gain saturation occurs regardless of the amount of channel separation [45–47]. Signal-gain spectra of a near-TWA in one-channel and two-channel amplification [47] are shown in Figure 8.9 in Chapter 8. In the presence of a saturating signal (two-channel amplification), the gain decreases because the carriers are consumed in providing gain for the saturating signal. Additionally, Fabry–Perot resonance wavelengths shift toward the longer-wavelength side because the refractive index of the active region increases as a result of the reduction in carrier density. This resonance effect is very strong in FPAs, while it can be neglected in TWAs with infinitesimal facet reflectivity.

Signal gains of a 1.5-$\mu$m TWA for two channels having the same input power are shown in Figure 7.34 as a function of the amplified output power [46]. There is no crosstalk between the two channels as long as the amplifier operates in the low-signal power region, that is, in the linear amplification region. In the high-signal region, on the other hand, simultaneous amplification in both channels suppresses the signal gain in each channel. This is the origin of saturation-induced interchannel crosstalk. The saturation output power $P_{3\ dB}$ for each channel is reduced by 3 dB from that in one-channel amplification. The signal gain saturation characteristics with respect to the total output power ($\blacktriangle$) coincide with those under one-channel operation ($\triangle$). This cross-saturation characteristic between the two signal channels reveals that the degree of gain saturation is uniquely determined by the *total* output power from both channels. That is, the $P_{3\ dB}$ for $N$-channel simultaneous amplification is reduced to $1/N$ of that for one-channel amplification. The underlying physical mechanism is the homogeneous gain saturation of the semiconductor laser gain over the entire gain spectrum (see Figure 7.24). Consequently, TWAs, which have higher saturation output characteristics, are suitable for multichannel amplification. To suppress the gain saturation-induced crosstalk, the amplified signal level should be set well below the $P_{3\ dB}$ for each channel. As another approach, electronic compensation of saturation-

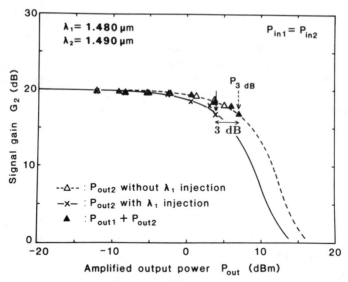

**Figure 7.34.** Signal gain of a TWA as a function of amplified output power. Solid and dashed lines represent theoretical results with and without signals in the other channel. Experimental results: × and △ denote signal gain $G_2$ of channel 2 versus amplified output power $P_{out2}$ with and without signal injection in channel 1; and ▲ indicates $G_2$ as a function of total amplified output power $P_{out1} + P_{out2}$ [46].

induced crosstalk has been demonstrated by using electrical feedforward linearization [122].

### 7.4.2. Nearly Degenerate Four-Wave Mixing

The crosstalk mentioned in Section 7.4.1 is considered to be absent in frequency shift-keying (FSK) or phase shift-keying (PSK) multiplexed transmission systems where the optical intensities of modulated signals are constant. However, even such a constant-powered FDM system, in which the signal channel separation is in the gigahertz range, suffers from interchannel crosstalk [123–126, 128] as a result of a nonlinear process in the semiconductor laser gain medium. Measured TWA output power spectra [129], when two signal light beams are injected, are shown in Figure 7.35. When two light beams of frequencies $f_0$ (pump) and $f_0 + \delta f$ (probe) are simultaneously injected, new light beams appear at frequencies of $f_0 - \delta f$ (signal 1) and $f_0 + 2\delta f$ (signal 2) in the TWA output. The new beams have the same order of output level as the injected beams. This phenomenon can be understood as nearly degenerate four-wave mixing (NDFWM) in the TWA [130–133]. That is, the incident beams at $f_0$ and $f_0 + \delta f$ are diffracted by the gain and refractive index gratings that are

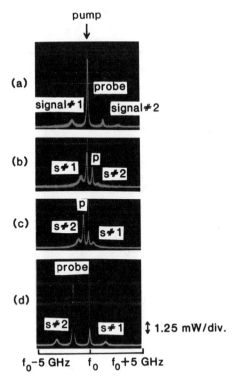

**Figure 7.35.** Nearly degenerate four wave mixing output power spectra of TWA. Pump and probe input powers are 240 $\mu$W and 166 $\mu$W, respectively. Vertical: 1.25 mW/division, Horizontal: 1 GHz/div. Probe detuning with respect to pump frequency ($f_0$) is positive in ($a$) and ($b$), and negative in ($c$) and ($d$) [129].

formed by the carrier-density modulation $\Delta n_\Omega$ at the beat frequency $\delta f$ $(= \Omega/2\pi)$.

Experimental output powers for the pump, probe, and NDFWM signal beams 1 and 2 at $I_{\text{out}}/I_s \simeq 1$ are shown in Figure 7.36 [129, 136] as a function of probe detuning $\delta f$. Calculated results are also shown as a function of normalized detuning $\Omega\tau_c$, using $\tau_c = 0.22$ ns and $\alpha$-parameter $= 6$. Pump and probe outputs exhibit strong asymmetry against the sign of probe detuning, while the NDFWM signals decrease symmetrically with probe detuning. Pump and probe input power levels are indicated by arrows on the vertical axis. Note that the conversion efficiency $\eta$, that is, the ratio of NDFWM signal output to probe input power, has a maximum of 8 dB around zero detuning, and has positive gain in the range of $\pm 6$ GHz. Furthermore, the observed probe asymmetry represents experimental verification of nonlinear additional gain, which was theoretically predicted by Bogatov et al. in 1975 [134]. Thus, the TWA is also used as an ideal device for investigating nonlinear interaction in a semiconductor laser gain medium.

Since the NDFWM signals are frequency-shifted from the incident signal frequencies by the beat frequency, they act as new crosstalk for the adjacent signal channels in closely spaced FDM systems, specifically,

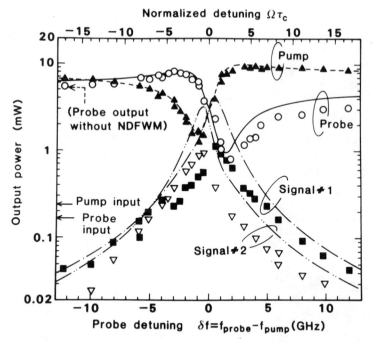

**Figure 7.36.** Experimental and calculated output powers for pump, probe, signal 1 and 2 beams as a function of probe detuning $\delta f$. Input power levels are the same as in Figure 7.35. Calculated results, obtained as a function of normalized detuning $\Omega \tau_c$, are shown for $\tau_c = 0.22$ ns [129].

intermodulation distortion (IMD). TWA operation well below the $I_s$ as well as a channel spacing of several gigahertz is required to prevent interference from the IMD [127]. The IMD can also be compensated by feedforward linearization [135]. In contrast to the crosstalk point of view, this new frequency generation due to efficient NDFWM interaction is attractive for other applications such as optical frequency exchange between two frequency channels [65] and wideband optical frequency conversion with a 4-THz conversion range [64].

Figure 7.37 shows how the NDFWM efficiency at around zero detuning is related to the TWA gain saturation characteristics [121, 136]. Figure 7.37a shows the usual signal-gain saturation characteristics for a single input beam. Here, $P_s$ denotes the TWA output power corresponding to the saturation intensity of the amplifier medium, specifically, $P_s = I_s S_e$. Figure 7.37b shows the NDFWM signal 1 output and its conversion efficiency $\eta$ as a function of total output power $P_{out}$. In the low-power region ($P_{out} \ll P_s$), NDFWM signal output increases with $P_{out}^3$, and conversion efficiency increases with $P_{out}^2$. In the extremely high-power region

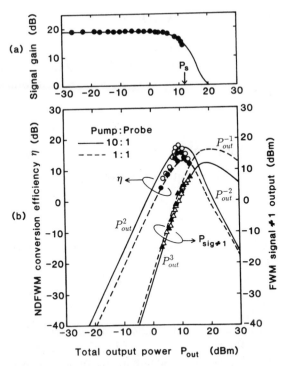

**Figure 7.37.** (a) Signal gain saturation characteristics of the TWA for the single input beam; $P_s$ denotes the TWA output power corresponding to the saturation intensity $I_s$. (b) FWM signal 1 output power and its conversion efficiency with respect to the probe input power versus the total output power of the TWA. Experimental: pump-to-probe input power ratio ranges from $2:1$ to $3:2$ (solid marks), pump:probe ranges from $5:1$ to $8:1$ (open marks) [121].

$(P_s \ll P_{out})$, that is, the completely gain-saturated region, on the other hand, NDFWM signal output and efficiency decrease with respective decreases in $P_{out}^{-1}$ and $P_{out}^{-2}$. Consequently, the NDFWM efficiency $\eta$ is maximum at around $P_s$. This is because the efficiency of the nonlinear operation in the amplifying medium is determined from both the signal gain and the third-order susceptibility [136]. TWA operation at higher signal gain is preferable for increasing the maximum NDFWM efficiency. The TWA is attractive and promising as a highly efficient nonlinear optical material and device because of the following features: (1) large optical gain ($\geq 20$ dB), (2) high optical intensity for small input power and long interaction length due to the waveguide structure, and (3) dominant contribution of refractive index grating due to the large $\alpha$-parameter (3–6).

## 7.5.  SHORT-PULSE AMPLIFICATION

One of the most important applications of TWAs is the direct amplification of high-speed signals. In this section we will formulate the saturation characteristics of pulse amplification and compare them with continuous-wave amplification.

### 7.5.1.  Theoretical Formulation

Let us consider energy conservation in a short segment of length $\Delta \hat{z}$ in the laser medium:

$$\frac{\partial}{\partial t}\left(\hat{S}(\hat{z},\hat{t})\frac{\Delta \hat{z}}{v_g}\right) = \hat{S}(\hat{z},\hat{t}) - \hat{S}(\hat{z}+\Delta \hat{z},\hat{t}) + \Gamma A_g \hat{N}(\hat{z},\hat{t})\hat{S}(\hat{z},\hat{t})\,\Delta \hat{z}$$

$$(7.26)$$

Here, variables with carets denote those in the laboratory coordinates. The three terms on the right-hand side of Eq. (7.26) respectively represent the photon density flow into the segment, the photon density flow out of the segment, and the photon density generated by stimulated emission within the segment. The rate equation for the instantaneous optical intensity $\hat{I}(\hat{z},\hat{t})$ carried by the pulse is given in the form

$$\frac{\partial \hat{I}(\hat{z},\hat{t})}{\partial \hat{t}} + v_g\frac{\partial \hat{I}(\hat{z},\hat{t})}{\partial \hat{z}} = \Gamma A_g v_g \hat{N}(\hat{z},\hat{t})\hat{I}(\hat{z},\hat{t}) \qquad (7.27)$$

On the other hand, the rate equation for the inverted carrier density $\hat{N}(\hat{z},\hat{t})$ in a four-level system is described as

$$\frac{\partial \hat{N}(\hat{z},\hat{t})}{\partial \hat{t}} = \frac{N_{00} - \hat{N}(\hat{z},\hat{t})}{\tau_c} - \frac{A_g}{h\nu}\hat{N}(\hat{z},\hat{t})\hat{I}(\hat{z},\hat{t}) \qquad (7.28)$$

Here, $N_{00}$ is the steady-state inverted carrier density without optical input signal. Equations (7.27) and (7.28) are the basic rate equations of the pulse amplification in SLAs. Converting the variables $z \equiv \hat{z}$ and $t \equiv \hat{t} - \hat{z}/v_g$ and rewriting $I(z,t) \equiv \hat{I}(\hat{z},\hat{t})$ and $N(z,t) = \hat{N}(\hat{z},\hat{t})$ transforms Eqs. (7.27) and (7.28) into *the moving coordinate system* which moves with the forward-traveling pulse:

$$\frac{\partial I(z,t)}{\partial z} = \Gamma A_g N(z,t) I(z,t) \tag{7.29}$$

$$\frac{\partial N(z,t)}{\partial t} = \frac{N_{00} - N(z,t)}{\tau_c} - \frac{A_g}{h\nu} N(z,t) I(z,t) \tag{7.30}$$

Note that the pulse written in this moving coordinate system is centered on $t = 0$ on every plane along the amplifier.

Introducing the idea of *the total number of inverted carriers* $N_{tot}(t)$ as the integration of $N(z,t)$ from the input to the output port of the amplifier reduces Eqs. (7.29) and (7.30) to equations in time $t$ only. This idea means that *the total number of inverted carriers* is condensed in the $z$ direction into a thin film and makes everything seem to happen simultaneously. This is correct because every segment of the pulse centered at a local time $t$ sees the corresponding space-integrated inverted carrier density. By assuming that the optical pulse duration is sufficiently short compared to the carrier lifetime ($\Delta t_p \ll \tau_c$), the first term on the right-hand side of Eq. (7.30) is eliminated in the following analysis.

The temporal gain $G(t)$ is derived [98] from the rate equations as

$$G(t) = \frac{G_i}{G_i - (G_i - 1)\exp[-U_{in}(t)/U_s]} \tag{7.31}$$

where the initial gain $G_i$ is the signal gain just before the pulse arrival and $U_{in}$ is the accumulated signal gain energy per unit area in the input pulse from starting time $t_0$ to time $t$. The saturation energy per unit area $U_s$ for a four-level system is given by

$$U_s = \frac{h\nu}{A_g} = I_s \tau_c \tag{7.32}$$

From Eq. (7.30), omitting the first term on the right-hand side, the inverted carrier density $N(z,t)$ is expressed in the simple form

$$N(z,t) = N_{00} \exp\left[-\frac{U(z,t)}{U_s}\right] \tag{7.33}$$

Therefore, $U_s$ is the pulse energy per unit area at which the probability of stimulated emission after the pulse amplification is $1/e$ of the probability before the pulse arrival. On the other hand, the *steady-state* inverted carrier density is also reduced from Eq. (7.30) in the form

$$N(z,t) = \frac{N_{00}}{1 + I(z,t)/I_s} \tag{7.34}$$

where $I(z,t) = h\nu v_g S(z,t)$. This confirms again that the probability of stimulated emission, that is, the gain coefficient, becomes half the unsaturated value when the signal intensity is equal to the saturation intensity $I_s$. The difference in saturation characteristics between CW and pulse amplification is clearly attributed to the carrier recovery effect.

The pulse energy gain $G_{pe}$ is given by the initial gain $G_i$ and the final gain $G_f$ [98] as

$$G_{pe} \equiv \frac{U_{out}}{U_{in}} = \frac{\ln[(G_i - 1)/(G_f - 1)]}{\ln[(G_i - 1)/(G_f - 1)] - \ln[G_i/G_f]} \tag{7.35}$$

The final gain $G_f$ is the final value of $G(t)$ after the pulse has passed, that is, the limit of $G(t)$ as $t \to \infty$, or

$$G_f = \frac{G_i}{G_i - (G_i - 1)\exp[-U_{in}/U_s]}. \tag{7.36}$$

Substituting Eq. (7.36) into Eq. (7.35), we obtain the pulse-energy gain $G_{pe}$ [137] as

$$G_{pe} = \frac{U_{out}}{U_s} \left[ \ln\left\{ \exp\left( \frac{U_{out}}{U_s} \right) + G_i - 1 \right\} - \ln G_i \right]^{-1} \tag{7.37}$$

where $U_{in}$ and $U_{out}$ are the total energy per unit area in the complete input and output pulses, respectively. Equation (7.37) demonstrates that the saturation characteristics of the pulse-energy gain are uniquely determined by the initial gain $G_i$ and the normalized output pulse energy $U_{out}/U_s$.

Theoretical gain saturation curves in CW and pulse amplifications for 20-dB unsaturated gains are compared in Figure 7.38. The ratio of the pulse energy gain $G_{pe,s}$, at which the output pulse energy becomes equal to the saturation energy, to the initial gain $G_i$ is obtained from Eq. (7.37) as

$$\frac{G_{pe,s}}{G_i} = \frac{1}{G_i \ln[1 + (e - 1)/G_i]}$$

$$\simeq \frac{1}{e - 1}, \qquad G_i \gg 1 \tag{7.38}$$

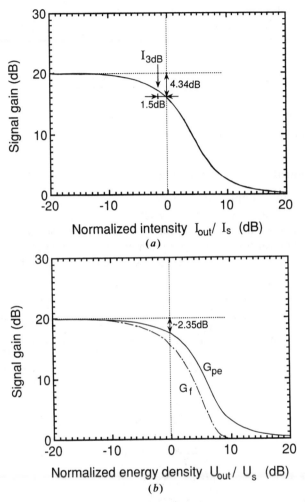

**Figure 7.38.** Theoretical signal gain saturation characteristics in (*a*) continuous-wave and (*b*) pulse amplification [144].

Therefore, the saturation energy per unit area $U_s$, which is also a physical parameter like $I_s$, is given as the output pulse energy per unit area at which the pulse gain is reduced to $1/(e-1)$, that is, by 2.35 dB, from the unsaturated value. This is in contrast to CW amplification in which the signal gain is decreased by 4.34 dB from the unsaturated value when the output signal intensity $I_{out}$ is equal to the saturation intensity $I_s$. This clearly implies that the carrier lifetime $\tau_c$ $(=U_s/I_s)$ cannot be obtained from the ratio of the saturation output energy per unit area $U_{3\,dB}$ to the saturation output intensity $I_{3\,dB}$. Note that the saturation curve of the final gain in the pulse amplification almost coincides with that of the signal gain in CW amplification. This is because the final gain $G_f$ is written in the form

$$G_f = G_i \exp\left[ -\frac{G_{pe} - 1}{G_{pe}} \frac{U_{out}}{U_s} \right] \qquad (7.39)$$

which is similar to Eq. (7.18).

### 7.5.2. Isolated-Pulse Amplification

In isolated-pulse amplification, where the pulse repetition period $T_r$ is much longer than the carrier lifetime $\tau_c$, the initial gain $G_i$ can be replaced by the unsaturated gain $G_0$. Output pulse shapes and the temporal gain variations for Gaussian input pulses [121] are shown in Figure 7.39. The refractive index dispersion effect of the material, which is

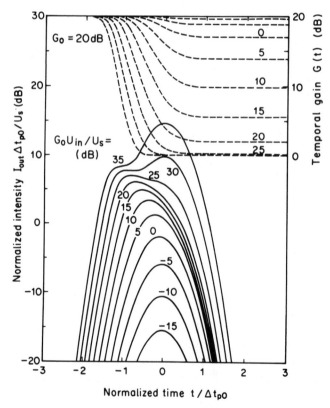

**Figure 7.39.** Pulse-shape distortion (solid lines) and temporal gain variation (dashed lines) in isolated pulse amplification for 20-dB unsaturated gain [121].

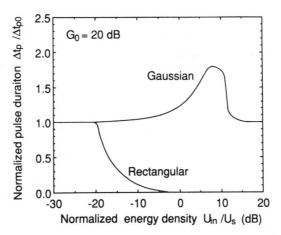

**Figure 7.40.** Output pulse duration of Gaussian and rectangular pulses for 20-dB unsaturated signal gain as a function of normalized input energy [121].

estimated to be a few tens of femtoseconds per nanometer for conventional device lengths [138], is not taken into account in this analysis. A pulse with an energy less than or comparable to the saturation energy can be amplified without distortion. However, in high-energy pulse amplification, the gain saturation during the pulse leads to a temporal gain difference between the leading edge and the trailing edge of the pulse. This advances the pulse peak in time and distorts the pulse shape. In the amplification of pulses with extremely high energy, the gain saturates completely on the leading edge. Therefore, the peak of the output pulse is the same as that of the input pulse, and the output pulse has a lump on the leading edge due to amplification.

The pulse duration variation due to gain saturation depends on the input pulse shape. The Gaussian pulse is broadened by amplification due to the above-mentioned distortion as shown in Figure 7.39. On the other hand, pulses are narrowed when they have a steep slope on the leading edge, as do rectangular pulses. Theoretical pulse duration variation for Gaussian and rectangular pulses with 20-dB unsaturated signal gains are shown in Figure 7.40 as a function of normalized input pulse energy per unit area [121]. When the normalized input pulse energy is less than $-20$ dB, where the output pulse energy equals to the saturation energy, the optical input pulse can be amplified without distortion regardless of the pulse shape. Above the $-20$-dB input pulse energy, the duration of the rectangular pulse is narrowed by the amplification, while the duration of the Gaussian pulse is broadened. The rectangular pulse is abruptly narrowed when the normalized input pulse energy becomes larger than $-20$ dB, and the pulse duration becomes infinitesimal when the input energy is larger than the saturation energy. In Gaussian pulse amplification, on the other hand, the pulse broadening becomes significant when the input pulse energy is larger than the saturation energy, and the output pulse duration becomes nearly twice the input pulse duration $\Delta t_{p0}$ when the input pulse energy is about six times the saturation energy. In the extremely-high-energy region, the pulse duration is again the same as that of the input pulse, because the main part of the output pulse is the nonamplified input pulse itself. These calculated results clearly explain the experimental results regarding both the distortionless pulse amplification reported by Wiesenfeld et al. [51] and the narrowing and broadening of the pulse duration reported by Marshall et al. [50].

Temporal gain variations due to dynamic carrier temperature changes in the low picosecond region have been observed by pump–probe methods [139, 140]. Saturation and recovery dynamics of an AlGaAs amplifier for gain, transparency, and absorption regions are shown in Figure 7.41. The probe transmissions are decreased by the pump pulse and subsequently recover after a few picoseconds to a certain level, which persists for nanosecond delays (band-to-band relaxation). In the transparency region, the pump pulse causes no net change in carrier density, so there is no nanosecond response. The response within a few picoseconds is caused by the carrier distributions relaxing to the lattice temperature. Kesler and Ippen have concluded [139] that the changes in dynamic carrier temperature mentioned above may provide the dominant contribution to the gain

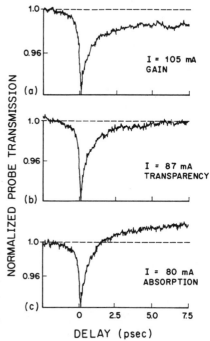

**Figure 7.41.** Saturation and recovery dynamics of a semiconductor laser amplifier in different regions. Dashed lines indicate the level of probe transmission without the pump pulse [139].

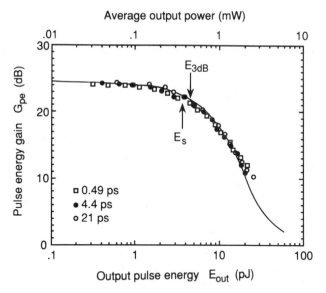

**Figure 7.42.** Pulse energy gain saturation characteristics of isolated pulses with durations of 0.49 ps ($\square$), 4.4 ps ($\bullet$), and 21 ps ($\bigcirc$). The solid line represents a theoretical curve calculated from Eq. (7.37) fitted to the experimental data [52].

compression factor $\varepsilon$ [141–143], which has been phenomenologically introduced in modeling the modulation characteristics of semiconductor lasers to characterize the degree of nonlinearity.

The pulse energy gains $G_{pe}$ of a 1.3-$\mu$m TWA for 0.49-, 4.4-, and 21-ps pulses [52] having a repetition period of 10 ns are shown in Figure 7.42 as a function of the output pulse energy $E_{out}$. The energy gain saturation characteristics are identical, within experimental uncertainty, even for the 0.49-ps pulses. In the amplification of isolated optical pulses, which have sufficiently short duration compared to the carrier lifetime, the pulse-energy gain saturation is independent of the pulse duration and is determined only by the pulse energy.

### 7.5.3. Repetitive-Pulse Amplification

When the pulse repetition period $T_r$ becomes comparable to or smaller than the carrier lifetime $\tau_c$, which is equal to the gain recovery time, the pulse-energy gain is reduced because the gain does not recover to the unsaturated value between pulses. The carrier density $N_i$ just before the pulse arrival, corresponding to $G_i$, is described as

$$N_i = N_f + (N_{00} - N_f)\left[1 - \exp\left(-\frac{T_r}{\tau_c}\right)\right] \tag{7.40}$$

where $N_f$ is the carrier density just after the pulse has completely passed through the amplifier. Equation (7.40) is transformed into a relation of the signal gain as

$$G_i = G_f \left( \frac{G_0}{G_f} \right)^{[1 - \exp(-T_r/\tau_c)]} \tag{7.41}$$

Various characteristics of repetitive pulse amplification can be calculated using this initial gain $G_i$ in Eqs. (7.31), (7.36), and (7.37).

Pulse-energy gain $G_{pe}$, initial gain $G_i$, and final gain $G_f$ for some repetition periods are shown in Figure 7.43 as a function of output pulse energy normalized by saturation energy [144]. When $\tau_c \ll T_r$, the three gain values differ significantly from each other in the high-energy region, while they are almost the same in the low-energy region. The pulse shape distortion originates from the difference between $G_i$ and $G_f$. When the pulse repetition period is much shorter than the carrier lifetime, the saturation curve of the pulse-energy gain coincides with the initial and final gains because gain recovery between pulses becomes incomplete. That is, neither pulse shape distortion nor pattern effects in coded pulse train amplification occur when $T_r \ll \tau_c$. However, the saturation output energy per unit area $U_{3\,dB}$ decreases with a decrease in repetition period. Here, $U_{3\,dB}$ is defined by the output energy per unit area at which the pulse-energy gain reduces to half the unsaturated value.

If we formally define the effective saturation energy per unit area $U_s'$ in repetitive pulse amplification as the output pulse energy at which $G_{pe}$ is decreased by 2.35 from $G_0$ (see Fig. 7.38b), then the relation between

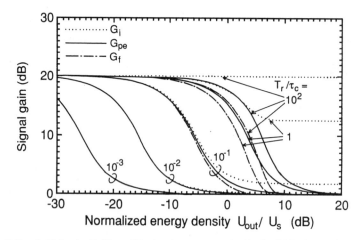

**Figure 7.43.** Initial gain $G_i$ (dotted line) pulse energy gain $G_{pe}$ (solid line), and final gain $G_f$ dash-dotted line) in repetitive pulse amplification as a function of normalized output pulse energy per unit area [144].

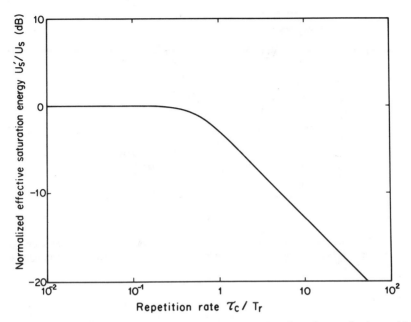

**Figure 7.44.** Normalized effective saturation energy as a function of normalized repetition rate [144].

$U_s'/U_s$ and normalized pulse repetition rate $\tau_c/T_r$ is obtained from Figure 7.42 and represented in Figure 7.44 [144]. When the repetition period becomes shorter than the carrier lifetime, $U_s'/U_s$ linearly decreases with $\tau_c/T_r$. This linear relation implies that the saturation of the pulse-energy gain at high repetition rates is determined by the saturation intensity $I_s$ in CW amplification. This fact is also understood as follows. The difference between the initial gain and the final gain also decreases as the repetition period $T_r$ decreases. Using Eq. (7.39) and an approximation such as $1 - \exp[-T_r/\tau_c] \simeq T_r/\tau_c$ in Eq. (7.41), the final gain $G_f$, which is approximately equal to $G_{pe}$ at the high repetition rates, is rewritten in the form

$$G_{pe} \simeq G_f$$

$$\simeq G_0 \exp\left[-\frac{G_{pe}-1}{G_{pe}}\frac{I_{\text{out}}}{I_s}\right], \qquad T_r \ll T_c \qquad (7.42)$$

Therefore, the saturation curve of the pulse energy gain $G_{pe}$ versus *the average output intensity* $I_{\text{out}}$ in high-repetition-rate pulse amplification coincides with that of the signal gain in CW amplification [compare Eq. (7.42) with Eq. (7.18)]. Consequently, the pulse-energy gain in high-repetition-rate pulse amplification is determined by the average power, that is, by the product of the input pulse energy and the repetition rate.

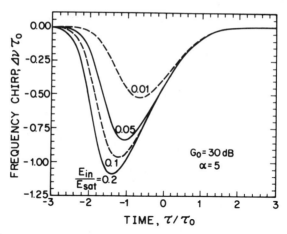

**Figure 7.45.** Frequency–chirp profiles of an amplified pulse for several input pulse energies when a Gaussian pulse is amplified in a semiconductor laser amplifier with an unsaturated gain of 30 dB [66].

### 7.5.4. Self-Phase Modulation

When a short high-energy pulses are amplified by SLAs, self-phase modulation occurs as a result of the nonlinear effect of the semiconductor gain medium. Calculated frequency-chirped profiles of the amplified pulse [66] are shown in Figure 7.45. The instantaneous frequency is down-shifted and the frequency shift is maximum at the pulse peak, which advances in time due to gain saturation. The chirped pulse can be compressed by being passed through a dispersive delay line with an anomalous group-velocity dispersion. Figure 7.46 represents the calculated results of the compressed pulse when a Gaussian input pulse is amplified and subsequently compressed by using a delay line of anomalous dispersion [66]. The input and chirped amplified pulses are also shown for comparison. The input pulse has been compressed by a factor of about 3 under the conditions of $G_0 = 30$ dB, $\alpha$-parameter $= 5$, and $E_{in}/E_s = 0.2$, where $E_{in}$ is the input pulse energy and $E_s$ the saturation energy. The compressed pulse has a broad pedestal on the leading side, because the energy contained in the leading edge disperses away with propagation. The experimentally compressed pulse is shown in Figure 7.47 together with the input pulse with a duration of 40 ps [66]. The compressed pulse duration is 23 ps, which corresponds to a compression factor of 1.7. Although the compression factor is small, its peak power is increased by a factor of about 13 dB as a result from the amplifier net gain. This scheme is able to compress weak picosecond pulses as low as 0.1 pJ, whereas this is difficult by other methods.

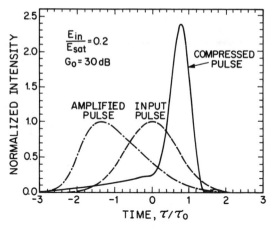

**Figure 7.46.** Compressed pulse when the amplified pulse is passed through a dispersive delay line. The dashed curve shows the input pulse, and the dash-dotted curve shows the amplified pulse before compression [66].

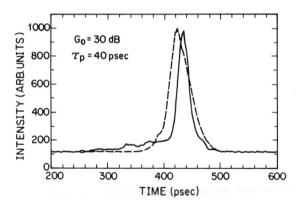

**Figure 7.47.** Experimental traces of the pulse shape obtained using a streak camera for the compressed pulse (solid curve) and the input pulse (dashed curve) [66].

## 7.6. SYSTEM APPLICATIONS OF SEMICONDUCTOR LASER AMPLIFIERS

### 7.6.1. Preamplifier, In-Line Repeater, and Booster Amplifier

Figure 7.48 shows three applications of linear optical amplifiers: ($a$) a preamplifier, ($b$) in-line repeaters, and ($c$) booster amplifiers.

An optical preamplifier (Figure 7.48$a$) is a front-end device of an optical receiver, in which a weak optical signal is amplified before photodetection to suppress the signal-to-noise ratio (SNR) degradation due to

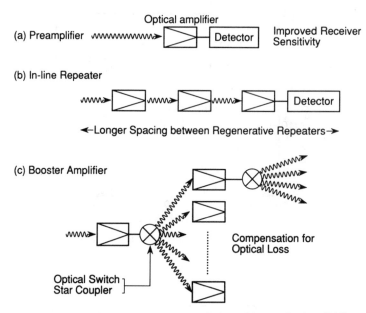

**Figure 7.48.** Configurations for linear optical amplifier applications [110].

thermal noise generated in the electronic circuit [26, 106]. Compared with other front-end devices, such as the avalanche photodiode or the optical heterodyne detector, an optical preamplifier produces a larger amplification factor and a broader bandwidth. A 1.3-$\mu$m TWA preamplifier receiver has recently achieved a 3.7-dB improvement in sensitivity over the best record for a conventional avalanche photodiode receiver at a bit rate of 8 Gbit/s [39].

An optical in-line repeater amplifier (Figure 7.48$b$) compensates for fiber loss and can extend the spacing between two regenerative repeaters [16, 26, 106]. Compared with a conventional repeater, which uses optical-to-electrical signal conversion and performs regenerating, reshaping, and retiming functions, an optical in-line repeater amplifier features versatility in data rate and modulation format. Furthermore, a TWA with a wide gain–bandwidth can simultaneously amplify wavelength-multiplexed or frequency-multiplexed optical signals. Optical amplifier repeaters are especially practical for coherent optical transmission systems [35], since they can directly amplify frequency- and/or phase-modulated optical signals. A nonregenerated transmission distance of 372 km has been demonstrated in a 400 Mbit/s, frequency shift-keying (FSK) coherent transmission experiment using four in-line 1.5-$\mu$m TWA repeaters [42].

An optical booster amplifier is used before sending an optical signal out in a lossy fiber and/or optical circuit or splitting it into several branches (Figure 7.48$c$). This partially overcomes the SNR degradation due to

**Table 7.1  Performance Requirements[a] in Amplifier Applications**

| Required Performance | (1) | (2) | (3) | (4) | (5) | (6) |
|---|---|---|---|---|---|---|
| Preamplifier | △ | △ | △ | × | ○ | ○ |
| In-line repeater | △ | △ | △ | ○ | ○ | △ |
| Booster amplifier | △ | △ | △ | ○ | ○ | △ |

[a]Key: ○, essential; △, useful; ×, unnecessary.

optical loss. Booster amplifiers enable us to construct large-scale, non-loss-limited optical signal processing systems, such as optical ICs, optical distribution networks, and optical switching systems.

The performance requirements for optical amplifiers are as follows:

1.  Sufficient small-signal gain
2.  Wide gain–bandwidth
3.  Polarization-insensitive signal gain
4.  High saturation output power
5.  Small noise figure
6.  Use of a narrowband optical frequency filter

The relative importance of each requirement is given in Table 7.1 for the amplifier applications described above.

Although the net fiber-to-fiber gain is reduced from the internal signal gain of the amplifier by the input and output coupling losses, the internal small-signal gain should be large enough to achieve a considerable net gain. The 3-dB signal gain bandwidth should be wide enough to remove the need for precise control of both the input signal and the amplifier gain peak frequencies. Wide gain–bandwidth also allows the simultaneous amplification of a large number of wavelength-multiplexed signals. If the signal gain is sensitive to the polarization state of the input signal, the effective bandwidth is reduced by the change in polarization direction at the fiber output, which randomly varies with time. The polarization-insensitive signal-gain characteristic is essential for practical amplifier applications in conventional intensity-modulated (IM) direct-detection systems without any polarization-control techniques.

When the saturation output power $P_{3\,dB}$ becomes large, the amplifier has wider dynamic range and is suitable for high-power operation. High power operation is especially important for in-line repeater and booster amplifier applications because it allows a larger number of cascaded amplifiers [26], as will be discussed in Section 7.6.2. A large $P_{3\,dB}$ is also required for multichannel amplification, because the saturation output power for each channel is given by $P_{3\,dB}/N$, where $N$ is the number of channels, due to the homogeneous nature of gain saturation (see Section

7.4.1). Note, however, that the preamplifier is free from gain saturation, since it is used to amplify a very weak signal.

Noise is the most important property in optical amplifiers, since it determines the SNR in system applications. In any amplifier application, the noise figure $F$, which determines the unavoidable SNR degradation due to the signal–spontaneous beat noise, should be as small as possible to achieve ultimate SNR performance. To suppress the additional SNR degradation due to the spontaneous–spontaneous beat noise arising from the ASE components over a wide gain spectrum, a narrowband optical frequency filter matched to the signal spectrum should be placed between the preamplifier output and the photodetector. For in-line repeater applications, optical filters are also required at the output of each or at least every few repeaters to avoid gain saturation resulting from the accumulated ASE power.

### 7.6.2. Signal-to-Noise Ratio in Cascaded Linear Amplifier Chain.

Let us consider the signal-to-noise ratio (SNR) of a cascaded amplifier chain in the PCM-IM direct-detection system to clarify how the system performance is determined by the amplifier device characteristics. To focus without complexity on the essential features of an optical amplifier chain, the SNR determined by the mean and variance values in the photon number is discussed, ignoring the thermal noise generated in the electronic circuit. In pursuing the ultimate performance, it is assumed that the input signal into the cascaded system is at the shot-noise limit, and that each amplifier operates at the signal–spontaneous beat noise limit, where the spontaneous–spontaneous beat noise stemming from the broadband ASE components is rejected by a narrowband optical frequency filter.

Consider the system of $k$ repeaters shown in Figure 7.49, where $L_i$ denotes the attenuation factor in front of the $i$th amplifier whose signal gain and noise figure are expressed by $G_i$ and $F_i$. The input and output SNRs for this system are defined at the transmitter output and at the $k$th repeater output, respectively. The total noise figure, $F_{tot}$, of the cascaded amplifier system is given by [1, 110]

$$F_{tot} \equiv \frac{(S/N)_0}{(S/N)_k} \tag{7.43}$$

$$= \frac{F_1}{L_1} + \frac{F_2}{L_1 G_1 L_2} + \frac{F_3}{L_1 G_1 L_2 G_2 L_3} + \cdots + \frac{F_k}{\{\Sigma_{i=1}^{k-1} L_i G_i\} L_k}. \tag{7.44}$$

First, consider the case of $L_i = 1$. Here, each amplifier is connected in tandem without any loss, to create a composite high-gain amplifier. Setting

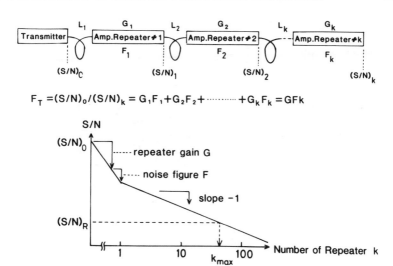

$$F_T = (S/N)_0/(S/N)_k = G_1F_1 + G_2F_2 + \cdots + G_kF_k = GFk$$

**Figure 7.49.** Signal-to-noise ratio in an amplifier repeater system [110].

$L_i = 1$ in Eq. (7.44), the total noise figure is given by

$$F_{\text{tot}} = F_1 + \frac{F_2}{G_1} + \frac{F_3}{G_1G_2} + \cdots + \frac{F_k}{G_1G_2 \cdots G_{k-1}} \qquad (7.45)$$

Here, $F_{\text{tot}}$ is dominated by $F_1$ of the first-stage amplifier and the contribution from the following amplifier noise figure is suppressed by the factor of gain products up to the preceding stages, in a manner similar to cascaded electrical amplifiers. Thus a low-noise high-gain amplifier can be constructed by combining a first-stage amplifier that has a low-noise figure with succeeding amplifiers that have high saturation output powers.

Next, consider the case of $L_iG_i = 1$. This is a cascaded in-line repeater system, where each transmission line loss $L_i$ is compensated for by the amplifier gain $G_i$. Setting $L_iG_i = 1$ in Eq. (7.44), the total noise figure of the system is given by

$$\begin{aligned} F_{\text{tot}} &= G_1F_1 + G_2F_2 + G_3F_3 + \cdots + G_kF_k \\ &= \sum_{i=1}^{k} G_iF_i \end{aligned} \qquad (7.46)$$

Equation (7.46) indicates that, in the compensated in-line repeater system, the noise figures for all amplifiers contribute equally to the total SNR degradation of the system. Furthermore, if the amplifier gains and noise figures of all the repeaters are identical with each other, then Eq. (7.46)

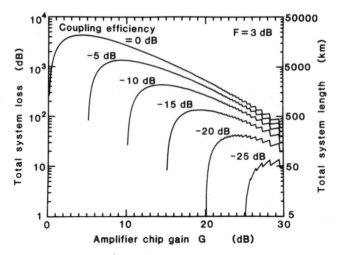

**Figure 7.50.** Total system loss and/or length versus amplifier chip gain as a function of fiber-to-amplifier coupling efficiency. In this figure amplifier noise figure is 3 dB, peak signal power at the amplifier output is 0 dBm, data rate is 10 Gbit/s, and fiber loss is 0.2 dB/km.

reduces to the simple formula

$$F_{tot} = GFk \qquad (7.47)$$

When allowable SNR degradation, $F_{tot}$ $[= (S/N)_0/(S/N)_R]$, is given for a system and both the amplifier gain $G$ and noise figure $F$ are known, the maximum number of cascaded repeaters $k_{max}$ can easily be calculated using Eq. (7.47). The SNR degradation in the cascaded linear amplifier repeater system is shown in the lower part of Figure 7.49. An abrupt SNR degradation occurs at the first repeater output, which amounts to $GF$, that is, the product of the repeater gain and the noise figure. Then it gradually decreases as $k^{-1}$ as the number of repeaters $k$ increases. Therefore, once in-line repeaters are introduced into a transmission line, the maximum number of repeaters (i.e., the maximum system length) is achieved by using the last amplifier as the preamplifier, instead of using as the in-line repeater, which follows the transmission fiber. It is important to reduce the amplifier noise figure $F$ when increasing $k_{max}$. Since the $(S/N)_0$ value is proportional to the signal power in the shot-noise limit, it is also important to operate amplifier repeaters in the high-power region, which is limited by the saturation output power $P_{3\,dB}$.

Figure 7.50 shows calculated results for total system loss and/or length versus the amplifier chip gain $G$ as a function of fiber-to-amplifier coupling efficiency $\eta$. Here, we assume an amplifier noise figure of $F = 3$ dB as an ultimate limit of optical amplifiers [110, 112]. Peak signal power at the amplifier output is set to 0 dBm, which is well below the saturation output power $P_{3\,dB}$ of the TWA [3]. The data-rate is 10 Gbit/s, and the fiber loss is assumed to be 0.2 dB/km at 1.55 $\mu$m. Since the interamplifier fiber loss is determined by $\eta G$, total system loss in decibels is given by

$$\text{Total system loss (dB)} = k_{\max} \cdot 10\log(\eta G) = \frac{F_{\text{tot}}}{FG} \cdot 10\log(\eta G) \quad (7.48)$$

Taking the derivative of Eq. (7.48) with respect to $G$, shows that the maximum system loss is achieved when $\eta G = e$, that is, when the amplifier chip gain is 4.3-dB larger than the coupling loss [145]. This optimization is not useful, however, because it requires impractically short fiber links: about 21 km. Since $k_{\max}$ decreases with an increase in $G$ according to Eq. (7.47), the total system loss decreases with increasing $G$. However, even with a suboptimal choice of $\eta G$, for instance, at $G = 20$ dB, a total system loss of 540 dB (i.e., system length of 2700 km) can be achieved with $\eta = 0$ dB. At a relatively large coupling loss of 10 dB, the total system loss reduces to 270 dB (i.e., 1350-km system length). Note that, since the SNR is inversely proportional to the baseband noise bandwidth, the above figures should be multiplied by 10 when the data rate is reduced to 1 Gbit/s.

If we regard $L_i$ in Eq. (7.44) and Figure 7.49 as the power-splitting loss for optical branching circuits in place of transmission fiber loss, then the previously mentioned SNR design for a cascaded in-line repeater system can be readily applied to that for the booster amplifier system shown in Figure 7.48c. The fan-out value $N_{\text{fan-out}}$, that is, the number of output ports, of a booster amplifier system is given by

$$N_{\text{fan-out}} = (\eta G)^{k_{\max}} \quad (7.49)$$

where $\eta G$ corresponds to the number of branches at each optical power divider and $k_{\max}$ denotes the number of cascaded stages of optical dividers. Note that the left ordinate of Figure 7.50 corresponds to $10\log(N_{\text{fan-out}})$. For example, $N_{\text{fan-out}} = 10^{27}$ is possible for $G = 20$ dB and $\eta = -10$ dB. Thus such a booster amplifier system is very attractive for constructing large-scale optical distribution networks.

More precise SNR calculations, which take into account many details not mentioned here, such as spontaneous–spontaneous beat noise, thermal noise, optical fiber bandwidth, and polarization sensitivity of signal gains between TE and TM modes, have been presented by Mukai et al. [1, 26], Loudon [146], and Lord and Stallard [147]. Theoretical calculations for sensitivity and power penalty in bit-error-rate characteristics have also been reported by Yamamoto [106], Simon [111], Olsson [148], and Fyath et al. [149].

## 7.7. FUTURE POSSIBILITIES

TWAs available at present still have problems that need to be solved. Polarization sensitivity of the signal gain is a severe problem for use in transmission systems. When TWAs are used as in-line repeaters, the signal-gain difference between polarization states for each amplifier is multiplied by the number of cascaded amplifiers. This problem will be solved to some extent by proper design of the device structure and by using multiamplifier configurations, which might also increase the net gain. When SLAs are used in photonic integrated circuits as optical amplifiers, optical modulators, or optical switches, polarization sensitivity will no longer be a disadvantage. Another problem is the coupling loss between amplifier and optical fiber. Low coupling efficiency reduces net gain and available output power and degrades practical noise characteristics [6]. This problem should be approached from both the amplifier side and the optical fiber side, including mode matching between them.

To obtain high-gain broad-bandwidth TWAs, the facet reflectivity must be reproducibly suppressed to as low as $10^{-4}$ for $G = 30$ dB over a wide wavelength range. This condition will be satisfied by structural contrivances such as an angled facet or a window facet combined with a multilayer AR coating. For high-gain amplifiers, the device length should be increased. However, a simple long device may not have high-gain characteristics because of the gain saturation due to ASE. If we can introduce a narrowband optical filter within the device, we can make a high-gain amplifier with a signal gain of more than 30 dB. This idea has already been demonstrated using two serially arranged discrete amplifiers with an optical filter between them [150].

Recently, a multiple-quantum well (MQW) TWA has been demonstrated [151] to have a saturation output power as large as 20.6 dBm, which is about 10 dB larger than usual TWAs. The MQW amplifier is suitable for use in photonic integrated circuits because it has a large polarization sensitivity, and reasonable signal gain can be obtained with a small bias current [115]. Even though an MQW structure has inherent polarization-sensitive gain, it will attract considerable attention as a future amplifier structure. SLAs have often been studied using devices intended as semiconductor laser *oscillators*. However, SLAs should be fabricated using wafers grown for amplifiers because the optimum structures for SLAs are different from those for laser oscillators, as discussed in Section 7.3.4.

## ACKNOWLEDGMENTS

The authors wish to express their gratitude for the guidance and encouragement received from Dr. Tatsuya Kimura and Dr. Hiroshi Kanbe of NTT Basic Research Laboratories.

## REFERENCES

1. T. Mukai, Y. Yamamoto, and T. Kimura, "Optical amplification by semiconductor lasers," in *Semiconductors and Semimetals*, R. K. Willardson and A. C. Beer, eds., Academic Press, New York, 1985, Vol. 22-E, pp. 265–319.

2. P. Urquhart, *IEE Proc.* **135-J**, 385–407 (1988).

3. T. Saitoh and T. Mukai, *IEEE J. Quantum Electron.* **QE-23**, 1010–1020 (1987).

4. J. C. Simon, *IEEE J. Lightwave Technol.* **LT-5**, 1286–1295 (1987).

5. G. Eisenstein, B. C. Johnson, and G. Raybon, *Electron. Lett.* **23**, 1020–1022 (1987).

6. M. J. O'Mahony, *IEEE J. Lightwave Technol.* **LT-6**, 531–544 (1988).

7. M. J. Coupland, K. G. Mambleton, and C. Hilsum, *Phys. Lett.* **7**, 231–232 (1963).

8. J. W. Crowe and R. M. Graig, Jr., *Appl. Phys. Lett.* **4**, 57–58 (1964).

9. R. Vuilleumier, N. E. Collins, J. M. Smith, J. C. S. Kim, and H. Raillard, *Proc. IEEE* **55**, 1420–1425 (1967).

10. W. F. Kosonocky and R. H. Cornely, *IEEE J. Quantum Electron.* **QE-4**, 125–131 (1968).

11. R. N. Hall, G. E. Fenner, J. D. Kingsley, T. J. Soltys, and R. O. Carlson, *Phys. Rev. Lett.* **9**, 366–368 (1962).

12. M. I. Nathan, W. P. Dumke, G. Burns, F. H . Dill, Jr., and G. Lasher, *Appl. Phys. Lett.* **1**, 62–64 (1962).

13. T. M. Quist, R. H. Rediker, R. J. Keyes, W. E. Krag, B. Lax, A. L. McWhorter, and H. J. Zeigler, *Appl. Phys. Lett.* **1**, 91–92 (1962).

14. I. Hayashi, M. B. Panish, P. W. Foy, and S. Sumski, *Appl. Phys. Lett.* **17**, 109–111 (1970).

15. Zh. I. Alferov, V. M. Andreev, D. Z. Garbuzov, Yu. V. Zhilyaev, E. P. Morozov, E. L. Portnoi, and V. G. Trofim, "Investigation of the influence of the AlAs–GaAs heterostructure parameters in the laser threshold current and the realization of continuous emission at room temperature," *Sov. Phys.–Semicond.* **4**, 1573–1575 (1971), transl. from *Fizika i Tekhnika Poluprovodnikov* **4**, 1826–1829 (1970).

16. G. Zeidler and D. Schicketanz, *Siemens Forsch. -u. Entwickl. -Ber.* **2**, 227–234 (1973).

17. G. Zeidler, *Siemens Forsch. -u. Entwickl. -Ber.* **2**, 235–238 (1973).

18. D. Schicketanz and G. Zeidler, *IEEE J. Quantum Electron.* **QE-11**, 65–69 (1975).

19. S. Kobayashi and T. Kimura, *Electron. Lett.* **16**, 230–232 (1980).

20. Y. Yamamoto and H. Tsuchiya, *Electron. Lett.* **16**, 233–235 (1980).

21. Y. Yamamoto, *IEEE J. Quantum Electron.* **QE-16**, 1047–1052 (1980).

22. V. N. Luk'yanov, A. T. Semenov, and S. D. Yakubovich, *Sov. J. Quantum Electron.* **10**, 1432–1435 (1980).

23. T. Mukai and Y. Yamamoto, *IEEE J. Quantum Electron.* **QE-17**, 1028–1034 (1981).

24. F. Favre, L. Jeunhomme, I. Joindot, M. Monerie, and J.-C. Simon, *IEEE J. Quantum Electron*. **QE-17**, 897–906 (1981).

25. J. C. Simon, *Electron. Lett*. **18**, 438–439 (1982).

26. T. Mukai, Y. Yamamoto, and T. Kimura, *IEEE J. Quantum Electron*. **QE-18**, 1560–1568 (1982).

27. I. S. Goldobin, V. N. Luk'yanov, A. F. Solodkov, V. P. Tabunov, and S. D. Yakubovich, *Sov. J. Quantum Electron*. **14**, 255–259 (1984).

28. R. H. Clarke, *Int. J. Electron*. **53**, 495–499 (1982).

29. G. Eisenstein, *AT&T Bell Lab. Tech. J*. **63**, 357–364 (1984).

30. T. Saitoh, T. Mukai, and O. Mikami, *IEEE J. Lightwave Technol*. **LT-3**, 288–293 (1985).

31. C. Vassallo, *Electron. Lett*. **21**, 333–334 (1985).

32. T. Saitoh and T. Mukai, "A low-noise 1.5-μm GaInAsP traveling-wave optical amplifier with high-saturation output power," in *Tenth IEEE International Semiconductor Laser Conference Post-Deadline Papers Tech. Digest*, Kanazawa, 1986, Paper PD-5.

33. G. Eisenstein and R. M. Jopson, *Int. J. Electron*. **60**, 113–121 (1986).

34. M. G. Öberg and N. A. Olsson, *Electron. Lett*. **24**, 99–100 (1988).

35. Y. Yamamoto and T. Kimura, *IEEE J. Quantum Electron*. **QE-17**, 919–935 (1981).

36. M. J. O'Mahony, I. W. Marshall, H. J. Westlake, and W. G. Stallard, *Electron. Lett*. **22**, 1238–1240 (1986).

37. N. A. Olsson and P. Garbinski, *Electron. Lett*. **22**, 1114–1116 (1986).

38. I. W. Marshall and M. J. O'Mahony, *Electron. Lett*. **23**, 1052–1053 (1987).

39. R. M. Jopson, A. H. Gnauck, B. L. Kasper, R. E. Tench, N. A. Olsson, C. A. Burrus, and A. R. Chraplyvy, *Electron. Lett*. **25**, 233–234 (1989).

40. I. W. Marshall, D. M. Spirit, and M. J. O'Mahony, *Electron. Lett*. **22**, 253–255 (1986).

41. M. J. O'Mahony, I. W. Marshall, H. J. Westlake, W. J. Devlin, and J. C. Regnault, "A 200 km 1.5 μm optical transmission experiment using a semiconductor laser amplifier repeater," in *Conference on Optical Fiber Communication Tech. Digest*, Atlanta, 1986, Paper WE5.

42. N. A. Olsson, M. G. Öberg, L. A. Koszi, and G. J. Przybylek, *Electron. Lett*. **24**, 36–38 (1988).

43. M. G. Öberg, N. A. Olsson, L. A. Koszi, and G. J. Przybylek, *Electron. Lett*. **24**, 38–39 (1988).

44. S. Ryu, H. Taga, S. Yamamoto, K. Mochizuki, and H. Wakabayashi, *Electron. Lett*. **25**, 1682–1684 (1989).

45. G. Großkopf, R. Ludwig, and H. G. Weber, *Electron. Lett*. **22**, 900–902 (1986).

46. T. Mukai, K. Inoue, and T. Saitoh, *Electron. Lett*. **23**, 396–397 (1987).

47. R. M. Jopson, K. L. Hall, G. Eisenstein, G. Raybon, and M. S. Whalen, *Electron. Lett*. **23**, 510–512 (1987).

48. H. J. Westlake and M. J. O'Mahony, *Electron. Lett*. **23**, 649–651 (1987).

49. I. W. Marshall, D. M. Spirit, and M. J. O'Mahony, *Electron. Lett.* **23**, 818–819 (1987).

50. I. W. Marshall and D. M. Spirit, "Observation of large pulse compression by a saturated travelling wave semiconductor laser amplifier," in *Eighth Conference on Lasers and Electro-Optics Tech. Digest*, Anaheim, CA, 1988, Paper TuM64.

51. J. M. Wiesenfeld, G. Eisenstein, R. S. Tucker, G. Raybon, and P. B. Hansen, *Appl. Phys. Lett.* **53**, 1239–1241 (1988).

52. T. Saitoh, H. Itoh, Y. Noguchi, S. Sudo, and T. Mukai, *IEEE Photon. Technol. Lett.* **1**, 297–299 (1989).

53. C. E. Zah, J. S. Osinski, C. Caneau, S. G. Menocal, L. A. Reith, J. Salzman, F. K. Shokoohi, and T. P. Lee, *Electron. Lett.* **23**, 990–992 (1987).

54. C. E. Zah, C. Caneau, F. K. Shokoohi, S. G. Menocal, F. Favire, L. A. Reith, and T. P. Lee, *Electron. Lett.* **24**, 1275–1276 (1988).

55. B. Mikkelsen, D. S. Olesen, K. E. Stubkjaer, Z. Wang, A. J. Collar, and G. D. Henshall, *Electron. Lett.* **25**, 357–359 (1989).

56. J. T. K. Chang and J. I. Vukusic, *J. Mod. Opt.* **35**, 355–364 (1988).

57. I. Cha, M. Kitamura, H. Honmou, and I. Mito, *Electron. Lett.* **25**, 1241–1242 (1989).

58. N. A. Olsson, R. F. Kazarinov, W. A. Nordland, C. H. Henry, M. G. Oberg, H. G. White, P. A. Garbinski, and A. Savage, *Electron. Lett.* **25**, 1048–1049 (1989).

59. M. Ikeda, *Electron. Lett.* **17**, 899–900 (1981).

60. J. Hegarty and K. A. Jackson, *Appl. Phys. Lett.* **45**, 1314–1316 (1984).

61. D. M. Annenkov, A. P. Bogatov, P. G. Eliseev, O. G. Okhotnikov, G. T. Pak, M. P. Rakhval'skiy, Yu. F. Fedorov, and K. A. Khairetdinov, *Sov. J. Quantum Electron.* **14**, 163–164 (1984).

62. J. Mellis, *Electron. Lett.* **25**, 679–680 (1989).

63. G. Großkopf, R. Ludwig, R. Schnabel, and H. G. Weber, "Semiconductor laser optical amplifier as phase modulator in a 140 Mbit/s DPSK transmission experiment," in *Seventh International Conference on Integrated Optics and Optical Fiber Communication Tech. Digest*, Kobe, Japan, 1989, Paper 21B4-2.

64. G. Großkopf, R. Ludwig, and H. G. Weber, *Electron. Lett.* **24**, 1106–1107 (1988).

65. K. Inoue, *Electron. Lett.* **25**, 630–632 (1989).

66. G. P. Agrawal and N. A. Olsson, *Opt. Lett.* **14**, 500–502 (1989).

67. T. Saitoh and T. Mukai, *IEEE J. Lightwave Technol.* **6**, 1656–1664 (1988).

68. B. W. Hakki and T. L. Paoli, *J. Appl. Phys.* **46**, 1299–1306 (1975).

69. F. K. Reinhart, I. Hayashi, and M. B. Panish, *J. Appl. Phys.* **42**, 4466–4479 (1971).

70. D. C. Krupka, *IEEE J. Quantum Electron.* **QE-11**, 390–400 (1975).

71. D. R. Kaplan and P. P. Deimel, *AT&T Bell Lab. Tech. J.* **63**, 857–877 (1984).

72. J. C. Simon, B. Landousies, Y. Bossis, D. Doussiere, B. Fernier, and C. Padioleau, *Electron. Lett.* **23**, 332–334 (1987).

73. C. Vassallo, *Electron. Lett.* **24**, 61–62 (1988).

74. L. Atternäs and L. Thylén, *IEEE J. Lightwave Technol.* **7**, 426–430 (1989).

75. T. Ikegami, *IEEE J. Quantum Electron.* **QE-8**, 470–476 (1972).

76. G. Eisenstein, G. Raybon, and L. W. Stulz, *J. Lightwave Technol.* **6**, 12–16 (1988).

77. A. Somani, P. A. Goud, and C. G. Englefield, *Appl. Opt.* **27**, 1391–1393 (1988).

78. G. Eisenstein and S. W. Stulz, *Appl. Opt.* **23**, 161–164 (1984).

79. K. Wakita and S. Matsuo, *Jpn. J. Appl. Phys.* **23**, L556–L558 (1984).

80. M. Serenyi and H.-U. Habermeier, *Appl. Opt.* **26**, 845–849 (1987).

81. Y. Katagiri and H. Ukita, "Ion beam sputtered $(SiO_2)_x(Si_3N_4)_{1-x}$ antireflection coatings on laser facets using $O_2$-$N_2$ discharges," in *Second Microoptics Conference/Eighth Topical Meeting on Gradient-Index Optical Imaging Systems Tech. Digest*, Tokyo, 1989, Paper D5.

82. G. Eisenstein, L. W. Stulz, and L. G. Van Uitert, *J. Lightwave Technol.* **LT-4**, 1373–1375 (1986).

83. H. D. Edmonds, C. DePalma, and E. P. Harris, *Appl. Opt.* **10**, 1591–1596 (1971).

84. T. Saitoh, T. Mukai, and Y. Noguchi, "Fabrication and gain characteristics of a 1.5 $\mu$m GaInAsP traveling-wave optical amplifier," in *First Optoelectronics Conference Post-Deadline Papers Tech. Digest* (The Institute of Electronics and Communications Engineers of Japan), Tokyo, 1986, Paper B11-2.

85. L. N. Kurbatov, S. S. Shakhidzhanov, L. V. Bystrova, V. V. Krapukhin, and S. I. Kolonenkova, *Sov. Phys. Semicond.* **4**, 1739–1744 (1971).

86. M. B. Holbrook and D. J. Bradley, *Appl. Phys. Lett.* **36**, 349–350 (1980).

87. J. Salzman, R. J. Hawkins, and T. P. Lee, *Opt. Lett.* **13**, 455–457 (1988).

88. D. Marcuse, *J. Lightwave Technol.* **7**, 336–339 (1989).

89. C. E. Zah, R. Bhat, S. G. Menocal, N. Andreadakis, F. Favire, C. Caneau, M. A. Koza, and T. P. Lee, *IEEE Photon. Technol. Lett.* **2**, 46–47 (1990).

90. S. Akiba, K. Utaka, K. Sakai, and Y. Matsushima, *IEEE J. Quantum Electron.* **QE-19**, 1052–1056 (1983).

91. K. Utaka, S. Akiba, K. Sakai, and Y. Matsushima, *IEEE J. Quantum Electron.* **QE-20**, 236–244 (1984).

92. R. M. Jopson, G. Eisenstein, K. L. Hall, G. Raybon, C. A. Burrus, and U. Koren, *Electron. Lett.* **22**, 1105–1107 (1986).

93. T. Saitoh and T. Mukai, *Opt. Quantum Electron.* **21**, S47–S58 (1989).

94. S. Cole, D. M. Cooper, W. J. Devlin, A. D. Ellis, D. J. Elton, J. J. Isaak, G. Sherlock, P. C. Spurdens, and W. A. Stallard, *Electron. Lett.* **25**, 314–315 (1989).

95. G. Großkopf, R. Ludwig, R. G. Waarts, and H. G. Weber, *Electron. Lett.* **23**, 1387–1388 (1987).

96. N. A. Olsson, *Electron. Lett.* **24**, 1075–1076 (1988).

97. Y. Yamamoto, *IEEE J. Quantum Electron.* **QE-19**, 34–46 (1983).

98. A. E. Siegman, *LASERS*, Oxford University Press, Oxford, 1986.

99  T. Saitoh and T. Mukai, *Electron. Lett.* **23**, 218–219 (1987).

100. T. Mukai, T. Saitoh, and O. Mikami, "1.5 μm InGaAsP Fabry-Perot cavity type laser amplifiers," *Trans. IECE Jpn.* **J69-C**, 421–432 (1986) (in Japanese), transl. in *Electronics and Communications in Japan* (Scripta Technica, Inc.), Part 2, Vol. 70, No. 2, pp. 421–432, 1987.

101. A. Sugimura, *IEEE J. Quantum Electron.* **QE-18**, 352–363 (1982).

102. G. Eisenstein, R. S. Tucker, J. M. Wiesenfeld, P. B. Hansen, G. Raybon, B. C. Johnson, T. J. Bridge, F. G. Storz, and C. A. Burrus, *Appl. Phys. Lett.* **54**, 454–456 (1989).

103. I. D. Henning, M. J. Adams, and J. V. Collins, *IEEE J. Quantum Electron.* **QE-21**, 609–613 (1985) and Errata, ibid. **QE-21** 1973 (1985).

104. K. Inoue, T. Mukai, and T. Saitoh, *Electron. Lett.* **23**, 328–329 (1987).

105. T. Mukai, K. Inoue, and T. Saitoh, *Appl. Phys. Lett.* **51**, 381–383 (1987).

106. Y. Yamamoto, *IEEE J. Quantum Electron.* **QE-16**, 1073–1081 (1980).

107. K. Shimoda, H. Takahashi, and C. H. Townes, *J. Phys. Soc. Jpn.* **12**, 686–700 (1957).

108. T. Mukai and Y. Yamamoto, *IEEE J. Quantum Electron.* **QE-18**, 564–575 (1982).

109. T. Mukai and T. Saitoh, *Electron. Lett.* **23**, 216–218 (1987).

110. Y. Yamamoto and T. Mukai, *Opt. Quantum Electron.* **21**, S1–S14 (1989).

111. J. C. Simon, *J. Opt. Commun.* **4**, 51–62 (1983).

112. H. A. Haus and J. A. Müllen, *Phys. Rev.* **128**, 2407–2413 (1962).

113. Y. Yamamoto and H. A. Haus, *Rev. Mod. Phys.* **58**, 1001–1020 (1986).

114. K. J. Vahala and C. E. Zah, *Appl. Phys. Lett.* **52**, 1945–1947 (1988).

115. P. J. Stevens and T. Mukai, *IEEE J. Quantum Electron.* **QE-26** (1990).

116. Y. Yamamoto, S. Saito, and T. Mukai, *IEEE J. Quantum Electron.* **QE-19**, 47–58 (1983).

117. T. Saitoh and T. Mukai, *Trans. IEICE* **E71**, 482–484 (1988).

118. H. C. Casey and F. Stern, *J. Appl. Phys.* **47**, 631–643 (1976).

119. F. Stern, *J. Appl. Phys.* **47**, 5382–5386 (1976).

120. A. Sugimura, *Appl. Phys. Lett.* **39**, 21–23 (1981).

121. T. Mukai and T. Saitoh, *Trans. IEICE* **E73**, 46–52 (1990).

122. T. E. Darcie, R. M. Jopson, and A. A. M. Saleh, *Electron. Lett.* **24**, 1154–1156 (1988).

123. G. P. Agrawal, *Electron. Lett.* **23**, 1175–1177 (1987).

124. K. Inoue, *Electron. Lett.* **23**, 1293–1295 (1987).

125. R. M. Jopson, T. E. Darcie, K. T. Gayliard, R. T. Ku, R. E. Tench, T. C. Rice, and N. A. Olsson, *Electron. Lett.* **23**, 1394–1395 (1987).

126. T. E. Darcie, R. M. Jopson and R. W. Tkach, *Electron. Lett.* **23**, 1392–1394 (1987).

127. R. M. Jopson and T. E. Darcie, *Electron Lett.* **24**, 1372–1374 (1988).

128. T. G. Hodgkinson and R. P. Webb, *Electron. Lett.* **24**, 1550–1552 (1988).

129. T. Mukai and T. Saitoh, "Existence of maximum frequency conversion efficiency in nearly degenerated four-wave mixing for a traveling-wave laser amplifier," *Sixteenth International Conference on Quantum Electronics*, Tokyo, 1988, Paper WD-1.

130. K. Inoue, T. Mukai, and T. Saitoh, *Appl. Phys. Lett.*, **51**, 1051–1053 (1987).

131. G. Großkopf, R. Ludwig, R. G. Waarts, and H. G. Weber, *Electron. Lett.* **24**, 31–32 (1988).

132. G. P. Agrawal, *J. Opt. Soc. Am. B* **5**, 147–159 (1988).

133. F. Favre, D. Le Guen, J.-C. Simon and P. Doussière, *Electron. Lett.* **25**, 272–273 (1989).

134. A. P. Bogatov, P. G. Eliseev, and B. N. Sverdlov, *IEEE J. Quantum Electron.* **QE-11**, 510–515 (1975).

135. A. A. M. Saleh, R. M. Jopson, and T. E. Darcie, *Electron. Lett.* **24**, 950–952 (1988).

136. T. Mukai and T. Saitoh, *IEEE J. Quantum Electron.* **26**, 865–875 (1990).

137. T. Saitoh, H. Itoh, Y. Noguchi, S. Sudo, and T. Mukai, "Gain saturation in subpicosecond and picosecond pulse amplification by a traveling-wave semiconductor laser amplifier," in *Seventh International Conference on Integrated Optics and Optical fiber Communication Tech. Digest*, Kobe, Japan, 1989, Paper 20C2-3.

138. J. P. van der Ziel and R. A. Logan, *IEEE J. Quantum Electron.* **QE-19**, 164–169 (1983).

139. M. P. Kesler and E. P. Ippen, *Appl. Phys. Lett.* **51**, 1765–1767 (1987).

140. K. L. Hall, E. P. Ippen, J. Mark, and G. Eisenstein, "Subpicosecond gain dynamics in InGaAsP diode laser amplifiers," in *Conference on Lasers and Electro-Optics 1989 Tech. Digest*, Baltimore, 1989, Paper THI1.

141. D. J. Channin, *J. Appl. Phys.* **50**, 3858–3860 (1979).

142. R. S. Tucker, *IEEE J. Lightwave Technol.* **LT-3**, 1180–1192 (1985).

143. J. E. Bowers, *Solid-State Electron.* **30**, 1–11 (1987).

144. T. Saitoh and T. Mukai, *J. Quantum Electron.* (in press).

145. P. S. Henry, *IEEE J. Quantum Electron.* **QE-21**, 1862–1879 (1985).

146. R. Loudon, *IEEE J. Quantum Electron.* **QE-21**, 766–773 (1985).

147. A. Lord and W. A. Stallard, *Opt. Quantum Electron.* **21**, 463–470 (1989).

148. N. A. Olsson, *IEEE J. Lightwave Technol.*, **7**, 1071–1082 (1989).

149. R. S. Fyath, A. J. McDonald, and J. J. O'Reilly, *IEE Proc.* **136-J**, 238–248 (1989).

150. N. A. Olsson and R. M. Jopson, *J. Lightwave Technol.* **7**, 791–793 (1989).

151. D. M. Cooper, M. Bagley, L. D. Westbrook, D. J. Elton, H. J. Wickes, M. J. Harlow, M. R. Aylett, and W. J. Devlin, "Broadband operation in InGaAsP-InGaAs GRINSCH MQW amplifiers with 115 mW saturated output power," in *Conf. of Optical Fiber Communication Tech. Digest*, San Francisco, 1990, Paper PD32.

# 8

# Semiconductor Laser Amplifiers in High-Bit-Rate and Wavelength-Division-Multiplexed Optical Communication Systems

R. M. JOPSON AND T. E. DARCIE

*AT & T Bell Laboratories, Crawford Hill Laboratory, Holmdel, New Jersey*

## 8.1. INTRODUCTION

In the latter half of the 1980s, a quiet revolution swept through the telecommunications industry. Optical fiber networks were installed throughout the developed world and submarine cables containing optical fibers were deployed to link continents. This vast change in the fundamental technology used in long-haul transmission was driven by several advantages of optical fiber as a transmission medium over the two microwave transmission media it supplanted: (1) optical fiber potentially has more than 10 thousand times the bandwidth of its predecessors; (2) it allows large repeater spacings; and (3) it is free from electromagnetic interference, fading, and regulatory bandwidth restrictions. The total bandwidth of the two "long-wavelength" windows in fiber transmission, 1.25–1.35 $\mu$m and 1.40–1.65 $\mu$m, is 50,000 GHz. However, since present systems use less than 0.1% of this bandwidth and have spans roughly ten times shorter than theoretical limits, the conversion to photonic communication systems is far from complete. To make better use of the potential of optical fiber, it will be necessary to improve receiver sensitivities and to develop practical wavelength-division-multiplexing (WDM) techniques to combine channels. Practical optical amplifiers are critical components in the development of both these capabilities.

*Coherence, Amplification, and Quantum Effects in Semiconductor Lasers*, Edited by Yoshihisa Yamamoto.
ISBN 0-471-512494 © 1991 John Wiley & Sons, Inc.

The sensitivity of direct-detection (nonheterodyne) receivers is limited by thermal noise in the receiver front end to one or two orders of magnitude worse than the quantum limit. Since the equivalent input noise of an optical amplifier can be less than that of a direct-detection receiver, optical preamplifiers can be used to increase receiver sensitivities. This is especially true as bit rates increase beyond 10 Gbit/s, the limit of usefulness for avalanche photodiodes (APDs).

Given the limits on the speed of the electronics used in lightwave transmitters and receivers, WDM will be essential for ultra-high-bit-rate systems. This approach increases the regenerator complexity significantly if the separate wavelength channels must be demultiplexed, separately regenerated, and multiplexed again. One method for greatly simplifying the regenerators used in WDM systems is to replace them with optical amplifier repeaters. In this way, the separate receiver and transmitter for each wavelength, the signal processing electronics, and the optical multiplexer and demultiplexer are all replaced with a single electrooptic device.

Optical amplifiers may also offer advantages in long-distance systems with modest bandwidths. A long span may contain up to a hundred or more regenerators, each of which must contain signal processing components operating at the transmission bit rate using a particular transmission protocol. It is attractive to consider the flexibility that can be obtained using an optical amplifier to provide broadband amplification directly, without the use of high-speed electronics.

Finally, the interaction between carriers and photons within the active layer of the amplifier creates dynamic nonlinearities that, although often problematic, can be useful. Nonlinear mixing can be used to generate new frequency components or shift the frequency of a modulated carrier. Intensity modulation passing through the amplifier can be detected as a modulation of bias voltage, providing an in-line telemetry monitor.

Although optical amplifiers will be of great value in these applications, they can degrade signal quality, primarily through optical noise and nonlinearities caused by dynamic saturation of the optical gain. In this chapter, we first discuss the general properties of the nonlinearity and noise of semiconductor optical amplifiers. We then discuss specific applications of these devices and outline the limits imposed by nonlinearity and noise on the use of amplifiers. Finally, we describe a feedforward technique for reducing the amplifier nonlinearity.

## 8.2. AMPLIFIER PROPERTIES

Some of the properties of semiconductor optical amplifiers have been discussed in previous chapters. In this chapter, we will concentrate on the noise and saturation properties of semiconductor optical amplifiers and the effect of these properties on the performance of high-capacity systems.

We will assume that the problems of polarization dependence and Fabry–Perot ripple in the gain have been solved so that the amount of gain available in a single chip is limited by noise-, saturation-, or system-related conditions rather than by the facet reflectivity or the polarization dependence of the gain.

### 8.2.1. Gain Saturation

Gain saturation in an optical amplifier limits the maximum amplifier output power for both continuous-wave (CW) and pulsed input signals and also causes dynamic nonlinearity that impairs the performance of communication systems using optical amplifiers. This nonlinearity arises from coupling, through stimulated emission, between the carrier density and the instantaneous optical intensity in the active region of the amplifier. This coupling is described by a rate equation for the local carrier density (see, for example, Refs. 1 and 2),

$$\frac{dN(z,t)}{dt} = \frac{I}{qV} - R[N(z,t)] - \frac{g[N(z,t)]}{h\nu}\left\langle |E(z,t)|^2 \right\rangle \quad (8.1)$$

the $z$ direction is parallel to the length of the action region, $I$ is the current injected into the active region (volume $V$), $q$ is the electronic charge, $h\nu$ is the photon energy, and $E(z,t)$ is the optical field in the waveguide, normalized such that $|E(z,t)|^2$ is the optical intensity in the active region. The material gain $g$ is defined by

$$g = a(N - N_t) \quad (8.2)$$

where $N_t$ is the carrier density at which the material becomes transparent and $a$ is the gain constant. In Eq. (8.1) we assume that the transverse waveguide dimensions are smaller than the carrier diffusion length. The carrier density is then constant across the waveguide and transverse effects are included [2] by averaging the optical intensity over the transverse coordinates (angle brackets). The optical intensity can be determined from the power in the mode

$$\left\langle |E(z,t)|^2 \right\rangle = \frac{\Gamma P(z,t)}{A} \quad (8.3)$$

where $A$ is the cross-sectional area of the active region and $\Gamma$ is the confinement factor, that is, the fraction of the optical power that overlaps the active layer [3].

In the active region of an amplifier, the carrier density is sufficiently high that Auger recombination may contribute significantly to the carrier recombination rate $R(N)$. The total recombination rate is generally expressed as a polynomial in $N$, with a linear term describing nonradiative recombination at defects or traps, and quadratic and cubic terms for spontaneous emission and Auger recombination, respectively [3]. Expanding this polynomial to first order in a Taylor series about the average carrier density $(N_0)$ and using an effective current and the effective carrier lifetime

$$I' \equiv I - qV(R|_{N_0} - N_0/\tau_c), \qquad \frac{1}{\tau_c} \equiv \frac{\partial R}{\partial N}\bigg|_{N_0} \qquad (8.4)$$

yields

$$\frac{I.}{qV} - R(N) \approx \frac{I'}{qV} - \frac{N_0}{\tau_c} \qquad (8.5)$$

The saturated gain for CW input signals follows directly from Eqs. (8.1), (8.2), and (8.5) by setting the time derivative to zero and solving for $N$:

$$g = \frac{g_0}{1 + P/P_s} \qquad (8.6)$$

The unsaturated signal gain is $g_0$ (units of inverse length), $P$ is the optical power in the amplifier, and the saturation power is defined by

$$P_s = I_s \frac{A}{\Gamma} \equiv \frac{h\nu}{a\tau_c} \frac{A}{\Gamma} \qquad (8.7)$$

where $I_s$ is the saturation intensity. The growth of optical power along the length of the amplifier is described by

$$\frac{dP}{dz} = (\Gamma g - \alpha_{\text{int}})P = \left(\frac{\Gamma g_0}{1 + P/P_s} - \alpha_{\text{int}}\right)P \qquad (8.8)$$

Here, the internal loss term, $\alpha_{\text{int}}$, includes only that portion of the internal loss that is independent of the carrier density [4]. We have incorporated the carrier-density-dependent part of the internal loss into $g$ and assumed that the carrier density is far enough above the transparency level that this internal loss saturates in the same manner as the gain. For $\alpha_{\text{int}}$ independent of $z$, Eq. (8.8) can be solved exactly to yield

$$P_{\text{out}} = P_{\text{in}}G_0 \left[\frac{g' - P_{\text{out}}/P_s}{g' - P_{\text{in}}/P_s}\right]^{g'+1} \qquad (8.9)$$

where, for an amplifier of length $L$, the *unsaturated* single-pass gain is given by $G_0 = e^{(\Gamma g_0 - \alpha_{int})L}$ and $g'$ is a normalized gain given by

$$g' = \frac{(\Gamma g_0 - \alpha_{int})}{\alpha_{int}} = \frac{\ln(G_0)}{\alpha_{int} L} \qquad (8.10)$$

The normalized gain ranges from zero in an amplifier with no gain to infinity in an amplifier with negligible internal loss. As $\alpha_{int}$ approaches zero, this intractable relation approaches a simpler one:

$$P_{out} \approx P_{in} G_0 \frac{e^{(P_{in}/P_s)}}{e^{(P_{out}/P_s)}} \qquad (8.11)$$

It can be seen that in both Eqs. (8.9) and (8.11), the rightmost factor gives the factor by which the small-signal gain is saturated. An estimate of the accuracy of Eq. (8.11), for nonzero $\alpha_{int}$, can be obtained numerically. In Figure 8.1, the ratio of the right side of Eq. (8.11) to that of Eq. (8.9) is plotted for various values of $g'$. Although the value of $\alpha_{int}$ in amplifiers is not well known, it is probably less than 10 cm$^{-1}$. Using this value, a 500-$\mu$m-long amplifier operating at a gain of $10^3$ has a value for $g'$ greater than 14. We see in Figure 8.1 that Eq. (8.11) overestimates the gain by less than 12% (0.5 dB) for $P_{out} \leq P_s$. Under these conditions, we find

$$P_{out} \approx P_{in} G_0 e^{-P_{out}/P_s} \qquad (8.12)$$

provided $P_{in} \ll P_s$. Note that while the approximation yielding Eq. (8.11) clearly becomes invalid as output powers exceed $P_s$, even Eq. (8.9)

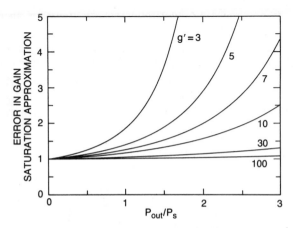

**Figure 8.1.** Factor by which the approximation [Eq. (8.11)] to the gain-saturation factor deviates from the exact solution. The curves are labeled by $g' = (\Gamma g_0 - \alpha_{int})/\alpha_{int}$.

becomes suspect as $P_{out}$ becomes much larger than $P_s$. The linear expansion of the carrier recombination rate, $R(N)$, about the unsaturated carrier density becomes inaccurate at high output power.

## 8.2.2. Noise

The gain in semiconductor optical amplifiers is provided by stimulated emission. In 1917, Einstein recognized that to maintain thermodynamic balance, stimulated emission must be accompanied by spontaneous emission occurring at a rate proportional to the cross section for stimulated emission [5]. After the invention of the maser, the noise in optical amplifiers was investigated theoretically [6, 7] and experimentally [8–10]. As the potential of optical fibers for use in communication systems was realized, the noise properties of semiconductor optical amplifiers were studied [11–13]. The spectral density of the optical noise in a single mode of one polarization emitted from the output facet of an amplifier is given by

$$S_{sp}(\nu) = h\nu\chi N_{sp}(G - 1)$$

$$\equiv h\nu\frac{F}{2}(G - 1) \tag{8.13}$$

where $h\nu$ is the photon energy, $G$ is the single-pass gain of the amplifier, and $F = 2\chi N_{sp}$ has been defined as the noise figure of the amplifier [14]. The noise figure [15] and gain are both functions of $\nu$. The spontaneous noise factor, $N_{sp}$, is equal to $N_e/(N_e - N_t)$, where $N_e$ and $N_t$ are the amplifier carrier densities at the operating point and at transparency, respectively; $N_{sp}$ is unity in an ideal optical amplifier and increases for incomplete population inversion. The excess noise factor, $\chi$ [11, 12] arising primarily from nonvanishing facet reflectivity, is unity for an ideal amplifier and is close to unity for useful amplifiers. Since $N_{sp}$ and $\chi$ are both unity for an ideal amplifier, the noise figure of an ideal amplifier is 3 dB. The lowest noise figure reported for semiconductor optical amplifiers is 5.2 dB [16]. Semiconductor optical amplifier noise figures typically range from this value to about 8 dB. The total noise emitted is determined by integrating Eq. (8.13) over the (optical) bandwidth of the noise:

$$P_{sp} = \int S_{sp}(\nu)\, d\nu$$

$$\equiv h\nu\frac{F}{2}(G - 1)\, \delta\nu \tag{8.14}$$

The terms $F$ and $G$ are evaluated at the frequency for which $S_{sp}(\nu)$ has its maximum value, $S_{peak}$. While Eq. (8.14) defines $\delta\nu$, the effective bandwidth of the noise spectrum, the full-width, half-maximum of the noise spectrum approximates $\delta\nu$ provided the noise spectrum is reasonably shaped. For anomalously shaped (multipeaked, pedestal, etc.) spectra, the effective bandwidth can be obtained from $\delta\nu = \int S_{sp}\, d\nu/S_{peak}$.

The deleterious effects of noise on a system's performance arise from three phenomena: (1) saturation of the gain of an amplifier by the amplified noise, (2) saturation of the gain in a chain of amplifiers by the accumulated noise in the chain, and (3) noise added to a receiver during the detection process by the presence of optical noise.

**8.2.2.1. Noise Saturation in a Single Amplifier.** We first consider the saturation of the gain of a single amplifier by amplified spontaneous emission noise. As optical noise at the output of the amplifier increases, the material gain, $g$, decreases, a result of the loss of carriers though emission stimulated by noise photons. This in itself does not impair the performance of the amplifier, provided the injection current in the amplifier can be increased to compensate for the loss in gain. Since this loss in carriers decreases the effective carrier lifetime, it actually increases the saturation power of the amplifier as measured by the signal power and it appears that there is no reason not to operate the amplifier in a regime where the output is heavily saturated by amplified spontaneous emission noise (ASE). However, assuming negligible facet reflectivity, there is as much ASE light at the input as at the output facet and the ASE light at the input reduces the carrier density at the input relative to the center of the amplifier. This decreases the population inversion at the input, increasing the noise figure of the amplifier for a given single-pass gain. Thus a limit on the maximum gain available from a single amplifier chip is set by the desired noise figure of that amplifier. Depletion of the carrier density will occur as the total ASE power approaches $P_s$. Using Eq. (8.14), to avoid increases in the noise figure, the gain should be limited to approximately

$$G - 1 \leq \frac{P_s}{h\nu F\, \delta\nu} \tag{8.15}$$

where the extra factor of 2 accounts for the presence of two polarizations. The total bandwidth of amplifiers operating at 1.3 and 1.5 $\mu$m is about 5000 and 6500 GHz, respectively. Using a noise figure of 6 dB and assuming a saturation power of $+5$ dBm, we find that the ASE limits the usable chip gain to about 30 dB.

To obtain more gain from a single amplifier, the saturation power can be increased, or the gain bandwidth can be reduced. A multiple-quantum well amplifier has achieved the former goal at the expense of the latter [17]; however, the net result was an increase in the allowable chip gain. If the required gain will saturate a single chip, then two amplifiers can be cascaded [18, 19] with a filter (and usually an optical isolator) between them to reduce the bandwidth of the spontaneous emission noise. Unless the filter is unusually narrow ($< 1$ nm), Eq. (8.15) is a good approximation, substituting the total module gain for $G - 1$ and using the filter bandwidth for $\delta\nu$. If a narrow filter is used, then the broadband noise generated by the second amplifier may dominate the amplified, filtered noise from the first amplifier. Then the gain and bandwidth of the second amplifier would be used in Eq. (8.15).

**8.2.2.2. Noise Saturation in a Chain of Amplifiers.** Consider a transmission system containing a chain of $N$ identical polarization-independent optical amplifier modules (Figure 8.2) with a peak gain of $G$, separated by fiber lengths of loss $L$. The amplifier modules contain one or more optical amplifiers with gain much greater than unity, and may contain an optical filter. It is convenient to include the amplifier/fiber coupling loss as part of $L$. Then the peak module gain, $G$, is the gain from the input facet of the first amplifier in the module to the output facet of the last amplifier in the module, measured at the peak of the module gain spectrum. If an optical filter follows the last amplifier, it is considered to be part of the module, but any filter loss at the peak in the filter spectrum is included in $L$. Assume that except for the initial and final amplifiers, the fiber loss is balanced by the peak amplifier gain ($GL = 1$). The losses (including coupling losses) of the first fiber and the final fiber are $L_T$ and $L_R$, respectively. An example of a gain spectrum of a single amplifier module is shown by the solid line in Figure 8.3. If the amplifier module contains a filter, this will be the filter shape, otherwise it will be the gain spectrum of the amplifier chip. In Figure 8.3, it is assumed that the amplifier module

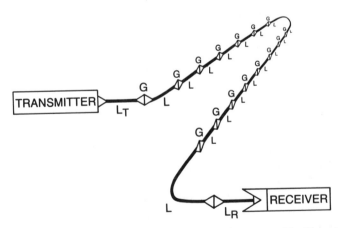

**Figure 8.2.** System containing $N$ identical amplifier modules separated by fiber with loss $L$. The gain of the amplifiers exactly compensates the fiber loss, $G = 1/L$.

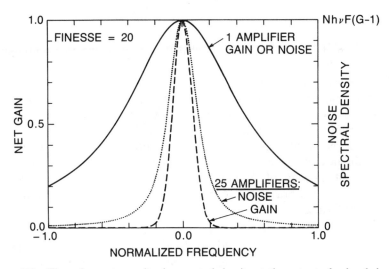

**Figure 8.3.** The gain spectra and noise spectral density at the output of a loss-balanced chain of optical amplifier modules. The frequency axis is normalized to the FWHM passband of a single amplifier module. Cascading 25 amplifiers narrows the net gain spectrum by a factor of 5.8, but the noise spectrum is narrowed by a factor of only 4.0. Each amplifier module contained a Fabry–Perot filter with a finesse of 20.

contains a Fabry–Perot filter with a finesse of 20. The ASE noise spectrum from the amplifier module has roughly the same shape as the gain spectrum. The ASE generated by an amplifier module in the chain undergoes successive cycles of attenuation and gain as it travels down the chain, but since the loss and gain are balanced at the wavelength of the gain peak, the peak spectral density of the ASE from any particular amplifier returns to the same value at the output of each amplifier module. Since the ASE from each amplifier is not coherent with that from any other amplifier, the peak spectral density of the optical noise at the output of the $N$th amplifier is just $N$ times that given in Eq. (8.13).

The widths of the noise spectrum and gain spectrum are not determined as easily since they depend on the shape of the module gain spectrum. If the filter passband is square, then the widths of the gain spectrum and the noise spectrum remain unchanged. However, in practice, these spectra do not have square sides and they therefore narrow as the number of amplifiers increases. For reasonably shaped (Fabry–Perot, Lorentzian, Gaussian, raised-parabola, etc.) spectra, the full-width half-maximum (FWHM) gain bandwidth after $N$ modules decreases approximately as $1/\sqrt{N}$. Therefore, the net gain spectrum after $N$ amplifier modules has a FWHM passband of $\delta\nu/\sqrt{N}$ where $\delta\nu$ is the FWHM passband of one amplifier module. The gain spectrum of a chain of 25 amplifier modules each containing a Fabry–Perot filter of finesse 20 is shown by the dashed line in Figure 8.3.

The noise from each amplifier module in the chain is amplified a different number of times before it reaches the output of the $N$th amplifier module. Thus the noise from the input of the first amplifier is amplified and filtered $N$ times and therefore has a bandwidth of roughly $1/\sqrt{N}$ times that of a single amplifier while the bandwidth of the noise from the final amplifier is unchanged. Since the noise from each amplifier is uncorrelated, total noise at the output of the $N$th amplifier is the sum of the contribution from each amplifier in the chain. The dotted line in Figure 8.3 shows the noise spectrum at the output of the final amplifier in the chain described earlier. To avoid excessive saturation in the final amplifier of a chain of $N$ polarization-independent *identical* amplifier modules, the total noise should (roughly) be less than $P_s$. Therefore, the gain should satisfy

$$P_s \geq h\nu FG\,\delta\nu \sum_{n=1}^{N} \frac{1}{\sqrt{n}} \tag{8.16}$$

Here, $\delta\nu$ is the effective bandwidth of the noise spectrum from one of the amplifiers in the chain, *including any filters*. The right side of Eq. (8.16) is the total noise power at the output of the final amplifier module, including any filters. This underestimates the noise at the output facet of the final amplifier when the modules contain a filter after the amplifier. Then, Eq. (8.16) must be modified in two ways so that the right side equals the noise power at the amplifier output facet. First, the noise from all previous amplifier modules pass through one less filter. Second, the noise from the final amplifier has the complete bandwidth of the amplifier, $\delta\nu_{\text{amp}}$. With these corrections, the gain for amplifier modules with a final filter should satisfy:

$$P_s(\nu) \geq h\nu FG\left[\delta\nu_{\text{amp}} + \delta\nu \sum_{n=1}^{N-1} \frac{1}{\sqrt{n}}\right] \tag{8.17}$$

**8.2.2.3. Noise-Induced Impairment in Receiver Sensitivity.** So far, we have considered the amplified spontaneous-emission noise only as a possible cause of saturation in the amplifier. Now we examine the effect of this noise on receiver sensitivity. The optical signal power at the end of a loss-balanced chain (Figure 8.2) of $N$ identical polarization-independent amplifiers is $P_T L_T L_R/L$, where $P_T$ is the transmitter power and the division by $L$ is a consequence of there being $N$ optical amplifiers (of gain $L^{-1}$) but only $N-1$ lengths of fiber (loss $L$) between them. At the wavelength of the gain peak, the received spectral density of the optical noise emitted by the amplifier chain *in a single polarization* is

$$S_{\text{ASE}} = Nh\nu\frac{F}{2}(G-1)L_R \tag{8.18}$$

We have included the loss, $L_R$, between the output facet of the final amplifier and the detector. The influence of this noise on a receiver performance depends on various signal parameters such as extinction ratio, chirp, intersymbol interference, modulation format, and the detection method.

We first discuss the case of coherent receivers. When designing a coherent receiver for systems without optical amplifiers, the local oscillator power, $P_{LO}$ is set high enough that the shot noise from it dominates the thermal noise of the receiver. Then the electrical signal-to-noise ratio (SNR) at the output of the receiver is (see for instance, Ref. 20):

$$SNR_c = \frac{\frac{1}{2}(i_s)^2}{i_n^2} = \frac{\frac{1}{2}\left(2R\sqrt{P_{LO}P_T L_{TR}}\right)^2}{2qRP_{LO}B_e} = \frac{RP_T L_{TR}}{qB_e} \qquad (8.19)$$

where the received signal power is the transmitted power, $P_T$, reduced by the fiber loss, $L_{TR}$; the receiver bandwidth is $B_e$; and $R$ is the detector responsivity, $\eta q/h\nu$. Further increases in $P_{LO}$ do not improve the SNR in the receiver because both the electrical signal power and the electrical noise power are proportional to $P_{LO}$. The introduction of optical amplifiers will change the SNR since in addition to the local oscillator (LO) shot noise, there are now several new noise terms, including one arising from beating between $P_{LO}$ and the ASE from the optical amplifiers:

$$\overline{i_{LO\text{-}sp}^2} = 2S_{ASE}P_{LO}R^2B_e \qquad (8.20)$$

We have assumed that the ASE has been filtered with a narrow optical filter that excludes the image band separated from the signal band by twice the IF frequency. We can neglect shot noise from the ASE, and noise terms arising from beating of the ASE against the signal and against itself (signal–spontaneous beat noise and spontaneous–spontaneous beat noise) because these terms are independent of $P_{LO}$ and therefore do not contribute to the SNR for large local oscillator power. Combining Eqs. (8.18), (8.19), and (8.20), we find that the SNR of a heterodyne receiver at the end of a chain of $N$ amplifiers is

$$SNR_c = \frac{2\eta P_T L_T L_R/L}{h\nu[2 + \eta NF(G-1)L_R]B_e} \qquad (8.21)$$

Once the SNR is known, the error rate performance of a system using a particular modulation format can be determined (see for example, Ref. 21). For instance, a system employing differential-phase-shift keying will require $SNR_c$ to be about 13 dB to achieve a $10^{-9}$ bit error rate while for frequency-shift keying, this increases to 16 dB.

An ideal direct-detection receiver is limited by shot noise in the signal. Thus, if a Poissonian photon distribution is assumed, an average of 10 photons per bit must be received to obtain an error rate of $10^{-9}$ (assuming the received bit stream is 50% spaces with no photons and 50% marks averaging 20 photons apiece). If direct-detection receivers approached this limit in practice, the use of optical amplifiers in a system would reduce receiver sensitivity since the noise emitted by the amplifiers would increase the number of errors. However, in practice, direct-detection receivers are limited not by photon statistics but by thermal noise. Measured sensitivities range from hundreds of photons per bit for subgigabit/second receivers to thousands of photons per bit for multigigabit/second receivers [22]. The addition of an optical amplifier in front of the receiver (optical preamplifier) can increase the sensitivity of these receivers.

In contrast to the ideal coherent receiver, the electrical signal-to-noise ratio at the output of an ideal direct-detection receiver differs from the optical signal-to-noise ratio at the photodetector input. This is a consequence of the square-law detection process. When a data signal and noise are detected by a PIN photodiode, sources of noise include detector dark current, shot noises, receiver thermal noise, and beat noises arising from beating between the signal and spontaneous emission noise as well as beating of the spontaneous emission noise against itself [10, 11, 13].

The receiver noise in a properly designed direct-detection system containing optical amplifiers will be dominated by signal–spontaneous beat noise and spontaneous–spontaneous beat noise when a mark is being received, and by spontaneous–spontaneous beat noise when a space is being received. For $G/(G - 1) \approx 1$, the electrical SNR for such a system is

$$SNR_d = \frac{(P_{in}/h\nu)^2}{\left[FP_{in}/h\nu + (F/2)^2 \, \delta\nu\right] B_e} \qquad (8.22)$$

where $P_{in}$ is the signal power at the amplifier input facet, $\delta\nu$ is the optical bandwidth of the ASE reaching the detector (usually limited by an optical filter), and we have assumed that a polarizer is placed between the amplifier and the detector. The conventional treatment [23] here is to determine the SNR necessary for the desired error rate. For marks, the noise is dominated by the first term, while for spaces, it is given by the second term. Using $F = 2$ and $B_e = \delta\nu = B$, the bit rate, one obtains an estimate of 42 photons per bit as the sensitivity of an ideal optically preamplified receiver for a bit error rate of $10^9$ [see, for example 13]. Although this method is commonly used, it assumes that the noise in the receiver has Gaussian statistics. This is true for the signal–spontaneous noise term which is the result of beating between the signal and optical noise with Gaussian statistics. However, the spontaneous–spontaneous beat noise has Gaussian-squared statistics. When the proper statistics are used, the sensitivity of an ideal optically preamplified detector is 38 photons per bit [24].

Equation (8.22) can be generalized for the chain of $N$ amplifiers (Figure 8.2)

$$SNR_d = \frac{[(P_T/h\nu)L_T]^2}{\left[NF(P_T/h\nu)L_T + (NF/2)^2\,\delta\nu + (h\nu N_{rec}/q\eta GL_R)^2\right]B_e}$$

(8.23)

where $N_{rec}$ is the receiver noise converted to equivalent detector noise current. For small $GL_R$, the receiver noise dominates. Calculations or measurements of receiver sensitivity for a chain of amplifiers [12, 25] are made by varying $L_R$ and plotting the bit error rate (BER) as a function of optical power at the receiver input. In such plots, for low optical power ($L_R \ll 1$), the denominator of Eq. (8.23) will be dominated by the receiver noise term and a conventional bit-error-rate curve will be obtained. For large $L_R$, the first two terms dominate the denominator of Eq. (8.23). This results in a floor in the BER curve determined by the SNR at the input to the final amplifier. As in the case for an optical preamplifier, the receiver noise in this chain has non-Gaussian statistics and in general, numerical techniques must be used to calculate error rates.

## 8.3. AMPLIFIERS IN SYSTEMS

In the previous section, the general noise and saturation properties of semiconductor optical amplifiers were described. In this section, we discuss the effect of noise and saturation power on system performance. The maximum performance obtainable from an optical amplifier generally depends on both the noise figure and the saturation power of the amplifier, but this dependence varies for different systems. We will separately discuss the use of optical amplifiers in three types of communication systems: direct-detection systems, wavelength-division-multiplexed systems using direct-detection, and wavelength-division-multiplexed systems employing coherent detection.

### 8.3.1. High-Speed Direct-Detection Systems

Optical amplifiers can perform three different functions in direct-detection systems: power amplifiers used to increase the transmitter power, in-line amplifiers used to reduce the effective fiber loss, and preamplifiers used to increase receiver sensitivity. In the power amplifier role, performance is usually limited by signal-induced gain saturation [26] and noise is not important. The reverse is usually true for optical preamplifiers, although as will be shown, a very-high-speed receiver may have such poor sensitivity that amplifier saturation becomes an issue. The maximum performance from an in-line amplifier is limited by both saturation and noise.

We have seen that the gain of optical amplifiers saturates at high output power levels according to Eq. (8.12). In an on–off-keyed communication system, this saturation results in dynamic gain changes that can degrade receiver sensitivity. During a long string of spaces, the carrier density builds up to the unsaturated value. The leading edge of the next mark experiences the unsaturated gain of the amplifier. After a long string of marks, the amplifier gain is fully saturated and the output power has changed to the saturated output power level. In general, the output power in a bit will vary between these two extremes. Changes in the carrier density not only change the gain of an amplifier, but are accompanied by shifts in the phase of the Fabry–Perot ripple in the amplifier gain spectrum. The degree of amplifier saturation and also its effect are then dependent on the exact wavelength of the signal relative to the Fabry–Perot peaks (see, for example, Ref. 27) and are therefore not predictable in a practical system employing Fabry–Perot amplifiers. For this reason, saturation in traveling-wave amplifiers only will be discussed.

The upper portion of Figure 8.4 (obtained from A. A. M. Saleh and I. M. I. Habbab) shows an eye diagram of the signal power at the output of an amplifier operating at various power levels with a bit period of half the effective carrier lifetime, $\tau_c$. This corresponds to a bit rate of 8 Gbit/s for

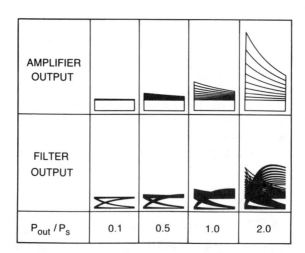

**Figure 8.4.** Calculated eye diagrams for four levels of amplifier saturation using a bit period of half the carrier lifetime (8 Gbit/s for $\tau_c = 250$ pc). The upper plots show the optical power at the output of the optical amplifier while the lower plots show the receiver voltage after a low-pass filter with an 3-dB rolloff frequency of 2.3 times the bit rate. In both cases each plot is normalized to the fully saturated level.

the typical amplifier lifetime of 250 ps. The plots are normalized to the minimum output power for a mark. If an ideal integrate-and-dump receiver is used, the threshold can be set to halfway between the power in the spaces and $P_{min}$, the power in a fully saturated mark. In this ideal receiver, error-free reception can be obtained at any level of amplifier saturation because the eye is never completely closed [28]. However, in a nonideal receiver, two problems arise: (1) it is not generally possible to set the detection threshold optimally; and (2) the filter in the receiver will introduce intersymbol interference. The threshold is usually set to the average of the received signal or, alternatively, to halfway between the extrema in the received signal or to some other easily determined level. It can be seen from the upper portion of Figure 8.4 that if the threshold is set at the average power level, it will approach the level of a fully saturated mark as the amplifier output power approaches and exceeds $P_s$. This will cause a floor in the bit-error-rate curve. The lower portion of Figure 8.4 shows the effect of filtering after detection with an $RC$ time constant of 0.43 times the bit period. Even though the influence of a mark decays to 10% of its value by the end of the following bit period, significant eye-closure occurs as the output power exceeds $P_s$.

Figure 8.4 shows that the amount of eye closure caused by saturation in an amplifier and the resultant bit error rate are dependent on the receiver design. In general, the evaluation of saturation-induced eye-closure requires extensive numerical computation; however, it is still possible to estimate reasonable amplifier output power levels. Results of numerical computations [26, 28] suggest that the difference between the maximum and minimum amplifier output level for marks be limited to about 3 dB. The maximum amplifier output power occurs during a mark that follows a long string of spaces. The leading edge of this mark experiences the full unsaturated gain of the amplifier and has a peak output power of $P_{out} = P_{in}G_0$. The minimum amplifier output power occurs during long strings of marks and is given by $P_{out} = P_{in}G_0 e^{-P_{out}/P_s}$. By setting the ratio between these powers to 3 dB, we find that the *minimum* output power must satisfy

$$P_{out} < \ln(2)P_s \qquad (8.24)$$

to avoid excessive receiver degradation.

**8.3.1.1. High-Speed Transmission Experiment.** Although optical preamplifiers for detectors have been studied for some time [8], it is only recently (for optical communication systems) that better performance was obtained from an optically preamplified receiver than could be obtained from an avalanche-photo-diode receiver [29]. This partially reflects improvement in optical amplifiers, but is also a result of the decrease in APD-receiver sensitivity at the multigigabit/s bit rates only recently attempted. It can be seen from Figure 8.5 that optically preamplified direct-detection receivers

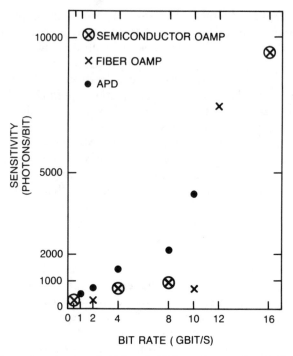

**Figure 8.5.** Comparison in the sensitivity of APD receivers and optically preamplified receivers. (Data obtained from Refs. 22, 29–35).

outperform conventional receivers, particularly at high bit rates. It is likely that all high-performance receivers operating faster than 10 Gbit/s will contain preamplifiers using either semiconductor amplifiers or Erbium-doped fiber amplifiers.

As an example of an optically preamplified receiver, we describe one that has been used to transmit a 16-Gbit/s pseudorandom bit pattern through 64 km of fiber [35]. The 1.3-$\mu$m optical amplifier improved the sensitivity of the high-impedance receiver by 3 dB. A diagram of the experimental system is shown in Figure 8.6. Eight copies of a pseudorandom 2-Gbit/s sequence (pattern length of $2^{15} - 1$ bits) were multiplexed to produce a 16-Gbit/s non-return-to-zero bit pattern that was used to modulate a 1.305-$\mu$m etched-mesa buried-heterostructure DFB laser. The receiver used a 250-$\mu$m-long optical preamplifier with an average facet reflectivity of $1 \times 10^{-4}$, a chip gain of about 20 dB, and an input coupling loss of 3.5 dB. The output of the preamplifier passed through a polarizer and a Fabry–Perot filter (50-nm free-spectral range and 1.8-nm passband) before being focused onto the detector. The receiver front end employed a commercial $p$–$i$–$n$ photodiode in a high-impedance circuit.

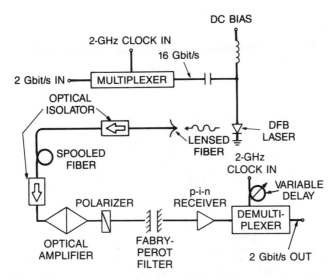

**Figure 8.6.** Apparatus used to transmit 16 Gbit/s over 64 km of fiber.

In addition to the overshoots and intersymbol interference visible in the portion of the received bit stream shown in Figure 8.7, two products of the optical preamplifier can be observed, noise and saturation. The noise in the marks is noticeably greater than the noise in the spaces. This is the result of greater signal–spontaneous beat noise during the marks [$P_{in}$ in Eq. (8.22) is larger for marks]. Saturation of the gain of the optical preamplifier causes the negative slope in groups of uninterrupted marks. Since the bit stream immediately preceding 0 ns is dominated by marks,

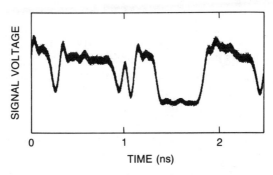

**Figure 8.7.** Section of the received 16-Gbit/s bit stream 40 bits long obtained with a sampling oscilloscope operating in the persistence mode. The increased noise and slope of the marks compared to that of the spaces can be seen. The amplifier output power was set above the optimal level to demonstrate the amplifier saturation.

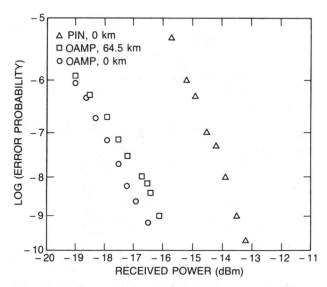

**Figure 8.8.** Error-rate performance of the 16-Gbit/s optical-preamplifier receiver with and without 64 km of fiber compared to the performance of the receiver without the optical preamplifier. The preamplifier improved the sensitivity by 3 dB.

the group of nine marks ending near 1 ns experience the saturated gain of the amplifier. During the next nanosecond, the gain recovers so that the group of marks at 2 ns experience higher initial gain, which again decreases as the amplifier goes into saturation.

Figure 8.8 shows the bit error rate (BER) plotted against the received power as measured in the input fiber for the worst 2-Gbit/s channel. The receiver sensitivity after 64.5 km of fiber (squares) was $-16.1$ dBm (10,100 photons/bit). Comparison with the baseline measurement (circles) shows that the dispersion penalty for the 64.5 km of (non-dispersion-shifted) fiber was 0.5 dB. The baseline receiver sensitivity obtained using the optical preamplifier, $-16.6$ dBm (9000 photons/bit), was 3.1 dB better than the sensitivity of the receiver without the optical preamplifier (triangles) but still 20.3 dB worse than the theoretical limit corrected for coupling loss. Of this 20.3-dB penalty, 4.5 dB was attributed to the 4:1 extinction ratio at the transmitter, 5.5 dB was attributed to the noise figure of the optical amplifier, and 10 dB was attributed to intersymbol interference. The intersymbol interference was a result of inadequate bandwidth and nonflat frequency response in several of the system components.

### 8.3.2. Multichannel Direct-Detection Systems

We have seen that in a single-channel system, the gain of an optical amplifier saturates as the output power of the amplifier approaches $P_s$.

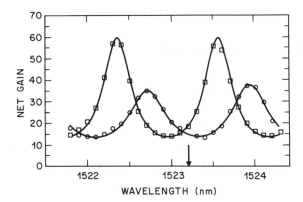

**Figure 8.9.** Small-signal gain spectrum of an amplifier with (circles) and without (squares) the presence of a saturating tone at the wavelength indicated by the arrow.

Since the amplifier gain is homogeneously broadened [36] for time scales longer than 1 ps [37], in a WDM system, saturation of the amplifier gain by any one channel will reduce the gain for all channels. In amplitude-modulated multiwavelength systems, this saturation can lead to crosstalk between channels [38–43]. Unlike the dynamic saturation effects discussed in Section 8.3.3, this saturation-induced crosstalk (SIC) does not decrease for channel spacings greater than the reciprocal of the carrier lifetime. Amplitude modulation in one channel changes the amplifier carrier density and thereby changes the gain in other channels. Since changes in the carrier density also change the amplifier refractive index, the phase of any Fabry–Perot ripple in the amplifier gain spectrum also changes. A portion of a small-signal optical-amplifier gain spectrum is shown in Figure 8.9 in squares. When a large-signal second channel with a wavelength shown by the arrow is added, the gain spectrum changes to that shown by the circles. It can be seen that the peak gain reduced to 60% of the unsaturated peak gain while the Fabry–Perot spectrum shifted by a third of the free-spectral range. Analysis shows that the output power of the second channel is only 18% of the saturation power of the amplifier and that at this level of saturation, the chip gain of the amplifier has decreased by only 17%. Since the presence of Fabry–Perot ripples in the gain spectrum greatly enhances the susceptibility of the amplifier to crosstalk, amplifiers used in WDM systems will need to have low gain ripple.

The allowable amplifier output power levels in a WDM system can be determined from an extension of the reasoning used for a single-channel system. In a WDM system as in a single-channel system, *for an ideal receiver*, error-free transmission can be obtained for arbitrarily high levels of saturation since the eye never closes completely. However, once again, in a practical receiver, the problem of setting the threshold and the presence of intersymbol interference intervene, and it is necessary to restrict amplifier output power levels to avoid a floor in the bit-error-rate curve. Assume that a traveling-wave amplifier is used in a system with $M$ amplitude-modulated channels of equal input power $P_{in}$, and perfect extinction ratio. As before, we require that the range of output power in each channel for marks be restricted to 3 dB. After a long string of spaces in all the channels, the leading edge of a mark will experience the unsaturated gain of the amplifier and will have an output power of $P_{out} = P_{in}G_0$. When all channels have a long string of marks, the output power of one channel will be

$$P_{out} = P_{in}G_0 e^{-P_{tot\ out}/P_s}$$

$$= P_{in}G_0 e^{-MP_{out}/P_s} \tag{8.25}$$

Setting the ratio between these powers to 3 dB, to avoid unacceptable receiver degradation in a WDM system with $M$ channels, the *minimum* output power per channel during a long string of marks in all channels should be limited to

$$P_{out} < \frac{\ln(2)}{M}P_s \tag{8.26}$$

This condition will be satisfied when the input power per channel satisfies

$$P_{in} < \frac{2\ln(2)}{M}\frac{P_s}{G_0}$$

### 8.3.3.  Multichannel Coherent-Detection Systems

Lightwave systems that use coherent-detection techniques could benefit most from the broadband gain available from semiconductor optical amplifiers. Multichannel coherent-detection WDM networks [44, 45] have been proposed in which tens to thousands of closely-spaced channels are combined and distributed through a passive optical star coupler. Similar systems using tunable narrowband optical filters to select and demodulate FSK-modulated channels have also been proposed [46]. Both detection techniques allow channel separations that are only a few times larger than the data rate of each channel. Optical amplifiers can be used to overcome the loss associated with distributing power from each transmitter to a large number of receivers, thereby increasing the number of users on the network.

**8.3.3.1.  Amplifier Intermodulation-Distortion Theory.** In the previously discussed direct-detection WDM systems the principle cause of nonlinearity was the intensity modulation of each source. Only the optical intensity, and not the spectral content of each modulated signal, was important in determining the severity of the impairment. In FSK and PSK (frequency shift-keyed and phase shift-keyed) systems the intensity of each signal is, ideally, constant. However, the instantaneous intensity of the total field within the amplifier is modulated by the interference between each pair of optical carriers and therefore the carrier density is modulated at the beat frequencies between these pairs. The resulting modulation of the gain modulates both the amplitude and phase of all signals within the amplifier in a nondegenerate four-wave mixing (NDFWM) process that can be described by a third-order optical susceptibility. As a result, unacceptable levels of third-order intermodulation distortion (IMD) can be generated even though the total output power may be well below the saturated output power.

IMD is of particular concern in multichannel systems since many intermodulation distortion products (IMPs) are generated within each channel. Each pair of frequencies $\omega_i$, $\omega_j$ leads to the generation of a pair of two-wave products at frequencies $2\omega_i - \omega_j$ and $2\omega_j - \omega_i$. Each group of three frequencies $\omega_i, \omega_j, \omega_k$ results in three three-wave products at $\omega_i + \omega_j - \omega_k$, $\omega_i - \omega_j + \omega_k$, and $-\omega_i + \omega_j + \omega_k$. For $N$ channels, the numbers of two-wave and three-wave products are proportional to $N^2$ and $N^3$, respectively. Since many products are generated within each channel, the total interference is many times the magnitude of each IMP. The total intermodulation interference in a multichannel system can be determined by summing the contributions from all possible two-wave and three-wave combinations, as described later in this section.

IMPs may be generated by several mechanisms in addition to the modulation of the carrier density. Since the response of the carrier-density modulation (CDM) is limited by the carrier lifetime $\tau_c$, the mixing efficiency decreases as the beat frequency $\Omega$ increases and vanishes for $\Omega\tau_c \gg 1$. For channel separations greater than 100 GHz, the small intrinsic third-order nonlinear susceptibility $\chi^{(3)}$ of the crystal will dominate. For separations less than about 100 GHz, IMPs may be generated through modulation of the intraband population, spectral hole-burning [2] or carrier heating effects [47]. However, CDM is currently the only amplifier nonlinearity with enough strength to be of interest or concern for lightwave systems.

Several investigations into IMD or NDFWM in semiconductor laser amplifiers have been reported, initiated by the realization that interacting optical fields in semiconductor amplifying media could produce dynamic carrier-density fluctuations [1, 48], and that these fluctuations could provide efficient NDFWM [49]. The theoretical formalism for gain and index variations for semiconductor materials [50] sparked a flurry of systems-related activity. Measurements of IMD [51–54], development of an analytic small-signal model [55] and numerical studies [56, 57] have quantified

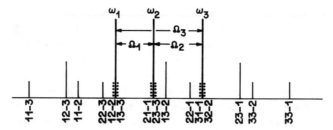

**Figure 8.10.** Two-tone and three-tone products generated by the interaction between three input frequencies, $\omega_1$, $\omega_2$, and $\omega_3$. In addition to the two-wave IMD products, like the one at $2\omega_1 - \omega_3$ (shown as 11-3 in the diagram), and three-wave products like $\omega_1 + \omega_2 - \omega_3$ (12-3 in the diagram), the amplitude and phase of the input frequencies are also modified by a coherent interaction of the type $\omega_1 + \omega_2 - \omega_2$ (12-2 in the diagram) shown as dotted lines.

the processes involved. Based on the magnitudes of two- and three-wave IMPs, the overall limitations imposed on multichannel systems have been calculated [58, 59]. Finally, IMD has been measured in coherent systems with many closely spaced channels [60–62].

We now describe the generation of IMPs using the small-signal approach in which it is assumed that the magnitude of each IMP is much smaller than each of the amplified input signals. With this assumption, the interaction can be treated analytically. Since the IMPs are generated by a third-order nonlinearity, multiwave interactions can be analyzed as the sum of all possible three-wave interactions. Then product-counting techniques, borrowed from treatments of distortion in electronic amplifiers [63], can be employed. For most system applications, which require total carrier-to-noise ratios greater than 15 dB, the resulting predictions are sufficiently accurate. However, for applications where one desires high conjugate reflectivities or to maximize the mixing efficiency, a large-signal approach is required.

Consider the interaction between the carrier density $N$ and any three equal-amplitude input signals, with frequencies $\omega_1$, $\omega_2$, and $\omega_3$, interacting with the carrier density to generate IMPs at frequencies shown in Figure 8.10. Neglecting modulation of each wave, the input field is expressed as

$$E(z,t) = \sum_{j=1}^{3} \frac{1}{2}\left[E_j(z)e^{-i\omega_j t} + \text{c.c.}\right] \qquad (8.27)$$

where c.c. refers to the complex conjugate. Transverse effects are included later using the confinement factor $\Gamma$. The $E_j(z)$ are the complex field amplitudes of the three input signals, normalized such that $|E_j(z)|^2/2$ is the optical intensity in the active region. The coupling between the optical field and the carrier density results in CDM at the beat frequencies between the input waves, $\Omega_1 = \omega_2 - \omega_1$, $\Omega_2 = \omega_3 - \omega_2$ and $\Omega_3 = \omega_3 - \omega_1$. If the CDM is small compared to the average carrier density $\overline{N}$, the rate equation for the carrier density, Eq. (8.1), can be solved [50, 58] to give

$$N = \overline{N} + (\overline{N} - N_t) \sum_{j=1}^{3} \left[ N_j e^{-i\Omega_j t} + \text{c.c.} \right] \qquad (8.28)$$

where

$$\overline{N} - N_t = (N_p - N_t)\left[ 1 - \sum_{j=1}^{3} \frac{E_j E_j^*}{2I_s} \right] \equiv (N_p - N_t)(1 - \varepsilon) \quad (8.29)$$

$$N_j = - \frac{E_m E_n^*}{2I_s(1 - i\Omega_j \tau_c)} \qquad (8.30)$$

and

$$N_p = \frac{I' \tau_c}{qV} \qquad (8.31)$$

The average carrier density $\overline{N}$ is reduced slightly from pump carrier density $N_p$ by the stimulated emission induced by the total input optical field. This DC gain saturation is included to first order by $\varepsilon(z)$ in Eq. (8.29), rendering the following analysis valid only when the total optical signal power within the amplifier is small compared to the saturation power, $P_s$. In Eq. (8.30), from the definition of $\Omega_j$, for $j = 1$, we use $m = 2$ and $n = 1$, while for $j = 2$, $m = 3$ and $n = 2$ and for $j = 3$, $m = 3$ and $n = 1$.

The polarization induced in the amplifier is given [1, 50] by the product of the carrier-density-dependent susceptibility $\chi(N)$ and the total electric field $E$:

$$\mathbf{P} = \chi(N)\varepsilon_0 E$$

$$= K(\alpha + i\zeta)(1 - \varepsilon)\left\{ \sum_{k=1}^{12} \left( E_k e^{-i\omega_k t} + \text{c.c.} \right) \right.$$

$$+ \sum_{k=1}^{3} \sum_{j=1}^{3} \left[ N_j \left( E_k e^{-i(\omega_k + \Omega_j)t} + \text{c.c.} \right) \right.$$

$$\left. \left. + N_j^* \left( E_k e^{-i(\omega_k - \Omega_j)t} + \text{c.c.} \right) \right] \right\} \qquad (8.32)$$

where

$$K = \frac{1}{M} \frac{\varepsilon_0 ncg_0}{4\pi\nu} \tag{8.33}$$

The speed of light in the medium is $c/n$, $\alpha$ is the linewidth-broadening factor, $\varepsilon_0$ is the permittivity of free space, and $g_0$ is the unsaturated material gain of the amplifier [Eq. (8.2)]. The optical frequency of any of the input signals can be used for $\nu$ with little loss in accuracy. $\zeta$ is a waveguide loss parameter $(0 < \zeta < 1)$ with a value of unity for a lossless waveguide:

$$\zeta = 1 - \frac{\alpha_{int}}{\Gamma g_0} \tag{8.34}$$

where $\alpha_{int}$ is the carrier-density-independent part of the total internal power absorption coefficient.

Equation (8.32) shows that in addition to the three input frequencies contained in the first term, polarization is induced at 12 frequencies in the second and third terms. Three of these frequencies overlap the input frequencies and result in changes in the amplitude and phase of the amplified input waves. The remaining nine of these frequencies correspond to the generation of three 3-wave and six 2-wave IMPs, shown in Figure 8.10. Higher-order products, resulting from mixing between input and output waves, are neglected. The generated IMPs are simply the phase and amplitude modulation sidebands induced on each input wave by CDM at each beat frequency. Each product generated, as well as the input signals, propagate along the active layer and experience gain described by the scalar wave equation

$$\frac{\partial^2 E(z,t)}{\partial z^2} - \frac{n^2}{c^2} \frac{\partial^2 E(z,t)}{\partial t^2} = \frac{\Gamma}{\varepsilon_0 c^2} \frac{\partial^2 P(z,t)}{\partial t^2} \tag{8.35}$$

where $\rho$ is the carrier-density-dependent portion of the polarization and the penetration of the transverse-mode profile into the unpumped waveguide cladding layers is incorporated using the confinement factor $\Gamma$.

Each frequency component can be treated separately by writing $E(z,t)$ and $P(z,t)$ as in Eq. (8.27) and extending the sum over all 12 frequencies

$$P(z,t) = \sum_{j=1}^{12} \frac{1}{2} \left[ P_j(z) e^{-i\omega_j t} + \text{c.c.} \right] \tag{8.36}$$

and similarly for $E_j$. Eq. (8.35) then can be written as

$$\frac{\partial^2 E_j(z)}{\partial z^2} + k_j^2 E_j(z) = -\frac{\Gamma}{\varepsilon_0} \frac{k_j^2}{n^2} P_j(z) \qquad (8.37)$$

where $k_j = n\omega_j/c$.

It is convenient to treat each of the types of interaction shown in Figure 8.10 separately. The performance of an $N$-channel system can be derived by considering these basic interactions only.

Consider, for example, the two-wave IMP, 11-3, generated at frequency $2\omega_1 - \omega_3$. Beating between signals at frequencies $\omega_1$ and $\omega_3$ induce CDM at beat frequency $\Omega_3$, which mixes with the input frequency $\omega_1$ to induce a polarization [Eq. (8.32)] given by

$$\mathbf{P}_{1+1-3}(z) = 2K(\alpha + i\zeta)(1 - \varepsilon)(N_3^* E_1 + E_{1+1-3}) \qquad (8.38)$$

In addition to the two-wave IMP just described, CDM at beat frequency $\Omega_3$ modulates $\omega_3$ to produce the two-tone IMP at $2\omega_3 - \omega_1$. The three-wave product at frequency $\omega_1 + \omega_2 - \omega_3$ is the combined result of CDM at $\Omega_2$ mixing with $\omega_1$ and CDM at $\Omega_3$ mixing with $\omega_2$. These two contributions add coherently resulting in an induced polarization:

$$\mathbf{P}_{1+2-3}(z) = 2K(\alpha + i\zeta)(1 - \varepsilon)(N_2^* E_1 + N_3^* E_2 + E_{1+2-3}) \qquad (8.39)$$

Finally, when input wave $\omega_1$, for example, is modulated by CDM at $\Omega_1$, the second modulation sideband is at frequency $\omega_2$. This results in a coherent interaction with the wave at $\omega_2$, as described by the polarization

$$\mathbf{P}_2(z) = 2K(\alpha + i\zeta)(1 - \varepsilon)[E_2 + N_1 E_1 + N_2^* E_3] \qquad (8.40)$$

Equation (8.37) can be solved for each input wave and each IMP [55, 58], given several assumptions. First, we assume that the magnitude of each product is small compared to that of each input wave ($\varepsilon \sim 0$). We then write

$$E_j(z) = \sqrt{I_s} A_j(z) e^{ik_j z} \qquad (8.41)$$

for each of the 12 frequency components and assume that the amplitude of each grows slowly in the $z$ direction. Also, for frequency separations less than 10 GHz and typical amplifier lengths the phase matching may be assumed to be perfect, such that

$$\frac{\partial^2 A}{\partial z^2} \sim 0 \qquad \text{and} \qquad e^{i(k_j - k_k)z} \sim 1 \qquad (8.42)$$

Expressions for the polarization components [e.g., Eqs. (8.38)–(8.40)] can then be inserted into Eq. (8.37), leading to first-order differential equations for the normalized field amplitudes, $A_j(z)$.

Under the condition of high single-pass gain, the power in each IMP (proportional to $A_j A_j^*$) for equal-amplitude input waves, is given by

$$P_{imd} = \frac{P_{out}}{4} \left( \frac{P_{out}}{P_s} \right)^2 \left( 1 + \frac{\alpha^2}{\zeta^2} \right) f(\Omega \tau_c) \qquad (8.43)$$

where, for two-wave interactions [from Eq. (8.38)]

$$f(\Omega \tau_c) = \left| \frac{1}{1 + i(\omega_i - \omega_j)\tau_c} \right|^2 \qquad (8.44)$$

and for three-wave interactions [from Eq. (8.39)]

$$f(\Omega \tau_c) = \left| \frac{1}{1 + i(\omega_j - \omega_k)\tau_c} + \frac{1}{1 + i(\omega_i - \omega_k)\tau_c} \right|^2 \qquad (8.45)$$

The amplitude and phase change induced on the input wave, $\omega_2$ for example, can be solved using Eq. (8.40):

$$\frac{A_2(L)}{A_2(0)} = \exp\left[ \frac{\Gamma}{2} g_0 L (\zeta - i\alpha)(1 - \delta) \right] \qquad (8.46)$$

where

$$\delta = \frac{P_{out}}{P_s} \frac{1}{\Gamma g_0 \zeta L} \left[ \frac{1}{1 - i\Omega_1 \tau_c} + \frac{1}{1 + i\Omega_2 \tau_c} \right] \qquad (8.47)$$

Equations (8.43)–(8.46) show that each of these interactions results from modulation of both the gain and refractive index within the active layer. Modulation of the refractive index, through $\alpha$, is responsible for most of the power in each IMP [Eq. (8.43)] and the strongest interaction with each input wave [Eq. (8.46)]. The magnitudes of the intermodulation products increase when the frequency separation of the generating waves decreases [Eqs. (8.44) and (8.45)] or when their output power ($P_{out} = |E(L)|^2 A/\Gamma$) increases [Eq. (8.43)]. The frequency dependence arises because the modulation of the carrier density is reduced as $\Omega \tau_c$ increases. The amplitude and phase changes described by $\delta$ result in asymmetric gain wherein the lowest frequency input wave ($\omega_1$) experiences the largest gain and $\omega_3$ the smallest [52, 56]. For FSK systems, the slight gain asymmetry and changes in phase and amplitude will not be a problem.

However, the type of interference described by Eq. (8.46) must be considered separately for ASK and PSK systems [64]. The impact of each of these nonlinear interactions on multichannel systems will be discussed later, but first we summarize the many experimental investigations that confirm the predicted behavior.

**8.3.3.2. Amplifier Intermodulation-Distortion Measurements.** Numerous experimental investigations of CDM-induced nonlinear mixing have been reported. Beginning with the demonstration of high conjugate reflectivity in GaAs lasers [49], several measurements of the generation of IMPs have been reported [51–54, 65]. Each shows a dependence of the mixing efficiency on pump power and pump frequency detuning (beat frequency between pump waves).

For systems applications, where the small-signal approximation is justified and where effects such as asymmetric gain and gain saturation are unimportant, the dependence described by Eqs. (8.43) and (8.44) has been demonstrated [53, 66]. The IMPs were measured by amplifying two equal-amplitude tones in the apparatus shown in Figure 8.11. The input tones were generated by two external-cavity lasers, L1 and L2, operating with optical frequencies $\nu_1$ and $\nu_2$, respectively. The amplifier output was heterodyne detected on a high-speed InGaAsP detector using most of the

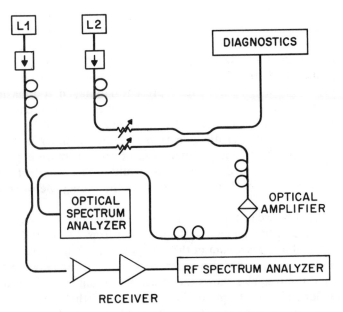

**Figure 8.11.** Apparatus used to measure two-tone intermodulation distortion. Lasers L1 and L2 are external cavity lasers; L1 also served as a local oscillator.

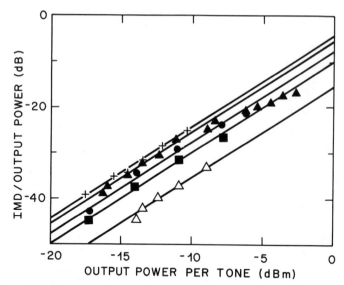

**Figure 8.12.** Increase in normalized IMD power with increase in output power per channel. Points represent data taken under conditions described in the text. The lines are slope 2 fits to the data.

power from L1 as the local oscillator (LO) to down-convert the IMP with frequency $2\nu_2 - \nu_1$ to $2|\nu_1 - \nu_2|$.

In Figure 8.12, the ratio of IMP output power to the output power of L1 is plotted against the output power of L1. The results were fit to a line of slope 2 as suggested by Eq. (8.43). All data were taken with a frequency separation, $\Omega$, of 300 MHz. The data shown in triangles were taken using the same injection current (130 mA) and nearly the same wavelength (1487 nm), except that the solid triangles represent data taken at a maximum in the amplifier FP spectrum while the open triangles represent data taken at a minimum in the FP spectrum. The difference in the IMP generated is about 9 dB, twice the 4.5 dB gain ripple of the amplifier at this wavelength. Although the effects of nonzero facet reflectivity were not considered in the derivation of Eq. (8.43), one would expect a perfect traveling-wave amplifier to exhibit IMD behavior near the average between the two curves.

Comparison between the solid triangles, circles and squares shows the effects of reducing the net gain of the amplifier from 12.0 to 10.5 to 7.5 dB, respectively, all at a maximum in the FP spectrum. Gain-ripple effects are the main source of the observed differences. Comparison between the plus symbols and the triangles shows the effect of shifting from the red side of the amplifier gain spectrum to the blue side of the spectrum. At the shorter wavelength, $\alpha$ is expected to be smaller [42, 67] leading to lower IMP levels, and the gain ripple was smaller in the amplifier measured

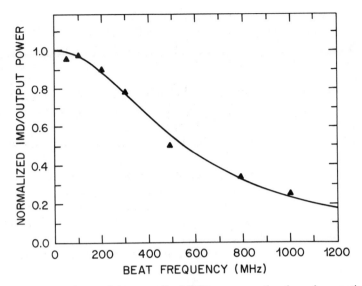

**Figure 8.13.** Dependence of the normalized IMD power on the channel separation. The line is a Lorentzian fit that yields an effective carrier lifetime $\tau_c$, of 280 ps.

(1 dB vs. 4.5 dB at $\lambda = 1287$ nm) thus also leading to lower IMP levels. The large IMD observed suggests that $P_s$ is also smaller at the shorter wavelength. This is consistent with the observed wavelength dependence of saturation in optical amplifiers [27, 68].

In Figure 8.13, the dependence of IMP generation on the frequency difference between the two tones can be seen. The data (triangles) were taken with a wavelength of 1287 nm, injection current of 130 mA, and with the input tones at a maximum in the amplifier FP spectrum. The line shows the expected [Eq. (8.44)] frequency dependence of $1/[1 + \Omega^2 \tau_c^2]$. The value of $\tau_c$ was adjusted for the best fit, yielding $\tau_c = 280$ ps. This is consistent with the expected value of $\tau_c$ for an amplifier operating with high gain or injection current levels [3]. By varying the bias current and temperature of the amplifier, values of $\tau_c$ ranging from 180 to 800 ps were observed.

**8.3.3.3.** *N*-Channel IMD. Each nonlinear interaction described previously presents an impairment to the performance of closely-spaced multifrequency lightwave systems. The total effect on the performance of an *N*-channel system can be calculated by considering all possible combinations of three channels that interact according to $(\omega_i, \omega_j, \omega_k; \omega_l)$, where $\omega_l = \omega_i + \omega_j - \omega_k$. This notation states that frequency $\omega_i$ interacts with CDM, induced by beating between $\omega_j$ and $\omega_k$, and generates a mixing product at $\omega_l$. Combinations of the types $(\omega_i, \omega_i, \omega_i; \omega_i)$ and $(\omega_j, \omega_i, \omega_i; \omega_j)$ represent (DC) saturation of the gain by the intensity of the wave at $\omega_i$.

These are included by the factor $\varepsilon$ in Eq. (8.32) and result in a reduction in available output power for constant-amplitude PSK or FSK systems. For ASK WDM systems with widely spaced channel frequencies, the modulation of this intensity leads to saturation-induced crosstalk (SIC), discussed in Section 8.3.2.

Combinations of the type $(\omega_i, \omega_j, \omega_i; \omega_j)$ result in coherent addition of power to the signal at $\omega_j$. This results in an altered amplitude and phase of the wave at $\omega_j$, as described by Eqs. (8.46) and (8.47), and leads to the previously noted gain asymmetry, wherein the lowest-frequency wave experiences a higher gain. For ASK systems, the amplitude and phase changes are time-dependent, resulting in beat-induced crosstalk (BIC, labeled CMI crosstalk in Ref. 58). Although both SIC and BIC are consequences of gain saturation, the BIC results from the beating between separate carriers and vanishes for large frequency separations. SIC results from intensity or amplitude modulation of each carrier and does not vanish for large frequency separations. FSK or PSK systems are relatively insensitive to either type of crosstalk.

The two remaining types of combinations include $(\omega_i, \omega_i, \omega_j; 2\omega_i - \omega_j)$, where the signal generated at $2\omega_i - \omega_j$ is a two-wave IMP, and the three-wave interaction $(\omega_i, \omega_j, \omega_k; \omega_i + \omega_j - \omega_k)$. This three-wave interaction adds in phase with $(\omega_j, \omega_i, \omega_k; \omega_j + \omega_i - \omega_k)$ to make one three-wave IMP. The total intermodulation interference for each channel is obtained by summing the intensities of all IMPs that fall within that channel, while considering all combinations of three input carriers. The intensities are added, rather than the optical fields, since the IMPs generated within each channel are generally uncorrelated, owing to an absence of coherence among the generating signals.

The calculated IMD levels across a band of channels are shown in Figure 8.14 [58], where the total output power is held constant ($NP_{out} = 0.1P_s$). Since $f(\Omega\tau_c)$, in Eq. (8.44), vanishes for large $\Omega\tau_c$, the ratio of total intermodulation power, $P_{imd}$ to $P_{out}$ is dominated by IMPs that result from CDM between closely-spaced channels. $P_{imd}/P_{out}$ is then approximately constant for all channels except those near the edges of the band of $N$ channels. As the channel separation $\Omega$ is decreased, CDM from more channels becomes significant and $P_{imd}/P_{out}$ increases. Figure 8.15 shows the change in band-center IMD levels as the number of channels is increased with the total power in the band held constant. A $10^{-9}$ error rate requires a carrier to noise ratio of about 13 dB for systems employing differential phase shift-keying and about 16 dB for coherent-detection systems employing frequency shifting-keying. It can be seen that depending on the number of channels and the channel spacing, the total output power of the amplifier may have to be less than $0.1P_s$ to achieve acceptable performance. While the exact calculation of IMD levels requires numerical summation of many contributions, analytic upper bounds have

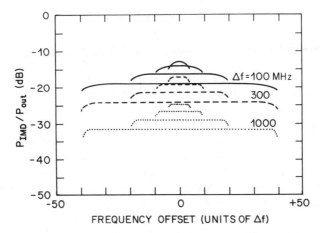

**Figure 8.14.** Relative intermodulation–distortion (IMD) power (ratio of IMD power in a channel to signal power in the channel) in angle-modulated systems for 10, 20, 40, and 80 channels and three frequency separations, $\Delta f$. Total output power is one-tenth the amplifier saturation power $P_s$, $\tau_c$ is 280 ps, and $\alpha/\zeta = 5$.

been derived that facilitate calculation of the effect of IMD on system performance [59].

The IMD levels plotted have been derived under the assumption that the carriers in the band are not coherently related. Under these conditions, the optical *powers* of the various intermodulation products are summed rather than the optical *fields*. The results plotted therefore represent the average IMD power. Transitory coherences between differ-

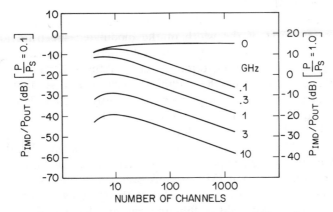

**Figure 8.15.** Relative IMD power in the center channel of a multichannel system. The total output signal power in the band is equal to $P_s$ (right scale) or $0.1P_s$ (left scale). The lines are labeled with the carrier spacing. The effective carrier lifetime is $\tau_c = 300$ ps, and $\alpha/\zeta = 10$.

ent IMD contributions occur randomly and the IMD power will fluctuate. The statistics of these fluctuations have not been studied; however, the problem is roughly similar to the multipath-signal-fading problem [69]. In addition, the IMD power can differ significantly from the plotted values if the optical carriers in the band are coherently related. This can occur if the carriers are derived from a small number of sources by nonlinear techniques or by filtering a single mode-locked source.

The impairment to FSK or PSK systems is caused by the accumulation of IMPs within each channel. Since the magnitude of each IMP decreases as the beat frequency increases, products generated by widely-spaced channels do not contribute. The carrier lifetime effectively introduces a filter to the mechanism for IMP generation so that the overall impairment becomes worse as the carrier lifetime or frequency separation decreases. For ASK systems, two additional types of crosstalk are introduced by the amplifier, one that vanishes for large channel separations and one that is not affected by channel separation. These impairments do not prohibit the use of semiconductor amplifiers in closely spaced lightwave systems, but they do place constraints on the allowed output power, number of channels, and frequency separation of the optical carriers.

**8.3.3.4. Intermodulation Distortion with Modulated Carriers.** Until this point, we have discussed the generation of nonlinear mixing products using unmodulated optical carriers only. The analysis becomes much more difficult when realistic modulation is considered and no such investigations have yet been reported. Nevertheless, the previously described approach of product counting using unmodulated carriers provides a reasonable estimate of the impairment in the presence of data. Also, measurements of intermodulation interference in coherent system demonstrations have been published [61–63, 70].

Effects related to modulation are most easily incorporated in two limiting cases. First, if the width of the modulated spectrum for each channel is less than a third of the separation between channels, then the previous results can be used directly. Each of the IMPs has a spectral width that is three times the width of each carrier so that roughly one-third of each product will impair the underlying data. This fraction depends on the type and index of modulation. The remaining power from the IMPs is outside the bandwidth of the detected channel and causes no impairment. In densely spaced coherent homodyne systems [71], the modulated spectrum can be comparable to the channel separation. In this case, the spectral width of each IMP extends over about three channels. The product counting approach is still valid. Although only one third of the power from each IMP falls within the central channel, the remaining power overlaps adjacent channels. The net result is roughly the same IMP interference as if no modulation were applied, since all the IMP power generated appears as an impairment in some channel.

Between these two extremes, one must consider the exact shape of the modulation spectrum and determine the appropriate fraction of IMP power that falls within the information-containing bandwidth of each channel. This has been investigated experimentally [72] using four FSK channels separated by 200 MHz and a 1.3-$\mu$m near-traveling wave optical amplifier. Difficulties in stabilizing and controlling frequencies in densely spaced networks generally result in the use of equally spaced channels that are locked to periodic resonances of, for example, a Fabry–Perot etalon [73]. Otherwise, one could consider spacing the channels according to a plan that minimizes the interference from intermodulation distortion.

Another complication arises from the dependence of the IMP power on the frequency separation of the beating waves. If both waves are modulated then the beat frequency becomes the distribution of frequencies given by the convolution of the two modulation spectra. In the limit where the largest frequency in this distribution is less than the reciprocal of $\tau_c$, this complication makes little difference. But for data-rates greater than a few hundred megabits per second, the generation of IMPs from the high-frequency components of the beat-frequency distribution ($\Omega \tau_c \gg 1$) is reduced. As a result, the predictions from the previous section overestimate the total interference in multicarrier systems with high data-rates.

This variation of the mixing efficiency from the interacting modulated spectra may cause problems in PSK systems. As described by Eqs. (8.46) and (8.47), CDM alters the phase of each of the carriers. With no modulation, this phase alteration is simply an additive constant. However, when modulation is applied to each carrier, this phase change becomes a function of the instantaneous beat frequency between uncorrelated modulated signals. The resulting phase noise could introduce an impairment that may be severe for powers approaching the saturated output power.

**8.3.3.5. Large-Signal Effects.** The small-signal model is a convenient tool for understanding the effect of the NDFWM on the performance of multichannel lightwave systems. Each category of impairment can be described by analytic solutions to linearized equations, and the approximations used are valid for the low power levels required to maintain acceptable error rates. However, many interesting and useful phenomena occur that cannot be described using the small-signal approach.

Even for small signals, shown by Eqs. (8.46) and (8.47), a gain asymmetry is created by the carrier-density modulation, resulting in higher gain for the lowest-frequency channels. Increasing the input powers increases the gain asymmetry and leads to asymmetry [52] in the magnitudes of the generated IMPs. For amplifier-based systems with just a few closely spaced channels operating near saturation of the amplifier, this gain asymmetry could be compensated for by preemphasis of the appropriate channels. Numerical studies have shown [56] that, for two channels, gain asymmetries as high as 10 dB may be obtained.

As IMPs propagate and grow along the active region, the carrier density is also modulated at the beat frequencies between the input and the generated frequencies. The higher-order IMPs generated by this cascaded mixing have been observed [51].

Finally, the mixing process can be used to shift the frequency of a modulated optical carrier over wide frequency ranges [74]. The CDM induced between a modulated carrier and an unmodulated pump wave induces sidebands on any other carrier within the gain–bandwidth of the amplifier. By introducing a second unmodulated signal at a desired frequency, the modulation originally at one frequency can be transferred to the second. The efficiency of this process depends on the efficiency of the mixing between the original signal and the pump, and a fiber-to-fiber efficiency of $-10$ dB was demonstrated.

## 8.4.   REDUCTION OF NONLINEARITY

We have seen that the nonlinearity inherent in semiconductor optical amplifiers can cause adverse system effects, such as intermodulation distortion (IMD) in densely spaced WDM systems, crosstalk in on–off-keyed WDM systems, and pulse distortion in on–off-keyed systems. While these effects can be made arbitrarily small by reducing the amplifier output power to a sufficiently low value, this can defeat the object of using such amplifiers in the first place. Each of these processes results from the coupling between the carrier density, hence the optical gain, and the optical intensity within the amplifier. In this section, we describe a feedforward linearization method that can reduce the nonlinearity of an amplifier by maintaining a constant carrier density.

### 8.4.1.   Theory of Linearizer

Consider a traveling-wave optical amplifier of length $L$ oriented along the $z$ direction. Let the bias current consist of the usual DC pump component, $I_{DC}$, and an additional time-dependent component, $I_1$, provided by a feedforward circuit with a transfer function to be determined. We average the rate equation for the carrier density [Eq. (8.1)] along the $z$ direction [75] to obtain

$$\frac{d\bar{N}}{dt} = \frac{I_1}{qV} - \frac{\bar{N} - \bar{N}_p}{\tau_c} - \frac{|E_{out}|^2 - |E_{in}|^2}{h\nu V} \tag{8.48}$$

where time dependence is implicit. The length-averaged carrier density is $\bar{N} = (1/L)\int_0^L N(z,t)\,dz$, $\bar{N}_p$ is the (time independent) averaged carrier density caused by $I_{DC}$, and $E_{in}$, and $E_{out}$ are, respectively, the input and

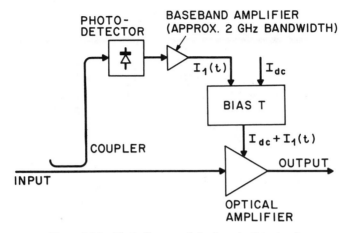

**Figure 8.16.** Block diagram of the linearization circuit.

output optical fields, which are normalized here such that $|E|^2$ represents optical power.

The scalar wave equation [Eq. (8.35)] gives the optical power gain as

$$G \equiv \frac{|E_{out}|^2}{|E_{in}|^2} = e^{\Gamma a[\bar{N} - \bar{N}_t]L} \tag{8.49}$$

where $\bar{N}_t$ is the averaged carrier density at transparency and the internal loss $\alpha_{int}$, has been ignored. The nonlinearity of the amplifier can be attributed to the time dependence of the optical gain, which is shown in Eq. (8.49) to be caused by the time variations of $\bar{N}$.

It is clear from Eq. (8.48) that, if one sets

$$I_1 = \frac{q}{h\nu}(G - 1)|E_{in}|^2 \tag{8.50}$$

then one obtains the solution: $\bar{N} = \bar{N}_p$; in other words, the average carrier density becomes constant. It follows from Eq. (8.49) that $G$ also becomes constant, and the amplifier becomes linear.

Figure 8.16 shows a feedforward circuit [76] that linearizes the optical amplifier by fulfilling Eq. (8.50). It consists of an input directional coupler (with power coupling factor, $C$), a photodetector (with quantum efficiency, $\eta$) and a baseband amplifier (with *current* gain, $G_c$). The amplifier's output current, $I_1$, is combined with the DC pump current, $I_{DC}$. If $K$ is the power coupling at the input facet of the optical amplifier, then Eq. (8.50) requires

$$G_c = \frac{K(1 - C)}{\eta C}(G - 1) \tag{8.51}$$

Note that, if the optical signal extends over hundreds of gigahertz of bandwidth, as would be the case in many lightwave communications systems, then Eq. (8.50) cannot be fulfilled, since no baseband amplifier can deliver such a signal. However, because of the nonzero value of the carrier lifetime, $\tau_c \approx 0.3$ ns [53], the carrier density cannot respond to variations faster than a few gigahertz (several times $1/2\pi\tau_c$). Thus, the linearizing current, $I_1$, and the baseband amplifier need not cover frequency components higher than a few gigahertz, irrespective of the optical bandwidth.

### 8.4.2. Linearizer Measurements

In a demonstration of the operation of the feedforward linearizer, the linearizer consisted of a 3-dB input coupler, a broadband PIN photodetector, and an (electrical) amplifier chain providing about 50 dB of gain with less than 2 dB of ripple. The 3-dB bandwidth of the chain and associated components was about 2 GHz. Not shown in Figure 8.16 is the variable attenuator and a variable delay line used to adjust the amplitude and delay of the feedforward bias current, $I_1$, to optimize the linearization.

The operation of the linearizer was studied for two cases of amplifier nonlinearity: (1) intermodulation distortion (IMD) induced by the amplification of two closely-spaced optical carriers [76] and (2) the crosstalk (SIC) induced by the amplification of on–off-keyed signals [77]. For the former case, using an apparatus similar to that shown in Figure 8.11, two external-cavity lasers generated equal-power signals with frequencies $f_1$ and $f_2$. The 1.3-$\mu$m optical amplifier had a chip gain of 24 dB, coupling loss of about 5 dB per facet, and facet reflectivities of about $1 \times 10^{-4}$, resulting in less than 1 dB of gain ripple. The optical spectrum after the amplifier was measured by mixing a portion of the light from one of the tunable lasers (frequency $f_1$) with the amplifier output. The resulting RF signal was displayed on a spectrum analyzer, which was used to measure the amplitudes of the signal at $f_2$ and the IMD product at $2f_2 - f_1$.

Figure 8.17 shows plots of the normalized IMD versus facet output power per tone without the linearizer (open circles) and with the linearizer (solid circles). The two-tone separation here was about 1000 MHz. The figure shows a reduction of the IMD by 12 dB. The solid and dashed lines in the figure represent the least-square-error, straight-line fits through the data points, with unconstrained slope and with a slope of 2, respectively. This slope shows that the third-order IMD product is proportional to the cube of the input power, which is expected, at least for the case without the linearizer [53–55, 58, 75]. Figure 8.17 shows that one can operate the linearized amplifier with a 6-dB increase in the output power compared to a nonlinearized amplifier operating with the same IMD-to-signal ratio.

A convenient method for observing the saturation-induced crosstalk with on–off-keyed signals is to amplify a sinusoidally intensity-modulated

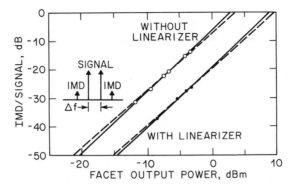

**Figure 8.17.** Normalized IMD power versus output power with and without linearizer. The frequency separation was about 1 GHz.

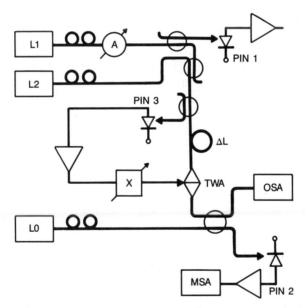

**Figure 8.18.** Apparatus used to measure linearizer suppression of saturation-induced crosstalk. Laser L1 was amplitude modulated, while CW laser L2 served as a probe. Amplifier nonlinearity was observed by measuring the modulation induced on the probe using local oscillator LO. The linearizer circuit consisted of a coupler, PIN 3, an amplifier, an attenuator and a phase shifter.

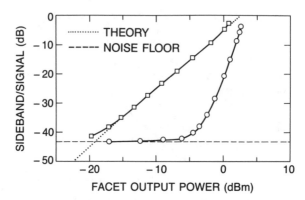

**Figure 8.19.** Measured crosstalk with (circles) and without (squares) linearizer. The dotted line is a theoretical prediction and the dashed line is the measurement noise limit.

signal and an unmodulated probe signal simultaneously, and measure the modulation sidebands induced on the probe [77]. Using the apparatus shown in Figure 8.18, modulated signals from a DFB laser (L1) with a wavelength of 1305 nm, and a probe beam from a grating-tuned external-cavity laser (L2) with a wavelength of 1290 nm, were amplified simultaneously in a traveling-wave amplifier that had a fiber-to-fiber gain of 14 dB with 1 dB of gain ripple. The amplified beam was heterodyne-detected using local-oscillator laser LO, and the result was displayed on a microwave spectrum analyzer (MSA). With full sinusoidal modulation applied to the modulated beam, the crosstalk induced on the probe beam, with and without the linearizer, is shown as a function of the output power of the modulated beam, on Figure 8.19. Also shown is the prediction (dotted line) from theory [77] using $\tau_c = 280$ ps and $P_s = 3$ dBm. The reduction in the amplifier nonlinearity is small for modulated beam powers that produce substantial gain saturation, but is at least 25 dB (limited by the measurement noise floor) for smaller output powers.

The usefulness of the feedforward circuit in reducing crosstalk in a digital system can be seen in Figure 8.20, which shows the heterodyne-detected spectrum of the amplified probe signal (centered at 1.5 GHz), when the other beam was intensity modulated with 200-Mb/s pseudorandom data. (The directly-detected spectrum of the modulated beam can be seen at frequencies less than 600 MHz.) Since modulation was not applied to the probe beam, the modulation evident in the heterodyne spectrum is the crosstalk induced by the presence of the modulated beam. With the linearizer, the crosstalk was reduced by about 20 dB. Also, as evident from the 6-dB difference shown at zero frequency in Figure 8.20, the effective gain of the amplifier is increased by 3 dB (optical).

It can be seen from these two demonstrations that a feedforward linearizer can reduce the nonlinearity of an optical amplifier by a signifi-

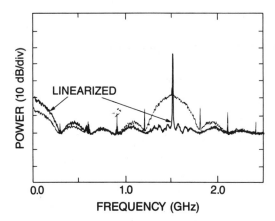

**Figure 8.20.** Induced modulation of probe beam, L2, with and without linearizer for 200 Mb/s pseudorandom on–off keying in saturating channel, L1.

cant amount. It remains to be seen whether this technique will be a practical solution to the problem since the gain and delay time of the feedforward circuit must be carefully matched to that of the signal circuit [78]. The linearizer can be viewed as a means of increasing the saturation power of the amplifier, which, in the demonstrations, increased by more than 7 dB. For effective linearization, the spectrum of the optical signal must be limited to a bandwidth over which the optical amplifier has a constant gain, which could be several hundreds of gigahertz in width. The bandwidth of the baseband *electronic* amplifier, on the other hand, does not need to be larger than a few gigahertz.

## 8.5. SUMMARY

We have described some of the noise and nonlinearity properties of InGaAsP optical amplifiers and outlined general constraints they impose on the use of amplifiers in high-bit-rate and multiwavelength communication systems. The maximum signal power is limited by amplifier saturation, while the minimum signal power is determined by the amplifier noise.

In a single-channel direct-detection system using on–off-keying, unacceptable saturation-induced degradation in receiver sensitivity will occur for amplifier output powers greater than $0.7P_s$. If more channels are added, this limit applies to the *total* output power. In a direct-detection system containing one (nonideal) optical amplifier, one would expect that a signal power at the amplifier input facet of about 80 photons/bit would be sufficient to obtain an error rate of less than $10^{-9}$. For a loss-balanced amplifier chain with $N$ amplifiers, the required signal input power is a factor of $N$ larger. In practice, the required input powers have been factors of 2–50 greater than expected. This difference can probably be attributed to extinction-ratio and intersymbol-interference penalties, both of which are worsened by optical amplifiers.

The power constraints imposed by optical amplifiers in multichannel coherent systems are not easily stated since they depend on the channel separation, channel packing, number of channels, linewidth broadening factor, and modulation format. As a worst case, an amplifier in a densely packed 10-channel frequency-shift-keying system with 100-MHz channel separation will generate unacceptable intermodulation distortion if the total output power in the band is greater than about 5% of $P_s$. On the other hand, if the channel spacing is increased to 10 GHz, the total output power can exceed $P_s$ (provided FM/AM conversion does not occur somewhere in the system with resultant saturation-induced crosstalk). The minimum input signal level required in a coherent detection system to obtain an error rate of $10^{-9}$ depends on the modulation format. It is expected that a nonideal amplifier will require about 80 photons per bit for DPSK modulation and about 150 photons per bit for FSK modulation at the input facet of the amplifier. These minimum power levels increase linearly with the number of amplifiers in the chain. The accumulation of saturation effects in a chain of amplifiers has not been extensively studied for either direct-detection or coherent-detection systems.

In this chapter, we have limited the discussion to semiconductor optical amplifiers. The performance of Erbium-doped fiber amplifiers is advancing rapidly and it is likely that they will be used in the first-generation optical amplifier systems. Most of the discussion in this chapter applies to fiber amplifiers with an important distinction. The saturation recovery time in Erbium amplifiers is about 0.2 ms [79], some six orders of magnitude slower than that of semiconductor amplifiers. For this reason gain-saturation-induced signal degradation ceases to be a problem for high-speed systems. Any saturation of the gain occurs over a time scale much slower than the bit rate so the entire bit stream experiences the same gain. Similarly, the amplifier averages over any beating between adjacent channels so intermodulation distortion ceases to be an issue. The respective roles of semiconductor amplifiers and Erbium-doped fiber amplifiers in future communication systems are not yet known. In comparison to semiconductor amplifiers, Erbium-doped fiber amplifiers are more easily made and packaged, they have much better dynamic saturation properties and have demonstrated higher gain. On the other hand, they only operate at wavelengths between 1.53 and 1.56 $\mu$m, they require high (tens of milliwatts) optical pump powers, they are relatively bulky and cannot be integrated into a PIC (photonic integrated circuit).

Telecommunication industries throughout the world have converted from fully electronic long-haul transmission systems to systems that are electronic–photonic hybrids. Optical amplifiers are the component that will allow conversion to fully photonic transmission systems. This conversion will make it easier to increase bit rates and to multiplex several wavelength channels onto a single fiber. By using optical amplifiers, unrepeated span lengths can be increased, as well as unregenerated span lengths. The performance of these devices is progressing rapidly and they will soon find application in commercial communication systems.

## REFERENCES

1. A. P. Bogatov, P. G. Eliseen, and B. N. Sverdlov, *IEEE J. Quantum Electron.* **QE-11**, 510–515 (1975).
2. G. P. Agrawal, *J. Opt. Soc. Am. B* **5**, 147–158 (1988).
3. G. P. Agrawal and N. K. Dutta, *Long-Wavelength Semiconductor Lasers*, Van Nostrand Reinhold, New York, 1986, Chapter 2.
4. U. Koren, B. I. Miller, Y. K. Su, T. L. Koch, and J. E. Bowers, *Appl. Phys. Lett.* **51**, 1744–1746 (1987).
5. A. Einstein, *Phys. Z.* **18**, 121–128 (1917).
6. K. Shimoda, H. Takahasi, and C. H. Townes, *J. Phys. Soc. Jpn.* **1**, 686–700 (1957).
7. H. Steinberg, *Proc. IEEE* **51**, 943 (1963).
8. W. B. Bridges and G. S. Picus, *Appl. Opt.* **3**, 1189–1190 (1964).
9. F. Arams and M. Wang, *Proc. IEEE* **53**, 329 (1965).
10. J. A. Arnaud, *IEEE J. Quantum Electron.* **QE-4**, 893–899 (1968).
11. Y. Yamamoto, *IEEE J. Quantum Electron.* **QE-16**, 1073–1081 (1980).
12. T. Mukai, Y. Yamamoto, and T. Kimura, *IEEE J. Quantum. Phys.* **QE-18**, pp. 1560–1568 (1982).
13. J. C. Simon, *J. Opt. Commun.* **4**, 51–62 (1983).
14. T. Saitoh and T. Mukai, *IEEE J. Quantum Electron.* **QE-23**, 1010–1020 (1987).
15. M. G. Oberg and N. A. Olsson, *Electron. Lett.* **24**, 99–100 (1988).
16. T. Mukai and T. Saitoh, *Electron. Lett.* **23**, 216–218 (1987).
17. G. Eisenstein, U. Koren, G. Raybon, T. L. Koch, J. M. Wiesenfeld, R. S. Tucker, and B. I. Miller, "A 1.5 $\mu$m quantum well amplifier with high saturation output power," *Post-Deadline Digest, Seventh International Conference on Integrated Optics and Optical Fiber Communication* (IOOC '89), pp. 62–63, Kobe (1989).
18. N. A. Olsson and R. M. Jopson, *Lightwave Technol.* **7**, 791–793 (1989).
19. I. W. Marshall and P. D. Constantine, "High-gain semiconductor laser amplifier package," *Digest: 15th European Conference on Optical Communication* (ECOC '89), pp. 54–57, Gothenburg (1989).
20. T. Okoshi and K. Kikuchi, *Coherent Optical Fiber Communications*, KTK Scientific Publishers, Tokyo (1988).
21. S. Benedetto, E. Biglieri, and V. Castellani, *Digital Transmission Theory*, Prentice-Hall, Englewood Cliffs, NJ, (1987).

22. B. L. Kasper and J. C. Campbell, *J. Lightwave Technol.* **LT-5**, 1351–1364 (1987).

23. S. D. Personick, *Bell Syst. Tech. J.* **52**, 843–874 (1973).

24. P. S. Henry, "Error-rate performance of optical amplifiers," *Digest: Optical Fiber Communication Conference* (OFC '89), Paper THK3, Houston (1989).

25. N. A. Olsson, *J. Lightwave Technol.* **LT-7**, 1071–1081 (1989).

26. A. A. M. Saleh and I. M. I. Habbab, "Performance of digital high-speed intensity-modulation lightwave systems utilizing semiconductor optical power amplifiers," *Optical Fiber Communication Conference* (OFC '90), Paper WM23, San Francisco (1990).

27. G. Eisenstein, K. L. Hall, R. M. Jopson, G. Raybon, and M. S. Whalen, "Two-color gain saturation in an InGaAsP near-traveling-wave optical amplifier," *Technical Digest*, OFC '87, Paper ThC4, Reno, NV (1987).

28. A. A. M. Saleh and I. M. I. Habbab, *IEEE Trans. Commun.* **38**, 839–846 (1990).

29. N. A. Olsson, M. G. Oberg, L. D. Tzeng, and T. Cella, *Electron. Lett.* **24**, 569–570 (1988).

30. N. A. Olsson and P. Garbinski, *Electron. Lett.* **22**, 114–1116 (1986).

31. C. R. Giles, E. Desurvire, J. L. Zyskind, and J. R. Simpson, "Near quantum-limited Erbium-doped fiber preamplifier with 215 photons/bit sensitivity at 1.8 Gb/s," *Post-Deadline Digest, Seventh International Conference on Integrated Optics and Optical Fiber Communication* (*IOOC* '89), *pp.* 20–21, *Kobe* (1989).

32. R. M. Jopson, A. H. Gnauck, B. L. Kasper, R. E. Tench, N. A. Olsson, C. A. Burrus, and A. R. Chraplyvy, *Electron. Lett.* **25**, 233–235 (1989).

33. K. Hagimoto, Y. Miyagawa, M. Miyamoto, M. Ohhashi, M. Ohhata, K. Aida, and K. Nakagawa, "A 10 Gb/s long-span fiber transmission experiment employing optical amplification technique and monolithic IC technology," *Post-Deadline Digest, Seventh International Conference on Integrated Optics and Optical Fiber Communication* (IOOC '89), pp. 22–23, Kobe (1989).

34. H. Nishimoto, I. Yokata, M. Suyama, T. Okiyama, M. Seino, T. Horimatsu, H. Kuwahara, and T. Touge, "Transmission of 12 Gb/s over 100 km using an LD-pumped Erbium-doped fiber amplifier and a Ti:LiNbO₃ Mach–Zender modulator," *Post-Deadline Digest, Seventh International Conference on Integrated Optics and Optical Fiber Communication* (IOOC '89), pp. 26–27, Kobe (1989).

35. A. H. Gnauck, R. M. Jopson, J. D. Evankow, C. A. Burrus, S.-J. Wang, N. K. Dutta, and H. M. Presby, "16-Gbit/s, 64-km optical-fiber transmission experiment using an optical-preamplifier receiver," *Post-Deadline Proceedings, 15th European Conference on Optical Communication* (ECOC '89), Vol. 3, pp. 25–28, Gothenburg (1989).

36. T. Mukai, K. Inoue, and T. Saitoh, *Appl. Phys. Lett.* **51**, 381–383 (1987).

37. K. L. Hall, J. Mark, E. P. Ippen, and G. Eisenstein, *Applied Phys. Lett.* **56**, 1740–1742 (1990).

38. R. P. Webb, "Evaluation of a semiconductor-laser amplifier for multiplexed coherent systems," *Proceedings of SPIE—The International Society for Optical Engineering*, Vol. 587, pp. 39–43 (1986).

39. G. Grosskopf, R. Ludwig, and H. G. Weber, *Electron. Lett.* **22**, 900–902 (1986).

40. T. Mukai, K. Inoue, and T. Saitoh, *Electron. Lett.* **23**, 396–397 (1987).

41. M. G. Oberg and N. A. Olsson, *IEEE J. Quantum Electron.* **QE-24**, 52–59 (1988).

42. R. M. Jopson, K. L. Hall, G. Eisenstein, G. Raybon, and M. S. Whalen, *Electron. Lett.* **23**, 510–512 (1987).

43. Y. K. Park, S. W. Granlund, L. D. Tzeng, N. A. Olsson, and N. K. Dutta, *Electron. Lett.* **24**, 475–477 (1988).

44. E. J. Bachus, R. P. Braun, C. Caspar, E. Grossman, H. Foisel, K. Heimes, H. Lamping, B. Strebel, and F. J. Westphal, *Electron. Lett.* **22**, 1002–1003 (1986).

45. B. S. Glance, K. Pollock, C. A. Burrus, B. L. Kasper, G. Eisenstein, and L. W. Stulz, *IEEE J. Lightwave Technol.* **6**, 67–72 (1988).

46. I. P. Kaminow, P. P. Iannone, J. Stone, and L. W. Stulz, *IEEE J. Lightwave Technol.* **6**, 1406–1414 (1988).

47. M. P. Kesler and E. P. Ippen, *Electron. Lett.* **24**, 1102–1104 (1988).

48. J.-L. Nishizawa and I. Katsuhiko, *IEEE J. Quantum Electron.* **QE-11**, 515–519 (1975).

49. H. Nakajima and R. Frey, *IEEE J. Quantum Electron.* **QE-22**, 1349–1354 (1986).

50. G. P. Agrawal, *Opt. Lett.* **12**, 260–262 (1987).

51. R. Nietzke, P. Fenz, W. Elsasser, and E. O. Göbel, *Appl. Phys. Lett.* **51**, 1298–1300 (1987)

52. K. Inoue, T. Mukai, and T. Saitoh, *Appl. Phys. Lett.* **51**, 1051–1053 (1987).

53. R. M. Jopson, T. E. Darcie, K. T. Gayliard, R. T. Ku, R. E. Tench, T. C. Rice, and N. A. Olsson, *Electron. Lett.* **23**, 1394–1395 (1987).

54. G. Grosskopf, R. Ludwig, R. G. Waarts, and H. G. Weber, *Electron. Lett.* **24**, 31–32 (1988).

55. T. E. Darcie, R. M. Jopson, and R. W. Tkach, *Electron. Lett.* **23**, 1392–1394 (1987).

56. I. M. I. Habbab and G. P. Agrawal, *IEEE J. Lightwave Technol.* **7**, 1351–1359 (1989).

57. F. Favre, D. LeGuen, J.-C. Simon, and P. Doussiere, *Electron. Lett.* **25**, 272–273 (1989).

58. T. E. Darcie and R. M. Jopson, *Electron. Lett.* **24**, 638–640 (1988).

59. R. M. Jopson and T. E. Darcie, *Electron. Lett.* **24**, 1372–1374 (1988).

60. R.-P. Braun, R. Ludwig, and R. Molt, *NTZ, Nachrichtentechnische Zeitschrift* **39**, 804–808 (1986).

61. B. Glance, G. Eisenstein, P. J. Fitzgerald, K. J. Pollock, and G. Raybon, *Electron. Lett.* **24**, 1157–1159 (1988).

62. S. Ryu, K. Mochizuki, and H. Wakabayashi, *J. Lightwave Technol.* **7**, 1525–1529 (1989).

63. M. T. Abuelma'atti, *IEEE Trans. Commun.* **COM-33**, 246–248 (1985).

64. G. P. Agrawal, *Electron. Lett.* **23**, 1175–1177 (1987).

65. K. Inoue, *Electron. Lett.* **23**, 1293–1295 (1987).

66. R. M. Jopson, T. E. Darcie, and K. Gayliard, "Observation of carrier-density mediated intermodulation distortion in an optical amplifier," *Technical Digest, Optical Fiber Communication Conference* 1988 (OFC '88), Vol. 1, p. 92, New Orleans (1988).

67. L. D. Westbrook, *Electron. Lett.* **21**, 1018–1019 (1985).

68. K. Inoue, T. Mukai, and T. Saitoh, *Electron. Lett.* **23**, 328–329 (1987).

69. M. Schwartz, W. R. Bennett, and S. Stein, *Communication Systems and Techniques*, McGraw-Hill, New York, 1966, Chapter 9.

70. G. Grosskopf and R. Ludwig, *Electron. Lett.* **24**, 1052–1054 (1988).

71. J. M. Kahn, *IEEE Photon. Technol. Lett.* **1**, 340–342 (1989).

72. B. S. Glance, G. Eisenstein, P. J. Fitzgerald, K. J. Pollock, and G. Raybon, *J. Lightwave Technol.* **7**, 759–765 (1989).

73. B. S. Glance, J. Stone, K. J. Pollack, P. J. Fitzgerald, C. A. Burrus, Jr., B. L. Kasper, and L. W. Stultz, *J. Lightwave Technol.* **6**, 1770–1781 (1988).

74. G. Grosskopf, R. Ludwig, and H. G. Weber, *Electron. Lett.* **24**, ·1106–1107 (1988).

75. A. A. M. Saleh, *Electron. Lett.* **24**, 835–837 (1988).

76. A. A. M. Saleh, R. M. Jopson, and T. E. Darcie, *Electron. Lett.* **24**, 950–952 (1988).

77. T. E. Darcie, R. M. Jopson, and A. A. M. Saleh, *Electron. Lett.* **24**, 1154–1155 (1988).

78. T. Durhuus, B. Mikkelsen, K. E. Stubkjaer, P. B. Hansen, Z. Wang, and D. S. Olesen, "Intermodulation distortion due to optical amplifiers in multichannel systems," *Proceedings of ECOC'89*, Vol. 1, pp. 46–49, Gothenburg (1989).

79. C. R. Giles, E. Desurvire, and J. R. Simpson, *Opt. Lett.* **44**, 880–882 (1989).

# 9

# Injection-Locked Semiconductor Laser Amplifiers

S. KOBAYASHI
*Photonic Integration Research, Inc., Columbus, Ohio*

## 9.1. INTRODUCTION

Injection-locking (IL) phenomena were theoretically verified by Adler in 1946, who developed a differential equation accounting for many of the observed phenomena of synchronization [1]. A practical IL amplifier was developed in the microwave circuits. The basic characteristics of IL for avalanche diodes were obtained by Shaw and Stover in 1966 [2]. Experimental examinations of IL repeaters for frequency-modulated (FM) signals were performed by Mastalli et al. in 1968 [3]. Sideband locking and IL amplification for phase-modulated (PM) signals in the millimeter wave range were verified by Fukatsu in 1969 [4]. Frequency-modulated (FM) noise reduction of an IMPATT oscillator using IL was experimentally verified by Isobe and Tokida [5], and these results could be explained by the theory drawn by Heines et al. in 1969 [6]. In 1969 Kohiyama developed an IL amplifier, with a low-noise Gunn diode as a master oscillator and an IMPATT (impact avalanche transit time) diode as a slave oscillator [7]. IL had been performed not only in the fundamental mode at a free-running frequency but also at harmonic and subharmonic frequencies. The harmonic locking was verified by Oltman and Nonnemaker [8] and the subharmonic locking by Daikoku and Mizushima in 1969 [9]. An FM demodulator employing an IL was analyzed by Ruthroff in 1968 [10]. An AM limiter was theoretically proposed by Osborn and Elemendorf in 1970 [11]. A PM wave was obtained using IL technique by Martin and Hobson in 1971 [12]. The basic characteristics of IL were theoretically verified by Kurokawa in 1973 [13].

---

*Coherence, Amplification, and Quantum Effects in Semiconductor Lasers*, Edited by Yoshihisa Yamamoto.
ISBN 0-471-512494 © 1991 John Wiley & Sons, Inc.

Laser frequency locking was first demonstrated by Stover and Steier in 1966, who directly injected a beam from an He–Ne laser into another laser [14]. The experimental results were found to be in qualitative agreement with Adler's classical frequency-locking analysis [1]. Hybrid IL of a high-power carbon dioxide ring laser was achieved with a stable low-power master laser by Buczek and Freiberg in 1972 [15]. Dye laser IL was performed by Vrehen in 1972. Helium–neon laser injection locking at 1.15 $\mu$m was precisely studied by Urisu et al. in 1980 [16]. The classical theory was extended to include locking in laser oscillators [17, 18]. Semi-classical laser theory was studied using the analogy of the van der Pol equation by Sargent et al. in 1974 [19]. A precise explanation of laser IL was also given by Shimoda and Yajima in 1975 [20]. Laser IL theory was clarified by Tang and Statz assuming the two-energy-level model in 1967 [18]. In a semiconductor laser, however, IL phenomena had not been observed until 1980 because of the oscillating frequency instability depen-dent on the mode competition in a laser oscillating in multilongitudinal modes. Coupling between two semiconductor lasers and the effect of reflected light on a semiconductor laser have been studied theoretically and experimentally [21–24]. Coupling between two lasers was tried in order to suppress relaxation oscillation [21, 25] or to measure the optical gain in semiconductor lasers [26]. To measure precise interaction with the injected laser beam, however, optical isolation would have to be equipped in order to avoid undesired feedback into the master laser. The theoreti-cal treatment of the semiconductor laser was achieved by Hirota and Suematsu using a semiclassical theory in 1979 [22]. Combining the rate equation with the van der Pol equation to verify semiconductor laser IL was performed by Otsuka and Trucha in 1981 [27]. The asymmetry of the locking curve was first explained by Lang and Kobayashi using the linewidth enhancement factor $\alpha$ in 1982 [28]. This treatment was authorized by Henry et al. in 1985 [29–31]. The AM and FM quantum noise of an IL laser amplifier was analyzed quantum-mechanically by Haus and Yamamoto in 1984 [32].

Semiconductor laser IL was verified experimentally by Kobayashi and Kimura as follows: the relation of the locking bandwidth to the locking gain and PM signal amplification in 1980 and 1981 [33–35], FM noise suppression and FM signal amplification in 1981 and 1982 [36, 37], and PM signal generation by IL in 1982 [38]. Locking bandwidth and relaxation oscillation of IL with AlGaAs lasers were also discussed by Petitbon et al. [39, 40]. The asymmetric locking curve in the 1.3-$\mu$m InGaAsP laser diode was measured by Kobayashi et al. [41]. The sideband injection locking with AlGaAs laser diodes and the microwave IL to Field-effect transistor (FET) by the beat signal between optically injection-locked upper and lower sidebands of the slave laser were performed by Goldberg et al. in 1982 [42, 43]. Goldberg's group also tried IL to an array semiconductor laser in 1985 [44]. Automatic frequency control (AFC) of the IL amplifier was performed by Kobayashi and Kimura in 1983 using the induced voltage change due to light injection [45]. The influence of the master laser noise on the slave laser was experimentally and theoretically clarified by Spano et al. in 1986 [46].

Coherent optical fiber transmission systems, in which coherent laser waves are used as optical carrier waves, are expected to improve system performance toward long repeater spacing and large information capacity [47]. Amplitude, frequency, or phase modulation of the optical carrier wave, and optical homodyne or heterodyne detection are employed as modulation and deomodulation schemes, in place of the intensity modulation (IM) and direct detection in conventional systems. Direct optical amplification is also regarded as an important technique for advanced systems [48]. Optical homodyne and heterodyne detection schemes are directly applied to frequency division multiplexing for further wideband communication [49, 50]. For further improvement in receiver performance, a narrow laser linewidth is indispensable. Optical phase shift-keying (PSK) signal generation by means of semiconductor laser IL is attractive for the development of the receiver performance that is expected in PSK hetero-dyne systems. Predictions of IL technology application to coherent optical systems were made by Yamamoto and Kimura in 1981 [47]. Amplitude shift-keying (ASK) homodyne systems using IL of an InGaAsP semicon-ductor laser with a He–Ne laser at 1.55 $\mu$m were verified in 1982 by Hodgkinson et al. [51] and Wyatt and Smith [52]. It is possible to combine a frequency-stabilized laser oscillator and an external phase modulator as a transmitter configuration for FSK heterodyne differential detection and PSK heterodyne–coherent-detection schemes. A conventional guided wave electrooptic modulator with, for example, LiNbO$_3$ [53], can be used as a phase modulator. Its insertion and coupling loss may be compensated for by an injection-locked semiconductor laser postamplifier. An IL repeater amplifier experiment was performed by Smith and Malyon [54]. In conven-tional intensity modulation–direct-detection systems, an injection-locked single-mode semiconductor laser with a frequency stabilized DC drive master laser was used at 2-Gbit/s modulation frequency by Yamada et al., who obtained repeater spacing of 44.3 km in 1981 [55]. An experiment was performed over 102 km of low-loss single-mode fiber at 140 Mbit/s using an IL source at 1.52 $\mu$m by Malyon and McDonna in 1982 [56]. A laser system with an IL DC–PBH as a slave and a DFB as a master was operated with 445.8-Mbit/s RZ signal over a 170-km single-mode fiber at 1.55 $\mu$m by Toba et al. in 1984 [57]. Olsson et al. performed the transmis-sion experiment over 82.5 km of fiber at 2 Gbit/s with DFB lasers without any measurable dispersion penalty [58].

In this chapter, Section 9.2 focuses on the basic theoretical model for semiconductor laser injection locking (IL). First the formulation of the van der Pol equation is discussed. Second, an asymmetric locking curve that is induced from the carrier density dependence of the refractive index will be presented. Section 9.3 describes the IL phenomena of AlGaAs semiconductor lasers, locking bandwidth, phase locking, and locked power saturation. Section 9.4 focuses on the optical FM signal amplification and FM noise suppression. Section 9.5 discusses phase modulated signals induced from IL. Section 9.6 provides system applications using semiconductor laser IL amplifiers.

## 9.2. THEORY OF INJECTION LOCKING

### 9.2.1. Formulation of the van der Pol Equation

The optical injection-locking phenomena shown in Figure 9.1 are explained here. When a weak narrow-bandwidth optical beam emitted from a master laser is coupled into a slave laser, the slave laser becomes a locked oscillator, forced to oscillate at the frequency $f_{in}$ of the incident light. The injection-locked laser (slave laser) emits its intense laser light at the same frequency $f_1$ as that $f_{in}$ of the incident signal. When the slave laser oscillates at the free-running frequency $f_0$, its frequency is switched $f_1 = (f_{in})$ by the master laser light injection. In this analysis, injection locking occurs when the optical isolator is used avoiding reverse light injection into the master laser.

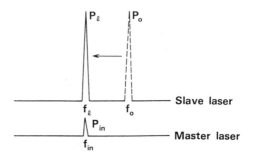

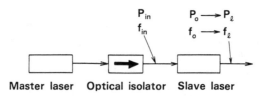

**Figure 9.1.** Schematic diagram of optical injection locking ($P_0$, free running power of slave laser; $P_1$, injection locked power; $P_{in}$, input power from master laser).

The basic equation for semiconductor laser injection locking is represented by the van der Pol equation, assuming that the laser oscillates in a single longitudinal mode. The van der Pol equation, including an injection wave, can be represented in the rotating-wave approximation as follows [20, 22]:

$$\frac{dE_1}{dt} - (g - \alpha - \beta|E_1|^2)E_1 = \frac{E_{in}}{2\tau_p} \tag{9.1}$$

where $E_1$ and $E_{in}$ represent the locked and injected complex electric fields, respectively. The gain $g$ is represented by $I/(2\tau_p I_{th})$; the loss $\alpha$ is represented by $1/(2\tau_p)$, where $\tau_p$ is photon lifetime; $\beta$ is represented by $1/(2\tau_p|E_{sat}|^2)$; and $I_{th}$ is the threshold current. Electric fields $E_1$ and $E_{in}$, which satisfy the rotating-wave approximation and saturation amplitude $|E_{sat}|^2$, are represented as follows:

$$E_1 = |E_1|\exp[j\{(\omega_1 - \Omega_0)t + \phi_1(t)\}] \tag{9.2}$$

$$E_{in} = |E_{in}|\exp[j\{(\omega_{in} - \Omega_0)t + \phi_{in}(t)\}] \tag{9.3}$$

$$|E_{sat}| = \frac{|E_0|}{(I/I_{th} - 1)^{1/2}} \tag{9.4}$$

where $\Omega_0$ is the center angular frequency of the resonator and $E_0$ is the free-running amplitude. By substituting Eqs. (9.2) and (9.3) into Eq. (9.1), the real and imaginary parts are given as follows:

$$\frac{d|E_1|}{dt} - (g - \alpha - \beta|E_1|^2)|E_1| = \frac{|E_{in}|\cos\theta}{2\tau_p} \tag{9.5}$$

$$|E_1|\left\{(\omega_1 - \Omega_0) + \frac{d\phi_1(t)}{dt}\right\} = \frac{|E_{in}|\sin\theta}{2\tau_p} \tag{9.6}$$

$$\theta = (\omega_{in} - \omega_1)t + (\theta_{in} - \phi_1) \tag{9.7}$$

When the phase of the locked wave is assumed to be fixed, that is, $d\phi_1(t)/dt = 0$, $\omega_1 - \Omega_0$ is represented by the locking half angular bandwidth $\Delta\omega = 2\pi\Delta f = \omega_{in} - \Omega_0$. When $\omega_{in} = \omega_1$, $P_{in} = |E_{in}|^2$, $P_1 = |E_1|^2$, and $\theta = \pm\pi/2$ in Eq. (9.6), the locking half-bandwidth is represented as follows:

$$\Delta f = \frac{1}{4\pi\tau_p}\left(\frac{P_{in}}{P_1}\right)^{1/2} \tag{9.8}$$

The relation between $P_1$, $P_{in}$, and $\Delta\omega$ in a steady state is obtained from

Eqs. (9.5) and (9.6) with $d|E_1|/dt = 0$, $d\phi_1(t)/dt = 0$, $|E_1|^2 = P_1$, and $|E_{in}|^2 = P_{in}$ as follows:

$$P_1\left[\left\{\left(\frac{1}{2\tau_p}\right)\left(\frac{I-I_{th}}{I_{th}}\right)\left(\frac{P_0-P_1}{P_0}\right)\right\}^2 + \Delta\omega^2\right] = \frac{P_{in}}{4\tau_p^2} \qquad (9.9)$$

where $P_0$ is free-running power. At zero detuning $\Delta\omega = \omega_{in} - \Omega_0 = 0$, $P_1$ has the maximum value given by Eq. (9.9). When the injected wave is detuned to the locking frequency boundary, $P_1$ is equal to the free-running power $P_0$ obtained from Eq. (9.9) with $P_1 = P_0$.

The photon lifetime $\tau_p$ in the laser cavity is represented as follows:

$$\tau_p = \left\{\frac{c}{n}\left[\alpha' + \left(\frac{1}{L}\right)\ln\left(\frac{1}{R}\right)\right]\right\}^{-1} \qquad (9.10)$$

where $c$ is the light velocity, $n$ is the refractive index, $\alpha'$ is the internal loss, $L$ is the cavity length, and $R$ is the laser facet reflectivity.

By differentiating Eq. (9.7) to obtain a phase condition, the derivative is as follows:

$$\frac{d\theta}{dt} = (\omega_{in} - \omega_1) + \frac{d(\phi_{in} - \phi_1)}{dt} \qquad (9.11)$$

By substituting Eq. (9.6) into Eq. (9.11) and integrating Eq. (9.11) with $\theta = \theta_0$ at $t = 0$,

$$t = \int_{\theta_0}^{\theta} \frac{d\theta}{(\omega_{in} - \omega_1) - (|E_{in}|\sin\theta)/(2\tau_p|E_1|)} \qquad (9.12)$$

The denominator of the right-hand side of Eq. (9.12) becomes 0 when the conditions are

$$|\omega_{in} - \omega_1| \leq \frac{|E_{in}|\sin\theta}{(2\tau_p|E_1|)} \qquad (9.13)$$

and $\theta = \theta_0$. Under this condition, $t \to \infty$ and $\theta$ reaches the constant value of $\theta_1$, the next relations are satisfied as follows:

$$\begin{cases} \omega_{in} = \omega_1 \\ \phi_{in} - \phi_1 = \theta_1 \end{cases} \qquad (9.14)$$

That is, the locking angular frequency equals the injected light angular frequency and the phase difference between the locked laser light and the

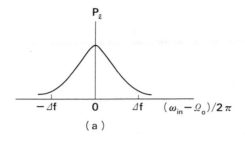

(a)

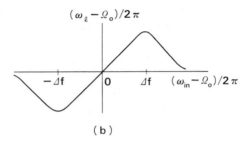

(b)

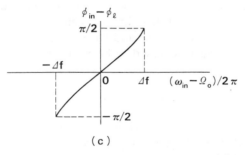

(c)

**Figure 9.2.** Schematic diagram of injection-locking phenomena: (a) locked power against injected light frequency, where $2\,\Delta f$ is locking bandwidth; (b) locked optical frequency against injected light frequency; (c) phase difference between injected and locked light against injected light frequency.

injected light electric fields becomes constant $\theta_1$. These relations are shown in Figure 9.2.

### 9.2.2. Asymmetric Detuning Due to Refractive Index Change

In an injection-locking tuning curve of semiconductor lasers, asymmetric characteristics are observed, but cannot be explained only by the theoretical expression drawn from the van der Pol equation. This asymmetric curve was analyzed by Lang and Kobayashi [28] taking into consideration the refractive index change due to the carrier-density change as a result of the light injection. Mogensen et al. follow a similar analysis using the linewidth enhancement factor $\alpha$ [31].

The basic differential equations are as follows:

$$\frac{dE_1(t)}{dt} - \left\{ j\omega(N) + \frac{1}{2}\left[ G(N) - \left(\frac{1}{\tau_p}\right) \right] \right\} E_1(t) = \eta f_d E_{in}(t) \quad (9.15)$$

where $\omega(N)$ and $G(N)$ are the angular optical frequency and the modal gain per second of the slave laser, which change as a result of the carrier density $N$; $f_d$ is the longitudinal-mode spacing; and $\eta$ is the coupling efficiency considering transmission and mode-matching losses in the slave laser.

The complex fields are represented by

$$E_1(t) = |E_0(t)|\exp\left[ j\{\omega_0 t + \phi_0(t)\} \right] \quad (9.16)$$

$$\eta E_{in}(t) = |E_{in}|\exp\left[ j\{\omega_{in} t + \phi_{in}(t)\} \right] \quad (9.17)$$

where $|E_0(t)|$, $|E_{in}|$, $\phi_0(t)$ and $\phi_{in}$ are real values and $\omega_0$ is the angular oscillation frequency of the slave laser without the injected signal; $|E_0(t)|$ is considered to be a constant here and $\phi_{in}$ can be set at zero. The master laser is assumed to be very stable and have a very narrow linewidth. The modal gain per second is assumed to be a linear function of the carrier density

$$G(N) = G_N(N - N_0) \quad (9.18)$$

where $G_N$ and $N_0$ are constants. Without light injection the carrier density is damped at the threshold value $N_{th}$, where

$$G(N_{th}) = \frac{1}{\tau_p} \quad (9.19)$$

The carrier-density change due to the external light injection can usually be regarded as small and the angular frequency can be expressed as follows:

$$\omega(N) = \omega_0 + \tfrac{1}{2}\alpha G_N \Delta N \quad (9.20)$$

where

$$\Delta N = N - N_{th} \quad (9.21)$$

and

$$\omega_0 = \omega(N_{th}) \quad (9.22)$$

The parameter $\alpha$ describes the coupling between the amplitude and phase of the electric field and is given by

$$\alpha = -\frac{2\omega_0}{n^*G_N}\frac{\partial n}{\partial N}\bigg|_{N=N_{th}} \tag{9.23}$$

where $n$ and $n^*$ are the effective phase and the group refractive indices, respectively, and $\alpha$ is the linewidth enhancement factor [8].

Using Eqs. (9.16), (9.17), and (9.18)–(9.23), Eq. (9.15) can be rewritten as the amplitude-phase representation:

$$\frac{d|E_0(t)|}{dt} = \frac{1}{2}G_N\,\Delta N(t)|E_0(t)| + f_d|E_{in}|\cos[\Delta(t)] \tag{9.24}$$

$$\frac{d\phi_0(t)|}{dt} = \frac{1}{2}\alpha G_N\,\Delta N(t) + f_d\frac{|E_{in}|}{|E_0|}\sin[\Delta(t)] \tag{9.25}$$

The carrier density is represented as follows:

$$\frac{dN(t)}{dt} = J - \frac{N(t)}{\tau_s} - G(N)|E_0(t)|^2 \tag{9.26}$$

where the optical power is represented by the squared amplitude of the electric field, $J$ is the pumping term, and $\tau_s$ is the spontaneous emission lifetime. In Eqs. (9.24) and (9.25)

$$\Delta(t) = \Delta\omega\,t - \phi_0(t) \tag{9.27}$$

and

$$\Delta\omega = 2\pi\,\Delta f = \omega_{in} - \omega_0 \tag{9.28}$$

is the angular frequency difference between the free-running lasers.

The stationary solutions without injection terms (index s) are represented using Eq. (9.19) and

$$\left|\widetilde{E_{0s}}\right|^2 = \tau_p J - \frac{N_{th}}{\tau_s} \tag{9.29}$$

$$\phi_{0s} = 0 \tag{9.30}$$

In the injection-locked state, the slave laser oscillates at the same frequency as the master laser, and the locked phase $\phi_1$ of the slave laser relative to that of the master laser can be shown as follows:

$$\phi_0(t) = (\omega_{in} - \omega_0)t + \phi_1 \tag{9.31}$$

and from Eq. (9.27)

$$\Delta(t) = -\phi_1 \tag{9.32}$$

The stationary solutions to Eqs. (9.24)–(9.26) are found by further setting $|E_0(t)| = |\widetilde{E_0(t)}|$ and $N(t) = N_{th} + \widetilde{\Delta N_i}$:

$$\widetilde{\Delta N_i} = -2\left(\frac{f_d}{G_N}\right)\frac{|E_{in}|}{|\widetilde{E_0}|}\cos\phi_1 \tag{9.33}$$

$$\Delta\omega = -f_d\frac{|E_{in}|}{|\widetilde{E_0}|}[\sin\phi_1 + \alpha\cos\phi_1] \tag{9.34}$$

$$\left|\widetilde{E_0(t)}\right|^2 = \frac{|\widetilde{E_{0s}}|^2 - \tau_p\widetilde{\Delta N_{in}}/\tau_s}{\left(1 + \tau_p G_N\widetilde{\Delta N_{in}}\right)} \tag{9.35}$$

Setting

$$\psi = \arctan\alpha \tag{9.36}$$

Eq. (9.34) can be satisfied only when

$$|\Delta\omega| \leqq \omega_1 \tag{9.37}$$

where the angular locking half-bandwidth is given by

$$\Delta\omega_1 = f_d\frac{|E_{in}|}{|\widetilde{E_0}|}(1 + \alpha^2)^{1/2} \tag{9.38}$$

The linewidth enhancement factor $\alpha$ gives rise to an increase in the locking bandwidth with a factor of $(1 + \alpha^2)^{1/2}$.

From Eq. (9.38) the locked phase can be determined as a function of the frequency detuning $\Delta\omega$. Confining $\phi_1$ to the internal $[-\pi; \pi]$, the phase characteristics become

$$\phi_1 = -\arcsin\left(\frac{\Delta\omega}{\Delta\omega_1}\right) - \arctan\alpha \tag{9.39}$$

The $\alpha$ parameter gives rise to an offset from zero, so that $\phi_1$ belongs to the interval $[-\pi/2 - \arctan\alpha; \pi/2 - \arctan\alpha]$. Since the interval for $\phi_1$ is not centered around 0, asymmetric locking characteristics will occur with a maximum in output power for negative detuning. Figure 9.3 shows the calculated variation of locked power and phase across the locking range. Both Figures 9.3a and 9.3b illustrate the asymmetric locking

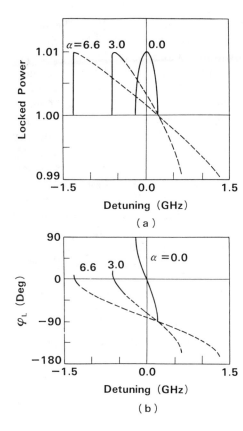

**Figure 9.3.** Locking characteristics for various $\alpha$ values. The injection level is $-40$ dB. The dynamically unstable parts are marked with dashed lines. (a) Locked power relative to free-running power; (b) locked phase. After Mogensen et al. [31].

characteristics resulting from a nonzero $\alpha$ value. The dynamic behavior and instability of injection locking was discussed by Lang and Kobayashi [28] and Mogensen et al. [31].

## 9.3. INJECTION-LOCKING PHENOMENA

Important basic characteristics for injection locking are frequency locking, phase locking, and locked power saturation. The relation between locking bandwidth and locking gain is the most fundamental characteristic. For future applications it is important to determine whether the relation obeys conventional classical injection locking theory [1].

### 9.3.1. Injection-Locking Setup and Procedures

A diagram of the experimental setup is shown in Figure 9.4. Figure 9.4a shows the main feature of the injection-locking experiment. Two AlGaAs double-heterostructure semiconductor lasers with identical cavity lengths operate in a single longitudinal mode at about 840 nm [60]. The master laser (ML) is mounted on a copper heat sink, which is placed in a chamber that is temperature-controlled to within ±0.025°C. Optical isolators are used to avoid undesired feedback from the slave laser (SL). Two isolators are used in tandem to secure more than 40 dB of isolation. The SL is stabilized to an accuracy of within ±0.1°C with a thermoelectric element.

Rough longitudinal-mode matching between the two lasers was accomplished by controlling the slave laser heat-sink temperature. Frequency fine-tuning was achieved by adjusting the master laser drive current. Locking phenomena were observed using the mode spectrum measurement setup shown in Figure 9.4b. A spectrometer, in combination with a TV camera and a display, is used to roughly identify longitudinal modes. The slave laser output is fed into a scanning Fabry–Perot longitudinal modes. The slave laser output is fed into a scanning Fabry–Perot (FP) interferometer and the mode spectrum is displayed on an oscilloscope. Figure 9.5 shows modes in injection locked and unlocked states observed with an FP interferometer. The locking state is also identified when the beat note is extinguished by controlling the oscillating frequency of ML with the drive current.

The coupling efficiency of an incident beam into a semiconductor laser is important to quantitatively evaluate the optical power required in

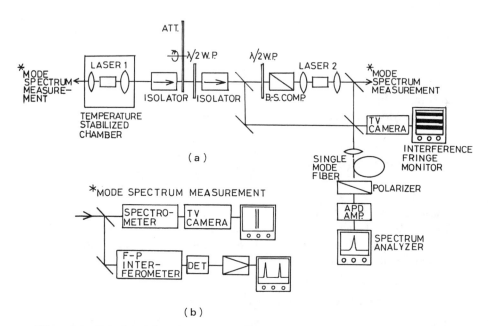

**Figure 9.4.** Experimental equipment setup for measuring optical injection locking: (a) injection-locking experimental setup; (b) mode spectrum measurement setup. After Kobayashi and Kimura [34].

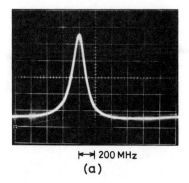

⊢—→ 200 MHz
(a)

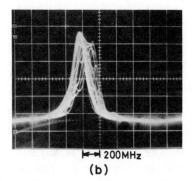

⊢—→ 200MHz
(b)

**Figure 9.5.** Fabry–Perot interferometer observation of (*a*) injection-locked laser output and (*b*) free-running laser output. After Kobayashi and Kimura [34].

injection locking. The coupling efficiency was measured by the same method as was employed in optical gain measurement. Figure 9.6 shows the gain measurement setup. In the amplifier experiment reported previously [61], the coupling efficiency was measured in the following way: the reference laser (LASER 1) was operated at a fixed frequency under a constant temperature and drive current. The amplified power was de-

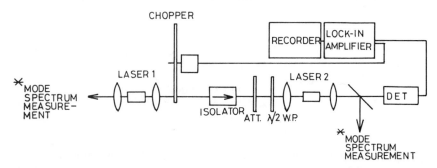

**Figure 9.6.** Gain and coupling efficiency measurement setup. After Kobayashi and Kimura [34].

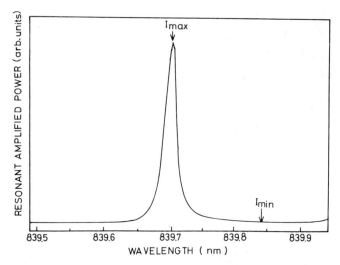

**Figure 9.7.** Resonant amplification gain versus input signal frequency. The $I_{max} : I_{min}$ ratio was 20. Input signal laser frequency shift was 0.08 nm/deg. Mode-jumping was not observed in the 0.4-nm wavelength range and 5°C temperature range. After Kobayashi and Kimura [34].

tected with a lock-in amplifier and chopper. The slave laser (LASER 2) was operated at a constant temperature, and its drive current was reduced to just below its oscillation threshold, so it could operate as an amplifier instead of an oscillator. The frequency of the reference laser was tuned by controlling its temperature and the amplified output was measured as shown in Figure 9.7. The single-path gain and resonant amplifier gain are derived from the peak resonant gain and the bottom nonresonant gain. The coupling efficiency is evaluated from the gain values as well as from the input and output power levels of the amplifier. When the ratio of the maximum power $I_{max}$ to the minimum power $I_{min}$ is represented by $v$, the single-path gain $G$ and the coupling efficiency $C$ are given by

$$G = \frac{v^{1/2} - 1}{(v^{1/2} + 1)R} \tag{9.40}$$

$$C = \frac{G_r(1 - GR)^2}{(1 - R)^2 G} \tag{9.41}$$

where $G_r$ is the resonant amplifier gain and $R$ is the amplifier facet reflectivity. The obtained minimum coupling efficiency $C$ was $-8.8$ dB, measured with power below $-25$ dBm injected into the laser cavity.

### 9.3.2. Locking Bandwidth

The locking state is identified when the beat note is extinguished by controlling the oscillating frequency of the master laser with the drive current. The beat note frequency observed on the RF spectrum analyzer, at which the beat note is extinguished, is regarded as one-half the locking bandwidth. The locking bandwidth was also measured with an FP interferometer. Figure 9.8 shows the half-width of the injection-locking frequency range as a function of the locking gain or the ratio of locked laser power to injected power $P_1/P_{in}$. Both the locked laser power $P_1$ and the injected power $P_{in}$ refer to values in the laser cavity. Symbols identify the relative drive current, as shown in the insets. Each group, from $a$ to $e$, corresponds to a constant injected power in the laser cavity. The locking half-bandwidth is calculated using the small-signal van der Pol equation as obtained in Eq. (9.8), which is shown as the solid line in Figure 9.8.

$$\Delta f = \frac{1}{4\pi\tau_p}\left(\frac{P_{in}}{P_1}\right)^{1/2} \tag{9.42}$$

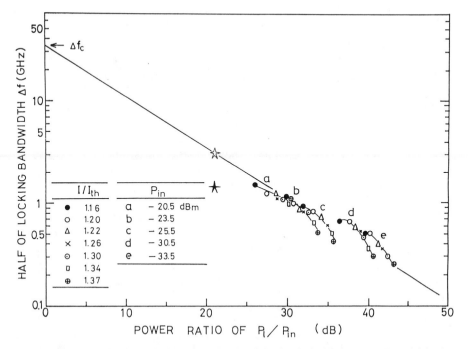

**Figure 9.8.** Locking bandwidth versus ratio of an injection-locked laser power $P_1$ to injected power $P_{in}$ in slave laser. Closed and open star symbols show the hysteresis characteristics at $1.03 \times I_{thSL}$ drive current and $-21$-dBm input power. They correspond to pull-in range and locking range half-widths, respectively. After Kobayashi and Kimura [34].

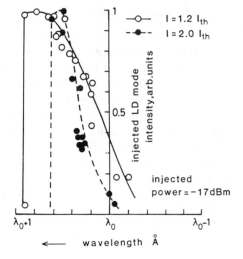

**Figure 9.9.** Relation between injected longitudinal-mode intensity and wavelength detuning. Refractive index dependence on carrier density causes strong asymmetry in detuning characteristics with strong signal injection [$I = 1.2 \times I_{th}$ (—o—); $I = 2.0 \times I_{th}$ (—●—)]. After Kobayashi et al. [41].

where $\Delta f$ is the locking half-bandwidth and $\tau_p$ is the photon lifetime. This relation was also represented by Adler using classical theory [1]. The obtained characteristics, shown in Figure 9.8, mean that injection-locking characteristics in semiconductor lasers are analogous to those for the classical theory and that similar applications can be expected for microwave and millimeter-wave injection locking. The maximum locking full bandwidth obtained here was 5.8 GHz at an 18-dB injection-locking gain. The maximum observed locking gain was 40 dB, when the locking bandwidth was 500 MHz and the minimum locking gain was $-29$ dBm. Here, the injection-locking gain was defined as the ratio of locked laser output to injection power, both measured *outside* the laser cavity. Hysteresis phenomena were observed in this experiment, especially when the oscillating power level was low. An example is shown by the star symbols. When the slave laser is biased at $I_{SL} = 1.03 \times I_{thSL}$ and the injected power is $-22.5$ dBm, 2.9-GHz locking-range half-width and 1.6-GHz pull-in-range half-width values were observed, as noted by the open and closed star symbols. With a relatively high input signal level, detuning characteristics show a strong asymmetry, as seen in Figure 9.9. The locked mode power reaches maximum at a wavelength longer than that in the free-running state. The peak power wavelength shift and detuning bandwidth are small for higher excitation. It has been suggested that this behavior coincides with the theoretical analysis, when refractive index dependence on the carrier density is taken into account [28, 31].

### 9.3.3. Phase Locking

Figure 9.10 shows a typical interference pattern between the input signal and injection-locked output beams when injection locking occurs. This was

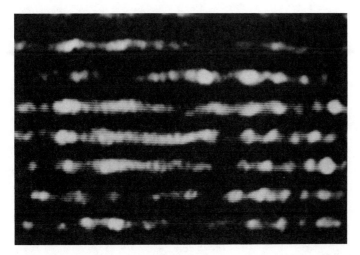

**Figure 9.10.** Interference pattern between laser beams when slave laser is locked by master laser. After Kobayashi and Kimura [34].

observed with a TV camera as shown in Figure 9.4. The interference fringes shift as the input signal phase is changed by a Babinet–Soleil compensator. The interference pattern is very stable, but vanishes in the unlocked state. Figure 9.11 shows intensity variation with injected beam phase change, which is measured using a short single-mode fiber with a 0.8-$\mu$m cutoff wavelength followed by a polarizer for wavefront matching. Visibility, defined as

$$V = \frac{I_{max} - I_{min}}{I_{max} + I_{min}} \tag{9.43}$$

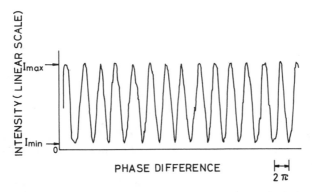

**Figure 9.11.** Intensity variation in interference fringes when the injecting wave phase is changed. Visibility was 0.87. After Kobayashi and Kimura [34].

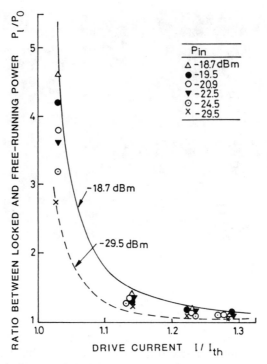

**Figure 9.12.** Ratio between locked laser power and free-running laser power. Various symbols show injected power levels. Solid and broken lines show the calculated values using the van der Pol equation, including injection wave, when injected powers are $-18.7$ and $-29.5$ dBm, respectively. After Kobayashi and Kimura [34].

is obtained as $V = 0.87$, when the master and the slave laser power values are adjusted to maximize the visibility. Figures 9.10 and 9.11 show the experimental proof of constant phase difference $\Delta\phi_c = \phi_{in} - \phi_1$ from Eq. (9.7). Figure 9.11 also shows that visibility is degraded from unity. The visibility degradation is attributed to (1) additional spontaneous emission power and stimulated emission power of spurious unlocked longitudinal modes and (2) unwanted power caused by reflection on the heat-sink plane of the injected laser [34].

### 9.3.4. Locked Power and Power Saturation

Figure 9.12 shows the ratio of locked mode power to central mode power as a function of drive current in the free-running state of the slave laser. The various symbols identify the injected power coupled into the slave laser cavity. The solid and broken lines show the values of the ratio calculated using Eq. (9.9) when the injected power was $-18.7$ and $-29.5$ dBm. The free-running central mode power of the slave laser evaluated inside the cavity was varied from 0.12 to 7.3 dBm. The locked power was measured at the center frequency of the oscillating longitudinal mode. As the drive current increases, the ratio of the locked power to the free-running power decreases further. When the drive current exceeds $1.3 \times I_{\text{thsl}}$, the locked mode power is hardly affected by the injection level. At the drive current of $1.28 \times I_{\text{thsl}}$, the slave laser power varies from 7.3 to 7.41 dBm, while the injected power varies from $-29.5$ to $-18.7$ dBm. These characteristics show that the injection-locked amplifier acts as an optical limiter amplifier.

## 9.4. OPTICAL FM SIGNAL AMPLIFICATION AND FM NOISE SUPPRESSION

Injection-locked amplification of an FM signal and FM noise suppression for a semiconductor laser were studied using a signal that was directly frequency-modulated. Analytical, basic formulas for frequency characteristics due to both effects are represented by the small-signal van der Pol equation.

### 9.4.1. FM Signal Amplification

The frequency-modulated (FM) optical wave is represented by

$$E = E_0 \exp\left[ j\{2\pi f_0 t + \beta \sin(2\pi f_m t)\}\right] \tag{9.44}$$

$$\beta = \frac{\Delta F}{f_m} \tag{9.45}$$

where $f_0$ is the center frequency, $\Delta F$ is the maximum frequency deviation, $f_m$ is the modulation frequency, and $\beta$ is the FM index. The optical field spectrum of the frequency-modulated wave in Eq. (9.44) can be expanded in terms of Bessel functions of the first kind,

$$
\begin{aligned}
E = {}& J_0(\beta) E_0 \sin(2\pi f_0 t) + J_1(\beta) E_0 \sin\{2\pi(f_0 + f_m)t\} \\
& - J_1(\beta) E_0 \sin\{2\pi(f_0 - f_m)t\} + \cdots + J_1(\beta) E_0 \\
& \cdot \sin\{2\pi(f_0 + 1f_m)t\} + (-1)^1 J_1(\beta) E_0 \\
& \cdot \sin\{2\pi(f_0 - 1f_m)t\} \tag{9.46}
\end{aligned}
$$

Figure 9.13 shows the power spectra for a directly modulated semiconductor laser measured with a Fabry–Perot interferometer [62]. Each observed sideband intensity corresponds to the square of the coefficient of the corresponding terms in Eq. (9.46). Figure 9.14 shows the normalized

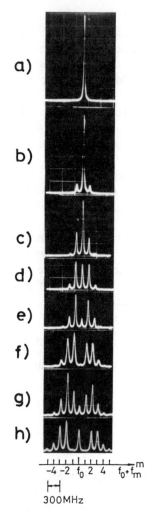

**Figure 9.13.** FM signal power spectra for a directly modulated GaAlAs (CSP) laser measured by Fabry–Perot interferometer. Modulation frequency $f_m$ is 150 MHz, $I = 1.4 \times I_{th}$, $I_{th} = 69$ mA; $\Delta I_{p-0} = $ (a) 0 mA, (b) 0.76 mA, (c) 1.4 mA, (d) 1.84 mA, (e) 2.24 mA, (f) 2.88 mA, (g) 3.44 mA, (h) 4 mA (FSR = 1.5 GHz). After Kobayashi et al. [62].

amplitude for the carrier and various sidebands as a function of modulation current. Marks corresponding to the sidebands are experimentally obtained by averaging the upper and lower sideband amplitudes to remove the AM component. Solid curves correspond to theoretical Bessel function values in Eq. (9.46), respectively. Figure 9.15 shows the maximum frequency deviation $\Delta F$ in the channel substrate planar (CSP)-type semiconductor laser as a function of modulation increases linearly with the modulation current.

Figure 9.16 shows the normalized suppression rate for the FM index in the case of FM signal amplification as a function of normalized modulation frequency. The suppression rate was obtained from the ratio of the frequency modulation indices, $\beta_0/\beta_{in}$, of the output to the input signal of

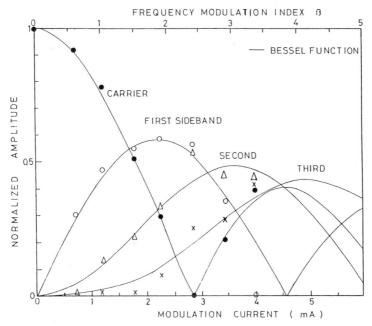

**Figure 9.14.** Normalized amplitude for various orders of sideband observed in the power spectra of Figures 9.13(a)–(h) versus modulation current and frequency modulation index. Symbols •, carrier; ○, first sideband; △, second sideband; ×, third sideband. Solid lines show theoretical values of Bessel function values of the first kind. After Kobayashi et al. [62].

the locked amplifier (slave laser). Modulation frequency $f_m$ is normalized by the locking half-bandwidth $\Delta f_L$. The locking bandwidth was measured by shifting the longitudinal mode of the master laser under CW operation. Frequency modulation indices $\beta_{in}$ for input signals are between 0.8 and 1.0. The solid line shows the drawn results calculated using the van der Pol equation. Phase and frequency relations are drawn from Eqs. (9.6) and (9.7).

To represent the frequency characteristics, first-order perturbation is assumed as follows:

$$\phi_1 = \phi_{10} + \Delta\phi_1 \tag{9.47}$$

$$\phi_{in} = \phi_{in0} + \Delta\phi_{in} \tag{9.48}$$

$$\Omega_a = \Omega_0 + \Delta\Omega \quad \left(\Delta\Omega = \frac{d\phi_a}{dt}\right) \tag{9.49}$$

$$\phi_a = \phi_{0a} + \Delta\phi_a \tag{9.50}$$

where index 1 represents the output locked state, in the input signal, and a the unlocked state output of the slave laser. When an input signal

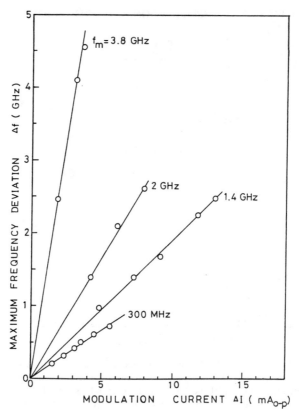

**Figure 9.15.** Maximum frequency deviation for a GaAlAs (CSP) laser versus modulation current. After Kobayashi et al. [62].

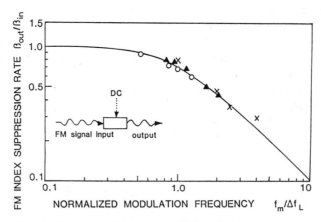

**Figure 9.16.** Suppression rate of FM index in FM signal amplification characteristics when FM-modulated signal is injected. Master laser, $I = 1.5 \times I_{th}$, slave laser; $I \times 1.1 \times I_{th}$; $\times$, $P_{in} = 42$ dBm, $G = 42$ dB, $\blacktriangle = P_{in} = -36.5$ dBm, $G = 34$ dB; $\circ$, $P_{in} = -25.5$ dBm, $G = 27$ dB. After Kobayashi et al. [36].

frequency is assumed to be the cavity center frequency and a small phase deviation is considered

$$\sin(\Delta\phi_{in} - \Delta\phi_1) \doteq \Delta\phi_{in} - \Delta\phi_1 \tag{9.51}$$

can be approximated. Substituting Eqs. (9.47)–(9.50) into Eq. (9.6), perturbation terms are represented as follows:

$$\frac{d\,\Delta\phi_1}{dt} - \frac{d\,\Delta\phi_a}{dt} = 2\pi\,\Delta f(\Delta\phi_{in} - \Delta\phi_1) \tag{9.52}$$

$\Delta f$ is the locking half-bandwidth represented by Eq. (9.8). Considering the sinusoidal input signal, Eq. (9.52) becomes

$$j\Omega\,\Delta\phi_1 - j\omega\phi_a = 2\pi\,\Delta f(\Delta\phi_{in} - \Delta\phi_1) \tag{9.53}$$

Various FM indices are represented as follows:

$$\beta_{in} = \Delta\phi_{in} = \Delta F_{in}\,f_m \tag{9.54}$$

$$\beta_{out} = \Delta\phi_1 = \Delta F_1\,f_m \tag{9.55}$$

$$\beta_a = \Delta\phi_a = \Delta F_a\,f_m \tag{9.56}$$

Here, $\Delta F_{in}$, $\Delta F_1$, and $\Delta F_a$ are maximum frequency deviations of various waves and $f_m$ is a modulation frequency.

In the case of FM signal injection, the condition of $\Delta\phi_{in} \geq \Delta\phi_a$ can be assumed as follows:

$$\Delta\phi_1 = \Delta\phi_{in}\left\{\left[1 + \left(\frac{f_m}{\Delta f}\right)^2\right]^{-1}\right\}^{1/2} \tag{9.57}$$

Here, $\omega = 2\pi f_m$. The solid line in Figure 9.16 is represented by FM indices as follows:

$$\left(\frac{\beta_{out}}{\beta_{in}}\right)_{ILA} = \left\{\left[1 + \left(\frac{f_m}{\Delta f}\right)^2\right]^{-1}\right\}^{1/2} = F_1 \tag{9.58}$$

This equation is the solution of the small-signal van der Pol equation with an external signal injection, when the FM noise for the locked amplifier itself and the amplitude change for the input signal are ignored. All experimental results can be well represented by the theoretical values and show that the bandwidth and the amplifier gains tend to be in a trade-off relation.

When the slave laser is operated as a Fabry–Perot amplifier (FPA) at or slightly below its oscillation threshold, linear gain is available at a sufficiently low input level. The amplification of a frequency-modulated optical signal is studied in FPA. The theoretical curves for an FPA with a small input FM index are as follows:

$$
\left(\frac{\beta_{out}}{\beta_{in}}\right)_{FPA} = \left\{\left[1 + \left(\frac{f_m}{\Delta f_{1/2}}\right)^2\right]^{-1}\right\}^{1/2} \tag{9.59}
$$

Here, $\Delta f_{1/2}$ is the half-width of half maximum value (HWHM) with CW optical injection in FPA. Figure 9.17 shows HWHM for an injection locked amplifier (ILA) and an FPA with both CW and FM signal input as a function of the amplifier gain. The broken and solid lines were obtained from Eqs. (9.58) and (9.59), respectively. Figure 9.17 shows that the

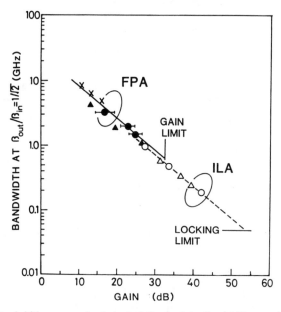

**Figure 9.17.** Bandwidth versus gain characteristics for injection-locking amplifier (ILA) and Fabry–Perot amplifier (FPA). Bandwidths measured with FM signal are indicated by $\circ$, ILA and $\bullet$ FPA. The bandwidth is depicted by $\beta_{out}/\beta_{in} = 1/\sqrt{2}$. Bandwidths measured by frequency-swept CW wave are indicated by $\triangle$, ILA and $\blacktriangle$, FPA ($P_{in} = 12$ dBm). Broken and solid lines are calculated values. Experimental and calculated results show $\sqrt{G} \cdot B = 25$ GHz for ILA and FPA. After Kobayashi and Kimura [37].

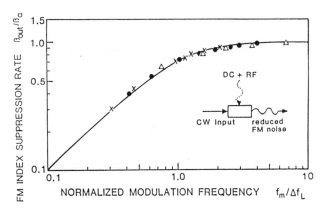

**Figure 9.18.** Suppression of FM index in FM signal generation when optical CW signal is injected. Master laser, $I = 1.64 \times I_{thML}$; slave laser, $I \times 1.25 \times I_{thSL}$; $\times$, $P_{in} = -26$ dBm, $G = 38$ dB, $\blacktriangle = P_{in} = -22.5$ dBm, $G = 34$ dB, $\circ$, $P_{in} = -20$ dBm, $G = 32$ dB. After Kobayashi et al. [36].

gain–bandwidth product for both ILA and FPA modes are given approximately by $\sqrt{G} B = 25$ GHz [37]. The FPA is suitable for relatively low gain and broadband operation, where the maximum gain is imposed by the gain around 30 dB at threshold. The ILA is suitable for high gain and relatively narrow bandwidth operation, where the bandwidth is limited by the center frequency fluctuation at around 50 MHz.

To obtain a large gain–bandwidth product $\sqrt{G} B$, small reflectivity $R$, and short cavity length $L$ are effective for both the ILA and the FPA. The cavity with reduced $Q$ value, however, may increase AM and FM noise in these amplifiers. Further development of optimum design criteria is necessary for optical amplifiers with respect to the gain–bandwidth product as well as noise.

### 9.4.2. FM Noise Suppression by Injection Locking

Frequency-modulated noise suppression simulation was undertaken by injecting coherent CW laser light into a directly frequency modulated semiconductor laser. Figure 9.18 shows the FM index suppression rate as a function of the normalized modulation frequency. The suppression rate is obtained by the ratio $\beta_{out}/\beta_a$ for the output locked by the CW input signal to the unlocked modulated output signals. The FM noise suppression rate is represented by $\overline{\Delta f_{out}^2}/\overline{\Delta f_a^2} = (\beta_{out}/\beta_a)^2$. Obtained locking gains are 38, 34, and 32 dB for $-26$, $-22.5$, and $-20$ dBm input levels, respectively. FM index $\beta_a$ without injection is 0.4–0.6. If the FM noise of the injected light is much smaller than that of the slave laser ($\Delta\phi_a \gtreqqless \Delta\phi_{in}$), the phase change in the slave laser is shown from Eq. (9.52) as follows:

$$\Delta\phi_1 = \Delta\phi_a \left\{ \frac{(f_m/\Delta f)^2}{\left[1 + (f_m/\Delta f)^2\right]} \right\}^{1/2} \tag{9.60}$$

The FM index ratio, which represents the FM noise suppression character-istics, is represented from Eq. (9.58) as follows:

$$\frac{\beta_{out}}{\beta_a} = \left[ \frac{(f_m/\Delta f)^2}{1 + (f_m/\Delta f)^2} \right]^{1/2} = F_2 \tag{9.61}$$

The solid line in Figure 9.18 shows the theoretical values calculated by Eq. (9.61). Equation (9.61) shows the frequency characteristics solving the small-signal van der Pol equation with an external signal, excluding ampli-tude change and frequency deviation for the input CW signal. Although the experiment was carried out using a modulated laser, the results can be interpreted as if the FM noise component were suppressed in the injection process. If we regard the modulation sideband as a component of angular modulation noise in the semiconductor laser, suppression of the FM index in FM signal generation can be regarded as FM noise reduction. The FM noise for an ILA is reduced to within the locking bandwidth by coherent-signal injection. This indicates that an optical FM signal with reduced FM quantum noise can be amplified in an IL semiconductor laser amplifier without serious distortion or S/N (signal-to-noise ratio) degradation. When the FM noise reduction is performed by ILA, FM → AM conver-sion occurs when the FM modulation index is large.

## 9.5. PHASE MODULATION BY INJECTION LOCKING

Phase modulation (PM) is expected to be the most effective modulation in coherent transmission systems, because the lowest minimum receiving power level can be achieved with a PSK homodyne system [47]. This section shows the proposed and experimentally realized optical phase modulation using injection locked semiconductor lasers. The model of PM by injection locking technology can be represented as shown in Figure 9.19. When the injected signal frequency is scanned in the locked state, the phase of the slave laser output changes from $-\pi/2$ to $\pi/2$ as shown in Figure 9.19a. However, the output frequency also changes in synchrony with the input signal. On the other hand, when the injected signal frequency is fixed at the center frequency of the slave laser cavity changes, a similar phase change is observed as shown in Figure 9.19b, but the

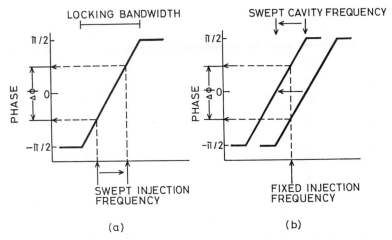

**Figure 9.19.** Schematic diagram of phase change by injection locking. (*a*) Phase change with injected light frequency shift; (*b*) phase change with locked laser cavity frequency shift.

output frequency does not change. The efficiency of the phase deviation becomes large when the locking bandwidth narrows.

### 9.5.1. Static Phase Change by Injection Locking

The experimental setup for measurement of phase-modulation characteristics is also shown in Figure 9.4. The Mach–Zehnder (MZ) interferometer configuration was used for PM measurement. The Babinet–Soleil (BS) compensator acts as a phase shifter to calibrate the phase deviation. The path difference between the two interferometer branches was reduced to less than 1.4 mm. When CW light is injected into the slave laser (SL), maximum frequency deviation (MFD) is suppressed. Simultaneously, the PM signal is induced with the injection. The induced PM signal is observed in the display on the sampling oscilloscope through the MZ interferometer. The optical phase bias is adjusted by the BS compensator to obtain a sinusoidal signal on the display. The maximum PM deviation (MPD) is calibrated by the BS compensator. The visibility of the MZ interferometer output is more than 95% when the BS compensator and the attenuator are adjusted at the SL cavity center in a locking state. The measured visibility was between 74% and 89% in relation to that of the MZ interferometer. The visibility becomes large when the locking bandwidth, which is determined by SL bias current, injected power, and cavity $Q$ becomes broad. Two lasers, the master laser (ML) and the SL, are driven by DC currents $I_{ML} = 1.5 \times I_{thML}$ and $I_{SL} = 1.54 \times I_{thSL}$, respectively. The injected CW light power in the SL is $P_{in} = 22$ dBm. A state

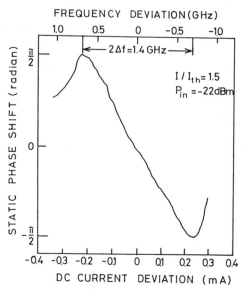

**Figure 9.20.** Static phase shift obtained by changing the slave laser cavity frequency with respect to that of the master laser through the bias current. After Kobayashi and Kimura [38].

phase shift of $\pi$ takes place when the SL bias current is changed by 0.48 mA, as shown in Figure 9.20. The full locking bandwidth $2 \Delta f$ is 1.4 GHz. The phase shift curve obtained between $-\pi/2$ and $\pi/2$ is nearly linear, rather than sinusoidal. This result shows that, in order to increase the phase shift per unit drive current in static operation, the locking bandwidth should be narrow.

The static phase relation is represented from Eq. (9.6) as follows:

$$\frac{|E_{in0}|}{|E_{10}|} \frac{\sin (\phi_{in0} - \phi_{10})}{2\tau_p} - \Delta\omega = 0 \qquad (9.62)$$

Here, index 0 means DC bias, $\Delta\omega = 2\pi \Delta f = \omega_{in} - \Omega_0 = \omega_1 - \Omega_0$ when injection-locked. When the injected light has a frequency $f_{in}$ that is identical to the slave laser cavity center frequency $f_c$ and a constant phase $\phi_{in0}$, which is arbitrarily chosen as 0, the place of the locked laser becomes

$$\phi_{10} = \sin^{-1}\left[\left(\frac{E_{10}}{E_{in0}}\right)4\pi\tau_p(f_0 - f_{in})\right] \qquad (9.63)$$

where $f_0$ is the free-running slave laser frequency without light injection. As $f_0$ is shifted by means of the drive current, $\phi_{10}$ is changed by $\pm\pi/2$

from 0, when the frequency shift $f_{in} - f_0$ is equal to the locking half-band-width:

$$f_{in} - f_0 = \frac{|E_{in0}|/|E_{10}|}{4\pi\tau_p} = \Delta f \tag{9.64}$$

At around $f_{in} = f_0$, the phase can be approximated by a linear function

$$\phi_{10} = \left(\frac{|E_{10}|}{|E_{in0}|}\right) 4\pi\tau_p (f_0 - f_{in}) \tag{9.65}$$

Experimental results in Figure 9.20 show the linear phase shift characteristics near the cavity center frequency.

### 9.5.2. Phase Modulation

When the SL was driven by an RF current and CW light was injected, a PM output is obtained. Figure 9.21 shows the frequency spectra for (a)

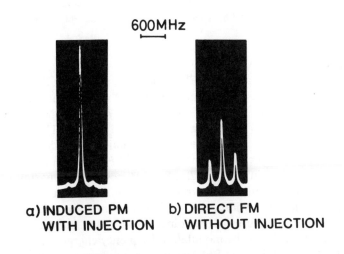

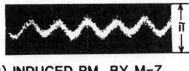

**Figure 9.21.** Frequency spectra and phase deviation. (a) PM spectrum with $-28$ dBm power injection. Maximum phase deviation $\Delta\phi = 0.17$ radians/mA. (b) FM spectrum without injection. FM index $\Delta f/f_m = 0.47$/mA. (c) PM signal measured by MZ interferometer; $\Delta\phi = 0.17$ radians/mA. After Kobayashi and Kimura [38].

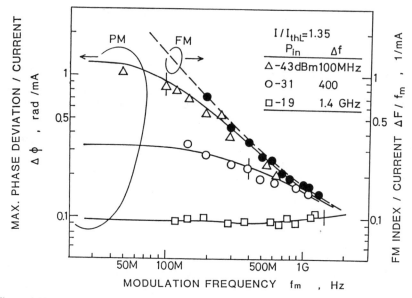

**Figure 9.22.** Maximum phase deviation characteristics. Bias current is $1.35 \times I_{th}$. Closed circles and broken lines show experimental and calculated results of maximum frequency deviation without injection. Solid lines show theoretical values for PM. After Kobayashi and Kimura [38].

PM modulation with and (b) FM modulation without light injection, measured with the MZ interferometer, and (c) a PM signal converted by the MZ interferometer. Modulation frequency $f_m$ is 300 MHz. DC drive currents are $I_{ML} = 1.6 \times I_{thML}$ and $I_{SL} = 1.34 \times I_{thSL}$. The modulation current is 2 mA$_{0-p}$. Measured frequency deviation is 130 MHz/mA. The modulation index, which is defined by $\beta = \Delta F / f_m$, is suppressed from 0.47 to 0.17 mA$^{-1}$ by injecting $-28$-dBm CW power. The maximum phase deviation (MPD) per unit drive current is 0.17 radian/mA for the $\Delta f = 800$ MHz locking half-bandwidth. The MPD measured by sideband amplitudes, as shown in Figure 9.21a, coincides with that indicated in Figure 9.21c. Figure 9.22 shows the modulation frequency characteristics between 100 MHz and 1.3 GHz for PM signals. The ordinate on the left shows MPD per unit drive current with injection. That on the right indicates the FM index per unit current without injection. The closed circles and the broken line represent experimental and theoretical FM indices per unit current [62]. Other symbols and lines show results of MPD per unit current for each locking half-bandwidth. Small vertical bars on the curves show the locking half-bandwidth for each experimental condition. The slope becomes gradual at frequencies higher than 500 MHz because of the carrier resonant phenomena [63]. The MPD curves tend to approach the FM index curve at high modulation frequencies, and the MPD becomes

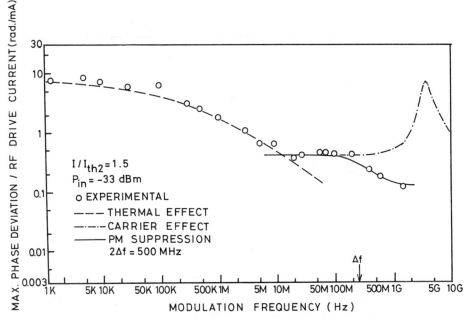

**Figure 9.23.** Frequency characteristics of the maximum phase deviation. After Kobayashi and Kimura [38].

constant at low modulation frequencies. As the locking bandwidth broadens, the MPD per unit RF current decreases and the cutoff frequency $f_{mc}$, at which the MPD approaches $1/\sqrt{2}$, increases. The MPD decreases as SL bias current increases.

Figure 9.23 shows frequency characteristics for MPD per unit current at $I_{SL} = 1.5 \times I_{thSL}$. The open circles show the experimental results. The solid-line curve shows the theoretical PM deviation. The broken line shows theoretical thermal and carrier effect curves over the wide modulation frequency range [64].

Phase deviation in the locked state is obtained as follows:

$$\Delta\phi_1 = \Delta\phi_{in}\left\{\frac{(f_m/\Delta f)^2}{\left[1 + (f_m/\Delta f)^2\right]}\right\}^{1/2} \qquad (9.66)$$

This relation is the same representation as the FM noise suppression. The FM index ratio is also obtained as follows:

$$\frac{\beta_{out}}{\beta_a} = \frac{\Delta\phi_1}{\Delta\phi_{in}} = \left[\frac{(f_m/\Delta f)^2}{1 + (f_m/\Delta f)^2}\right]^{1/2} \qquad (9.67)$$

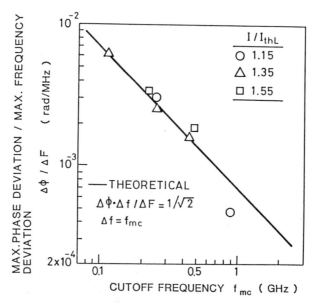

**Figure 9.24.** Maximum phase modulation normalized by maximum frequency as a function of the cutoff frequency. After Kobayashi and Kimura [38].

From $\Delta\phi_{in} = \beta_a = \Delta F/f_m$, Eq. (9.67) becomes

$$\Delta\phi_1 = \frac{\Delta F}{\left[\Delta f^2 + f_m^2\right]^{1/2}} \tag{9.68}$$

where $\Delta F$ is MFD; $\Delta\phi_1$ is constant at $\Delta F/\Delta f$ for $f_m$ lower than the locking half-bandwidth $\Delta f$, while it tends to be small at $f_m > \Delta f$. The solid-line curves in Figures 9.18 and 9.23 show the value calculated using Eq. (9.68). Phase modulation can be induced in a modulation frequency range lower than the locking half-bandwidth $\Delta f$. Figure 9.24 shows experimental results equal to the phase shift value of $\Delta F/\Delta f$ in a lower frequency region than the 250-MHz locking half-bandwidth and shows the MPD normalized by the MFD at cutoff modulation frequency $f_{mc}$ as a function of cutoff frequency. The solid line shows the theoretical results calculated using Eq. (9.68) when $f_{mc} = \Delta f$. The experimental results at a low cutoff frequency are in good agreement with the theoretical values. In high $f_{mc}$, however, MFD is affected by the carrier resonant characteristics [62]. When the MFD is constant, the phase deviation and the cutoff frequency are in inverse relation. Therefore, optimum values have to be selected, or the phase shift and cutoff frequency product is expected to increase. To obtain large product values, a low-$Q$ cavity will be required. The phase modulation obtained in this study has such advantages for future coherent optical transmission systems as (1) narrow carrier linewidth due to frequency-stabilized light injection and (2) high frequency-modulation capability within a wide locking bandwidth.

## 9.6. INJECTION LOCKED AMPLIFIER APPLICATIONS

### 9.6.1. High-Speed Fiber Transmission Systems

Single-mode optical fiber transmission systems have long repeater spacing and large capacity in the long-wavelength region because of low loss at 1.55 $\mu$m and low dispersion at 1.3 $\mu$m. Semiconductor lasers, directly modulated at high data-rate, oscillate in multilongitudinal modes. After transmission through long-span single-mode fibers, the multimode oscillation gives rise to the following serious problems: (1) the transmission bandwidth becomes narrow because of the finite spectrum width of the

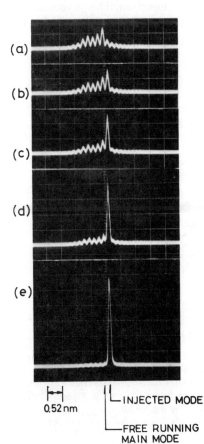

(a)

(b)

(c)

(d)

(e)

0.52 nm

INJECTED MODE

FREE RUNNING
MAIN MODE

**Figure 9.25.** Spectra of AlGaAs heterostructure laser modulated by 500 Mbit/s RZ pseudorandom pulse patterns. DC bias current, $I = 0.93 \times I_{th}$, $I_{th} = 81$ mA. Pulse signal current 46 mA (zero peak); $P_{in} = (a)$ free-running spectrum without light injection, $(b) = -43.2$ dBm, $(c)$ $-36.3$ dBm, $(d)$ $-32.3$ dBm, $(e)$ $-30.5$ dBm. After Kobayashi et al. [64].

laser output, and (2) intensity noise appears as a result of the fluctuation of each oscillating mode in lasers and the chromatic delay differences in fibers. In the 1.5-$\mu$m band, silica fibers have minimum loss, but a finite fiber dispersion degrades system performance, for instance, the optical receiving level. Single-mode fibers, with zero dispersion wavelength, are expected to improve the receiving performance in the 1.5-$\mu$m-band transmission systems. An alternative method is to realize a single-longitudinal-mode operation in lasers even under modulation.

Modulated single-longitudinal-mode operation has been achieved by injection locking first in AlGaAs lasers [64]. The single-mode behavior by light injection is shown in Figure 9.25. When a CW signal above $-20$ dBm is injected into a longitudinal mode of the laser modulated at 500 Mbit/s, the power in the spurious modes is reduced to less than 10% of the total power. The intensity fluctuation noise, or mode partition noise, in the slave laser is improved significantly by laser light injection [65, 66]. Figure 9.26 shows the improved noise spectrum by the light injection with 1.3-$\mu$m master and slave lasers [67]. At 1.55 $\mu$m with 100 Mbit/s, transmitted pulses through 44.3 km were improved by light injection. This is verified by the error rate characteristics as shown in Figure 9.27 [57]. Degradation of

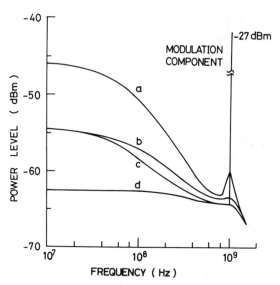

**Figure 9.26.** Noise spectra modulated by an RZ all mark pattern at 1 GHz/s. (*a*) Central dominant mode extracted from slave laser without optical injection; $I_{SL} = 0.8 \times I_{thSL}$, $I_{thSL} = 72$ mA. Pulse signal current 60 mA (zero peak). (*b*) Central dominant mode extracted from slave laser with optical injection. Drive condition is the same as in Figure 9.26(*a*). (*c*) Dominant mode extracted from continuously operating master laser; $I_{ML} = 1.4 \times I_{thML}$, $I_{thML} = 47$ mA. (*d*) All modes in slave laser. The drive condition is the same as in (*a*). After Yamada et al. [67].

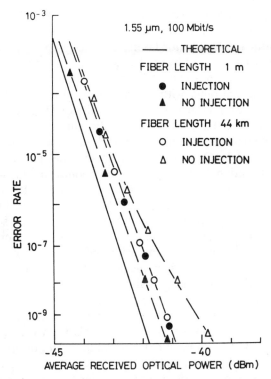

**Figure 9.27.** Plot of 100-Mbit/s error-rate characteristics at 1.55 $\mu$m. Solid-line curve shows theoretical results. Symbols ▲, ●, △, and ○ show experimental results after 1-m transmission without and with light injection and after 44.3-km transmission without and with injection, respectively. Light injection power is $-15$ dBm. After Yamada et al. [55].

the received optical power could be compensated by 1.4 dB at a $10^{-9}$ error rate with the light injection.

A higher bit rate of 450 Mbit/s and a longer distance of 106 km were achieved by using the injection-locked oscillator in the system [68]. After these experiments, DFB lasers were used as master and slave lasers in the 1.5-$\mu$m region and long distance and high-bit-rate transmission system experiments were performed [57, 58].

## 9.6.2. Coherent Optical Transmission Systems

Coherent optical transmission systems are foreseen as future systems that will feature increased repeater spacing and information transmission capacity in comparison to conventional optical-fiber transmission systems. The contributions of injection locking technology to coherent optical transmission systems are expected to be (1) stabilized and narrow-linewidth oscillators as coherent sources, (2) postamplifier after an external modulator, (3) repeater or preamplifiers, and (4) local oscillators in FM or PM signals with homodyne or heterodyne detection systems. Narrow-linewidth semiconductor lasers for a coherent source and a local oscillator have been realized by (1) an external grating feedback [69] and (2) injection locking with a very stable master laser. Helium–neon lasers emitting in the 1.5-$\mu$m range with a narrow linewidth of less than 1.5 MHz were used as a master laser and a local oscillator [70]. As a transmitter configuration for the FSK heterodyne differential detection and PSK heterodyne coherent detection schemes shows, the combination of a frequency-stabilized laser oscillator and an external phase modulator is possible. A conventional guided wave electrooptic modulator with, for example, $LiNbO_3$, can be used as a phase modulator. Its insertion and coupling loss may be compensated for by the injection-locked semiconductor–laser postamplifier. The semiconductor laser could be used to repeat optical phase- or frequency-modulated signals. In addition the injection-locked repeater becomes a limiting amplifier and may have greater immunity from signal fading than the equivalent system using intensity modulation with linear optical repeaters. An optical fiber link operating over 10 km of single-mode fiber using an injection-locked semiconductor laser as a repeater was demonstrated with a 1.5-$\mu$m InGaAsP semiconductor laser by Smith and Malyon [71]. FM signals from the repeater were demodulated by using an optical delay technique. The future potential of semiconductor laser injection-locked repeaters will ultimately depend on their noise performance. Experimental frequency characteristics of suppression rates in FM signal amplification and FM signal generation in injection-locked amplifiers are in good agreement with the simple Eqs. (9.58) and (9.61), which are basically the same as those of microwave solid-state devices [6]. FM noise properties in cascade locked amplifier repeaters of a coherent FSK system are considered here. The FM noise of the locked amplifiers is obtained by combining Eqs. (9.58) and (9.61):

$$\left(\overline{\Delta f_{\text{out}}^2}\right)_N = \left(\overline{\Delta f_a^2}\right)_N \times F_2^2 + \left(\overline{\Delta f_i^2}\right)_N \times F_1^2 \qquad (9.69)$$

$$\left(\overline{\Delta f_i^2}\right) = \left(\overline{\Delta f_{\text{out}}^2}\right)_{N-1} \qquad (9.70)$$

where $\overline{\Delta f_a^2}$ is the FM noise in the $N$th locked amplifier, $\overline{\Delta f_i^2}$ is the FM noise of the signal output from the $N$th amplifier, and $N$ denotes the number of cascade locked amplifiers. Figure 9.28 shows the FM noise power calculated using Eqs. (9.69) and (9.70) when the locked amplifiers are connected in series, assuming that each locked amplifier has the same FM noise power. The top broken line shows the FM noise of the locked amplifier; and the bottom broken line, that of the input signal. The ratio of these FM noise powers is assumed to be $4 \times 10^3$. The FM noise of the semiconductor laser is caused by (1) a thermal and mechanical cavity

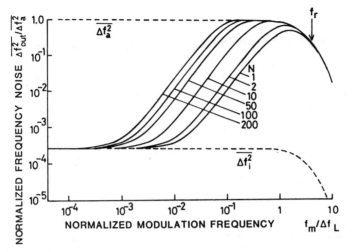

**Figure 9.28.** Cumulative FM noise increase with cascade locked amplifiers. Symbols: $\overline{\Delta f_a^2}$, FM noise of the locked amplifiers; $\overline{\Delta f_{in}^2}$, FM noise of injected signal; $f_r$, 3 dB down frequency of $\overline{\Delta f_a^2}$ Lorentzian distribution; $N$, number of locked amplifiers. After Kobayashi et al. [36].

length change, (2) a refractive index change caused by temperature and carrier density fluctuation, and (3) phase fluctuation by mixed spontaneous emission light in the lasing mode [72]. In Figure 9.28 the frequency characteristics of FM noise power are assumed to be flat up to the 3-dB down cutoff frequency $f_r$. Calculated results show that the FM noise of the injection-locked amplifier is reduced to the input FM noise power if the locking half-bandwidth $\Delta f_l$ is larger than the frequency bandwidth of the baseband signal bandwidth. Residual FM noise power, however, increases as the number of locked amplifiers increases. Optical FM transmission with low FM noise, that is, narrow linewidth, is important for a coherent FSK system. Injection-locked amplifiers with a wide locking bandwidth employing low cavity $Q$ are promising for repeaters in a coherent FSK system, since the FM noise of the amplifier is well suppressed by injecting coherent input waves, and amplified FM signals are not distorted when there is a sufficient locking bandwidth.

Automatic gain and frequency control techniques are indispensible for the optical amplifier repeater systems. Figure 9.29 illustrates the principle of the AFC in an injection-locked amplifier, which was achieved through using the terminal voltage change induced by light injection. The junction voltage decreases with the injection of tuned light into the semiconductor laser when the light frequency is shifted as shown in Figure 9.29$a$. When the temperature of the locked amplifier is sinusoidally modulated at $f_0$ and coherent light is injected at a frequency ($f_A$, $f_B$, or $f_c$ in Figure 9.29$b$), the induced differential voltage changes depending on the optical

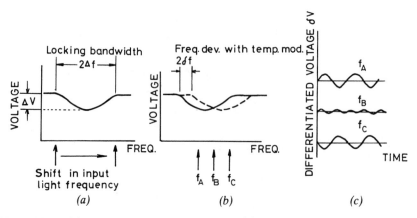

**Figure 9.29.** Voltage change with injection locking. (*a*) Voltage change with shifting input light frequency. (*b*) Light injection at $f_A$, $f_B$, or $f_c$ frequencies. (*c*) Differentiated voltage signals as control signals for AFC at $f_A$, $f_B$, $f_c$. After Kobayashi and Kimura [45].

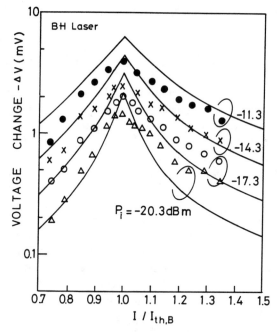

**Figure 9.30.** Voltage change with light injection. Solid lines show values calculated by the rate equation using GHL bandtail model. After Kobayashi and Kimura [45].

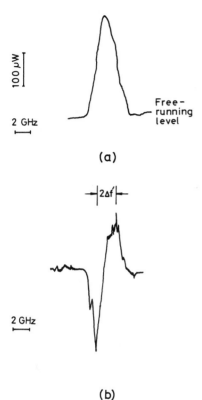

100 μW

2 GHz

Free-running level

(a)

2Δf

2 GHz

**Figure 9.31.** Injection-locking characteristics with temperature modulation. Tuning curve for (a) amplified excess power and (b) first derivative of voltage change ($P_{in} = -15.7$ dBm; $I/I_{thSL} = 1.16$). Slave laser central frequency deviation is $2\,\delta f = 24$ MHz. After Kobayashi and Kimura [45].

(b)

frequency tuning as shown in Figure 9.29$c$). The phases of the first derivatives for $f_A$ and $f_C$ are opposite. The second harmonic signal output is obtained for central tuning $f_B$. The phase difference is detected by means of a phase discriminator. Phase information in the induced voltage change is fed back to control the laser temperature and the tuning is executed. The voltage change induced in an AlGaAs BH laser by injecting a coherent laser light is shown in Figure 9.30. The voltage change has a maximum value at the threshold bias current. AFC can be realized at the center frequency in the tuning curve of the injection-locked amplifier. An amplified excess power (AEP) curve was obtained as shown in Figure 9.31$a$ when the input light frequency was swept. The control signal, which was obtained as the first deviation of voltage change, was measured as shown in Figure 9.31$b$. Figure 9.32 shows the first derivative measured for the input power levels of curves ranging from $-33.2$ to $-21.2$ dBm. Asymmetry in the tuning curves increases as the input power increases. The slopes in the linear region in the first derivative had nearly the same value at $5 \times 10^{-10}$ [V/(MHz)$^2$]. To examine the stability of the feedback system against ambient temperature changes, a running AFC test was

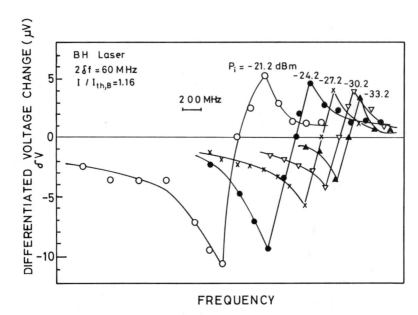

**Figure 9.32.** First derivative curves of voltage change by injection locking. Slopes are constant as $3 \times 10^{-8}$ (V/MHz) in linear regions at $2\,\delta f = 60$ MHz. After Kobayashi and Kimura [45].

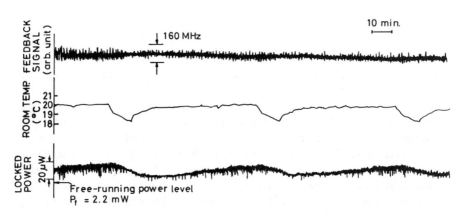

**Figure 9.33.** Results of test running injection locking amplifier with AFC in GaAlAs BH laser for 3 h. At top is control signal represented by first derivative in voltage. Center is ambient temperature. Bottom is locked output power level ($P_{in} = 42$ dBm). Bias current of locked laser $I/I_{thSL} = 1.16$; of master laser, $I/I_{th, ML} = 1.3$. After Kobayashi and Kimura [45].

carried out over 3 h, as can be seen in Figure 9.33. The room temperature changed by 2°C three times during the test. The locked laser output changed by 0.3% with a 2°C temperature change. The temperature change caused changes in the threshold current and free-running power level. The noise level, which determines the control limit, is due to the FM noise and temperature modulation amplitude. The lowest input power was decided by the linear region bandwidth for the first voltage derivative to be $-49$ dBm, which corresponds to a 53-dB locking gain and a locking bandwidth of 100 MHz.

The control sensitivity was dependent on the carrier intensity of the modulated signal. With a low-pass filter, a low-frequency voltage change signal by temperature modulation would be obtained even with the modulated input signal. In homodyne and heterodyne detection, a narrow linewidth and a frequency-stabilized local oscillator is needed. The injection-locked amplifier (ILA) has the potential to satisfy the above requirements. The ILA synchronized to the carrier frequency extracted from the transmitted modulated signal is very useful as a local oscillator in homodyne detection. Sideband injection locking is one of the tools for the local oscillator in heterodyne detection systems. When the carrier frequency extracted from the transmitted modulated signal can lock the sideband of the ILA, which is already modulated with the IF frequency, the heterodyning by detected beat signal on the quadratic detector becomes the IF frequency. Using sideband locking, frequency stabilized beat signals can always be realized by the frequency-stabilized microwave oscillator during the stabilized locking state.

The heterodyne detection experiments using a sideband injection-locked oscillator have been performed by Goldberg et al. who also obtained microwave injection locking to the FET with the beat signal between the ILA locked to the upper and lower sidebands of the master oscillator [42, 43].

## 9.7. SUMMARY

Semiconductor laser injection locking (SLIL) shows the possibility of application to coherent optical transmission and high-bit-rate intensity-modulation systems. Nonlinear phenomena have been specifically observed in semiconductor lasers compared with other laser IL because of high gain and low cavity $Q$. One of the remarkable SLIL phenomena is the asymmetric locking curve, which can be explained by the basic theory analyzed by Lang, taking the carrier density dependence of the active region refractive index into account. This asymmetry leads to optical bistability, which was observed at a low bias level and high-power light injection [73].

SLIL has the potential for use in optical amplifiers as a repeater, a booster, a postamplifier, and a limiter amplifier [48]. The repeater and limiter amplifiers have been discussed in this chapter. The possibility of using SLIL as a postamplifier occurs in coherent transmission systems, which is the IL high-power semiconductor laser array [44]. The laser array has been used as a transmitter for coherent optical transmission systems, where the transmitter power was 310 mW at 830 nm with an AlGaAs semiconductor laser [74]. Another feature of the SLIL is that it is a low-noise amplifier, compared to the Fabry–Perot (FP) type and the traveling-wave (TW)-type amplifiers because the locking bandwidth is very narrow and the operation is performed in the saturated gain condition at a bias level over threshold. In the future, broadcasting services will have the possibility of expanding into optical communication systems, in which case high-power booster amplifiers will be required with a $1 \times N$ low-loss optical splitter.

If there is an elegant technology for inducing a carrier signal from the transmitted FM- or PM-modulated signal, SLIL can operate as a local oscillator locking to the carrier frequency in a homodyne detection system. Also, heterodyne detection will be possible using sideband locking when the transmitted signal carrier frequency is located in the sideband of a modulated SLIL at an IF frequency. However, the detection level in coherent transmission systems is very low, so these techniques need not be applicable only to communication systems but also to sensing for other optical measurements. Automatic frequency control (AFC) technology introduced in this chapter was achieved through using the terminal voltage change induced by light injection. This technique is applicable to a line monitor because the input power change can be detected by the terminal voltage change as an on-line monitor. This chapter has mainly concerned itself with SLIL with AlGaAs semiconductor lasers, but SLIL is also applicable to long-wavelength systems employing InGaAsP.

## REFERENCES

1. R. Adler, *Proc. IRE* **34**, 351 (1946).
2. H. C. Shaw and H. L. Stover, *Proc. IEEE* (*Lett.*) **54**, 710 (1966).
3. P. Mastalli, S. Randi, and G. Vannuchi, *Alta Frequenza* (Engl. Iss.) **2**, 415 (1968)
4. Y. Fukatsu and H. Kato, *ECR Rev.* (in Japanese) **18**, 1361 (1969).
5. T. Isobe and M. Tokida, *IEEE J. Solid-State Circuits* **SC-4**, 400 (1969).
6. M. Heines, J. C. Collinet, and J. G. Ondria, *IEEE Trans. Microwave Theory Tech.* **MIT-16**, 738 (1968).
7. K. Kohiyama and K. Monma, *Proc. IEEE* **57**, 1205 (1969).
8. H. G. Oltman and C. H. Nonnemaker, *IEEE Trans. Microwave Theory Tech.* **MIT-17**, 728 (1969).

9. K. Daikoku and Y. Mizushima, *Int. Electron.* **31**, 279 (1971).

10. C. L. Ruthroff, *Bell Syst. Tech. J.* **47**, 1653 (1968).

11. T. L. Osborne and C. H. Elemendorf, *Proc. IEEE* **57**, 214 (1969).

12. B. Martin and G. S. Hobson, *Electron. Lett.* **7**, 399 (1971).

13. K. Kurokawa, *Proc. IEEE* **61**, 1386 (1973).

14. H. L. Stover and W. H. Steier, *Appl. Phys. Lett.* **8**, 91 (1966).

15. C. J. Buczek and R. J. Freiberg, *IEEE J. Quantum Electron.* **QE-8**, 641 (1972).

16. T. Urisu, T. Sugeta, and Y. Mizushima, *J. Appl. Phys.* **52**, 3154 (1981).

17. R. H. Pantell, *Proc. IEEE* **53**, 474 (1965).

18. C. L. Tang and H. Statz, *J. Appl. Phys.* **38**, 323 (1967).

19. M. S. Sargent III, M. O. Scully, and W. E. Lamb, Jr., *Laser Physics*, Addison-Wesley, Reading, MA, 1974.

20. K. Shimoda and T. Yajima, *Quantum Electronics*, Reading, Shokabo, Tokyo, 1972 (in Japanese).

21. G. Arnold, K. Petermann, P. Russer, and F. J. Berlec, *AEU* **34**, 129 (1978).

22. O. Hirota and Y. Suematsu, *IEEE J. Quantum Electron.* **QE-15**, 142 (1979).

23. T. Kanada and K. Nawata, *IEEE J. Quantum Electron.* **QE-15**, 559 (1979).

24. R. Lang and K. Kobayashi, *IEEE J. Quantum Electron.* **QE-16**, 347 (1980).

25. R. Lang and K. Kobayashi, *IEEE J. Quantum Electron.* **QE-12**, 194 (1979).

26. R. Salathe, C. Voumard, and H. Weber, *Phys. Status Solidi (a)* **23**, 675 (1974).

27. K. Otsuka and S. Trucha, *IEEE J. Quantum Electron.* **QE-17**, 1515 (1981).

28. R. Lang and K. Kobayashi, *IEEE J. Quantum Electron.* **QE-18**, 976 (1982).

29. C. H. Henry, N. A. Olsson, and N. K. Dutta, *IEEE J. Quantum Electron.* **QE-21**, 1152 (1985).

30. P. Gallion and G. Debarge, *Electron. Lett.* **21**, 264 (1985).

31. F. Mogensen, H. Olesen, and G. Jacobsen, *IEEE J. Quantum Electron.* **QE-21**, 784 (1985).

32. H. A. Haus and Y. Yamamoto, *Phys. Rev. A.* **29**, 1261 (1984).

33. S. Kobayashi and T. Kimura, *IEEE J. Quantum Electron.* **QE-16**, 915 (1980).

34. S. Kobayashi and T. Kimura, *IEEE J. Quantum Electron.* **QE-17**, 681 (1981).

35. S. Kobayashi and T. Kimura, *Electron. Lett.* **16**, 668 (1980).

36. S. Kobayashi, Y. Yamamoto, and T. Kimura, *Electron. Lett.* **17**, 849 (1981).

37. S. Kobayashi and T. Kimura, *IEEE J. Quantum Electron.* **QE-18**, 575 (1982).

38. S. Kobayashi and T. Kimura, *IEEE J. Quantum Electron.* **QE-18**, 1662 (1982).

39. I. Petitbon, P. Gallion, G. Debarge, and C. Chabran, *Electron. Lett.* **22**, 889 (1986).

40. I. Petitbon, P. Gallion, G. Debarge, and C. Chabran, *IEEE J. Quantum Electron.* **QE-24**, 148 (1988).

41. K. Kobayashi, H. Nishimoto, and R. Lang, *Electron. Lett.* **18**, 54 (1982).

42. L. Goldberg, H. F. Taylor, and J. F. Weller, *Electron. Lett.* **18**, 1019 (1982).

43. L. Goldberg, C. Rauscher, J. F. Weller, and H. F. Taylor, *Electron. Lett.* **19**, 848 (1983).

44. L. Goldberg, H. F. Taylor, and J. F. Weller, *Appl. Phys. Lett.* **46**, 236 (1985).

45. S. Kobayashi and T. Kimura, *IEEE J. Lightwave Technol.* **LT-1**, 394 (1983).

46. P. Spano, S. Piazzora, and M. Tamburrini, *IEEE J. Quantum Electron.* **QE-22**, 427 (1986).

47. Y. Yamamoto and T. Kimura, *IEEE J. Quantum Electron.* **QE-17**, 919 (1981).

48. T. Mukai, Y. Yamamoto, and T. Kimura, *Semiconductor–Semimetal*, Academic Press, Orlando, FL, 1985, Vol. 22, Part E.

49. K. Inoue, H. Toba, and K. Nosu, *Electron. Lett.* **21**, 387 (1985).

50. H. Toba, K. Inoue, and K. Nosu, *Electron. Lett.* **21**, 656 (1985).

51. T. G. Hodgkinson, D. W. Smith, and R. Wyatt, *Electron. Lett.* **18**, 929 (1982).

52. R. Wyatt and D. W. Smith, *Electron. Lett.* **18**, 292 (1982).

53. D. J. Malyon, T. G. Hodgkinson, D. W. Smith, R. C. Booth, and B. E. Daymond-John, *Electron. Lett.* **19**, 145 (1983).

54. D. W. Smith and D. J. Malyon, *Electron. Lett.* **18**, 43 (1982).

55. J. Yamada, S. Kobayashi, H. Nagai, and T. Kimura, *IEEE J. Quantum Electron.* **QE-17**, 1006 (1981).

56. D. J. Malyon and A. P. McDonna, *Electron. Lett.* **18**, 445 (1982).

57. H. Toba, Y. Kobayashi, H. Nagai, and M. Nakahara, *Electron. Lett.* **20**, 371 (1984).

58. N. A. Olsson, H. Temkin, R. A. Logan, L. F. Johnson, G. J. Dolan, J. P. van der Ziel, and J. C. Campbell, *IEEE J. Lightwave Technol.* **LT-3**, 63 (1985).

59. C. H. Henry, *IEEE J. Quantum Electron.* **QE-18**, 259 (1982).

60. M. Nakamura, K. Aiki, N. Chinone, and J. Umeda, *J. Appl. Phys.* **49**, 4644 (1978).

61. S. Kobayashi and T. Kimura, *Electron. Lett.* **16**, 230 (1980).

62. S. Kobayashi and T. Kimura, *IEEE J. Quantum Electron.* **QE-18**, 582 (1982).

63. S. Kobayashi, Y. Yamamoto, and T. Kimura, *Electron. Lett.* **17**, 350 (1981).

64. S. Kobayashi, J. Yamada, S. Machida, and T. Kimura, *Electron. Lett.* **16**, 746 (1980).

65. K. Ogawa, *Fourth International Conference on Integrated Optics and Optical Fiber Communication* (IOOC '83), 29B1-6, 1983.

66. K. Iwashita and S. Nakagawa, *IEEE J. Quantum Electron.* **QE-18**, 1669 (1982).

67. J. Yamada, S. Kobayashi, S. Machida, and T. Kimura, *Jpn. J. Appl. Phys.* **16**, L689 (1980).

68. H. Nishimoto, H. Kuwahara, and M. Motegi, *Electron. Lett.* **19**, 509 (1983).

69. S. Saito, O. Nilsson, and Y. Yamamoto, *IEEE J. Quantum Electron.* **QE-18**, 961 (1982).

70. T. G. Hodgkinson, R. Wyatt, and D. W. Smith, *Electron. Lett.* **18**, 523 (1982).

71. D. W. Smith and D. J. Malyon, *Electron. Lett.* **18**, 44 (1982).

72. Y. Yamamoto, S. Saito, and T. Mukai, *IEEE J. Quantum Electron.* **QE-19**, 47 (1983).

73. K. Otsuka and S. Kobayashi, *Electron. Lett.* **19**, 262 (1983).

74. M. Lucente, E. S. Kintzer, S. B. Alexander, J. G. Fujimoto, and V. W. S. Chan, *Electron. Lett.* **25**, 112 (1989).

# 10

# Photon Statistics and Mode Partition Noise of Semiconductor Lasers

Pao-Lo Liu

*Department of Electrical and Computer Engineering, State University of New York at Buffalo, Buffalo, New York*

## 10.1.  INTRODUCTION

The optical beam plays an important role in optical communications systems. Most of us are participating in the generation, transmission, and detection of optical beams. For a strong coherent optical beam, we can use the classical picture, namely, a sinusoidal electromagnetic field with certain amplitude, to describe it. After transmitting through a long lossy medium, the power of beam is attenuated. When the beam is weak, the quantum nature of photons becomes important. The number of photons in a weak optical beam fluctuates. We describe the effect of such fluctuations by the shot noise when we consider the system performance. As a matter of fact, this quantum nature affects not only the receiver but also the laser at the transmitter end. The ultimate sensitivity of optical communications systems is limited by the quantum nature of the photon generation and absorption processes.

To characterize a collection of photons, we use statistical distributions. For example, we can repetitively sample the output power of a light beam. Because of the quantum nature of the optical beam, the amount of energy detected fluctuates. We can plot the probability of occurrence as a function of optical energy detected per window period. Such a distribution is called *photon* or *photoelectron statistics* [1]. If the beam is coherent, we obtain the Poisson distribution. If the beam is totally incoherent, we obtain an exponential distribution. Photon statistics of conventional lasers indeed had been studied and experimentally measured [2, 3]. It is a vital part of the theory of optical coherence [4–11].

*Coherence, Amplification, and Quantum Effects in Semiconductor Lasers,* Edited by Yoshihisa Yamamoto.
ISBN   0-471-512494   ©1991 John Wiley & Sons, Inc.

Both phase fluctuations and amplitude fluctuations of semiconductor lasers affect the performance of optical communications systems. Phase fluctuations determine the linewidth [12]. The linewidth is a very important parameter for coherent communications systems. Amplitude fluctuations appear in both the total output and the output levels of individual longitudinal modes. The total output power is partitioned among all longitudinal modes. Since they share the same pool of carriers for optical gain, one mode may grow at the expense of others. The density of carriers only stabilizes the total output. It does not stabilize the power of individual longitudinal mode. There are variations in how the power is partitioned. Coupled with a dispersive transmission medium, such as a fiber or wavelength-selective element, the effect of such fluctuations, recognized as mode partition noise, can severely limit the system performance [13–16]. It is of great interest to understand the origin of mode partition and, one step further, to eliminate its effect on system performance.

Fluctuations of the total output and of individual modes were first studied in the frequency domain by measuring their RF spectra [17, 18]. The spectral power density, that is, the relative-intensity noise (RIN), is frequency-dependent. For the total output, there is a strong peak occurring at the relaxation frequency of the semiconductor laser. The RIN is very low for frequencies well below the relaxation frequency. For individual modes, however, the RIN is high from DC all the way to the relaxation frequency.

System engineers take a different approach to describe the mode partition noise. The most important information for system performance is the bit error rate as a function of optical power received. The best error-rate performance is obtained directly from the output of a semiconductor laser without going through any dispersive medium. In optical communications systems, it usually requires more optical power to reach the same error rate. Such an increase is recognized as the power penalty. Mode partition noise is characterized by the power penalty [13, 14, 16]. Sometimes it may even make the error rate saturate at an intolerably high level. The error rate simply cannot be reduced further from this error-rate floor by increasing the optical power [13, 14, 16].

The present author's involvement in photon statistics of semiconductor lasers started in 1982, with an interest in short-pulse generation, high-speed modulation and transient phenomena of semiconductor lasers driven by a fast-rising current step. At that time, the effect of mode partition on system performance was developed by K. Ogawa [14, 15]. J. A. Copeland was writing a computer program using the Monte Carlo method to simulate fluctuations of injection lasers [19]. One day E. A. J. Marcatili came to this author's laboratory and asked whether it was possible to record individual, mode-resolved turn-on events. He and S. E. Miller were convinced by J. A. Copeland, that there should be fluctuations in turn-on events. Using an oscilloscope with a fast-writing screen, this author was able to observe directly mode partition in real time.

At about the same time, very fast Ti:LiNbO$_3$ modulators became available [20]. They can be modulated beyond relaxation frequencies of semiconductor lasers. On the other hand, amplitude fluctuations of semiconductor lasers peak at the relaxation frequency. The performance of high-data-rate systems using fast external modulators may be limited by amplitude fluctuations. To study the effect of amplitude fluctuations, this author developed the experimental technique to measure the photon statistics of the total output [21]. The technique was later used to study photon statistics of individual modes during the transient stage as well as in the steady state [22, 23].

In addition to photon statistics of injection lasers, there are many experimental and theoretical efforts contributing to our understanding of mode partition noise. The correlation measurements among individual longitudinal modes [24, 25], dropout rate measurements [26, 27], analytical solutions to rate equations with Langevin noise terms [28, 29], the Monte Carlo method [19, 30], numerical solutions to rate equations with Langevin noise terms [31], and analytical solution for nearly single-mode lasers have been demonstrated [32]. From these studies, we now have a good understanding of the mode partition noise. To eliminate mode partition noise, semiconductor lasers with very large side-mode-suppression ratio were developed leading to a series of record-setting experiments [33].

As system performance further advances, we will eventually be limited by the quantum nature of the optical beam. Photon statistics may help us to understand such quantum limitations. The following sections of this chapter contain reviews of the mode partition noise, experimental techniques used in measuring photon statistics of semiconductor lasers, direct observation of fluctuations in real time, photon statistics of the total output of injection lasers, photon statistics of individual longitudinal modes both in the steady state and during the turn-on transient, transient fluctuations of highly single-mode lasers, and theoretical modeling of low-probability turn-on events.

## 10.2.  MODE PARTITION NOISE

To evaluate the system performance, we typically measure the error rate as a function of optical power received. In early system experiments, a power penalty was commonly observed. Sometimes, an error-rate floor was also observed. Both the power penalty and the error-rate floor can be explained by fluctuations of the laser spectrum and the dispersion in optical fibers [13, 14].

In the theory of mode partition noise developed by Ogawa [14] the spectrum, that is, power of the individual mode, is considered to be fluctuating. A probability function, $p(a_1, a_2, \ldots, a_N)$, which represents the joint probability of having the $i$th mode with a power level of $a_i$, is used to describe such fluctuations. It is also assumed that the total output remains constant for all pulses;

$$\sum_i a_i = 1 \tag{10.1}$$

The average spectral distribution, $p(\lambda_i)$, can be described as a time or ensemble average

$$p(\lambda_i) = \int a_i p(a_1 a_2 \dots a_N) \, da_1 \, da_2 \dots da_N \tag{10.2}$$

After going through the dispersive transmission medium, there is a differential time delay, $\Delta\tau_i$, between each longitudinal mode and the center mode. By assuming that the eye pattern can be represented by $\cos(\pi Bt)$, where $B$ is the data-rate, the total amplitude fluctuation at the decision time is found to be

$$\Delta = \tfrac{1}{2}(\pi B)^2 \sum (\Delta\tau_i)^2 a_i \tag{10.3}$$

The average noise power due to fluctuation can be obtained from the average value and the variance of $\Delta$ as follows:

$$\sigma_{pc}^2 = \langle \Delta^2 \rangle - \langle \Delta \rangle^2 \tag{10.4}$$

where $\langle \ \rangle$ represents the ensemble average.

Mode partition noise is worst when longitudinal modes are mutually exclusive. In other words, all the output power appears in one mode. However, this power-carrying mode fluctuates from pulse to pulse. On the other hand, if the spectral distribution remains constant from pulse to pulse, there is no mode partition noise. To accommodate such variations, a mode partition coefficient, $k$, was introduced [15]. The value of $k$ ranges from zero to unity. The worst case corresponds to $k = 1$. The value of $k$ varies from laser to laser and is a function of the data rate; $k$ is related only to the mean value and variance of $a_i$ as

$$k^2 = \frac{\langle a_i^2 \rangle - \langle a_i \rangle^2}{\langle a_i \rangle - \langle a_i \rangle^2} \tag{10.5}$$

It is assumed that the same $k$ value applies to all modes of a laser. By introducing the partition parameter we can calculate the effect of mode partition noise without requiring to know the probability distribution. Only the mean value and the variance of $a_i$ are used to characterize mode partition.

The signal-to-noise ratio due to mode partition becomes

$$\frac{S}{N_{pc}} = \frac{1}{\sigma_{pc}^2 k^2} \tag{10.6}$$

Further simplification can be achieved by assuming the average spectrum to be Gaussian with a center wavelength, $\lambda_c$

$$p(\lambda_i) = \frac{1}{\sqrt{2\pi\sigma^2}} \exp\left[-\frac{(\lambda_i - \lambda_c)^2}{2\sigma^2}\right] \tag{10.7}$$

and by using the following approximation for the dispersion of the group delay

$$\frac{d\tau}{d\lambda} = \frac{s(\lambda - \lambda_0)}{c\lambda^2} \tag{10.8}$$

where $\lambda_0$ represents the minimum dispersion wavelength, $c$ is the speed of light, and $s = 0.05$ for silica glass fibers. Therefore, we have

$$\frac{N_{pc}}{S} = \frac{1}{2}(\pi B)^4\left[A_1^4\sigma^4 + 48A_2^4\sigma^8 + 42A_1^2A_2^2\sigma^6\right] \tag{10.9}$$

where

$$A_1 = \frac{s(\lambda_c - \lambda_0)z}{c\lambda_c^2} \tag{10.10}$$

$$A_2 = \frac{sz}{2c\lambda_c^2} \tag{10.11}$$

and $z$ is the length of the fiber.

If we assume that the noise is approximately a Gaussian random variable, the error rate can be calculated from

$$P_{\text{error-rate}}(Q) = \int_Q^\infty e^{-x^2/2}\,dx \tag{10.12}$$

In order to achieve a certain error rate determined by the value $Q$, the signal-to-noise ratio must satisfy

$$\frac{S}{N_{pc}} \geq Q^2 \tag{10.13}$$

For example, the signal-to-noise ratio shall be $6^2$ in order to obtain an error rate of $10^{-9}$.

Both the receiver noise and the mode partition noise contribute to the overall signal-to-noise ratio of the system according to

$$\frac{N}{S} = \sigma_{pc}^2 k^2 + \frac{N_{receiver}}{S} \le \frac{1}{Q^2} \qquad (10.14)$$

The first term represents the mode partition noise. The second term represents the receiver noise. By increasing the signal level at the receiver, we can suppress the contribution of the receiver noise and improve the signal-to-noise ratio of the system. The mode partition noise, however, is independent of the signal level. Both the error-rate floor and the power penalty can be explained by Eq. (10.14). If the mode partition noise is already dominant, the signal-to-noise ratio of the system cannot be improved by increasing the signal level; thus, there is an error-rate floor. Because of the presence of the mode partition noise, we must increase the signal level, that is, suppress the receiver noise term, in order to have the same error rate in a system without mode partition noise. Assuming that the error rate is $10^{-9}$, the power penalty, $\alpha$ dB, due to mode partition can

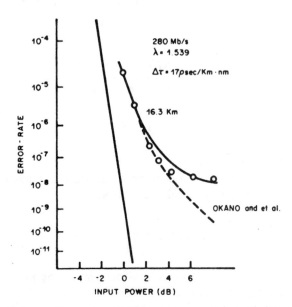

**Figure 10.1.** Mode partition noise. Circles represent experimental data measured in a 16.3-km-long, 280-Mbit/s system operating at 1.52 $\mu$m with a spectral width of 2 nm. The horizontal axis represents the power penalty. The dashed curve is the theoretical result of Ref. 13. The solid line is the theoretical result of Ref. 14. (© 1982 IEEE)

be calculated from

$$\alpha = 5 \log \frac{Q^2}{Q^2 - 6^2} \qquad (10.15)$$

where $Q^2 = 1/(\sigma_{pc}k)^2$.

In addition to power penalty, the bit rate–distance product can also be calculated by the theory. Comparison of early experimental measurements to theoretical results is shown in Figure 10.1. The model can explain experimental data very well.

## 10.3. EXPERIMENTAL SETUP

To thoroughly understand the mode partition noise, we must know the probability distribution function, that is, the photon statistics of semiconductor lasers. To measure it, we need a sampling window that is narrow enough so that we can record extremes of fluctuations. If the sampling window is too long, fluctuations are reduced because of time averaging. Photon statistics of gas lasers was measured in the 1960s using the photon counting technique. The response time of photomultiplier tubes (PMT) is short enough to record extremes of fluctuations [2, 3]. For semiconductor lasers, however, the RF spectrum of amplitude fluctuations extends to the relaxation frequency that is in the gigahertz region [17, 18, 34]. The photon counting technique using PMT does not have sufficient bandwidth for measuring photon statistics of semiconductor lasers. The alternative to the digital photon counting technique is to use an analog circuit to sample the instantaneous output power [21]. Fast semiconductor detectors, including $p$–$i$–$n$ and avalanche photodiodes (APD) and wideband amplifiers, are readily available. We can use a fast detector in series with an amplifier. The output of the amplifier can be monitored in real time by a wideband oscilloscope or a transient digitizer. It can also be sampled by a sampling head. The sampling head performs like a sample-and-hold circuit. The sampling gate is open for a very short time, typically 25 ps. The level of the analog signal is detected, and an output directly proportional to the instantaneous input signal is generated. The output is held constant until the next sampling is performed. Depending on the sampling head used, the sampling rate is in the range 33–60 kHz. In other words, we can accumulate an ensemble of $10^9$ samples in approximately 4 h.

The original experimental setup for measuring photon statistics of injection lasers is shown in Figure 10.2. The output of an injection laser is collimated by a microscope objective. The beam is coupled into a monochromator. The output of the monochromator is focused by another microscope objective onto a fast detector. The output of the detector is

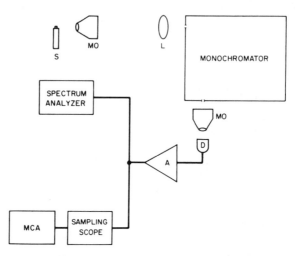

**Figure 10.2.** A schematic diagram of the experimental setup; S represents a semiconductor laser; MO represents microscope objectives; L is a focusing lens; D is a detector; A represents the broadband amplifier; MCA stands for multichannel analyzer. The monochromator selects one longitudinal mode from the output of the laser.

amplified by a broadband amplifier. The amplified signal is monitored by a sampling head. The output of the sampling head is processed by a multichannel analyzer (MCA) operating in the pulse-height-analysis mode. If the photon statistics of the total output is to be measured, the monochromator is removed. In the original measurements, we used InGaAs $p$–$i$–$n$ detector with a full-width-at-half-maximum (FWHM) pulse response of 85 ps for the total output and Si or Ge APD for mode-resolved measurements. The FWHM pulse response of APDs was measured to be approximately 450 ps. A broadband amplifier, such as B & H7000 or B & H8000 with 7- or 8-GHz bandwidth, was used to amplify the signal to a level of approximately 100 mV. The amplified signal was sampled by a sampling head, Tektronix S-6 or S-4. A Nuclear Data model ND62 MCA was used to obtain the statistical distribution. A gating signal synchronized to the sampling gate was obtained from the oscilloscope to control the MCA. For real-time observations, we used a 1-GHz oscilloscope, the Tektronix 7104. Its screen has a built-in microchannel plate, therefore, can provide a clear trace of very fast event.

The analog sampling technique, of course, is substantially noisier than the digital photon counting technique. The main source of noise is the sampling head. The sampling head is typically used to monitor repetitive waveforms. We were using it to monitor nonrepetitive fluctuating events. We carefully adjusted the dot response to minimize over- or undershoot during sampling. The background noise level of the sampling head was measured to be 5 mV. Thus, a signal level around 100 mV is needed to ensure that the background noise can be neglected.

To study transient events, a fast-rising current step is required to drive semiconductor lasers. This was derived from the Tektronix pulse generator, S-52. It put out a step waveform with a risetime of less than 25 ps. A B & H broadband amplifier was used to reverse the polarity and to boost the level of the output. The output after the amplifier had a risetime of 85 ps and a duration of 500 ns. The pulse was combined with the DC bias in a Hewlett-Packard bias T to drive injection lasers.

Since we are interested in monitoring fluctuations during the transient stage, the jitter of the trigger must be minimized. The electrical output of S-52 pulse generator was used to ensure that the jitter was negligible in comparison with fluctuations of semiconductor lasers. The instrumentation jitter was less than 50 ps.

## 10.4. OBSERVATION OF TURN-ON FLUCTUATIONS IN REAL TIME

The effect of mode partition can be seen in many experiments, including power penalty, bit-error-rate floor, dropout rate, and correlation among individual modes. The most direct evidence, however, can be obtained by monitoring the output of semiconductor lasers in real time [35, 36].

Many semiconductor lasers, especially those with built-in wavelength-selective structures, such as distributed feedback (DFB) or distributed Bragg reflector (DBR), exhibit single-longitudinal-mode output under DC operation above threshold. A commonly used parameter to gauge single-mode lasers is the ratio of the power of the highest side mode to that of the main mode, that is, the side-mode-suppression ratio (SMSR). The typical SMSR has been improved from $1/500$ in the early 1980s to $1/40,000$ or more at present. However, when the laser is biased below or near threshold, side modes containing substantial amount of power are present in all semiconductor lasers [37, 38]. The output builds up from the bias condition when a laser is driven by a current step. We expect to see an evolution from the multimode regime to the single-mode regime during the transient. By monitoring the output in real time, we can study fluctuations during the turn-on transient. We can also learn how long it takes for semiconductor lasers to evolve into the steady state.

Many lasers were used in the original study. Results of two lasers are presented here. They represent typical examples of laser studied. One laser is a ridge-waveguide InGaAsP/InP laser with as-cleaved facets and a cavity length of 250 $\mu$m [39]. The other laser is an oxide-stripe short-cavity laser with a cavity length of 51 $\mu$m [40]. One facet of the short-cavity laser was gold-coated to reduce the threshold. The mode spacing of short-cavity laser is wider; therefore, the side mode is smaller. According to the

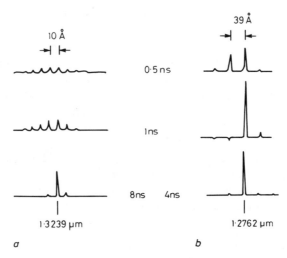

**Figure 10.3.** Time-resolved, average spectra of (*a*) multimode ridge-waveguide laser; (*b*) short-cavity laser. The time indicates the delay between the turn-on and the moment when spectrum was taken.

standard in early 1980s, it was in the category of single-mode lasers. Judging from today's standard, however, it is not quite single-mode.

The average spectra of these lasers were monitored during the transient. The time resolution was obtained by controlling the delay time of the sampling head with respect to the turn-on of the current step. The spectral resolution was obtained by scanning the monochromator. The average time-resolved spectra are shown in Figure 10.3. Both lasers start as multimode. For the 250-$\mu$m-long laser, the spectrum evolves into a single dominant mode after an elapsed time of 6 ns. The center of the emission peak shifts to longer wavelengths during the transient stage. However, the output of the short-cavity laser under similar driving conditions evolves into a single dominant mode within 1 ns. The apparently negative spectral features are artifacts caused by a minor ringing in spectrum measuring circuit.

The output of these lasers, at discrete wavelengths and in real time, is shown in Figure 10.4. The evolution of four individual shots (clean traces) and an accumulation of several thousand pulses (smeared traces) are shown. Note that for the 250-$\mu$m-long laser, the individual pulse of any one mode can start at different times and go through different evolution paths. The secondary modes decay while the dominant mode increases to its steady-state value. A similar display of the output of the short-cavity laser shown in Figure 10.4*b* shows far less pulse-to-pulse variation.

From this real-time study, we realize that the mode partition noise is a severe problem during the transient. We cannot depend on the average

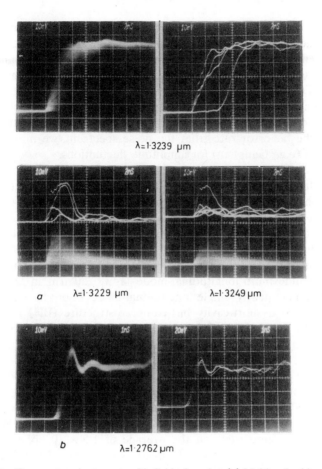

**Figure 10.4.** Turn-on transient events of individual modes. (*a*) Multimode ridge-waveguide laser; (*b*) short-cavity laser. The wavelength indicates the monochromator setting at which traces were recorded.

spectra. In fact, a mode that is rather small on the average can still be the dominant mode during the transient stage. In some cases, the side mode may last for nanoseconds.

## 10.5. AMPLITUDE FLUCTUATIONS OF THE TOTAL OUTPUT

Amplitude fluctuations of semiconductor lasers may limit the system performance at high data rate. One convenient approach to characterize amplitude fluctuations is to measure the noise spectrum with a fast detector and an RF spectrum analyzer [17]. Using this method, the noise spectra of semiconductor lasers have been found to have a large peak at the relaxation frequency. For bit rates substantially lower than the relaxation frequency, fluctuations are averaged during the time slot of a single bit. The effect of amplitude fluctuations on system performance is reduced. At very high bit rates, obtained, for example, by externally modulating a DC-biased semiconductor laser, amplitude fluctuations may introduce a serious increase in bit error rate. The noise spectrum, however, cannot directly provide information on error rate. In order to calculate the bit error rate from the noise spectrum, we must assume a distribution (e.g., Gaussian) for amplitude fluctuations.

We measured the photon statistics of the total output of several InGaAsP lasers driven above threshold by DC current [21]. The statistical distributions were obtained by repetitively sampling the output of the laser with a time resolution of 120 ps. Measurements of the photon statistics of semiconductor lasers are very valuable. In addition to the coherent properties of the laser, we can predict the bit error rate.

Three different InGaAsP lasers emitting at 1.3 $\mu$m were studied: (1) a ridge-waveguide multi-longitudinal-mode laser, 250 $\mu$m in length; (2) a ridge-waveguide nearly single-longitudinal-mode laser, also 250 $\mu$m in length; and (3) a short-cavity buried-heterostructure (BH) laser with a cavity length of 70 $\mu$m. All lasers were biased with DC current above threshold. The output of the power supply was shunted with a 2500-$\mu$F capacitor to eliminate any fluctuations in drive current. A $p$–$i$–$n$ InGaAs detector and a broadband amplifier were used [41]. The FWHM pulse response was measured to be 120 ps. The detector was purposely misaligned. Feedback from external reflectors may cause instability of the light output. The detector current was approximately 200 $\mu$A. Within the 120-ps window there were on the average of $1.5 \times 10^5$ photoelectrons. The output of the amplifier was sampled by a sampling head. The output of the sampling head was digitized and processed by a computer or a MCA. The computer or the MCA categorized the power levels and recorded the number of events occurring in each power interval. The separation between two adjacent intervals, which we call *channel width*, is an experimental parameter. We need the value of the channel width to fit theoretical calculations to experimental data.

The distributions for the single-mode laser at average power levels of 1.2 and 3 mW are shown in Figures 10.5$a$ and 10.5$b$, respectively. In each figure, the data points represent the probability of recording light output levels falling within each channel. The sum of probabilities of all data points in each plot is, of course, unity. In order to see the deviation from the average directly, both the light output and the channel width were normalized to the average power. The normalized channel width is 0.011. A similar plot for a multimode laser at 3 mW is shown in Figure 10.6. The normalized channel width is also 0.011. The distribution of a short-cavity BH laser at an output of 1.7 mW is shown in Figure 10.7. The normalized channel width in this case is 0.015.

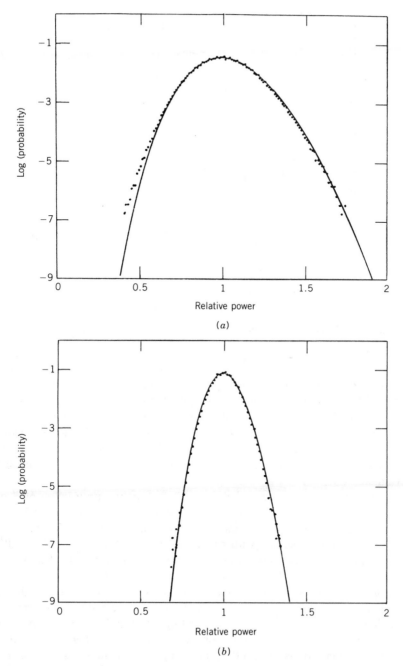

**Figure 10.5.** Photon statistics of the total output of a nearly single-mode InGaAsP laser at (*a*) 1.2 mW; (*b*) 3 mW. The horizontal axis represents power sampled in a 120-ps window normalized to the average power. The solid curves are theoretical results assuming that the output of the laser can be represented as the superposition of a coherent electromagnetic field and Gaussian noise.

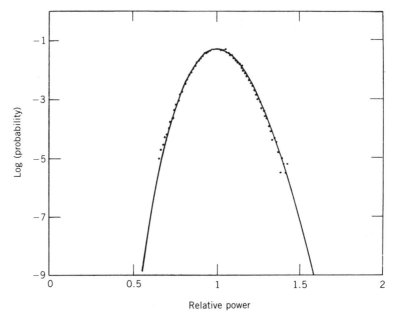

**Figure 10.6.**   Photon statistics of a multimode laser. The solid curve is the theoretical fit.

From these plots, we realize that amplitude fluctuations of the total output of semiconductor lasers are quite large. In fact, they are much larger than what can be expected from the shot noise. Shot noise can be described by the Poisson distribution. With more than $10^5$ photoelectrons detected in each sampling window, the Poisson distribution is very narrow. The measured distributions are much wider, and there is a distinct asymmetry common to all measured results. According to the 1.2-mW, single-mode laser data, the probability of having light output below 50% of the average is $1.45 \times 10^{-5}$. The probability of having light output below 50% of the average for the same laser at 3 mW output is substantially below $10^{-9}$. In comparison with single-mode laser, the output of a multimode laser has more fluctuations. In comparison with regular lasers, the short-cavity laser is noisier.

From these experimental measurements, we realize that amplitude fluctuations may indeed contribute significantly to error rate for externally modulated semiconductor lasers operating at high data rates. At low output power, lasers are not that coherent. The output may contain large amplitude fluctuations. The effect of amplitude fluctuations can be reduced by operating at lower data rates or by increasing the output power.

The statistical distribution of amplitude fluctuations of a laser biased above threshold can be described by either the Fokker–Plank equation or a phenomenological model [28, 29, 42–45]. The Fokker–Plank equation results in a Gaussian distribution that is symmetrical and therefore can fit only the central portion of our experimental data. It cannot explain the asymmetry in the experimental data.

The phenomenological model is actually quite simple. We consider the total output of a semiconductor laser to be the superposition of a coherent electromagnetic field and a narrowband Gaussian noise. The probability of finding a net electromagnetic field amplitude, $R$, lying between $R$ and $R + dR$ is given by

$$p(R)\, dR = \frac{R}{\psi} \exp\left[ -\frac{R^2 + P^2}{2\psi} \right] I_0\left( \frac{RP}{\psi} \right) dR \qquad (10.16)$$

where $P$ is the amplitude of the coherent field, $\psi$ is the mean-square value of the in-phase and the out-of-phase Gaussian noise, and $I_0$ is the modified Bessel function. The quantity measured by the detector is the power that is proportional to $R^2$. The probability of finding a power level lying between $I$ and $I + dI$ is given by

$$p(I)\, dI = \frac{1}{2\psi} \exp\left[ -\frac{I + P^2}{2\psi} \right] I_0\left( \frac{\sqrt{I}\, P}{\psi} \right) dI \qquad (10.17)$$

By normalizing all parameters with respect to the amplitude of the coherent field $P$:

$$\psi' = \frac{\psi}{P^2}, \qquad I' = \frac{I}{P^2}$$

we obtain

$$p(I')\, dI' = \frac{1}{2\psi'} \exp\left[ -\frac{I' + 1}{2\psi'} \right] I_0\left( \frac{\sqrt{I'}}{\psi'} \right) dI' \qquad (10.18)$$

By using measured channel width as $dI'$ and $\psi'$ as an adjustable parameter, we can compare experimental results to the phenomenological model. The solid curves in Figures 10.5–10.7 represent the best fits to experimental data. Values of $\psi'$ are $\frac{1}{235}$, $\frac{1}{1100}$, $\frac{1}{525}$, and $\frac{1}{250}$, respectively. Theoretical results fit experimental data covering five orders of magnitude quite well. There is only a small discrepancy on the lower-power side. It is probably caused by the residual over- and under-shoot in the sampling circuitry and by, possibly, noise with characteristics different from those of the narrowband Gaussian noise.

The adjustable parameter can actually be measured by using a microwave power meter or by integrating the RF spectrum of the RIN over the detection bandwidth. We believe that using data fitting in determining

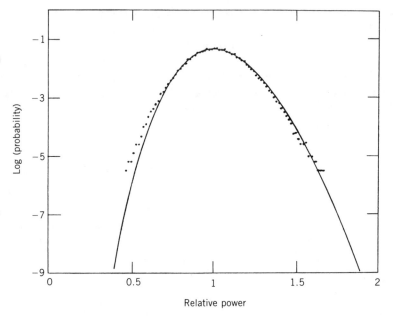

**Figure 10.7.**   Photon statistics of a short-cavity laser. The solid curve is the theoretical fit.

$\psi'$ is more accurate. The noise spectrum, however, does provide direct information on how $\psi'$ varies with the duration of the sampling window, that is, the bandwidth of the system.

As shown in Figures 10.5–10.7, amplitude fluctuations of solitary semiconductor lasers are rather large. There are several contributing factors. First, the volume of the semiconductor laser is small; therefore, the number of photons stored in the cavity is small. The small volume also implies a large acceptance angle for the spontaneous emission. In addition, the density of carriers does not change immediately as the light output changes. Carriers and photons interact dynamically rather than instantaneously. One practical method to reduce fluctuations is to increase the cavity length by using an external resonator [46]. Both the volume and the round-trip time are increased. Such external cavity lasers have been proven to have linewidth as narrow as 10–15 kHz. Experimental observations indicate that the output power of such a laser is stable without pulsation or fluctuations.

We have constructed a simple composite-cavity laser consisting of an antireflection (AR)-coated InGaAsP gain medium in an optical fiber resonator. A sputtered $Si_3N_4$ AR coating was applied to a 250-$\mu$m-long, 1.3-$\mu$m ridge-waveguide laser. The measured reflectivity of the AR coating was $4 \times 10^{-4}$. A microlens was formed at the tip of a 50-$\mu$m-core-diameter multimode fiber using a combination of chemical etching and arc

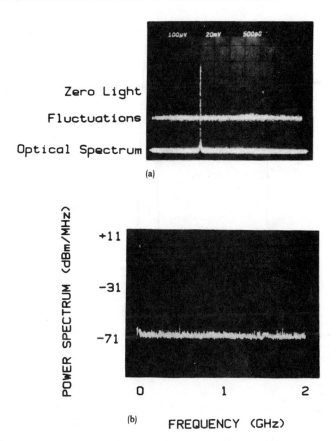

**Figure 10.8.** Amplitude fluctuations of a composite-cavity laser operating in a single mode. (*a*) Output power and spectrum; (*b*) RF spectrum of the noise.

melting. The fiber was cleaved to an arbitrary length of 6.8 cm. A gold high-reflection coating was applied to the cleaved end of the fiber to provide feedback. The lens was aligned to the AR-coated facet of the InGaAsP ridge-waveguide gain medium. The threshold current of the InGaAsP laser before coating was 68 mA. The threshold current of the composite cavity laser was 103 mA. Biased at 160 mA, the output power from the uncoated facet of the ridge-waveguide structure was 2.5 mW. The output of the laser was monitored by a monochromator, an RF spectrum analyzer, and our photon statistics measurement setup using an InGaAs *p–i–n* detector.

By carefully optimizing the alignment, we achieved single-mode operation. Amplitude fluctuations recorded by the sampling oscilloscope along with the optical spectrum and RF spectrum are shown in Figure 10.8. Fluctuations of the laser output, shown in the oscillogram, are comparable

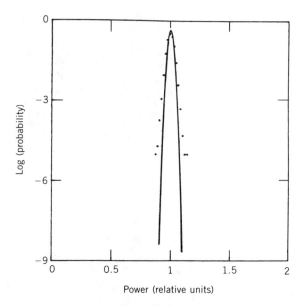

**Figure 10.9.** Photon statistics of a composite-cavity laser operating in a single mode. Dots represent experimental data. The solid curve is obtained by deconvolving the baseline fluctuations of the instrument. The distribution is substantially narrower than that of a solitary laser operating at the same power level.

to the baseline fluctuations of the instrument. The probability distribution measured directly from the output of the sampling oscilloscope is shown in Figure 10.9 as a series of points. The horizontal scale represents the output power normalized to the average power, namely, 2.5 mW. The continuous curve is the probability distribution obtained after deconvolving the baseline fluctuations of the instrument. The distribution is four times narrower than that of a 250-$\mu$m-long, single-mode, ridge-waveguide laser operating at 3 mW.

By adjusting the alignment of the fiber to the gain medium, this composite-cavity laser could also operate with several longitudinal modes. Under multimode operation, the RF spectrum of the amplitude fluctuations showed a peak at 1.45 GHz. Such a peak is a result of beating among longitudinal modes of the composite cavity. Because of mode beating, we did not measure the statistical distribution.

From this study, we conclude that amplitude fluctuations of semiconductor lasers can be reduced by using a long, composite cavity. However, it must operate in a single mode of the composite cavity. Otherwise, the output power has a sinusoidal undulation resulting from beating of longitudinal modes of the composite cavity. A wavelength-selective structure, such as DFB or DBR, can be used to ensure single-mode operation.

Surface emitting lasers with very short cavity are available. The packing density of such miniature lasers in two-dimensional laser arrays can be very high (e.g., $10^6/cm^2$). The heat dissipation is a rather severe limitation. It is advantageous to operate them at very low power. The photon statistics of such lasers at low power is yet to be measured.

## 10.6. STEADY-STATE PHOTON STATISTICS OF INDIVIDUAL MODES

The partition of output power among longitudinal modes of a semiconductor laser is a noise-driven process [47–53]. One way to characterize it is to measure the statistical distribution. Instead of describing fluctuations by a mean-square value, we can characterize fluctuations by using photon statistics. Such distributions can be used to understand the coherent properties of the optical beam. They are also very valuable in characterizing low-probability (e.g., $10^{-9}$) events.

The statistical distribution of the power contained in one longitudinal mode was first measured by Henning [24]. He used a transient digitizer with a bandwidth of a few hundred megahertz to measure the statistical distribution of the output power contained in one longitudinal mode of a modulated multimode laser. Since it has been observed that the RF spectrum of mode partition extends to the relaxation oscillation frequency of the semiconductor laser, a bandwidth comparable to the relaxation frequency is needed to observe extremes of fluctuations. We used our experimental setup to measure the photon statistics of individual longitudinal modes.

There are several noise sources in our measurement setup, namely, the excess noise of the APD, the noise of the broadband amplifier, and the baseline noise of the sampling head. In order to calibrate the instrumentation noise, we measured the distribution of an incoherent light source, that is, the total output of a light-emitting diode (LED). The distribution was a very sharp Gaussian distribution. We concluded that the instrumentation noise in our setup could be neglected in analyzing experimental data.

For mode partition in the steady state, we biased a one-mode dominant AlGaAs transverse junction stripe (TJS) laser above threshold to provide an output power of 3 mW [54]. The spectrum is shown in Figure 10.10a. The dominant mode contained on the average 92% of the total output power. Distributions of the total output power, the dominant mode, and one of the side modes are shown in Figure 10.10b. The horizontal scale was normalized by the average total output power. The distribution of the total output is peaked around unity with a finite width representing amplitude fluctuations of the total output. The distribution of the dominant mode has an extended tail on the low-power side. On the other hand,

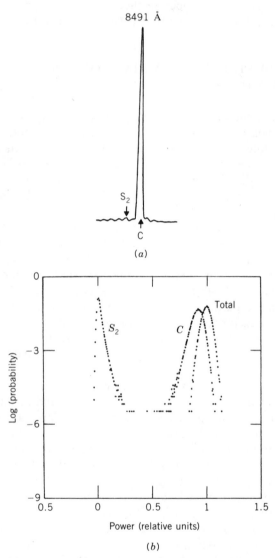

**Figure 10.10.** Mode partition of an AlGaAs TJS laser dc-biased at 3-mW output. (*a*) Average spectrum. (*b*) Photon statistics of the side mode, main mode, and the total output. For the side mode, there is a tail extending to the high-power side. The main mode, on the other hand, has a tail extending to the low-power side.

the distribution of the side mode, containing on the average only 1.6% of the total output power, has an extended tail on the high-power side. They clearly indicate that mode partition is present.

In order to see how far the tail of the side mode could extend, we measured a buried-crescent InGaAsP laser in which we could vary the side-mode power by slightly changing the driving current. The average spectra are shown in Figure 10.11a. Distributions corresponding to a side mode containing on the average 2 and 4% of the total output power are shown in Figure 10.11b. Because of the baseline fluctuations of the instrument, there is a slight spread below the zero mark in distributions of the side mode. The main feature of side-mode distributions is the exponential tail extending to the high-power side. The decay of the exponential function, however, is related to the average side-mode power. A side mode containing on the average 4% of the total output power may occasionally (i.e., with a probability of $10^{-6}$) become the dominant mode containing 50% of the total output power. The photon statistics reveals the characteristics of the side mode. Although the injection laser itself is biased above threshold, small, individual side modes still behave like a polarized chaotic light source, that is, a laser biased below threshold. The complex electromagnetic field of the side mode contains many spontaneously emitted photons. We can apply the central limit theorem and obtain a probability distribution for the instantaneous field amplitude $|V|$ [1]:

$$p(|V|) \, d|V| = \frac{|V|}{\sigma_c^2} \exp\left[ -\frac{|V|^2}{2\sigma_c^2} \right] d|V| \qquad (10.19)$$

where $\sigma_c$ is a measure of the width of the distribution.

In addition to the probability distribution of the field amplitude, the second-order correlation of the field is also required to fully characterize the statistical properties of the chaotic incoherent field. With the second-order correlation, one can calculate correlations of higher orders. The second-order correlation for the complex field is defined as [1]

$$G^{(2)}(x_1, x_2) = \langle V^*(x_1) V(x_2) \rangle \qquad (10.20)$$

where $x$ represents the spatial as well as the time coordinates at which the field is measured and $\langle \; \rangle$ represents the ensemble average. In the case of a stationary stochastic process that has no spatial variation, the second-order correlation for the field amplitude becomes a function that depends only on the time difference between measurements:

$$G^{(2)}(x_1, x_2) = G^{(2)}(t_1 - t_2) \qquad (10.21)$$

In order to comprehend the form of the second-order correlation explicitly, one must take a closer look at the system under consideration.

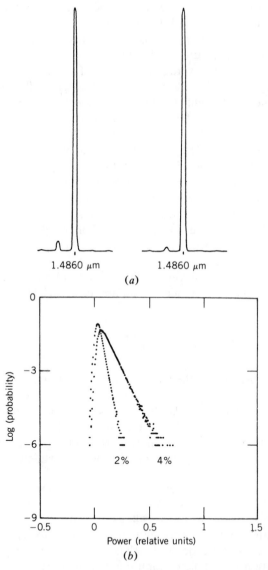

**Figure 10.11.** (*a*) Spectrum. (*b*) Photon statistics of a BH laser. By adjusting the current, the average power of the side mode can be controlled. With more average power, the tail of the side mode extends further to the high power side.

For a laser biased below threshold, one may use a damped harmonic oscillator with a Langevin noise term to describe the system. From that equation, one can deduce the second-order correlation for the field

$$G^{(2)}(t_1 - t_2) = G^{(2)}(0) \cdot \exp(-\beta_c|t_1 - t_2|) \qquad (10.22)$$

where $\beta_c$ is related to the coherence time $\tau_c$ by $\tau_c = 1/\beta_c$ and $G^{(2)}(0)$ is a normalization factor. The Fourier transform of the second-order field correlation is the power spectrum of the electromagnetic field. The spectrum corresponding to an exponential correlation function is a Lorentzian function.

The quantity of interest to the optics communications community is the power or intensity, which is proportional to $|V|^2$. From the characteristics of the field, one can deduce the statistical distribution of the intensity as well as the second correlation of the intensity. The statistical probability distribution for the instantaneous intensity, $I$, is given by

$$p(I)\, dI = \frac{1}{\langle I \rangle} \exp\left(-\frac{I}{\langle I \rangle}\right) dI \qquad (10.23)$$

where $\langle I \rangle$ represents the average intensity. The second-order correlation for the intensity is a special case of the fourth-order correlation of the field and is given by

$$G_I^{(2)}(t_1 - t_2) = G_I^{(2)}\left[\exp(-2\beta_c|t_1 - t_2|) + 1\right] \qquad (10.24)$$

where $G_I^{(2)}(0)$ is the normalization factor. The Fourier transform of the second-order intensity correlation is the RF power spectrum of the output power fluctuations. If we drop the time-independent term in Eq. (10.24), that is, consider only the deviation of th output power from its average value, we obtain the following Lorentzian function for the RF spectrum of the output power fluctuations

$$S(\omega) = \frac{S(0)}{1 + (\omega/2\beta_c)^2} \qquad (10.25)$$

where $S(0)$ represents the normalization factor.

The probability distribution of the instantaneous output power of the laser can, in principle, be measured experimentally. However, in reality, any detection system has a limited bandwidth. What is actually measured is the integral of the output power over a certain sampling window, $T$, rather than the instantaneous power. Only when the bandwidth is very large, that is, when the sampling window is very short, the measured distribution becomes an exponential function described by Eq. (10.23).

The longer the sampling window, the greater the averaging effect. This is one of the reasons why mode partition noise is not as serious for low data-rate systems as for high-data-rate systems.

The statistical distribution of $W$, defined as

$$W = \int_0^T I\, dt \qquad (10.26)$$

where $T$ is the duration of the sampling window, can be calculated for a Gaussian field with Lorentzian spectrum. The distribution is given by

$$p(W) = \frac{\exp(\gamma_c)}{\langle W \rangle} \sum_k (-1)^k \frac{2y_k^2}{y_k^2 + 2\gamma_c + \gamma_c^2} \cdot \exp\left[ -\frac{\gamma_c^2 + y_k^2}{2\langle W \rangle \gamma_c} \right] \qquad (10.27)$$

where $\gamma_c = \beta_c T$ and $y_k$ are solutions to the following equations:

$$y_k \cdot \tan\left( \frac{y_k}{2} \right) = \gamma_c \qquad (10.28)$$

$$y_k \cdot \cot\left( \frac{y_k}{2} \right) = -\gamma_c \qquad (10.29)$$

If $y_k$ satisfies Eq. (10.28), a positive sign is used in Eq. (10.27). If $y_k$ satisfies Eq. (10.29), a negative sign is used.

To compare theoretical predictions to experimental results, we measured the RF spectrum and the statistical distributions of the side mode of a ridge-waveguide laser biased at approximately 3 mW. The side mode contained on the average 3.7% of the total output power. The RF spectrum of the side mode is shown in Figure 10.12. There is no pronounced relaxation resonance. The noise spectrum is a decreasing func-

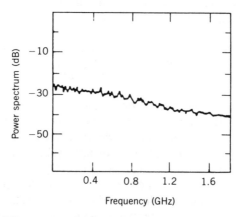

**Figure 10.12.** The RF spectrum of the output power fluctuations of a side mode of the semiconductor laser.

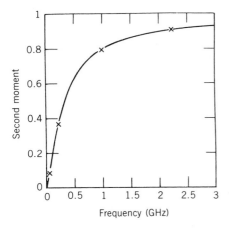

**Figure 10.13.** The second-order moment or the mean-square fluctuations of the side mode as a function of the bandwidth of the detection circuit. Experimental data are marked as ×. The solid curve represents calculated results from a Lorentzian spectrum with a −3-dB frequency of 330 MHz.

tion of frequency. The −3-dB point is approximately at 300–350 MHz. Because of the ripples on the spectrum, it is difficult to pinpoint the −3-dB point and the actual shape of the spectrum.

The noise power below certain cutoff frequencies was also measured with a microwave power meter. Such a measurement corresponds to the integration of the noise spectrum from low frequency to the cutoff frequency and is proportional to the mean-square fluctuations or the second moment. Results are shown in Figure 10.13. The calculated second-order moment from a Lorentzian spectrum with a −3-dB frequency of 330 MHz is also shown. The fit is excellent.

Statistical output power distributions were measured with several band-limiting filters. In Figure 10.14, we present results measured with (a) full bandwidth of Ge–APD and (b) a 250-MHz filter in series with the detector. By limiting the bandwidth, the distribution becomes narrower.

For comparison with Eq. (10.29), we must choose the parameter $\gamma_c$ for each band-limiting filter. We calculated the mean-square fluctuations, that is, the second-order moment, of the distribution given by Eq. (10.29) as a function of $\gamma_c$. Results are shown in Figure 10.15. For each band-limiting filter, we mapped out the second-order moment from Figure 10.13. From the second-order moment, we found the appropriate $\gamma_c$. Using the $\gamma_c$ value, we calculated the statistical distribution from Eq. (10.29). The series in Eq. (10.29) converges very quickly. In our computer program, we arbitrarily used 25 pairs of $y_k$. Calculated distributions are shown as continuous curves in Figure 10.14. Theoretical calculations match experimental results extremely well. The only region in which they differ is in the vicinity of the zero-power point. As a result of baseline fluctuations of the instrument, the intrinsic fluctuations of the laser cannot be measured accurately.

From RF spectrum and statistical distributions of the output of side mode, we confirm that side modes, containing on the average few percent

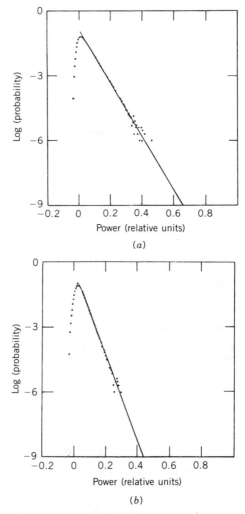

**Figure 10.14.** Measured and calculated photon statistics of the side mode of a semiconductor laser. (*a*) Full bandwidth of the detector. (*b*) With a 250-MHz bandpass filter. The amplitude fluctuations are reduced by using a smaller bandwidth. Solid lines are theoretical results.

of the total output power, behave like chaotic light source, or a laser that is biased below threshold.

Sometimes, the measured distributions of side modes deviate from the exponential distribution. This deviation is more pronounced when the average power of the side mode increases. It starts first with low-probability events. The power level of these low-probability events is more than what the exponential function predicts. This is an indication that the side

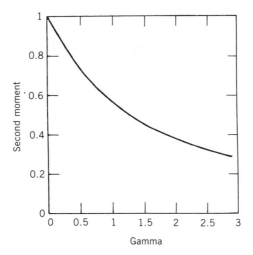

**Figure 10.15.** The second-order moment or the mean-square fluctuations calculated for a Gaussian field with Lorentzian spectrum versus a parameter $\gamma_c$, which is related to the duration of the time window or the bandwidth of the detection circuit.

mode is getting above the threshold. As the power of the side mode increases, the distribution becomes very wide, ranging from zero power to full power.

We measured statistical distributions of individual longitudinal modes of a multimode ridge-waveguide laser. The laser was DC-biased at 3 mW. There are three main modes with average power levels of 62, 27, and 8% as shown in Figure 10.15a. Distributions of these modes are shown in Figure 10.15b. All three modes have distributions extending from 0 to 100% of the total output power.

## 10.7.  TIME-RESOLVED PHOTON STATISTICS OF INDIVIDUAL MODES

Direct modulation is the most convenient method to code information onto the output of semiconductor lasers. Typically, a laser is DC-biased around threshold, a current signal modulates the laser to generate a coded optical beam. The side-mode-suppression ratio at threshold is usually very low. When the laser is driven well above threshold, the SMSR becomes large. During the transient stage, the laser is turned on from threshold to well above threshold [55]. The spectrum of the output of the semiconductor laser evolves from multimode to single-mode.

To study properties of semiconductor laser during the transient stage, we measured mode-resolved statistical distributions. An AlGaAs TJS laser

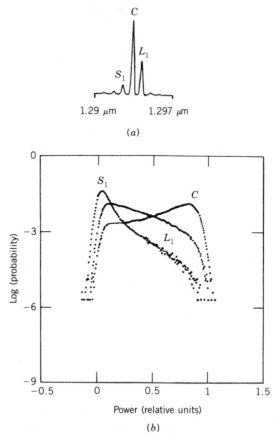

**Figure 10.16.** (*a*) Spectrum and (*b*) mode-resolved photon statistics of a multimode laser. There are three major modes with average power levels of 62, 27, and 8%. Distributions for these modes cover from zero to full power. In other words, one mode can be zero or can have the full output power momentarily.

was biased at threshold, 22 mA, and pumped by a step current with a risetime of 85 ps to provide 3-mW output. The average, time-resolved spectra are shown in Figure 10.16. At 100 ps after the turn-on, there are several longitudinal modes. The laser gradually evolves into a single-dominant-mode condition. The evolution of the dominant mode is shown in Figure 10.17*a*. At 100 ps, this mode is not the dominant mode. It has a statistical distribution corresponding to that of a polarized chaotic light beam. Gradually, it evolves toward the steady state in which this mode is dominant. Even at 6 ns after the turn-on, there still is a large probability to have low output power in this so-called dominant mode. At 8 ns, the laser reaches the steady state. The distribution of the dominant mode

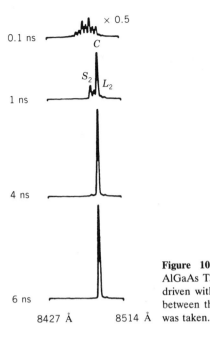

**Figure 10.17.** Time-resolved transient spectra of an AlGaAs TJS laser. The laser was biased at threshold and driven with a step current. The time indicates the delay between the turn-on and the moment when the spectrum was taken.

becomes identical to that of a DC-biased laser shown in Figure 10.10. As shown in Figure 10.16, the second side mode on the shorter-wavelength side of the eventually dominant mode is quite strong at 100 ps after the turn-on. It becomes negligibly small in the steady state. The evolution of its statistical distribution is shown in Figure 10.17b. The distribution at 1 ns is quite broad. There is a substantial probability for this mode to carry a major portion of the total power. The distribution evolves toward that of the chaotic light later on. On the other hand, the side mode on the longer-wavelength side of the eventually dominant mode never builds up. The evolution of its statistical distribution is shown in Figure 10.17c. This mode remains to be a chaotic light.

We also measured transient photon statistics of DFB lasers modulated by random pulses at 2 Gbit/s [56]. Laser A had an average side-mode-suppression-ratio of −30 dB under modulation. It had a SMSR of −9 dB at the bias point. This laser had an error-rate floor in system experiments. The error-rate floor improved by biasing the laser slightly above threshold. Laser B had an average SMSR of −35 dB under modulation. The SMSR at the bias point was −17 dB [38]. There was no observable error-rate floor.

Results of photon statistics measurements for these two lasers under modulation are shown in Figure 10.18. At 100 ps after the turn-on, the side mode of laser A has an extended tail indicating the presence of mode

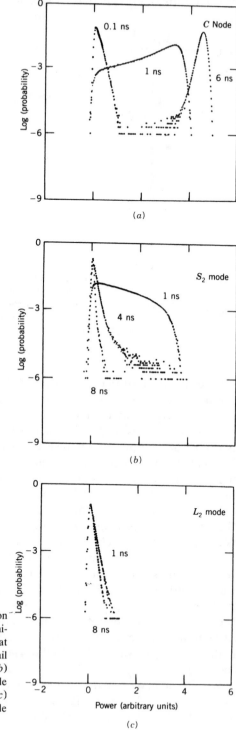

**Figure 10.18.** (*a*) Time-resolved photon statistics of the main mode. The dominant mode builds up from noise. Even at 6 ns after the turn-on, it still has a tail extending to the low-power side. (*b*) Time-resolved photon statistics of a mode on the shorter-wavelength side. (*c*) Time-resolved photon statistics of a mode on the longer-wavelength side.

partition. Laser B, on the other hand, is free from any mode partitioning. The broadening in main-mode distribution for laser A is a result of relaxation oscillations during the turn-on stage. From system experiments as well as photon statistics measurements, we realize that dynamic single-mode laser must pass very stringent tests. An average SMSR of $-30$ dB may not be sufficient.

From results of photon statistics measurements and real-time oscillograms shown in Figure 10.4, we understand the turn-on of semiconductor lasers. Biased below or at threshold, semiconductor lasers have several longitudinal modes. This is true even for injection lasers with high mode selectivity, such as DFB or DBR lasers. The average power levels of these modes are quite comparable. Individually, they all behave like polarized chaotic light, that is, a laser biased below threshold. The instantaneous power in each mode fluctuates. After being driven by a current step, the output starts to build up from noise. The mode that has the highest instantaneous power at the moment of turn-on tends to build up first. The mode selectivity eventually determines which mode is the dominant mode. If the dominant mode happens to be the mode having the highest instantaneous power at the moment of turn-on, it is turned on immediately after the current step. If, however, one of the side modes has the highest instantaneous power at the moment of turn-on, the dominant mode has to build up from a lower power level and to compete with that side mode for gain. The mode selectivity eventually favors the dominant mode. However, during the transient stage, there may be a delay in the turn-on of the dominant mode. Since which mode has the highest instantaneous power fluctuates, the path of evolution for each turn-on event is different.

On the other hand, the total output has far less pulse-to-pulse fluctuations. Although, at the bias point, individual modes fluctuate, the sum of all modes is actually quite stable. This can be understood by considering the central limit theorem of statistics. Side modes have uncorrelated fluctuations. By summing them together, fluctuations are averaged out. The turn-on of the total output is much more reproducible than that of individual modes.

To eliminate fluctuations, we must increase the mode selectivity and stabilize the initial condition. Large mode selectivity can be provided by lasers with built-in filters, such as, DFB or DBR. To stabilize the initial condition, we need a strong SMSR of the order of $-17$ dB. This can be achieved by using lasers with very high mode selectivity or by biasing lasers slightly above threshold or use the injection-locking scheme. By biasing laser above threshold, however, the extinction ratio is reduced. The dispersion penalty due to transient mode partition and the penalty of reduced extinction ratio must be compromised to get the best system performance.

## 10.8.   TURN-ON JITTER OF A SINGLE-MODE LASER

The mode selectivity of semiconductor lasers has been substantially improved over the passed few years. Lasers with a DC side-mode-suppression ratio in the −30- to −45-dB range are readily available. In some system experiments, such as, bit-error-rate measurements, the dispersion penalty is almost entirely eliminated. It is still interesting to know how such lasers perform during the turn-on transient. By understanding the

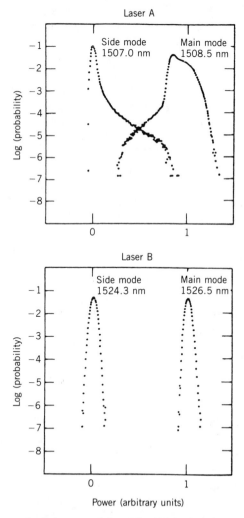

**Figure 10.19.**   Time-resolved photon statistics of two highly single-mode lasers modulated at 2 Gbit/s. Data were taken 100 ps after the turn-on. Laser A, which had a SMSR of −30 dB, still has mode partition. Laser B, which had a SMSR of −35 dB, is free from mode partition.

evolution of the output of such lasers, we may predict performance of future systems.

We studied a highly single-mode 1.3-$\mu$m DFB laser [57]. The laser was mounted in a stripline. The output was monitored by our photon statistics measurement setup. The laser was either biased at $2 \times I_{th}$ and modulated by sinusoidal signal at 1.5 GHz or biased at threshold and modulated with a fast-rising current step. When the modulation depth was slightly less than 100%, the laser had a SMSR of $-32$ dB. By setting the sampling window at the peak of the sinusoidal waveform, we measured the statistical distributions of the side mode and the main mode. Results are shown in Figure 10.19. The distribution of the side mode is actually instrumentation-limited. The distribution of the main mode is indistinguishable from the statistical distribution of the total output of the same laser DC-biased at $2 \times I_{th}$. There is no observable mode partition.

As a more stringent test, the DFB laser was overmodulated so that the output sinusoidal waveform became clipped. The laser was driven to below threshold. The sampling oscilloscope showed a smeared trace during the first 50 ps after the laser was turned on from below threshold. Statistical distributions of the main mode and the side mode measured at the leading edge are shown in Figure 10.20. The distribution of the main mode has a tail extending to the low-power side. However, in contrast to what mode partition normally gives, the side mode does not have any tail extending to the high-power side. In fact, the time-resolved distribution of the total output power also displays a low-power tail.

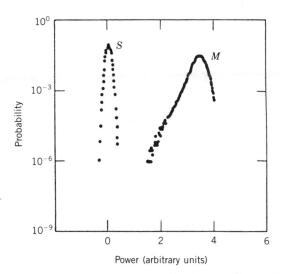

**Figure 10.20.** Time-resolved photon statistics of the side mode and the main mode of a single-mode laser. The tail of the main-mode distribution can be explained by a jitter in turn-on time.

To further clarify the cause of such an extended tail in distribution, we drove the laser with a fast-rising current step. When the laser was biased below or at threshold, we obtained similar distribution with tail extending to the low-power side during the first 50 ps of the turn-on transient. The distribution evolved into one with narrow width but no tail later on.

From our observations, we conclude that the turn-on time of such highly single-mode lasers fluctuates when biased below or at threshold. There is a turn-on jitter of the order of 50 ps. This jitter can be reduced by biasing the laser above threshold. The jitter can be understood by considering the turn-on process. Biased below or at threshold, the output of the laser is essentially a polarized chaotic light source. If a laser has several modes at the bias point, fluctuations of these modes average out. The turn-on process for the total output is highly reproducible. If side modes are very small at bias point, the instantaneous power of the main mode is almost identical to the instantaneous power of the total output. It fluctuates. Depending the instantaneous power at the moment of turn-on, there is a jitter in turn-on time. Mode selectivity alone cannot eliminate turn-on fluctuations entirely. The initial condition must be stabilized.

## 10.9.  MODELING OF LOW-PROBABILITY TURN-ON EVENTS

The dynamics and fluctuations of semiconductor lasers can be modeled by using rate equations with Langevin noise terms [28, 29, 31, 42, 43]:

$$\frac{dn}{dt} = \frac{J}{ed} - \frac{n}{\tau_{sp}} - \frac{c}{\eta n_g}\sum_\nu g_\nu S_\nu + F(t) \qquad (10.30)$$

$$\frac{dS_\nu}{dt} = \frac{\gamma}{\tau_{sp}} D_\nu n + \frac{c}{n_g}(g_\nu - \alpha_\nu)S_\nu + G_\nu(t) \qquad (10.31)$$

with

$$g_\nu = \eta\frac{n_g}{c}A(D_\nu n - n_0)\bigg/\bigg(1 + \sum_\nu S_\nu/\chi\bigg) \qquad (10.32)$$

The notations and their values used in our simulations are:

$\eta$    Mode confinement factor, 0.5
$n$    Carrier density
$n_0$    A parameter determining the threshold
$J$    Current density
$e$    Electronic charge, $1.6 \times 10^{-19}$ C
$d$    Thickness of the active layer, 0.3 $\mu$m
$D$    Width of the laser stripe, $2\mu$m
$\tau_{sp}$    Spontaneous lifetime, 2ns

$c$     Speed of light in vacuum, $3 \times 10^{10}$ cm/s

$n_g$    Group index, 4.0

$S_\nu$    Density of photons in the $\nu$th mode

$g_\nu$    Gain of the $\nu$th mode

$\alpha_\nu$    Loss of the $\nu$th mode

$D_\nu$    Lineshape factor

$\chi$     Saturation parameter, a nominal value of 20 times the steady-state photon density is used

$A$     Stimulated emission factor

$\gamma$     Fraction of spontaneous emission coupled into one mode, $2.2 \times 10^{-5}$

A gain saturation factor, $\chi$, is included in our calculations [58, 59]. It is used to bring relaxation oscillations in line with experimental observations. Here, we have followed the conventional laser theory. For semiconductor lasers, this saturation phenomenon, sometimes is described in terms of a gain compression factor, $k$. Instead of using $(1 + \Sigma_\nu S_\nu/\chi)$ in the denominator, $(1 - k\Sigma_\nu S_\nu)$ in the numerator of Eq. (10.32) is used. Factors $F(t)$ and $G_\nu(t)$ are Langevin noise terms that have an average value of zero. These Langevin noise terms are characterized by correlation functions. They represent the quantum nature of the photon generation and absorption processes. In order to obtain photon statistics for semiconductor lasers, we must solve these coupled rate equations numerically. By counting the number of occurrence of various events, we can obtain the photon statistics.

Two numerical methods have been developed. In the first approach, instead of dealing with Langevin noise terms directly, the evolution of photon packets is considered. For example, a photon packet containing $n$ photons is launched into a semiconductor gain medium. As the photon packet propagates, it may gain more photons through stimulated emission or it may lose photons through absorption. Because of the quantum nature of the absorption and emission processes, the number of photons at the output end may vary. Instead of an average gain factor, a statistical distribution is needed to describe the output. Indeed, there is a possibility that the output may have no photon at all. All the input photons happen to be absorbed by going through the semiconductor gain medium. The output may also have a large number of photons much more than what can be expected from the average gain. The probability distributions for photon packets with few photons, say, $1, 2, 4, \ldots, 64$, can be calculated. Results are stored in computer memory. We can then simulate a semiconductor laser with a large number of photons by using the Monte Carlo method and these pre-calculated results [19]. In the second approach, properly selected mathematical functions are used to represent the Langevin noise terms [31]. Using a suitable random number generator routine, we can simulate semiconductor laser by solving coupled rate equations driven by random noises. Typically, it takes approximately 30 min on the Cray I supercomputer to simulate an injection laser over a time period of 1 $\mu$s. In other words, if the data rate is 2 Gbit/s, we can simulate 2000 turn-on events [31].

Using the above-mentioned methods, computer simulations of semiconductor lasers have been done [60–65]. Photon statistics in the steady-state and during the transient stage has been obtained. Results are in good agreement with experimental data. The only limitation of computer simulations is the size of the ensemble. The number of events is limited. It is unlikely that low-probability events down in the $10^{-9}$ region can be covered in computer simulations.

We have taken a different approach. In this approach, the evolution of photon density during the transient stage is calculated by using rate equations without the Langevin noise terms. However, the initial photon density is not set at its average value [66]. Instead, it is calculated, for a given probability, according to the photon statistics. For example, the statistical distribution of a side mode is given by Eq. (10.23). In other words, the effect of Langevin noise terms is lumped into initial conditions. This assumption is applicable for very-low-probability events. It is highly efficient in simulating low-probability events with large fluctuations.

Since we are interested in nearly single-mode lasers, we have used rate equations with only two modes. When the mode spacing is very small in comparison with the lineshape of the gain curve, we have $D_1 = D_2 = 0.5$, $A = 5.62 \times 10^{-6}$ cm$^3$/s, and $n_0 = 6.8 \times 10^{17}$ cm$^{-3}$. The mode selectivity can be simulated through the loss parameter, $\alpha_\nu$. The loss of the main mode is chosen to be 50.3 cm$^{-1}$. It corresponds to a laser cavity having an internal loss of 20 cm$^{-1}$, an effective reflectivity of 29.8%, and an effective cavity length of 400 $\mu$m. An additional loss for the side mode ranging from 1 to 15 cm$^{-1}$ is used in our simulations.

We assumed that the laser was biased at its threshold. Modulated by a current step of $0.2 \times J_{th}$, the steady-state output power at $\lambda = 1.52$ $\mu$m was calculated to be 1.6 mW per facet. Modulated by $0.5 \times J_{th}$, the output power was 3.9 mW and the relaxation frequency, $f_R$, was 3.4 GHz. Modulated by $1.0 \times J_{th}$, the power was 8.0 mW and $f_R$ was 5.1 GHz. The dynamic quantum efficiency was 31% per facet. These calculated parameters match performance of typical devices, therefore, the values chosen for different parameters are justified.

We have calculated many turn-on transient events by using different laser parameters and driving conditions. Results can be summarized as follows: (1) the transient side-mode suppression ratio can be several orders of magnitude worse than that of the steady state; (2) there is a turn-on jitter for the main mode even when the side mode is absent; and (3) the slower the risetime of the output of the laser, the less is the transient mode partitioning.

Numerical results of specific examples are presented in the following figures. In Figure 10.21, we show that during the transient turn-on,

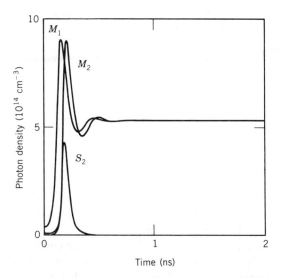

**Figure 10.21.** The evolution of both the main mode ($M_1$ and $M_2$) and the side mode ($S_2$). Curve $M_1$ was obtained with a large main-mode photon density at $t = 0$. Curves $M_2$ and $S_2$ were obtained with a large side-mode photon density at $t = 0$. The side mode was assumed to have an additional loss of 2.5 cm$^{-1}$ in comparison with the main mode. The laser was assumed to be biased at the threshold and driven by a modulation current of $0.5 \times J_{th}$.

although happening with a probability of $10^{-9}$, the side mode may contain a major portion of the total output. The mode selectivity, that is, additional loss of the side mode, used is 2.5 cm$^{-1}$. The modulation current is $0.5 \times J_{th}$. Both the main mode (curves $M_1, M_2$) and the side mode (curve $S_2$) are shown. Curve $M_1$ was calculated by using initial photon densities: $S_m(0) = 3.87 \times 10^{13}$ cm$^{-3}$ and $S_s(0) = 0$. In this case, the side mode is always too small to be seen. Curves $M_2$ and $S_2$ were obtained from initial photon densities: $S_m(0) = 0$ and $S_s(0) = 7.87 \times 10^{12}$ cm$^{-3}$. Indeed, the side mode contains a substantial amount of power during the transient stage. Assuming that all initial photons are in the side mode, we calculated the transient SMSR of these $10^{-9}$ events for different mode selectivities. Results are summarized in Table 10.1. Depending on the mode selectivity, the transient SMSR can sometimes be more than three orders of magnitude worse than that of the steady state.

The turn-on delay between curves $M_1$ and $M_2$ shown in Figure 10.21 persists even when the side mode is absent. In Figure 10.22, we show the evolution of the main mode assuming that the side mode has an additional loss of 15 cm$^{-1}$. Curve $M_1$ was obtained by having all initial photons in the main mode. Curve $M_2$ was obtained when initial photons are all in the negligibly small side mode. In either case, there is no side mode observable during the transient turn-on. Nevertheless, there is a jitter in turn-on time by 37 ps.

**Table 10.1   Transient SMRR of $10^{-9}$ Events for Different Mode Selectivities**

| $\Delta\alpha$ (cm$^{-1}$) | Side-Mode Suppression Ratio | | | | |
|---|---|---|---|---|---|
| | 100 ps | 250 ps | 500 ps | 1 ns | 2 ns |
| 1 | 15.1 | 4.82 | 0.734 | $1.92 \times 10^{-2}$ | $2.3 \times 10^{-3}$ |
| 2.5 | 2.94 | 0.185 | $2.52 \times 10^{-3}$ | $9.1 \times 10^{-4}$ | $9.0 \times 10^{-4}$ |
| 5 | 0.357 | $1.72 \times 10^{-3}$ | $4.5 \times 10^{-4}$ | $4.5 \times 10^{-4}$ | $4.5 \times 10^{-4}$ |

The effect of the saturation parameter can be seen by comparing Figure 10.21 to Figure 10.23. All parameters used are the same except that the saturation parameter is reduced by a factor of 3 in Figure 10.23. The relaxation oscillations are damped. The side mode contains less power. At the same time, the risetime is slowed down. By comparing several DFB lasers with different amounts of relaxation oscillations, it has been found experimentally that lasers with lower relaxation are less subject to transient mode partitioning. Our simulation confirms the experimental observation.

The effect of driving condition is shown in Figure 10.24. All parameters used are the same as that used in Figure 10.21 except the modulation current is set at $0.2 \times J_{th}$. A slower risetime again leads to a smaller side mode.

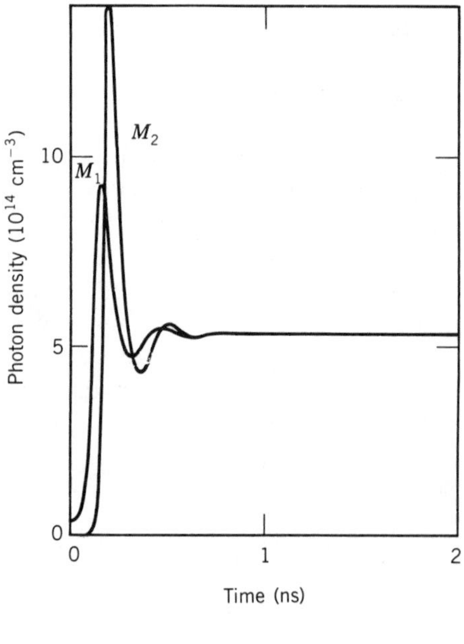

**Figure 10.22.** Evolution of the main mode. Curve $M_1$ represents an event with a large main-mode photon density at $t = 0$. Curve $M_2$ represents an event with a large side-mode photon density at $t = 0$. The side mode is too small to be seen because it has an addition loss of 15 cm$^{-1}$.

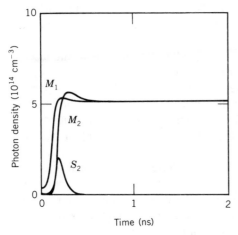

**Figure 10.23.** Evolution of the main mode ($M_1$ and $M_2$) and the side mode ($S_2$). The gain saturation parameter is reduced by a factor of 3. Other conditions are the same as in Figure 10.21.

From this study, we confirm that, biased at or below threshold, lasers behave like noise. The photon densities fluctuate. Such fluctuations in initial photon densities lead to transient mode partitioning and jitter in turn-on time. The SMSR can be orders of magnitude worse than that of the steady state. The turn-on jitter of the main mode always exists even when the side mode is essentially absent. Transient mode partition can be reduced by keeping the laser as close to equilibrium as possible. This can be achieved, for example, by using a current step with a longer risetime.

Besides mode selectivity, the spontaneous emission factor, $\gamma$, is the most crucial parameter in determining the SMSR. In our simulations, the transient mode partition can be eliminated by increasing the mode selective to 10 cm$^{-1}$. Even for those 10$^{-9}$ events, the power in the side mode never reaches beyond 2% of the steady-state power of the main mode.

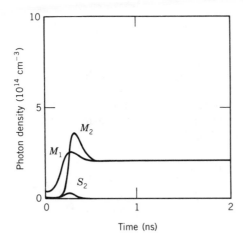

**Figure 10.24.** Evolution of the main mode ($M_1$ and $M_2$) and the side mode ($S_2$). The modulation current is $0.2 \times J_{th}$. Other conditions are the same as in Figure 10.21.

The steady-state SMSR is $-39$ dB. Beyond transient mode partition, eventually system designers will face the turn-on jitter for systems operating in the 10-Gbit/s range.

In order to minimize the dispersion penalty, the DC bias of the laser is raised to above the threshold. With a DC bias above the threshold, the extinction ratio becomes finite. There is a trade-off between the dispersion penalty due to mode partition and the finite extinction ratio. The DC bias usually is optimized empirically. There is still no theoretical analysis that can provide quantitative data on the effect of the bias level to transient mode partition.

Based on photon statistics, we have developed an analytical approach showing why the above threshold bias can eliminate the transient mode partition [67]. It can also provide data on the bias level needed and on the extinction ratio if the mode selectivity of the laser is known. With nearly single-mode lasers in mind, in order to simplify the mathematical procedures, we assume that only one side mode is present.

The side mode behaves like a laser biased below threshold. According to Eq. (10.23), its statistical distribution is given by

$$P_s(n_s) = \frac{1}{\langle n_s \rangle} \exp\left(-\frac{n_s}{\langle n_s \rangle}\right) \tag{10.33}$$

where $n_s$ represents the number of side-mode photons and $\langle n_s \rangle$ represents its average value. The most probable photon number for the side mode is zero. The distribution decays exponentially toward large photon numbers. If the statistical distribution of the total output, $P_0(n_s + n_m)$, where $n_m$ represents the number of main-mode photons, is known we can calculate the statistical distribution of the main mode according to

$$P_m(n_m) = \int_0^\infty P_s(n_s) P_0(n_s + n_m) \, dn_s \tag{10.34}$$

In the simplest case, the total output is assumed to be free of fluctuations; therefore, we have

$$P_0(n_s + n_m) = \delta(n_s + n_m - n_t) \tag{10.35}$$

where $n_t$ is not a variable. It is a number representing the total number of photons. The distribution of the main mode can be easily found by carrying out the integration in Eq. (10.34). We have

$$P_m(n_m) = \begin{cases} \frac{1}{\langle n_s \rangle} \exp\left[-\left(\frac{n_t - n_m}{\langle n_s \rangle}\right)\right], & n_m < n_t \\ 0, & n_m > n_t \end{cases} \tag{10.36}$$

The main mode also has an exponential distribution. The most probable photon number is $n_t$. It decays exponentially on the low-photon number side with the same decay rate as the side mode.

In fact, the number of photons in the total output does fluctuate. Its distribution as derived from the Fokker–Planck equation is given by

$$P_0(n_m + n_s) = N_0 \exp\left[-\frac{\beta}{2n_{sp}}(n_m + n_s) \cdot \left(n_m + n_s - \frac{2Rn_{sp}}{\beta}\right)\right]$$

(10.37)

with $\beta$ given by

$$\beta = \frac{A\tau_{sp}}{V_0}$$

(10.38)

where $\tau_{sp}$ is the spontaneous lifetime of carriers, $A$ is the stimulated emission factor, and $V_0$ is the volume of the optical mode. It is related to the volume of the active medium, $V_e$, and mode confinement factor, $\eta$, by $V_0 = V_e/\eta$. The term $R$ in Eq. (10.37) represents how high above threshold the bias is, specifically, $R = I/I_{th} - 1$. The term $n_{sp}$ is the population inversion parameter and $N_0$ is the normalization constant. It is given by

$$N_0^{-1} = \sqrt{\pi}\,\kappa[1 + \text{erf}(\kappa R)]\exp(\kappa^2 R^2)$$

(10.39)

where erf represents the error function and $\kappa$ is defined as

$$\kappa = \sqrt{\frac{n_{sp}}{2\beta}}$$

(10.40)

The population inversion parameter is related to several parameters commonly used in the rate equations by

$$n_{sp} = 1 + A\eta n_0 \tau_p$$

(10.41)

where $n_0$ is a parameter determining the threshold and $\tau_p$ is the photon lifetime. By using Eq. (10.37) and carrying out the integration in Eq. (10.34), we obtain

$$P_m(n_m) = \frac{N_0\sqrt{\pi}}{\langle n_s\rangle 2\sqrt{U}} \exp\left(\frac{n_m}{\langle n_s\rangle}\right)\exp\left(\frac{V^2}{4U}\right)\text{erfc}\left[\sqrt{U}\left(n_m - \frac{V}{2U}\right)\right]$$

(10.42)

where

$$U = \frac{\beta}{2n_{\text{sp}}} \qquad (10.43)$$

$$V = R - \frac{1}{\langle n_{\text{s}} \rangle} \qquad (10.44)$$

where the function erfc represents the complementary error function.

The Fokker–Planck approach used to derive Eq. (10.37) is based on the van der Pol equation. Instead of handling the depletion of carriers dynamically like the rate equations, the gain saturation is assumed to be instantaneous. Nevertheless, the average steady-state photon numbers calculated from both approaches usually match with each other.

From the bias condition and numerical values of various parameters, we can determine the average number of photons in the side mode from the steady-state rate equations. We can then determine the statistical distribution of the side mode and of the main mode according to Eqs. (10.33) and (10.42), respectively. The side mode is assumed to have a moderate, additional loss of 0.5 cm$^{-1}$. The average number of photons in the side mode and in the main mode can be calculated from the bias condition. At a bias of $1.05 \times I_{\text{th}}$, we have $\langle n_{\text{s}} \rangle = 1096$ and $\langle n_{\text{m}} \rangle = 2.44 \times 10^4$. At a bias of $1.1 \times I_{\text{th}}$, we have $\langle n_{\text{s}} \rangle = 1123$ and $\langle n_{\text{m}} \rangle = 4.98 \times 10^4$. The distribution of the main mode can be calculated according to Eq. (10.42); $\tau_{\text{p}}$ can be calculated from values of other parameters used in rate equations. It is $2.65 \times 10^{-12}$ s.

Distributions of the side mode and the main mode for two different bias levels, $1.05 \times I_{\text{th}}$ and $1.1 \times I_{\text{th}}$, are shown in Figure 10.25. At the higher bias, the peak of the main-mode distribution shifts toward a larger photon number. Its tail on the low-photon-number side is also exponential in nature. It comes from the $\exp(n_{\text{m}}/\langle n_{\text{s}} \rangle)$ factor in Eq. (10.42). In fact, for low-probability events that we are interested in, the simpler formula, Eq. (10.36), gives essentially identical results. The decay in probability on the high-photon-number side is governed by the erfc function. Since we used an integration routine with a finite accuracy to calculate the error function, only a portion of the distribution on the high-photon-number side is calculated. All distributions are normalized in the photon-number representation. The most probable event for the side mode happens with a probability of $1/\langle n_{\text{s}} \rangle$, or approximately $10^{-3}$. In the following discussions, we consider events happening with a relative probability of $10^{-9}$ in comparison with the most probable event.

At the higher bias, as shown in Figure 10.25$b$, the number of photons in the main mode is always larger than the number of photons in the side mode all the way from the peak of the distribution to $10^{-9}$ below the peak. On the other hand, at the lower bias as shown in Figure 10.25$a$, the

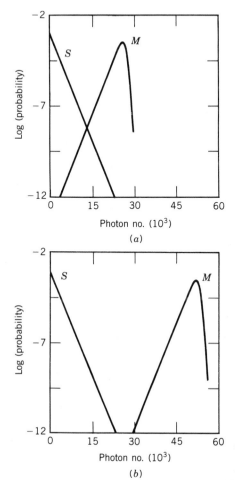

**Figure 10.25.** Statistical distributions of the main mode ($M$) and the side mode ($S$). (a) Biased at $1.05 \times I_{th}$, (b) Biased at $1.1 \times I_{th}$. Other parameters are cited in the text.

side mode may have more photons than the main mode for certain low-probability events. In fact, with a probability of $10^{-9}$ down from the peak of the distribution, the side-mode photon number can be 6.46 times the main-mode photon number.

The evolution of photon densities for the side mode and for the main mode was calculated from rate equations using initial photon densities determined from the statistical distributions. Photon densities were calculated from the photon numbers and the optical mode volume. We assumed that a current step was applied at time zero to drive the laser from the bias level to $1.5 \times I_{th}$. A gain saturation parameter at 20 times the steady-state photon density was used to bring the relaxation oscillations in line with experimental observations. Results are shown in Figure 10.26. At the lower bias, with a probability of $10^{-9}$, the side mode can dominate

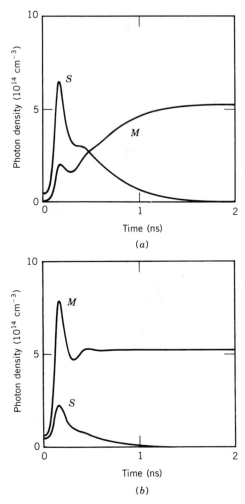

**Figure 10.26.** Evolution of the main mode ($M$) and the side mode ($S$). The initial conditions are determined from the statistical distributions. ($a$) A turn-on event with a probability of $10^{-9}$ down from the peak of the distribution. The bias was assumed to be $1.05 \times I_{th}$. ($b$) Same as ($a$) except that the bias was assumed to be $1.1 \times I_{th}$.

over the main mode during the transient stage as shown in Figure 10.26$a$. Transient mode partition is present at such a bias level. On the other hand, at the higher bias, the side mode is always substantially smaller than the main mode. Transient mode partition is avoided at the higher bias level. The steady-state SMSR is determined from the mode selectivity and how far above threshold the laser is driven. The steady-state SMSR calculated from values of parameters cited earlier in the chapter is $-23$

dB. Assuming that the higher biased level is used, therefore, the extinction ratio is 6.8 dB.

Transient mode partition can be eliminated by either using lasers with very large mode selectivity or by biasing lasers above threshold. In our approach, quantitative data on SMSR during the transient stage can be calculated for any given low-probability event. As long as the initial side-mode photon number is smaller than the initial main-mode photon number, the main mode will always dominate even during the transient stage. No mode partition penalty is expected. For a moderate mode selectivity (e.g., 0.5 cm$^{-1}$), transient mode partition can be avoided by biasing the laser at $1.1 \times I_{th}$. From the amplitude of the modulating signal, the extinction ratio can also be calculated.

If we neglect fluctuations of the total output, then a simple rule can be used in determining the bias level. For low probability events, the statistical distribution of the main mode is essentially an exponential distribution with a decay factor identical to that of the side mode. With a probability of $10^{-9}$ below the peak, the side mode may have 20.7 times its average number of photons. To ensure that the starting number of main-mode photons is always greater than that of the side mode even for those $10^{-9}$ events, a bias that can provide a main : side-mode ratio of 41.4, that is, an SMSR of $-16$ dB, is needed. From this rule one can easily determine the bias level for any given mode selectivity. For example, with a mode selectivity of 1 cm$^{-1}$, the bias would be $1.05 \times I_{th}$. If the on-state is at $1.5 \times I_{th}$, the extinction ratio becomes 10 dB.

## 10.10. CONCLUSION

We have characterized the amplitude fluctuations of semiconductor lasers by using statistical distributions, specifically, photon statistics. Such fluctuations appear both in the total output and in the output of individual modes. From detailed experimental measurements and theoretical modeling, we know that the quantum nature of the photon generation and absorption processes is the cause of these fluctuations. Biased at or below threshold, semiconductor lasers behave like a noise source. Biased above threshold, the total output is stabilized by carriers. However, the power of individual longitudinal modes can still fluctuate. A side mode may contain a large portion of the total output momentarily at the expense of the main mode. Because of such fluctuations and the fact that photon–carrier interaction is dynamic rather than instantaneous, the amplitude fluctuations of the total output are quite large. We can describe the total output of a semiconductor laser as the superposition of a coherent electromagnetic field and narrowband Gaussian noise. Amplitude fluctuations can be reduced by increasing the output power, by using a composite cavity, and by operating in single mode. The amplitude fluctuations of the total output may affect the performance of externally modulated semiconductor lasers.

When a current pulse is applied to drive a laser from the bias point to above threshold, the output builds up. Biased at or below threshold, the laser behaves like noise. The output must evolve from noise toward the steady state. Because the initial conditions fluctuate from pulse to pulse, a side mode sometimes may dominate during the transient stage. In the steady state, because of mode selectivity, the main mode dominates.

To avoid the transient mode partition, we must use semiconductor lasers with very large mode selectivity. From our experiences, the SMSR under modulation shall be better than $-32$ dB. The bias may also have to be raised to slightly above threshold. Raising the bias can stabilize the initial conditions, therefore, can ensure that the side mode is always small in comparison with the main mode. A simple rule can be applied to determine the bias level; namely, we need a SMSR of $-16$ dB or better at the bias level. With a truly single-mode laser, there may still be fluctuations in the turn-on time. Such a turn-on jitter, of the order of 50 ps, may eventually affect system performance at very high data rates.

Amplitude fluctuations of semiconductor lasers can be simulated by solving rate equations using the Monte Carlo method or by including Langevin noise terms in simulations. Alternatively, we can use initial conditions determined by photon statistics to study the behavior of very-low-probability events. Results are in good agreement with experimental data.

In addition to providing information on the cause of mode partition and amplitude fluctuations, photon statistics or modeling of the laser can be used to predict system performance [68–71].

## ACKNOWLEDGMENTS

Some of the results presented here were obtained while I was at Bell Telephone Laboratories and Bell Communications Research, Inc. I would like to acknowledge my collaborators, including C. A. Burrus, Jr., M. M. Choy, G. Einstein, I. P. Kaminow, J. S. Ko, T. P. Lee, and K. Ogawa. I would also like to acknowledge J. A. Copeland, E. A. J. Marcatili, and S. E. Miller for stimulating discussions. Research carried out in University at Buffalo is supported by the National Science Foundation.

## REFERENCES

1. B. E. A. Saleh, *Photoelectron Statistics*, Springer-Verlag, Berlin, Heidelberg, 1978.
2. C. Freed and H. A. Haus, *IEEE J. Quantum Electron.* **QE-2**, 190 (1966).

3. W. A. Smith, *Opt. Commun.* **12**, 236 (1974).

4. M. Lax and W. H. Louisell, *IEEE J. Quantum Electron.* **QE-3**, 47 (1967).

5. M. Lax and W. H. Louisell, *Phys. Rev.* **185**, 568 (1969).

6. E. R. Pike, "Some problems in the statistics of optical fields," in *Quantum Optics: Proceedings of the International School of Physics*, "*Enrico Fermi*," R. J. Glauber, ed., Academic Press, New York, 1969.

7. J. Perina, *Coherence of Light*, Van Nostrand Reinhold, London, 1971.

8. R. J. Glauber, "Photon statistics"; H. Haken, "The theory of coherence, noise, and photon statistics of laser light"; F. T. Arecchi and V. Degiorgio, "Measurement of the statistical properties of optical fields," in *Laser Handbook*, Vol. 1, F. Arecchi and E. O. Schulz-Dubois, eds., North-Holland, Amsterdam, 1972.

9. Murray Sargent III, Marlan O. Scully, and Willis E. Lamb, Jr., *Laser Physics*, Addison-Wesley, Reading, MA, 1974.

10. D. E. McCumber, *Phys. Rev.*, **141**, 306 (1966).

11. J. A. Armstrong and A. W. Smith, "Experimental studies of intensity fluctuations in lasers," in *Progress in Optics*, Vol. VI, E. Wolf, ed., North-Holland, Amsterdam, 1967.

12. C. H. Henry, *J. Lightwave Technol.* **LT-4**, 298 (1986).

13. Y. Okano, K. Nakagawa, and T. Ito, *IEEE Trans. Commun.* **COM-28**, 238 (1980).

14. K. Ogawa, *IEEE J. Quantum Electron.* **QE-18**, 849 (1982).

15. K. Ogawa and R. W. Vodhanel, *IEEE J. Quantum Electron.* **QE-18**, 1090 (1982).

16. S. Yamamoto, H. Sakaguchi, and N. Seki, *IEEE J. Quantum Electron.* **QE-18**, 264 (1982).

17. H. Jäckel and G. Guekos, *Opt. Quantum Electron.* **9**, 233 (1977).

18. T. Ito, S. Machida, K. Nawata, and T. Ikegami, *IEEE J. Quantum Electron.* **QE-13**, 574 (1977).

19. J. A. Copeland, *J. Appl. Phys.* **54**, 2813 (1983).

20. T. Sueta and M. Izutsu, *J. Opt. Commun.* **3**, 53 (1982).

21. P.-L. Liu, L. E. Fencil, J.-S. Ko, I. P. Kaminow, T. P. Lee, and C. A. Burrus, Jr., *IEEE J. Quantum Electron.* **QE-19**, 1348 (1983).

22. P. L. Liu and K. Ogawa, *J. Lightwave Technol.* **LT-2**, 44 (1984).

23. Pao-Lo Liu, *J. Lightwave Technol.* **LT-3**, 205 (1985).

24. I. D. Henning, *Electron Lett.* **18**, 368 (1982).

25. I. M. Joindot and C. Y. Boisrobert, *IEEE J. Quantum Electron.* **QE-23**, 1059 (1987).

26. R. A. Linke, B. L. Kasper, C. A. Burrus, Jr., I. P. Kaminow, J.-S. Ko, and T. P. Lee, *J. Lightwave Technol.* **LT-3**, 1303 (1985).

27. G. L. Abbas and T. K. Yee, *IEEE J. Quantum Electron.* **QE-21**, 1303 (1985).

28. Y. Yamamoto, *IEEE J. Quantum Electron.* **QE-19**, 34 (1983).

29. Y. Yamamoto, *IEEE J. Quantum Electron.* **QE-19**, 47 (1983).

30. S. E. Miller and D. Marcuse, *IEEE J. Quantum Electron.* **QE-20**, 1032 (1984).

31. D. Marcuse, *IEEE J. Quantum Electron*. **QE-20**, 1139 (1984).

32. C. H. Henry, P. S. Henry, and M. Lax, *J. Lightwave Technol*. **LT-2**, 209 (1984).

33. A. H. Gnauck, C. A. Burrus, Jr., S.-J. Wang, and N. K. Dutta, *Electron. Lett*. **25**, 1356 (1989).

34. J. W. M. Beisterbos and H. W. M. Salemink, *Electron. Lett*. **18**, 300 (1982).

35. I. D. Henning and D. A. Frisch, *J. Lightwave Technol*. **LT-1**, 202 (1983).

36. P. L. Liu, T. P. Lee, C. A. Burrus, Jr., I. P. Kaminow, and J.-S. Ko, *Electron. Lett*. **18**, 904 (1982).

37. M. Nakamura, K. Aiki, N. Chinone, R. Ito, and J. Umeda, *J. Appl. Phys*. **49**, 4644 (1978).

38. S. Sasaki, M. M. Choy, and N. K. Cheung, *Electron. Lett*. **24**, 26 (1988).

39. I. P. Kaminow, L. W. Stulz, J.-S. Ko, A. G. Dentai, R. E. Nahory, J. C. DeWinter, and R. L. Hartman, *IEEE J. Quantum Electron*. **QE-19**, 1312 (1983).

40. T. P. Lee, C. A. Burrus, Jr., J. A. Copeland, A. G. Dentai, and D. Marcuse, *IEEE J. Quantum Electron*. **QE-18**, 1101 (1982).

41. T. P. Lee, C. A. Burrus, Jr., K. Ogawa, and A. G. Dentai, *Electron. Lett*. **17**, 431 (1981).

42. H. Haug, *Z. Phys*. **200**, 57 (1967).

43. H. Haug, *Phys. Rev*. **184**, 338 (1969).

44. O. Hirota and Y. Suematsu, *Trans. IECE Japan*, **65**, 94 (1982).

45. S. O. Rice, *Bell Syst. Tech. J*. **24**, 46 (1945).

46. P. L. Liu, G. Eisenstein, R. S. Tucker, and I. P. Kaminow, *Appl. Phys. Lett*. **44**, 481 (1984).

47. K. Iwashita and K. Nakagawa, *IEEE J. Quantum. Electron*. **QE-18**, 2000 (1982).

48. T. M. Shen and G. P. Agrawal, *Electron. Lett*. **21**, 1220 (1985).

49. S. Yamamoto, H. Sakaguchi, and N. Seki, *J. Lightwave Technol*. **LT-4**, 672 (1986).

50. E. E. Basch, R. F. Kearns, and T. G. Brown, *J. Lightwave Technol*. **LT-4**, 516 (1986).

51. W. R. Throssell, *J. Lightwave Technol*. **LT-4**, 948 (1986).

52. N. H. Jensen, H. Olesen, and K. E. Stubkjaer, *IEEE J. Quantum Electron*. **QE-23**, 71 (1987).

53. K.-Y. Liou, M. Ohtsu, C. A. Burrus, Jr., U. Koren, T. L. Koch, *J. Lightwave Technol*. **LT-7**, 632 (1989).

54. M. Nagano and K. Kasahara, *IEEE J. Quantum Electron*. **QE-13**, 632 (1977).

55. M. Sargent III, M. O. Scully, and W. E. Lamb, Jr., *Appl. Opt*. **9**, 2423 (1970).

56. M. M. Choy, P. L. Liu, and S. Sasaki, *Appl. Phys. Lett*. **52**, 1762 (1988).

57. M. M. Choy, P. L. Liu, P. W. Shumate, and T. P. Lee, *Appl. Phys. Lett*. **47**, 448 (1985).

58. D. J. Channin, *J. Appl. Phys*. **50**, 3858 (1979).

59. R. S. Tucker and D. J. Pope, *IEEE J. Quantum Electron*. **QE-19**, 1179 (1983).

60. S. Miller, *IEEE J. Quantum Electron.* **QE-21**, 1644 (1985).

61. S. E. Miller, *IEEE J. Quantum Electron.* **QE-22**, 16 (1986).

62. S. E. Miller, *J. Quantum Electron.* **QE-23**, 1071 (1987).

63. D. Marcuse, *IEEE J. Quantum Electron.* **QE-20**, 1148 (1984).

64. D. Marcuse, *IEEE J. Quantum Electron.* **QE-21**, 154 (1985).

65. D. Marcuse, *IEEE J. Quantum Electron.* **QE-21**, 161 (1985).

66. P. L. Liu and M. M. Choy, *IEEE J. Quantum Electron.* **QE-25**, 1767 (1989).

67. P. L. Liu and M. M. Choy, *IEEE J. Quantum Electron.* **QE-25**, 854 (1989).

68. B. Hellstrom, *J. Opt. Commun.* **6**, 132 (1985).

69. J. C. Campbell, *J. Lightwave Technol.* **LT-6**, 564 (1988).

70. G. P. Agrawal, P. J. Anthony, and T. M. Shen, *J. Lightwave Technol.* **LT-6**, 620 (1988).

71. J. C. Cartledge, *J. Lightwave Technol.* **LT-6**, 626 (1988).

# 11

# Squeezed-State Generation by Semiconductor Lasers

Y. YAMAMOTO AND S. MACHIDA
*NTT Basic Research Laboratories, Nippon Telegraph and Telephone Corporation, Musashino-shi, Tokyo, Japan*

OLLE NILSSON
*Department of Microwave Engineering, The Royal Institute of Technology, Stockholm, Sweden*

## 11.1. INTRODUCTION

Quantum-statistical properties of laser light have been extensively studied for the last 30 years by using an operator Langevin equation, a density operator master equation, and a quantum-mechanical Fokker–Planck equation. It was generally accepted among physicists and quantum electronics engineers that an ideal laser operating at far above the threshold generates a coherent state of light. Various experimental facts such as the Poissonian photoelectron statistics, the shot-noise-limited photocurrent fluctuations, and the Gaussian distributions of optical homodyne detector output seem to support this. However, recent careful studies on a semiconductor laser have revealed that there are fundamental errors in the conventional reservoir theory and that a semiconductor laser does not necessarily produce a coherent state of light [1, 2]. Instead, a semiconductor laser generates a number-phase squeezed state, in which the photon number noise is smaller than the standard quantum limit (shot-noise limit or Poisson limit). This chapter reviews the theoretical and experimental aspects of number-phase squeezed-state generation by a semiconductor laser.

*Coherence, Amplification, and Quantum Effects in Semiconductor Laser*, Edited by Yoshihisa Yamamoto.
ISBN 0-471-512494 © 1991 John Wiley & Sons, Inc.

The chapter is organized as follows. In Section 11.2, a squeezed state of light is briefly reviewed. There are the two kinds of squeezed states, a quadrature amplitude squeezed state and a number-phase squeezed state. The properties and generation schemes of the squeezed states are discussed. Section 11.3 introduces the quantum-mechanical reservoir theory of a semiconductor laser and demonstrates the breakdown of the conventional theory. It is shown that the convention theory does not describe the external output field and does not take proper account of the pump noise. The quantum-mechanically self-consistent new theory predicts that a pump-noise-suppressed laser generates a number-phase squeezed state. The theory of squeezed-state generation by a semiconductor laser is presented in Section 11.4. The microscopic theory of the carrier transport processes in a $p$–$n$ junction is developed to demonstrate the principle of high-impedance suppression of the pump current noise. The noise-equivalent circuit is presented to give a physical insight into this new physics. The experimental evidence for squeezed state generation is finally delineated in Section 11.5.

## 11.2. BRIEF REVIEW OF A SQUEEZED STATE OF LIGHT

The Heisenberg uncertainty principle for two conjugate observables $\hat{O}_1$ and $\hat{O}_2$ is the direct consequence of the commutation relation between them, as given by

$$[\hat{O}_1, \hat{O}_2] = i\hat{O}_3 \tag{11.1}$$

The minimum uncertainty product is readily obtained by applying the Schwartz inequality [3]

$$\langle \Delta\hat{O}_1^2 \rangle \langle \Delta\hat{O}_2^2 \rangle \geq \tfrac{1}{4}\langle \hat{O}_3 \rangle^2 \tag{11.2}$$

A minimum uncertainty state that satisfies the equality in Eq. (11.2) can be defined as an eigenstate of an operator [4]

$$\hat{O}(r) = e^r\hat{O}_1 + ie^{-r}\hat{O}_2 \tag{11.3}$$

A parameter $r$ determines the distribution of quantum uncertainty (noise) between $\hat{O}_1$ and $\hat{O}_2$. This can be represented by

$$\langle \Delta\hat{O}_1^2 \rangle = \langle \Delta\hat{O}_1^2 \rangle_{r=0} \, e^{-2r}$$
$$\langle \Delta\hat{O}_2^2 \rangle = \langle \Delta\hat{O}_2^2 \rangle_{r=0} \, e^{2r} \tag{11.4}$$

A quadrature amplitude squeezed state of light [5–7] is one of the minimum uncertainty states, in which $\hat{O}_1$ and $\hat{O}_2$ correspond to the two quadrature ($\cos \omega t$ and $\sin \omega t$) components $\hat{a}_1$ and $\hat{a}_2$ of an electromagnetic wave; $\hat{O}_3$ is then a $c$ number ($= \frac{1}{2}$). In the special case of $r = 0$, $\hat{O}(r)$ becomes a photon annihilation operator $\hat{a} = \hat{a}_1 + i\hat{a}_2$, and the two quadrature components share the same amount of quantum noise, i.e., $\langle \Delta \hat{a}_1^2 \rangle = \langle \Delta \hat{a}_2^2 \rangle = \frac{1}{4}$. This is a coherent state [8]. The specialty of a coherent state among the general family of a quadrature amplitude squeezed state lies in the fact that a vacuum state $\hat{c}$ is in a coherent state with equal amounts of quantum mechanical zero-point fluctuations $\langle \Delta \hat{c}_1^2 \rangle = \langle \Delta \hat{c}_2^2 \rangle = \frac{1}{4}$. Therefore, a coherent state preserves its quantum statistical properties when it encounters a loss process. This fact makes a coherent state a unique and special quantum state of light.

A quadrature amplitude squeezed state features less fluctuation than a coherent state, if one quadrature component is measured by an optical homodyne detector. A signal-to-noise ratio for a given photon number is improved by using a squeezed state. However, the quantum noise of one quadrature cannot be reduced to zero if the photon number is finite, because the increased quantum noise of the other quadrature requires the infinite photon number. There exists an optimum squeezing parameter $r$, which maximizes the signal-to-noise ratio for a given photon number $\langle \hat{n} \rangle$ [9]. The signal-to-noise ratio achieved by the optimum squeezed state is

$$\left( \frac{S}{N} \right)_{ss} = 4\langle \hat{n} \rangle (\langle \hat{n} \rangle + 1) \tag{11.5}$$

which is larger than that achieved by coherent state $(S/N)_{cs} = 4\langle \hat{n} \rangle$.

A quadrature amplitude squeezed state is generated by a phase-sensitive amplification–deamplification process in various nonlinear optical interactions [10]. A four-wave mixer and degenerate parametric amplifier are the two most successful schemes for generating a quadrature amplitude squeezed state [11].

A number-phase squeezed state [12, 13] is also one of the minimum uncertainty states, in which $\hat{O}_1$ and $\hat{O}_2$ correspond to the respective photon number and the phase of an electromagnetic wave. Quantum-mechanical definition of a phase is, however, not unique. One mathematical definition of a number-phase squeezed state is based on [14]

$$\hat{O}_1 = \hat{n} \tag{11.6}$$

$$\hat{O}_2 = \hat{S} = \frac{1}{2i} \left[ (\hat{n} + 1)^{-1/2} \hat{a} - \hat{a}^\dagger (\hat{n} + 1)^{-1/2} \right] \tag{11.7}$$

and

$$\hat{O}_3 = \hat{C} = \frac{1}{2} \left[ (\hat{n} + 1)^{-1/2} \hat{a} + \hat{a}^\dagger (\hat{n} + 1)^{-1/2} \right] \tag{11.8}$$

When a squeezing parameter $r$ is equal to $-\frac{1}{2}\ln(2\langle\hat{n}\rangle)$, the photon number noise is equal to that of coherent state $\langle\Delta\hat{n}^2\rangle = \langle\hat{n}\rangle$. It is possible to reduce the photon-number noise to below this coherent state limit by accepting an increase of phase noise. The photon-number noise can ultimately be reduced to zero without requiring an infinite photon number because the enhanced phase noise does not require any photon (energy) at all. This is a photon-number eigenstate. Even though the quantum noise of a photon-number state is zero, the photon number takes only a discrete value. Therefore, the information communicated by a photon-number state having a given average photon number is not infinite. Nevertheless, a photon-number state realizes the maximum channel capacity in optical communications [15].

A number-phase squeezed state is generated by a photon-number-dependent phase shift (i.e., quantum phase diffusion) in a third-order nonlinear medium [13]. However, the most promising scheme for generating a number-phase squeezed state is a constant current-driven semiconductor laser [1, 2, 16–18] which is the immediate subject of this chapter.

In Figure 11.1, noise distributions ($Q(\alpha)$ function) in the $a_1$–$a_2$ phase space of a coherent state, quadrature amplitude squeezed state, quadrature amplitude eigenstate, number-phase squeezed state, and photon-number eigenstate are plotted schematically [13].

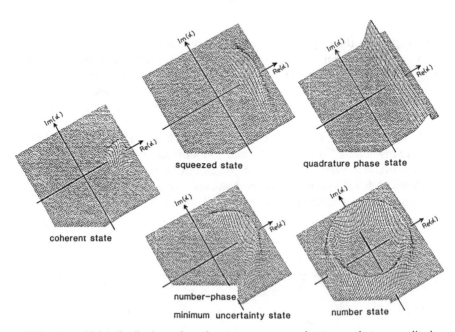

**Figure 11.1.** Noise distributions of a coherent state, squeezed state, quadrature amplitude eigenstate, number-phase squeezed state, and photon-number eigenstate.

## 11.3. RESERVOIR THEORY OF A SEMICONDUCTOR LASER

A laser is a nonequilibrium open system, in which the macroscopic coherence of light and matter is established by the subtle balance between a system ordering force (gain saturation) and fluctuating forces from external reservoirs. A conventional reservoir theory of a laser [19–21] treats only three degrees of freedom as "systems": a photon field inside a cavity, population inversion, and a dipole moment. The dipole moment couples the photon field and the population inversion, which serves as the system ordering force to establish the macroscopic coherence. All the other degrees of freedom are considered as "reservoirs," which are eliminated by use of the (quantum-mechanical) fluctuation–dissipation theorem. The photon field inside a laser cavity leaks into an external open space via an output coupling mirror. As a reverse process, the vacuum field fluctuation in an external space penetrates into the cavity and disturbs the internal photon field. The population inversion is usually produced by an external pump laser (or incandescent light source), which carries the shot-noise-limited fluctuation in most cases. The population inversion randomly decays by spontaneous emission, which is the consequence of vacuum field fluctuations surrounding the excited atom. The dipole moment randomly decays (loses its phase memory) due to collisions with electrons, holes, phonons, and impurity atoms. These collisions are inherently multibody problems, but the reservoir theory simplifies the problems by separating the system from the reservoirs.

Because these reservoirs usually consist of many degrees of freedom with continuous spectra, they do not have any memory effects in a time scale during which the above-mentioned three system parameters change appreciably. This approximation is called a *Markoffian process*, and it

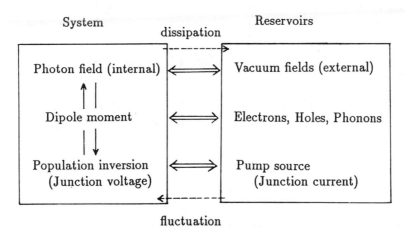

**Figure 11.2.** Model of a laser in the conventional reservoir theory.

holds for most of laser systems. Consequently, the fluctuation intensity from the reservoirs is the only important parameter, and it is proportional to the rate of dissipation (i.e., rate-in and rate-out). Figure 11.2 schematically shows the relation between the three system parameters and the associated reservoirs for a laser oscillator.

### 11.3.1.  Quantum-Mechanical Langevin Equations

The Heisenberg equations of motion for the three system operators are obtained directly from the interaction Hamiltonians between the system operators and between the system and reservoir operators [19–21].

The cavity internal field operator $\hat{A}$ is described by the equation [Eqs. (11.28)–(11.40)] of Ref. 21:

$$\frac{d}{dt}\hat{A} = -\frac{\gamma}{2}\hat{A} - igN\hat{\Sigma} + \sqrt{\gamma}\,\hat{S}_A \tag{11.9}$$

where $\gamma$ is the photon decay rate, $g$ is a gain factor proportional to the matrix element of the dipole moment, $N$ is the total number of atoms, $\hat{\Sigma}$ is the dipole moment operator, and $\hat{S}_A$ is the (reservoir) vacuum field fluctuation operator incident on the output coupling mirror. Operator $\hat{S}_A$ is normalized such that $\langle \hat{S}_A^\dagger \hat{S}_A \rangle$ represents the average photon flux incident on the laser cavity. The equation for the dipole moment operator is

$$\frac{d}{dt}\hat{\Sigma} = -\Gamma\hat{\Sigma} + ig\hat{\sigma}\hat{A} + \sqrt{\frac{2\Gamma}{N}}\,\hat{S}_\Sigma \tag{11.10}$$

where $\hat{\sigma}$ is the population inversion operator, $\Gamma$ is the dipole moment decay rate, and $\hat{S}_\Sigma$ is the (reservoir) noise operator, which is normalized to have the same correlation function as $\hat{S}_A$. The equation for the population inversion operator is

$$\frac{d}{dt}\hat{\sigma} = \Lambda - \frac{\hat{\sigma}}{\tau_{sp}} + 2ig(\hat{A}^\dagger\hat{\Sigma} - \hat{\Sigma}^\dagger\hat{A}) + \hat{S}_\sigma \tag{11.11}$$

where $\Lambda$ is the pump rate per atom, $\tau_{sp}$ is the spontaneous lifetime, and $\hat{S}_\sigma$ is the (reservoir) noise operator; $\hat{S}_\sigma$ is split into the two independent noise operators

$$\hat{S}_\sigma = \sqrt{\frac{\Lambda}{N}}\,\hat{S}_\Lambda + \sqrt{\frac{\langle\hat{\sigma}\rangle}{N\tau_{sp}}}\,\hat{S}_\tau \tag{11.12}$$

where $\hat{S}_\Lambda$ and $\hat{S}_\tau$ are associated with the pump and spontaneous decay processes, respectively.

To calculate the noise spectrum, we must find the correlation functions for noise operators $\hat{S}_A$, $\hat{S}_\Sigma$, $\hat{S}_\Lambda$, and $\hat{S}_\tau$. In the Markoffian approximation mentioned above, the correlation functions are the $\delta$ functions proportional to rate-in and rate-out:

$$\langle \hat{S}_{A1}(t)\hat{S}_{A1}(t+\tau)\rangle = \langle \hat{S}_{A2}(t)\hat{S}_{A2}(t+\tau)\rangle = \tfrac{1}{4}\delta(\tau) \quad (11.13)$$

$$\langle \hat{S}_{\Sigma1}(t)\hat{S}_{\Sigma1}(t+\tau)\rangle = \langle \hat{S}_{\Sigma2}(t)\hat{S}_{\Sigma2}(t+\tau)\rangle = \tfrac{1}{4}\delta(\tau) \quad (11.14)$$

$$\langle \hat{S}_\Lambda(t)\hat{S}_\Lambda(t+\tau)\rangle = \delta(\tau) \quad (11.15)$$

and

$$\langle \hat{S}_\tau(t)\hat{S}_\tau(t+\tau)\rangle = \delta(\tau) \quad (11.16)$$

Subscripts 1 and 2 indicate the respective Hermitian real and imaginary parts of the noise operator, for instance, $\hat{S}_{A1} = (\hat{S}_A + \hat{S}_A^\dagger)/2$ and $\hat{S}_{A2} = (\hat{S}_A - \hat{S}_A^\dagger)/2i$. The mutual correlation functions between any two noise operators are zero.

The physical interpretations of Eqs. (11.13)–(11.16) are as follows. The interpretation of Eq. (11.13) is that the field that is incident from an external space and is coupled into the laser mode is in a vacuum (coherent) state that carries equal amounts of quantum noise $\langle \Delta\hat{S}_{A1}^2\rangle = \langle \Delta\hat{S}_{A2}^2\rangle = \tfrac{1}{4}$ in the two quadrature components. The interpretation of Eq. (11.14) is that the reservoirs, which are responsible for the phase decay process of the dipole moment, are also at ground states and impose equal amounts of quantum noise on the two quadrature components of the dipole moment. The interpretation of Eq. (11.15) is that the pump process is a self-exciting Poisson point process (shot-noise-limited) due to the very short memory effect of a pump light. The interpretation of Eq. (11.16) is that the spontaneous decay of the excited atom also has a shot-noise character because of the very short memory effect of continuous vacuum field fluctuations, which are the origin of spontaneous emission.

### 11.3.2. Electrical Circuit Theory of a Laser

The quantum-statistical property of photon field $\hat{A}$ calculated by Eqs. (11.9)–(11.11) is that of the cavity internal field. However, what is measured in real experiments is that of the external output field. The quantum noises of the internal field and external field are not necessarily the same. The external output field $\hat{r}$ consists of the transmitted internal field and the reflected vacuum field fluctuation [22, 23]

$$\hat{r} = -\hat{S}_A + \sqrt{\gamma}\,\hat{A} \quad (11.17)$$

This relation is obtained from the time-reversal argument [24] or from the direct analysis of waves bouncing back and forth in a Fabry–Perot cavity

[25]. Later we will show that Eq. (11.17) corresponds to the (quantum-mechanical) Kirchhoff current law in electrical circuit theory [26]. This fact that the internal field fluctuation and the external field fluctuation are not the same has been widely accepted in microwave oscillator theory [22], but has been overlooked in laser oscillator theory until very recently.

One reason that the quantum theory of laser oscillators has never covered the external field fluctuations stems from the inherent difficulty of quantum mechanics. Quantum-mechanical theory usually treats a system as a whole world and sets up a Hamiltonian function. All the remaining subsystems, such as the external space outside the laser cavity, are treated as reservoirs. Thus, the quantum theory can describe the development of the system under influence of the reservoirs. However, the development of the reservoirs cannot be described in this way. Such difficulties do not appear in electrical circuit theory. A laser oscillator with an output coupling mirror followed by free space can be described by an electrical circuit connected to an infinite transmission line. The use of electrical circuit language to describe laser oscillator noise was pioneered by Lax [19]. He analyzed an equivalent electrical circuit containing positive and negative resistances with their own noise sources, but calculated only the cavity internal field fluctuations.

Based on the research of Kurokawa [22], we describe here the laser oscillator in terms of the equivalent electrical circuit shown in Figure 11.3a. The internal field in the real laser oscillator shown in Figure 11.3b is described by the peak amplitude $A$ and phase $\phi$ of the current $I$ flowing in the active element. The terms $e_e$ and $e_i$ are noise voltages of external and internal origin, respectively. We use the classical theory to describe the electrical circuit shown in Figure 11.3a. After linearization, that is, splitting an operator into its mean value and small fluctuation parts, the difference between the quantum theory and the classical theory disappear. Here, impedance $-Z(A)$ describes the active element, the real part of which is split into negative resistance $-R_i(A)$, representing a gain, and positive resistance $R_a$, indicating internal loss of the active element. The frequency-dependent part of the active-element impedance is, for convenience, attributed to the load.

Resistance $R_L$ is the load resistance and is equal to $R_i(A) - R_a$ under steady-state oscillation conditions. The imaginary part of $Z(A)$ is equal to cavity reactance $X(\omega)$, which is usually given by $\omega L - 1/\omega C \simeq 2L(\omega - \omega_0)$, where $\omega$ is the actual oscillation (angular) frequency, and $\omega_0 = 1/\sqrt{LC}$ is the cold cavity resonance (angular) frequency.

Referring to Figure 11.3a, the circuit equation reads

$$[R_L + jX(\omega) - R(A) + jX(A)] \cdot I = e_e + e_i \equiv e \qquad (11.18)$$

We write the current $I$ as

$$I = Re\{(\overline{A} + \Delta A)e^{j(\overline{\omega}t + \phi + \Delta\phi)}\}$$

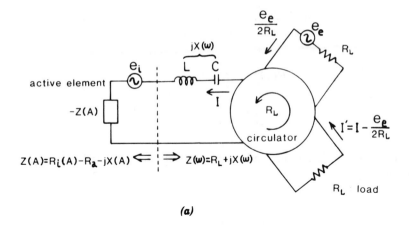

*(a)*

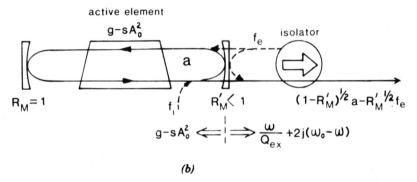

*(b)*

**Figure 11.3.** A microwave oscillator (*a*) and a laser oscillator (*b*) with circulator configurations.

allowing $\Delta A$ and $\Delta \phi$ to be slowly varying real quantities and using $\overline{A}$ and $\overline{\omega}$ to denote the respective noise-free amplitude and frequency given by

$$R_L - R(\overline{A}) = 0, \quad X(\overline{A}) = -X(\overline{\omega}) \tag{11.19}$$

Since $\Delta A$ and $\Delta \phi$ are slowly varying quantities, we can find a first-order differential equation by expanding $X(\omega)$ as

$$X(\omega) = (\omega - \omega_0) \frac{\partial X}{\partial \omega} = 2(\omega - \omega_0)L \tag{11.20}$$

and replacing $j\omega$ by $d/dt$ operating on the current. Furthermore, assuming that $|\Delta A/\overline{A}| \ll 1$, we may linearize $Z(A)$ according to

$$R(A) = R(\overline{A}) + \Delta A \cdot \frac{\partial R}{\partial A} \qquad (11.21)$$

$$X(A) = X(\overline{A}) + \Delta A \cdot \frac{\partial X}{\partial A} \qquad (11.22)$$

Using these modifications and neglecting products of small quantities, we obtain

$$Re\left\{\left[\left(j\frac{d\phi}{dt} + \frac{1}{A}\frac{d\Delta A}{dt}\right) \cdot 2L - \Delta A\frac{\partial R}{\partial A} + j\Delta A\frac{\partial X}{\partial A}\right] \cdot \overline{A}\, e^{j(\omega t + \phi)}\right\} = e(t) \qquad (11.23)$$

or

$$\left(\frac{d\Delta A}{dt} \cdot 2L - \Delta A\overline{A}\frac{\partial R}{\partial A}\right)\cos(\overline{\omega}t + \phi)$$

$$-\left(\overline{A}\frac{d\phi}{dt} \cdot 2L + \Delta A\overline{A}\frac{\partial X}{\partial A}\right)\sin(\overline{\omega}t + \phi) = e(t) \qquad (11.24)$$

Multiplying in turn by $\cos(\overline{\omega}t + \phi)$ and $\sin(\overline{\omega}t + \phi)$ and integrating over one period $2\pi/\overline{\omega}$, and keeping in mind that $\Delta A$ and $\Delta\phi$ can be regarded as constant during such a short time interval, we obtain

$$\frac{d\Delta A}{dt} \cdot 2L - \Delta A \cdot \overline{A}\frac{\partial R}{\partial A}$$

$$= e_c(t) \equiv \frac{\overline{\omega}}{\pi}\int_{t-\pi/\overline{\omega}}^{t+\pi/\overline{\omega}} e(u)\cos(\overline{\omega}u + \phi)\, du \qquad (11.25)$$

and

$$\overline{A} \cdot \frac{d\Delta\phi}{dt} \cdot 2L + \Delta A \cdot \overline{A}\frac{\partial X}{\partial A}$$

$$= -e_s(t) \equiv -\frac{\overline{\omega}}{\pi}\int_{t-\pi/\overline{\omega}}^{t+\pi/\overline{\omega}} e(u)\sin(\overline{\omega}u + \phi)\, du \qquad (11.26)$$

Since, in the following, we will be concerned only with variations in $e_s(t)$ and $e_c(t)$, which are slow compared to light angular frequency $\omega$, it is consistent with Eqs. (11.25) and (11.26) to write

$$e(t) = e_c(t)\cos(\bar{\omega}t + \phi) + e_s(t)\sin(\bar{\omega}t + \phi) \qquad (11.27)$$

We now introduce saturation parameters $s$ and $r$ through

$$s \equiv \frac{-\bar{A}}{R(\bar{A})} \cdot \frac{\partial R(A)}{\partial A} \qquad (11.28)$$

$$r \equiv \frac{\bar{A}}{R(\bar{A})} \cdot \frac{\partial X(A)}{\partial A} \qquad (11.29)$$

to write Eqs. (11.25) and (11.26) as

$$2L \cdot \frac{d\,\Delta A}{dt} + sR_L\,\Delta A = e_c(t) \qquad (11.30)$$

$$-2L \cdot \frac{d\phi}{dt} - rR_L\frac{\Delta A}{\bar{A}} = \frac{e_s(t)}{\bar{A}} \qquad (11.31)$$

where Eq. (11.19) is used to substitute $R_L$ for $R(\bar{A})$.

Since we are considering modulation frequencies very much lower than the oscillation frequency, $\Omega \ll \omega$, the solutions of Eqs. (11.30) and (11.31) in the frequency domain are

$$\Delta A_\Omega = \frac{e_{c\Omega}}{2j\Omega L + sR_L} \qquad (11.32)$$

$$\Delta\phi_\Omega = -\frac{1}{2j\Omega L\bar{A}}\left[e_{s\Omega} - e_{c\Omega}\frac{rR_L}{2j\Omega L + sR_L}\right] \qquad (11.33)$$

Here the Fourier series analysis with a period $T$ is used, for instance

$$e_{c\Omega} = \sqrt{\frac{2}{T}}\int_0^T e_c(t)e^{-j\Omega t}\,dt$$

If $e_c(t)$ and $e_s(t)$ are assumed to be uncorrelated, and the power spectra of the noise voltages are defined by

$$P_c(\Omega) \equiv \lim_{T\to\infty}|e_{c\Omega}|^2$$

we obtain

$$P_{\Delta A}(\Omega) = \frac{P_c(\Omega)}{4R_L^2(s/2)^2} \cdot \frac{1}{1 + (2\Omega L/sR_L)^2} \tag{11.34}$$

$$P_{\Delta\phi}(\Omega) = \frac{1}{4(\Omega L\bar{A})^2}\left[P_s(\Omega) + P_c(\Omega)\frac{r^2/s^2}{1 + (2\Omega L/sR_L)^2}\right] \tag{11.35}$$

With $Q_{\text{ex}} = \omega_0 L/R_L \sim \omega L/R_L$ and the average output power is given by

$$P_0 = \tfrac{1}{2}R_L\bar{A}^2$$

Eqs. (11.34) and (11.35) are often written as

$$P_{\Delta A}(\Omega) = \frac{P_c(\Omega)}{4R_L^2(s/2)^2} \cdot \frac{1}{1 + (2Q_{\text{ex}}\Omega/s\omega)^2} \tag{11.36}$$

and

$$P_{\Delta\phi}(\Omega) = \frac{1}{8P_0R_L} \cdot \frac{1}{(Q_e\Omega/\omega)^2}\left[P_s(\Omega) + P_c(\Omega)\frac{r^2/s^2}{1 + (2Q_{\text{ex}}\Omega/s\omega)^2}\right] \tag{11.37}$$

From Eq. (11.37) it is seen that the phase fluctuation spectrum is enhanced by a factor $1 + r^2/s^2$ at low frequences due to the reactive saturation parameter $r$, if $P_s(\Omega) = P_c(\Omega)$. For semiconductor lasers, $r^2/s^2$ is important and is generally referred to as the *linewidth enhancement factor* $\alpha^2$ (see Chapter 2). As for the resistive saturation parameter, $s$ is equal to 2 if amplitude $\bar{A}$ is adjusted (by choice of $R_L$) to give maximum generated power $(1/2)\bar{A}^2R(\bar{A})$. Furthermore, $s$ is equal to 2 in a strongly pumped laser oscillator independent of the exact matching conditions, provided the optical loss is small.

### 11.3.3. Breakdown of the Conventional Reservoir Theory (I): Internal Noise versus External Noise

Next we will calculate the amplitude and phase fluctuation in the outgoing wave. External current $I'$ is related to internal current $I$ by Kirchhoff's law:

$$I' = I - \frac{e_e(t)}{2R_L}$$

$$= (\bar{A} + \Delta A)\cos(\bar{\omega}t + \phi + \Delta\phi) - \frac{e_e(t)}{2R_L} \tag{11.38}$$

This equation is equivalent to Eq. (11.17). By combining Eqs. (11.38) and (11.27), it is concluded that amplitude fluctuation $\Delta A_e$ in the outgoing wave is

$$\Delta A_e = \Delta A - \frac{e_{ec}(t)}{2R_L} \tag{11.39}$$

or with Eq. (11.32) in the frequency domain

$$\Delta A_{e\Omega} = e_{ec\Omega}\left[\frac{1}{2j\Omega L + sR_L} - \frac{1}{2R_L}\right] + e_{ic\Omega}\frac{1}{2j\Omega L + sR_L} \tag{11.40}$$

Here, $e_{c\Omega}$ is divided into noise sources of external $e_{ec\Omega}$ and internal origin $e_{ic\Omega}$. From Eqs. (11.38) and (11.27), it is apparent that there is a phase shift $\Delta\phi'$ between $I'$ and $I$, given by

$$\Delta\phi' = \frac{1}{2R_L} \cdot \frac{e_{es}(t)}{\overline{A}} \tag{11.41}$$

When Eq. (11.33) is used, phase $\Delta\phi_e$ of the outgoing wave in the frequency domain is

$$\Delta\phi_{e\Omega} \equiv \Delta\phi_\Omega + \Delta\phi'_\Omega = -\frac{1}{2j\Omega L\overline{A}}\left[e_{es\Omega}\left(1 - \frac{j\Omega L}{R_L}\right) + e_{is\Omega}\right.$$
$$\left. -(e_{ec}\Omega + e_{ic}\Omega)\frac{rR_L}{2j\Omega L + sR_L}\right] \tag{11.42}$$

Since $e_e$ and $e_i$ as well as $e_c$ and $e_s$ are assumed to be uncorrelated, the power spectra corresponding to Eqs. (11.40) and (11.42) become

$$P_{\Delta A_e}(\Omega) = \frac{P_{ec}(\Omega)\left[\left(1 - \tfrac{1}{2}s\right)^2 + (\Omega L/R_L)^2\right] + P_{ic}(\Omega)}{4R_L^2\left(\tfrac{1}{2}s\right)^2\left[1 + (2\Omega L/sR_L)^2\right]} \tag{11.43}$$

$$P_{\Delta\phi e}(\Omega) = \frac{1}{4(\Omega L\overline{A})^2}\left[P_s(\Omega) + P_c(\Omega)\frac{r^2/s^2}{1 + (2\Omega L/sR_L)^2}\right] + \frac{P_{es}(\Omega)}{4R_L^2\overline{A}^2} \tag{11.44}$$

Here, as before, $P_{c,s}(\Omega) = P_{ec,s}(\Omega) + P_{ic,s}(\Omega)$.

We will now compare these results with those of internal fields Eqs. (11.34) and (11.35). First, we note that $P_{\Delta\phi e}(\Omega)$ differs from $P_{\Delta\phi}(\Omega)$ by only a frequency-independent term

$$P_{\Delta\phi e}(\Omega) = P_{\Delta\phi}(\Omega) + \frac{P_{es}(\Omega)}{4R_L^2 \overline{A^2}} = P_{\Delta\phi}(\Omega) + \frac{P_{es}(\Omega)}{8P_0 R_L} \quad (11.45)$$

Since $P_{\Delta\phi}(\Omega)$ approaches zero at high frequencies, $P_{\Delta\phi e}(\Omega)$ approaches the constant value $P_{es}(\Omega)/8P_0 R_L$, provided that $P_e(\Omega)$ [and $P_i(\Omega)$] has a flat spectrum.

From Eq. (11.43) we see that the high-frequency limit of $P_{\Delta A_e}(\Omega)$ is $P_{ec}(\Omega)/4R_L^2$. This contrasts with $P_{\Delta A}(\Omega)$, which approaches zero at high frequencies. Thus, both the amplitude and phase noise spectra of the outgoing wave extend to frequencies far beyond cavity cutoff frequency $sR_L/2L$. The reason for this, of course, is that the noise wave generated by the load is simply reflected by the oscillator, since the impedance of the oscillator is very large for frequencies outside its cold circuit bandwidth.

In the low-frequency limit, on the other hand, $P_{\Delta A_e}(\Omega)$ is given by

$$P_{\Delta A_e}(\Omega)\big|_{\Omega \to 0} = \frac{P_{ec}(\Omega)}{4R_L^2} \frac{(1 - s/2)^2}{(s/2)^2} + \frac{P_{ic}(\Omega)}{4R_L^2(s/2)^2} \quad (11.46)$$

Thus, if $s = 2$, the external noise source does not contribute at all to the low-frequency fluctuations in the outgoing wave. In this way the oscillator resembles a "matched load." Note, however, that this analogy does not hold for the phase fluctuations induced by the external noise source.

To apply the above theory to a laser, we must find the proper expressions for the internal and external noise sources $e_i$ and $e_e$, the negative resistance $R(A)$, and the saturation parameters $s$ and $r$. Figure 11.3b and Table 11.1 summarize the proper expressions for these parameters.

Figures 11.4a and 11.4b compare the amplitude and phase noise spectra of the internal field and external field of a highly saturated laser oscillator for three cases such as (1) where the internal noise source $e_i$ is twice as large as the external noise source $e_e$, (2) where the internal noise source is equal to the external noise source, and (3) where the internal noise source is zero [26]. The amplitude noise spectra of the internal field are all Lorentzian with a cutoff frequency $\Omega_c$ equal to $\gamma$, as is predicted by the conventional theory. However, the amplitude noise spectra of the external field has a white spectrum above the cutoff frequency $\Omega_c$ and features more complicated behavior below $\Omega_c$. These are respectively due to the facts that the vacuum field fluctuation does not penetrate into the cavity, but is simply reflected back above $\Omega_c$, and that the transmitted internal field fluctuation and the partly reflected vacuum field fluctuation destructively interfere because of the $\pi$-phase difference between the two fields below $\Omega_c$.

**Table 11.1  Comparison of Microwave Oscillator and Laser Oscillator Terminology**

| | Electrical Terminology | (Relation) | Laser Terminology |
|---|---|---|---|
| Oscillator field strength | Current amplitude | $A \longleftrightarrow$ $A_0 = A\sqrt{L/2\hbar\omega}$ | Photon amplitude $A_0$ |
| Active element | Complex impedance $Z(A) = R_i(A) - R_a - jX(A)$ | $G(A_0) = \dfrac{Z(A)}{L}$ | Complex gain factor $G(A_0) = \dfrac{\omega\chi_i(A_0)}{\mu^2} - \dfrac{\omega}{Q_0} - \dfrac{j\omega\chi_r(A_0)}{\mu^2}$ |
| Load | Complex impedance $Z(\omega) = R_L + j2L(\omega - \omega_0)$ | $G_0(\omega) = \dfrac{Z(\omega)}{L}$ | Complex photon decay rate $G_0(\omega) = \dfrac{\omega}{Q_{ex}} + j2(\omega - \omega_0)$ |
| Stored energy | $\frac{1}{2}L\overline{A}^2$ | $===$ | $\hbar\omega\overline{A}_0^2$ |
| Power flow to a load | $\frac{1}{2}R_L\overline{A}^2$ | $===$ | $\hbar\omega\overline{A}_0^2\left(\dfrac{\omega}{Q_{ex}}\right)$ |
| External noise source | Noise voltage $e_e(t)$ | $e_e(t) \longleftrightarrow$ $f_e = e_e\sqrt{1/R_L\hbar\omega}$ | Field fluctuation $f_e(t)$ |
| Commutator bracket | $[e_{e+}(\Omega_k), e_{e+}^\dagger(\Omega_{k'})] = \hbar\omega R_L \delta_{kk'}$ | $\longleftrightarrow$ | $[f_{e\pm}^\dagger(\Omega_k), f_{e\pm}(\Omega_{k'})] = \delta_{kk'}$ |
| Quantum noise spectral density | $S_e^{(q)} = 2\hbar\omega R_L$ | | $S_{fe}^{(q)} = 2$ |
| Thermal noise spectral density | $S_e^{(t)} = \dfrac{4\hbar\omega R_L}{\exp(\hbar\omega/kT) - 1}$ | $\longleftrightarrow$ | $S_{fe}^{(t)} = \dfrac{4}{\exp(\hbar\omega/kT) - 1}$ |

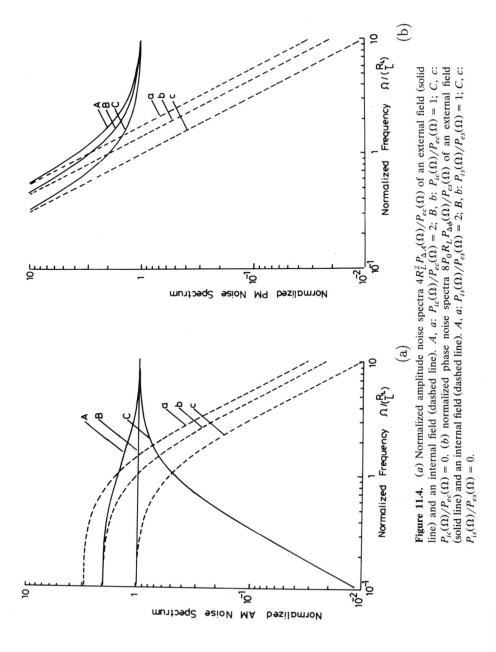

**Figure 11.4.** (a) Normalized amplitude noise spectra $4R_L^2 P_{\Delta A}(\Omega)/P_{ec}(\Omega)$ of an external field (solid line) and an internal field (dashed line). A, a: $P_{ic}(\Omega)/P_{ec}(\Omega) = 2$; B, b: $P_{ic}(\Omega)/P_{ec}(\Omega) = 1$; C, c: $P_{ic}(\Omega)/P_{ec}(\Omega) = 0$. (b) normalized phase noise spectra $8P_0 R_L P_{\Delta\phi}(\Omega)/P_{es}(\Omega)$ of an external field (solid line) and an internal field (dashed line). A, a: $P_{is}(\Omega)/P_{es}(\Omega) = 2$; B, b: $P_{is}(\Omega)/P_{es}(\Omega) = 1$; C, c: $P_{is}(\Omega)/P_{es}(\Omega) = 0$.

In the special case (2), that is, equal internal and external noise sources, the internal field amplitude noise power, which is given by the integrated amplitude noise spectrum, is equal to that of a coherent state $\langle \Delta \hat{A}^2 \rangle = \frac{1}{4}$. This is a "theorist's coherent state." On the other hand, the external field amplitude noise spectrum is equal to a shot-noise level $P_{\Delta,r}(\Omega) = \frac{1}{2}$. This is an "experimentalist's coherent state." The most striking result is obtained for case (3). Even though the internal field has a finite fluctuation at the low-frequency limit, the fluctuation of the external field completely disappears at the low-frequency limit. The amplitude noise power of the internal field is half that of a coherent state $\langle \Delta \hat{A}^2 \rangle = \frac{1}{8}$. The external field, on the other hand, features complete amplitude squeezing at the low-frequency limit. This surprising result stems from the fact that the residual internal field fluctuation and the reflected vacuum field fluctuation have exactly the same magnitude and opposite phase to cancel each other out. At far above the threshold, the internal noise sources except the pump noise are suppressed [1]. Therefore, whether the laser output is in a coherent state or in a squeezed state depends solely on the pump noise.

### 11.3.4. Breakdown of the Conventional Reservoir Theory (II): Pump Noise Suppression

If the pump light is sub-Poissonian light or a photon-number eigenstate and the quantum efficiency of optical pumping is close to unity, the pump noise becomes smaller than the shot-noise level given by Eq. (11.15). The phase noise of the pump light does not contribute to the laser noise at all because the pump process is actually a photoelectron emission process. Golubev and Sokolov were the first to point out this possibility [27]. However, this solution of how to suppress the pump noise is not interesting from our viewpoint, because the generation of sub-Poissonian light itself is the goal.

Preparation of a sub-Poissonian light or a photon-number state is not easy, but preparation of a sub-Poissonian electron beam or an electron-number state is easily achieved using mutual electron interaction in a space-charge-limited beam or in a simple resistor. It is expected that an electron (beam) excitation scheme suppresses the pump noise in a realistic way.

Figure 11.5 shows the scheme for generating a sub-Poissonian electron beam in a space charge limited vacuum tube [28]. When the electron emission rate increases above its average, the number of space charges near the cathode increases, and the potential minimum between the cathode and the anode becomes more negative owing to the Coulomb repulsion effect between space electrons. Therefore, more electrons are reflected back to the cathode as a result of the insufficient initial velocity. When the electron emission rate decreases below its average, the potential

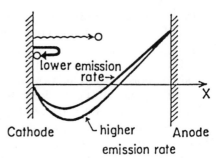

**Figure 11.5.** Sub-Poissonian electron beam generation by Coulomb blockade in a space-charge-limited vacuum tube.

minimum becomes less negative, and less electrons are reflected back. In this way, the modulation in the potential minimum regulates the number of electrons passing through the potential minimum. It is a natural feedback stabilization mechanism. The electron arrival process at the anode becomes strongly sub-Poissonian, and the anode current fluctuation features a sub-shot-noise level. In fact, using this principle, Teich and Saleh produced the sub-Poissonian light from the space-charge-limited vacuum tube containing mercury atoms [29] even though the light is not a laser emission but a weak spontaneous emission.

The current noise generated in a resistor is not shot noise, $2eI_0$, but is thermal noise, $4k_BT/R$, independent of the current $I_0$ precisely due to such a feedback effect. Therefore, if the voltage bias $V = I_0R$ across the resistor is larger than the equivalent thermal voltage $V_T = 2k_BT/e$, the current noise is smaller than shot noise.

Figure 11.6 shows the scheme for generating an electron-number state in a microtunnel junction using the principle of Coulomb blockage [30]. When the junction capacitance $C$ becomes very small (on the order of $10^{-15}$ F or less), the Coulomb energy produced by a single electron $e^2/2C$ becomes larger than the thermal background energy $k_BT$ at a cryogenic temperature ($T \sim 1$ K). Then, the electron tunneling is inhibited during the charging process until the charge of one electrode becomes $+e/2$ from $-e/2$. This is because, before reaching that charge, the electron tunneling would result in an enormous increase of the electrostatic energy, which cannot be provided by thermal phonons, since $e^2/2C \gg k_BT$. Consequently, single-electron tunneling is separated by the time constant $\tau = e/I$, where $I$ is the DC current flowing in an external circuit. This is called *single-electron tunneling* oscillation, which is nothing but a single-electron state generation. In fact, the correlated single-electron tunneling has been observed recently in an ultrasmall tunnel junction [31].

Next, let us consider the quantum-mechanical consistency of the two new discoveries, mentioned above, namely the difference between the internal and external field noise and the pump noise suppression.

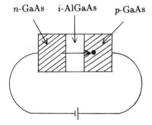

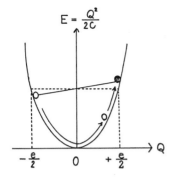

**Figure 11.6.** Single-electron tunneling regulated by Coulomb blockade in a microtunnel junction at a cryogenic temperature.

### 11.3.5. Commutator Bracket Conservation

The creation and annihilation operators of the waves incident on the laser and reflected (output) from it must obey the proper commutator bracket in free space. This fact alone, regardless of the medium in the cavity, imposes commutator brackets on the internal noise sources [32]. The quantum-mechanical consistency of any new physical idea can be checked by this commutator bracket conservation test.

When the dipole decay rate $\Gamma$ is much larger than both the photon decay rate $\gamma$ and the population decay rate $1/\tau_{sp}$ (which is the case for a semiconductor laser), the dipole operator can be eliminated adiabatically from Eqs. (11.9)–(11.11). The resulting equations are

$$\frac{d}{dt}\hat{A} = -\frac{\gamma}{2}\hat{A} + \frac{g^2}{\Gamma}\hat{N}\hat{A} + \sqrt{\gamma}\,\hat{S}_A - ig\sqrt{\frac{2N}{\Gamma}}\,\hat{S}_\Sigma \qquad (11.47)$$

and

$$\frac{d}{dt}\hat{N} = P - \frac{\hat{N}}{\tau_{sp}} - \frac{2g^2}{\Gamma}\hat{N}\hat{S} + \sqrt{P}\,\hat{S}_\Lambda$$

$$+ \sqrt{\frac{\langle\hat{N}\rangle}{\tau_{sp}}}\,\hat{S}_\tau + ig\sqrt{\frac{2N}{\Gamma}}\left(\hat{A}^\dagger\hat{S}_\Sigma - \hat{S}_\Sigma^\dagger\hat{A}\right) \qquad (11.48)$$

Here, $\hat{N} = N\hat{\sigma}$ is the total population inversion operator, $P = N\Lambda$ is the total pump rate, and $\hat{S} = \hat{A}^{\dagger}\hat{A}$ is the internal photon-number operator.

In the linear amplification regime, the population fluctuation does not couple with the field fluctuation so that $\hat{N}$ can be replaced by its average value $\langle\hat{N}\rangle = N\langle\hat{\sigma}\rangle$ in Eq. (11.47). With defining the stimulated emission rate $E_{cv} = (2g^2/\Gamma)N\langle\hat{\sigma}\rangle$, the Fourier transform of Eq. (11.47) results in

$$\tilde{A} = \frac{\sqrt{\gamma}\,\tilde{S}_A - ig\sqrt{(2N/\Gamma)}\,\tilde{S}_\Sigma}{i\Omega + \frac{1}{2}(\gamma - E_{cv})} \tag{11.49}$$

Here, the tilde indicates the Fourier transform (Fourier series analysis of period $T$) of the operator, according to

$$\tilde{A}(\Omega) = \sqrt{\frac{2}{T}} \int_{-T/2}^{T/2} \hat{A}(t)e^{i\Omega t}\,dt \tag{11.50}$$

When boundary condition Eq. (11.17) is used, the Fourier transformed external field operator is

$$\tilde{r} = \frac{[-i\Omega + \frac{1}{2}(\gamma + E_{cv})]\tilde{S}_A - ig\sqrt{(2N\gamma/\Gamma)}\,\tilde{S}_\Sigma}{i\Omega + \frac{1}{2}(\gamma - E_{cv})} \tag{11.51}$$

Assuming the commutator bracket of the incident vacuum field fluctuation $[\tilde{S}_A, \tilde{S}_A^{\dagger}] = 1$, which is self-consistent with Eq. (11.13), we can easily prove that the commutator bracket of the output field is properly conserved:

$$[\tilde{r}, \tilde{r}^{\dagger}] = \frac{1}{\Omega^2 + \frac{1}{4}(\gamma - E_{cv})^2}\left\{\left[\Omega^2 + \frac{1}{4}(\gamma + E_{cv})^2\right]\right.$$

$$\left.\times[\tilde{S}_A, \tilde{S}_A^{\dagger}] + \frac{2\gamma g^2 N}{\Gamma}[\tilde{S}_\Sigma, \tilde{S}_\Sigma^{\dagger}]\right\}$$

$$= 1 \tag{11.52}$$

Here, the commutator bracket of dipole noise operators $[\tilde{S}_\Sigma, \tilde{S}_\Sigma^{\dagger}] = -\langle\hat{\sigma}\rangle$ and the relation $E_{cv} = (2g^2N/\Gamma)\langle\hat{\sigma}\rangle$ are used. The atomic systems are represented by Pauli's spin operators $\hat{\Sigma} = \begin{pmatrix} 0 & 0 \\ 1 & 0 \end{pmatrix}$, $\hat{\Sigma}^{\dagger} = \begin{pmatrix} 0 & 1 \\ 0 & 0 \end{pmatrix}$, and $\hat{\sigma} = \begin{pmatrix} 1 & 0 \\ 0 & -1 \end{pmatrix}$. The commutator bracket is properly preserved by the joint contributions of incident vacuum field fluctuation $\hat{S}_A$ and dipole noise operator $\hat{S}_\Sigma$.

In the saturated oscillator regime, the nonlinear coupling between the field and the population inversion must be taken into account. Both the field operator and the population inversion operator can be divided into

$c$-number average values and small fluctuation operators: $\hat{A} = A_0 + \Delta\hat{A}_1 + i\,\Delta\hat{A}_2$, and $\hat{N} = N_0 + \Delta\hat{N}$. Without loss of generality, we can assume that $A_0$ is a real number. The equations for the small fluctuation operators are

$$\frac{d}{dt}\,\Delta\hat{A}_1 = \frac{g^2}{\Gamma}A_0\,\Delta\hat{N} + \sqrt{\gamma}\,\hat{S}_{A1} + g\sqrt{\frac{2N}{\Gamma}}\,\hat{S}_{\Sigma 2} \tag{11.53}$$

$$\frac{d}{dt}\,\Delta\hat{A}_2 = \sqrt{\gamma}\,\hat{S}_{A2} - g\sqrt{\frac{2N}{\Gamma}}\,\hat{S}_{\Sigma 1} \tag{11.54}$$

$$\frac{d}{dt}\,\Delta\hat{N} = -\left(\frac{1}{\tau_{sp}} + \frac{1}{\tau_{st}}\right)\Delta\hat{N} - \frac{2g^2}{\Gamma}N_0\,\Delta\hat{S} + \sqrt{P}\,\hat{S}_{\Lambda} + \sqrt{\frac{N_0}{\tau_{sp}}}\,\hat{S}_{\tau}$$

$$+ ig\sqrt{\frac{2N}{\Gamma}}\,A_0(\hat{S}_{\Sigma} - \hat{S}_{\Sigma}^{\dagger}) \tag{11.55}$$

Here, $\tau_{st} = (\Gamma/2g^2A_0^2)$ is the stimulated emission lifetime, $\Delta\hat{S} = 2A_0\,\Delta\hat{A}_1$ is the photon-number fluctuation operator, and the products of small fluctuation operators are neglected. At far above the threshold, $\tau_{st}$ becomes much shorter than $\tau_{sp}$ and the photon lifetime $\tau_p = 1/\gamma$. Therefore, the population fluctuation operator $\Delta\hat{N}$ can also be adiabatically eliminated from Eq. (11.53), as shown by

$$\frac{d}{dt}\,\Delta\hat{A}_1 = -\gamma\,\Delta\hat{A}_1 + \sqrt{\gamma}\,\hat{S}_{A1} + \frac{\sqrt{P}}{2A_0}\hat{S}_{\Lambda} \tag{11.56}$$

where the spontaneous emission terms are neglected because $\tau_{st} \ll \tau_{sp}$. The Fourier-transformed external field fluctuation operator satisfies

$$[\Delta\tilde{r}, \Delta\tilde{r}^{\dagger}] = \frac{1}{\Omega^4 + \Omega^2\gamma^2}\{-i\Omega^2(\Omega^2 + \gamma^2)[\tilde{S}_{A1}, \tilde{S}_{A2}]$$

$$+ i\Omega^2(\Omega^2 + \gamma^2)[\tilde{S}_{A2}, \tilde{S}_{A1}]\}$$

$$= 1 \tag{11.57}$$

Here, $[\tilde{S}_{A1}, \tilde{S}_{A2}] = i/2$ is used. The commutator bracket of the output field is preserved properly only by the incident vacuum field fluctuation. The dipole noise operator (at least, its in-phase component) is suppressed by the gain saturation, so that it does not contribute to the commutator bracket conservation.

The important observations made here are that the external field given by Eq. (11.17) conserve the commutator bracket properly and that the pump noise does not contribute to the commutator bracket conservation at all. This means that the shot-noise-limited pump noise [Eq. (11.15)] is not fundamental in a quantum-mechanical first principle. The boundary condition, Eq. (11.17), and the pump noise suppression schemes in the two previous subsections do not contradict the quantum-mechanical self-consistency.

### 11.3.6.  Origin of Standard Quantum Limit

Let us consider here the standard quantum limits of the amplitude and phase noise of a laser oscillator having a shot-noise-limited pump noise. The power spectrum of the fluctuation operator $\hat{g}(t)$ is calculated by Wiener–Khintchin's theorem as

$$P_{\hat{g}}(\Omega) = \lim_{T \to \infty} \left\langle \tilde{g}^{\dagger}(T, \Omega)\tilde{g}(T, \Omega)\right\rangle \tag{11.58}$$

where

$$\tilde{g}(T, \Omega) = \sqrt{\frac{2}{T}} \int_{-T/2}^{T/2} \hat{g}(t)e^{i\Omega t} \, dt \tag{11.59}$$

From Eqs. (11.53)–(11.55), we can readily obtain the Fourier transforms of amplitude noise $\Delta \hat{A} \equiv \Delta \hat{A}_1$ and phase noise $\Delta \hat{\phi} \equiv \Delta \hat{A}_2/A_0$ as

$$\Delta \tilde{A}(\Omega) = \frac{\left\{A_3\left[\sqrt{P}\tilde{S}_\Lambda + \sqrt{(N_0/\tau_{sp})}\,\tilde{S}_\tau\right] + (A_1 - i\Omega)\sqrt{\gamma}\,\tilde{S}_{A1} + (A_1 - A_3 A_0 - i\Omega)g\sqrt{(2N/\Gamma)}\,\tilde{S}_{\Sigma 2}\right\}}{(A_2 A_3 + \Omega^2) + i\Omega A_1} \tag{11.60}$$

$$\Delta \tilde{\phi}(\Omega) = \frac{\sqrt{\gamma}\,\tilde{S}_{A2} - g\sqrt{(2N/\Gamma)}\,\tilde{S}_{\Sigma 1}}{i\Omega A_0} \tag{11.61}$$

Here coefficients $A_1$–$A_3$ are

$$A_1 = \left(\frac{1}{\tau_{sp}} + \frac{1}{\tau_{sp}}\right) = \frac{1}{\tau_{sp}}(1 + n_{sp}R_p) \tag{11.62}$$

$$A_2 = \frac{4g^2}{\Gamma}N_0 A_0 = 2\gamma A_0 \tag{11.63}$$

$$A_3 = \frac{g^2}{\Gamma}A_0 = \frac{n_{sp}R_p}{2A_0\tau_{sp}} \tag{11.64}$$

where $n_{sp} = [2 - (N/N_0)]^{-1}$ is the population inversion parameter, $R_p = P/P_{th} - 1$ is the normalized pump rate, $\gamma = (2g^2/\Gamma)N_0$ is the threshold condition, $\beta = 2g^2\tau_{sp}/\Gamma$ is the quantum efficiency of spontaneous emission coupled into a single-laser mode (spontaneous emission coefficient), and $A_0^2 = R_p n_{sp}/\beta$. The power spectra of the noise operator can be calculated by correlation functions Eqs. (11.13)–(11.15). The power spectrum is defined here as the single-sided (unilateral) spectral density per cycles per second. The power spectra corresponding to Eqs. (11.60) and (11.61) are

$$P_{\Delta\hat{A}}(\Omega) = \left\{ 2A_3^2 \left( P + \frac{N_0}{\tau_{sp}} \right) + (A_1^2 + \Omega^2)\frac{\gamma}{2} + \frac{Ng^2}{\Gamma}\left[ (A_1 - A_3 A_0)^2 + \Omega^2 \right] \right\}$$

$$/\left[ (A_2 A_3 + \Omega^2)^2 + \Omega^2 A_1^2 \right] \tag{11.65}$$

$$P_{\Delta\hat{\phi}}(\Omega) = \frac{\gamma/2 + Ng^2/\Gamma}{\Omega^2 A_0^2} \tag{11.66}$$

Figures 11.7a and 11.7b show characteristic examples of the normalized amplitude and phase noise spectra. In Figure 11.7b, the finite linewidth enhancement factor $\alpha$ is assumed. The modification of Eq. (11.66) due to the $\alpha$ parameter is given by Eq. (11.37). The numerical examples were chosen for a typical semiconductor diode laser. When the pump rate is far above the threshold, Eq. (11.65) is reduced to the Lorentzian

$$P_{\Delta\hat{A}}(\Omega) = \frac{\gamma}{\Omega^2 + \gamma^2} \tag{11.67}$$

which is shown by the dashed curve in Figure 11.7a. The variance of the amplitude is calculated by Parseval's theorem as

$$\langle \Delta\hat{A}^2 \rangle \equiv \int_0^\infty P_{\Delta\hat{A}}(\Omega) \cdot \frac{d\Omega}{2\pi} = \frac{1}{4} \tag{11.68}$$

which is equal to that of a coherent state, as already mentioned in Section 11.3.2. Half of the noise power stems from pump noise $\hat{S}_\Lambda$, and the other half is due to incident vacuum field fluctuation $\hat{S}_{41}$. Spontaneous emission noise $\hat{S}_\tau$ and dipole moment noise $\hat{S}_{\Sigma2}$ are suppressed at such a high pump rate. When the population inversion parameter is equal to one and the linewidth enhancement factor $\alpha$ is zero, the phase noise spectrum is reduced to the Schawlow–Townes limit

$$P_{\Delta\hat{\phi}}(\Omega) = \frac{\gamma}{A_0^2 \Omega^2} \tag{11.69}$$

which is shown by the dashed line in Figure 11.7b. Half of the noise power

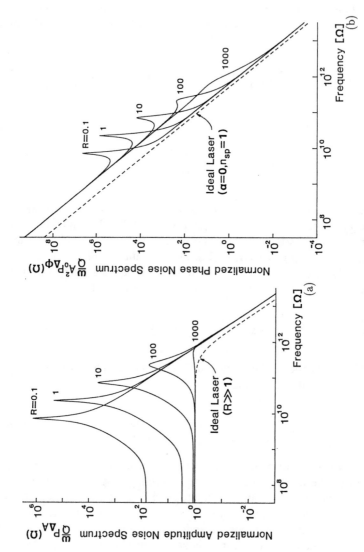

**Figure 11.7.** (*a*) Internal amplitude-noise spectra of a laser ($\omega/Q_0 = 0$). (*b*) Internal phase-noise spectra of a laser ($\omega/Q_0 = 0$).

stems from dipole moment noise $\hat{S}_{\Sigma 1}$, and the other half is due to incident vacuum field fluctuation $\hat{S}_{A2}$. Even though quadrature dipole moment noise component $\hat{S}_{\Sigma 2}$ is suppressed by gain saturation, in-phase dipole moment noise component $\hat{S}_{\Sigma 1}$ is not.

The power spectra of the external output field are calculated by using boundary condition, Eq. (11.17). The results are

$$\Delta \tilde{r}(\Omega) = \left\{ \sqrt{\gamma} A_3 \left[ \sqrt{P} \tilde{S}_\Lambda + \sqrt{\frac{N_0}{\tau_{sp}}} \tilde{S}_\tau \right] \right.$$

$$+ \sqrt{\gamma} (A_1 - A_3 A_0 - i\Omega) g \sqrt{\frac{2N}{\Gamma}} \tilde{S}_{\Sigma 2}$$

$$+ \left[ (\gamma A_1 - A_2 A_3 - \Omega^2) - i\Omega(\gamma + A_1) \right] \tilde{S}_{A1} \right\}$$

$$\Big/ \left[ (A_2 A_3 + \Omega^2) + i\Omega A_1 \right] \tag{11.70}$$

$$P_{\Delta \tilde{r}}(\Omega) = \left\{ 2\gamma A_3^2 \left( P + \frac{N_0}{\tau_{sp}} \right) + \frac{\gamma N g^2}{\Gamma} \left[ (A_1 - A_3 A_0)^2 + \Omega^2 \right] \right.$$

$$+ \left[ (\gamma A_1 - A_2 A_3 - \Omega^2)^2 + \Omega^2 (\gamma + A_1)^2 \right] \cdot \frac{1}{2} \right\}$$

$$\Big/ \left[ (A_2 A_3 + \Omega^2)^2 + \Omega^2 A_1^2 \right] \tag{11.71}$$

$$\Delta \hat{\psi}(\Omega) = \frac{(\gamma - i\Omega) \tilde{S}_{A2} - g\sqrt{(2N\gamma/\Gamma)} \, \tilde{S}_{\Sigma 1}}{i\Omega r_0} \tag{11.72}$$

$$P_{\Delta \hat{\psi}}(\Omega) = \frac{\frac{1}{2}(\gamma^2 + \Omega^2) + (N\gamma g^2/\Gamma)}{\Omega^2 r_0^2} \tag{11.73}$$

Figures 11.8a and 11.8b are characteristic examples of the respective amplitude and phase noise spectra of the external output field. Numerical parameters are the same as for Figure 11.7.

At the pump rate far above the threshold ($R_p \gg 1$), the amplitude noise spectrum, Eq. (11.71), is reduced to a broadband coherent state's white spectrum of

$$P_{\Delta \hat{r}}(\Omega) = \frac{1}{2} \tag{11.74}$$

which corresponds to the shot-noise-limited photon flux spectrum of

$$P_{\Delta \hat{N}}(\Omega) \equiv 4r_0^2 P_{\Delta \hat{r}}(\Omega) = 2r_0^2 \tag{11.75}$$

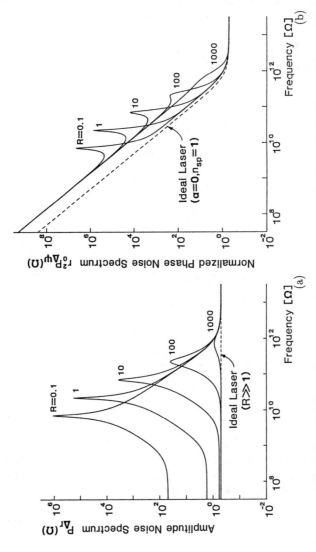

**Figure 11.8.** (a) External amplitude-noise spectra of a laser ($\omega/Q_0 = 0$). (b) External phase-noise spectra of a ($\omega/Q_0 = 0$).

as already mentioned in Section 11.3.2. The amplitude noise spectrum, Eq. (11.71), below cavity cutoff frequency $P_{\Delta \hat{r}}(\Omega < \gamma)$ is due to pump noise $\hat{S}_{\Lambda}$. The amplitude noise spectrum above cavity cutoff frequency $P_{\Delta \hat{r}}(\Omega > \gamma)$ stems from vacuum field fluctuation $\hat{S}_{A1}$. One-half of the internal amplitude noise spectrum is contributed by the vacuum field fluctuation. This part of the internal amplitude noise is completely suppressed by the destructive interference because reflected vacuum field fluctuation $-\hat{S}_{A1}$ has the same magnitude and opposite phase of transmitted internal field fluctuation $\sqrt{\gamma} \, \Delta \hat{A}_1$. In the frequency region above the cavity cutoff frequency, the internal amplitude noise is absent, leaving only the reflected vacuum field fluctuation.

If population inversion parameter $n_{sp}$ is one, the phase noise spectrum [Eq. (11.73)] is reduced to

$$P_{\Delta \hat{\psi}}(\Omega) = \frac{1}{2r_0^2}\left( \frac{2\gamma^2}{\Omega^2} + 1 \right) \tag{11.76}$$

Above the cavity cutoff frequency, the external phase noise is a broadband coherent state's white spectrum of $P_{\Delta \hat{\psi}}(\Omega > \gamma) = (2r_0^2)^{-1}$. However, below the cavity cutoff frequency, the Schawlow–Townes phase diffusion noise due to $\hat{S}_{\Sigma 1}$ and $\hat{S}_{A2}$ survives. The destructive interference between the transmitted internal field and the reflected vacuum field does not work for the phase noise, because of the quadrature phase difference between the two waves.

### 11.3.7. Amplitude Squeezing in a Pump-Noise-Suppressed Laser

The external amplitude and phase noise spectra above the cavity cutoff frequency pumped at levels far above the threshold satisfy minimum uncertainty product

$$P_{\Delta \hat{r}}(\Omega) P_{\Delta \hat{\psi}}(\Omega) = \frac{1}{4r_0^2} \tag{11.77}$$

or

$$P_{\Delta \hat{N}}(\Omega) P_{\Delta \hat{\psi}}(\Omega) = 1 \tag{11.78}$$

The quantum noises are equal to that of a broadband coherent state. They are simply imparted on the coherent laser emission by the reflected vacuum field fluctuation. The amplitude and phase noise spectra below the cavity cutoff frequency do not reach that limit yet, because of the phase diffusion noise shown in Figure 11.9a. On the other hand, the commutator bracket conservation requires vacuum field fluctuation $\hat{S}_A$ and dipole moment fluctuation $\hat{S}_{\Sigma}$, both of which constitute the Schawlow–Townes phase diffusion noise. Pump noise $\hat{S}_{\Lambda}$ is the only noise source that

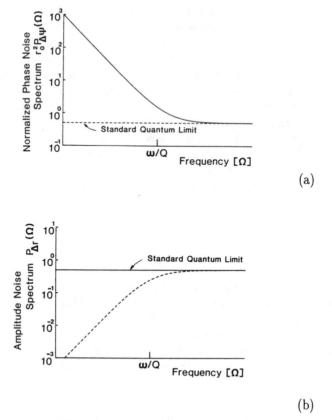

**Figure 11.9.** (*a*) Normalized phase-noise spectrum $r_0^2 P_{\Delta\psi}(\Omega)$ and (*b*) amplitude-noise spectrum $P_{\Delta r}(\Omega)$ of a pump-noise-suppressed laser oscillator biased at well above threshold ($r_0^2$ is the average photon flux and $\omega/Q$ is the cavity bandwidth).

can be eliminated without violating the commutator bracket conservation. If we assume zero pump noise, the amplitude noise spectrum at levels far above the threshold becomes

$$P_{\Delta\hat{r}}(\Omega) = \frac{\Omega^2}{2(\Omega^2 + \gamma^2)} \tag{11.79}$$

As shown in Figure 11.9*b*, the amplitude noise is squeezed to below the standard quantum limit. The phase diffusion noise compensates for the squeezed amplitude noise and satisfies the uncertainty product. In fact, the uncertainty product is twice as large as the minimum product. This is considered as a true quantum limit of an ideal laser oscillator.

Recently, similar conclusions were obtained by M. O. Scully et al. [33] and by D. F. Walls et al. [34], both using the density operator master equations and the quantum Langevin equations. In the special case of the three-level laser system with equal atomic decay rates for upper and lower states, reduction of the internal photon-number noise is not 50 but 25%. Consequently, the external field amplitude is squeezed by only 50%. As will be discussed next, a semiconductor laser is an exception and its amplitude can be completely squeezed.

## 11.4. THEORY OF SQUEEZED-STATE GENERATION IN SEMICONDUCTOR LASERS

The pump process of a semiconductor laser is the carrier injection process across a $p-n$ junction. The population inversion fluctuation is related to the junction voltage fluctuation via the quasi-Fermi level separation. The pump fluctuation is the junction current fluctuation. The junction voltage fluctuation and the junction current fluctuation couple, and they affect each other through the source resistance in an external circuit. Thus, the pump source cannot be considered a reservoir. Quantum-statistical properties of a reservoir must be independent of those of a system by definition, but junction current fluctuation is dependent on junction voltage fluctuation. To calculate the pump noise of a semiconductor laser correctly, the carrier injection process must be examined, taking into account the mutual coupling of current and voltage fluctuations.

### 11.4.1. Current Noise of a Single-Heterojunction Diode

This section discusses the fluctuation of a minority-carrier transport in a single heterojunction. Buckingham's theory of current noise in a $p-n$ homojunction [35] is extended to the case of a $p-N$ single heterojunction and to include the effect of a finite diode's series resistance. We treat only the low-frequency part of the current noise, so the cutoff characteristic due to the carrier lifetime is not discussed here. We also treat the situation $\hbar\Omega \ll k_B T$, where $\Omega$ is the current fluctuation frequency. Zero-point fluctuation in a resistive element can be safely neglected.

The following are four assumptions for the discussion in this section [36]:

1. The depletion layer has abrupt boundaries; built-in potential and applied voltage are supported within the two boundaries so that outside the depletion layer, the semiconductor is neutral.
2. Carrier densities may be represented by an exponential approximation of the Fermi–Dirac function.
3. Injected minority carrier densities are small compared to majority carrier concentration.
4. There is no generation or recombination current in the depletion layer.

**11.4.1.1. Current–Voltage Characteristic.** Consider the $p^+$–$N$ single-heterojunction diode shown in Figure 11.10. The heavily $p$-doped semiconductor has a narrow band gap, and the lightly $n$-doped semiconductor has a wide band gap. The junction current is carried by electrons injected from the wide-band-gap $N$-type semiconductor to the narrow-band-gap $p$-type semiconductor, because the band-gap discontinuity blocks the hole injection from the $p^+$ side to the $N$ side.

Built-in potential $V_D$, which is determined by band-gap energies and impurity concentrations of the two semiconductors, is shared by the $p$ side and $N$ side in accordance with relationships [36, 37]

$$V_{D_p} = \frac{V_D}{K} \tag{11.80}$$

and

$$V_{DN} = \left[1 - \frac{1}{K}\right] V_D \tag{11.81}$$

where

$$K = 1 + \frac{\epsilon_1(N_{A1}^- - N_{D1}^+)}{\epsilon_2(N_{D2}^+ - N_{A2}^-)} \tag{11.82}$$

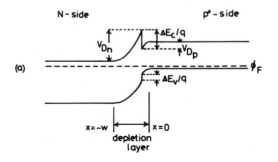

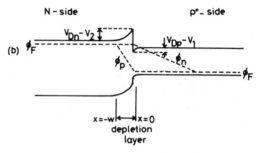

**Figure 11.10.** Band diagrams of a $p^+$–$N$ single heterojunction. ($a$) Zero-bias case $V = 0$; ($b$) forward-bias case $V > 0$.

where $\epsilon_1$ and $\epsilon_2$ are the dielectric constants of the $p$ side and $N$ side, respectively, $N_{A1}^- - N_{D1}^+$ is the hole concentration in the $p$ side, and $N_{D2}^+ - N_{A2}^-$ is the electron concentration in the $N$ side under the thermal equilibrium condition. Since a $p^+$-$N$ junction is assumed, most of the built-in potential is supported in the $N$ side ($K \gg 1$). The potential barrier for the electron flux from the $N$ side to $p^+$ side is $V_{DN} \simeq V_D$ at thermal equilibrium, while the potential barrier for the electron flux from the $p$ side to $N$ side is $\Delta E_c/q - V_{D_p} \simeq \Delta E_c/q$, as shown in Figure 11.10a, where $\Delta E_c$ is the conduction-band discontinuity.

When forward bias $V$ is applied to the junction, potential barriers become $V_{DN} - V_2 \simeq V_D - V$ and $\Delta E_c/q - (V_{D_p} - V_1) \simeq \Delta E_c/q$, as shown in Figure 11.10b, where $V_1 = V/K$ and $V_2 = (1 - 1/K)V$ are the junction voltages supported in the $p^+$ side and $N$ side. Electron density at $x = 0$ (the edge of the depletion layer) is given by the expression [37]

$$n_p = X n_{N0} \exp\left[ -\frac{V_D - V}{V_T} \right] \qquad (11.83)$$

where $X$ represents the transmission coefficient of electrons across the heterojunction interface, $n_{N0}$ is the electron concentration in the $N$ side at thermal equilibrium, and $V_T = k_B T/q$ is the thermal voltage. Since electron density $n_p$ at zero applied voltage is equal to the electron concentration in the $p^+$ side at thermal equilibrium, the relationship

$$n_{p0} = X n_{N0} \exp\left[ -\frac{V_D}{V_T} \right] \qquad (11.84)$$

holds.

Electrons injected into the $p^+$ side diffuse toward the $p$-side metal contact. The continuity equation for electron density $n(x, t)$ in the $p^+$ side is [36]

$$\frac{\partial}{\partial t} n(x, t) = -\frac{n(x, t) - n_{p0}}{\tau_n} - \frac{1}{q} \frac{\partial}{\partial x} i_n(x, t) \qquad (11.85)$$

Here, $n_{p0}$ is the electron concentration in the $p^+$ side at thermal equilibrium and $\tau_n$ is the electron lifetime. Because of assumption (i), electron current density $i_n(x, t)$ is given only by the diffusion current and is

$$i_n(x, t) = q D_n \frac{\partial}{\partial x} n(x, t) \qquad (11.86)$$

where $D_n$ is the electron diffusion constant.

The steady-state solution of Eq. (11.85) with boundary conditions [Eq. (11.83)] at $x = 0$ and $n_p(x \to \infty) = n_{p0}$ is

$$n(x) = n_{p0} + (n_p - n_{p0})e^{-x/L_n} \tag{11.87}$$

where $L_n = \sqrt{D_n \tau_n}$ is the electron diffusion length. The electron current density is obtained from Eqs. (11.86) and (11.87) as the expression

$$i_n(x) = \frac{q D_n}{L_n}(n_p - n_{p0})e^{-x/L_n} \tag{11.88}$$

and junction current density $i$ across the $p^+$–$N$ junction is equal to $i_n(x = 0)$:

$$i \equiv i_n(0) = \frac{q D_n}{L_n}(n_p - n_{p0}) = \frac{q D_n n_{p0}}{L_n}\left[\exp\left[\frac{V}{V_T}\right] - 1\right] \tag{11.89}$$

The diode's differential conductance $G$ is defined by

$$G \equiv A\frac{di}{dV} = \frac{Aq D_n X n_{N0}}{L_n V_T}\exp\left[-\frac{V_D - V}{V_T}\right] \simeq \frac{I}{V_T} \quad \text{(strong forward bias)} \tag{11.90}$$

Here, $A$ is the cross section of the junction and $I$ is the total junction current. The diode's diffusion capacitance $C$ is defined by

$$C \equiv Aq\frac{d}{dV}\left[\int_0^\infty [n(x) - n_{p0}]\,dx\right] = \frac{AqL_p X n_{N0}}{V_T}\exp\left[-\frac{V_D - V}{V_T}\right] \tag{11.91}$$

**11.4.1.2. Thermal Fluctuation of Minority-Carrier Flow.** Here we assume the following. The source resistance and the bulk resistance of the diode are negligibly small compared with the diode's differential resistance. The junction voltage supported between $x = 0$ and $x = -W$ is kept constant because of this assumption, so the electron density at $x = 0$ is fixed. The electron density at $x = w$ (the $p$-side metal contact) is also fixed. As a result, the electron distribution fluctuates only between $x = 0$ and $x = w$. (In this and following sections, we will calculate the current noise of such a "constant-voltage-driven" $p^+$–$N$ junction due to the electron distribution fluctuations.) Later, this assumption will be removed, which corresponds to taking the finite source resistance and associated thermal fluctuations of majority-carrier flow into account.

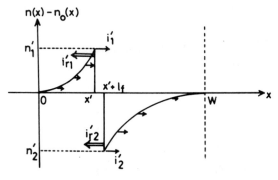

**Figure 11.11.** Perturbed electron distribution and relaxation currents due to thermal fluctuation of electron flows.

There are two mechanisms responsible for the current noise of a constant-voltage-driven $p^+$–$N$ junction. One is the thermal fluctuations of the electron flow (diffusion), and the other is the generation and recombination noise of electrons. These two mechanisms cause the departure of the electron distribution from the unperturbed distribution. These perturbations are immediately followed by electron relaxation flows so that equilibrium is restored. This relaxation current is the one that can be observed as the current noise in the external circuit. The electron diffusion process in the $p^+$ semiconductor is actually the random motion of electrons due to thermal agitation. Each action of this thermal fluctuation is represented by the transit of a single electron over the distance $l_f$ between collisions, where $l_f$ is an electron's mean free path. This corresponds to a current flow of $q\,\delta(t)$ over this very short distance (on the order of 100 Å). Perturbed electron distribution $n'(x)$, which is triggered by this initial action, is illustrated in Figure 11.11, where $n'(x)$ is the deviation from the unperturbed electron distribution. The electron is assumed to move in the $-x$ direction at $x = x'$ in the case of Figure 11.11. This perturbation immediately results in relaxation electron currents throughout the entire region between $x = 0$ and $x = w$, as shown in Figure 11.11.

The Fourier transform of continuity equation, Eq. (11.85), with Eq. (11.86) becomes

$$\frac{\partial^2}{\partial x^2} n'(j\Omega) = \frac{1}{L_n^2} n'(j\Omega) \tag{11.92}$$

The solution of Eq. (11.92) for boundary conditions $n'(j\Omega) = 0$ at $x = 0$ and $n'(j\Omega) = n_1'$ at $x = x'$ results in the relationship

$$n'(j\Omega) = \frac{n_1'}{e^{x'/L_n} - e^{-x'/L_n}} (e^{x/L_n} - e^{-x/L_n}) \quad (0 \leq x \leq x' - 0) \tag{11.93}$$

The Fourier transform of relaxation current density $i_1'(j\Omega)$ at $x = x' - 0$ is

$$i_1'(j\Omega) = qD_n \frac{\partial n'(j\Omega)}{\partial x}\bigg|_{x=x'} = k_1 n_1' \tag{11.94}$$

where

$$k_1 = \frac{qD_n}{L_n}\coth\left[\frac{x'}{L_n}\right] \tag{11.95}$$

The plus sign of Eq. (11.94) indicates that this current flows in the $+x$ direction. The perturbed electron distribution between $x = x' + l_f$ and $x = w$, and the Fourier transform of the relaxation current density $i_2'(j\Omega)$ at $x = x' + l_f$ are similarly obtained as expressions

$$n'(j\Omega) = \frac{n_2'}{e^{(w-x')/L_n} - e^{-(w-x')/L_n}} \times \left(e^{(w-x)/L_n} - e^{-(w-x)/L_n}\right) \tag{11.96}$$

$$i_2'(j\Omega) \equiv qD_n \frac{\partial n'(j\Omega)}{\partial x}\bigg|_{x=x'+l_f} = -k_2 n_2' \tag{11.97}$$

and

$$k_2 = \frac{qD_n}{L_n}\coth\left[\frac{w - x'}{L_n}\right] \tag{11.98}$$

Since $n_2'$ is negative, current $i_2'$ also flows in the $+x$ direction. Constants $n_1'$ and $n_2'$ remain to be determined.

There are two return currents, $i_{r1}'(j\Omega)$ at $x = x' + 0$ and $i_{r2}'(j\Omega)$ at $x = x' + l_f - 0$. Since $l_f \ll L_n$, these two currents become equal, resulting in

$$i_{r1}'(j\Omega) = i_{r2}'(j\Omega) = -\frac{qD_n}{l_f}(n_1' - n_2') \tag{11.99}$$

The minus sign of Eq. (11.99) indicates that these two return currents flow in the minus $x$ direction, as shown in Figure 11.11.

Since there can be no accumulation of electrons in the $p^+$ semiconductor, current continuity must be imposed both at $x = x'$ and $x = x' + l_f$. Consequently, equations

$$i_1'(j\Omega) + i_{r1}'(j\Omega) + q = 0 \tag{11.100}$$

and

$$i_2'(j\Omega) + i_{r2}'(j\Omega) + q = 0 \tag{11.101}$$

hold true. Then, from Eqs. (11.94), (11.97), and (11.99), relationships

$$n_1' = \frac{l_f}{D_n} \frac{k_2}{k_1 + k_2} \tag{11.102}$$

and

$$n_2' = -\frac{l_f}{D_n} \frac{k_1}{k_1 + k_2} \tag{11.103}$$

can be obtained.

The total outflow from the event is equal to the difference between the two Fourier-transformed current densities at $x = 0$ and $x = w$. This outflow causes a departure from charge equilibrium that is restored by majority-carrier flow through the metal contact, that is, the external circuit current pulse. It is described by

$$i_T'(j\Omega) = i_0'(j\Omega) - i_w'(j\Omega) \tag{11.104}$$

where

$$i_0'(j\Omega) \equiv q D_n \frac{\partial n'(j\Omega)}{\partial x}\bigg|_{x=0} = \frac{l_f}{D_n} \frac{k_0 k_2}{k_1 + k_2} \tag{11.105}$$

$$i_w'(j\Omega) \equiv q D_n \frac{\partial n'(j\Omega)}{\partial x}\bigg|_{x=w} = \frac{l_f}{D_n} \frac{k_w k_1}{k_1 + k_2} \tag{11.106}$$

Here,

$$k_0 = \frac{q D_n}{L_n} \operatorname{cosech}\left[\frac{x'}{L_n}\right] \tag{11.107}$$

and

$$k_w = \frac{q D_n}{L_n} \operatorname{cosech}\left[\frac{w - x'}{L_n}\right] \tag{11.108}$$

As indicated by Eqs. (11.105) and (11.106), $i_0'(j\Omega)$ and $i_w'(j\Omega)$ are caused mainly by the events occurring near $x = 0$ and $x = w$, respectively. This is because the electron relaxation far from the junction edge or the metal contact is entirely compensated for by majority carrier flow. Later, we will examine the current noise due to this majority-carrier flow separately.

The average number of thermal motion events per second in the small volume formed by $x = x'$ and $x = x' + \Delta x$ planes and by cross-sectional area $A$ is

$$\gamma_T = \frac{n(x') A \Delta x}{\tau_f} \tag{11.109}$$

where $\tau_f$ is the electron mean free time between collisions. From Carson's theorem [38], the power spectrum of the current noise due to the events in this small volume follows the relationship

$$\Delta P_{i'}(\Omega) \equiv 2\gamma_T |i_T(j\Omega)|^2$$

$$= 2\frac{n(x')A\,\Delta x}{\tau_f}\left[\frac{l_f}{D_n}\right]^2\left[\frac{k_0 k_2 - k_w k_1}{k_1 + k_2}\right]^2$$

(single-sided or unilateral)   (11.110)

Since thermal fluctuations in the different volumes have no correlation, the power spectrum of the total current noise is given by integrating Eq. (11.110) over the range from $x = 0$ to $x = w$, resulting in

$$P_{i'}(\Omega) = \frac{4A}{D_n}\int_0^\infty n(x)\left[\frac{k_0 k_2 - k_w k_1}{k_1 + k_2}\right]^2 dx$$

$$\simeq \frac{4Aq^2 D_n}{L_n}\left[\frac{n_p - n_{p0}}{3} + \frac{n_{p0}}{2}\right] \quad (w \gg L_n) \quad (11.111)$$

Here, Eq. (11.87) and the Einstein relationship [39]

$$D_n = \frac{l_f^2}{2\tau_f} \tag{11.112}$$

are used.

**11.4.1.3.   Generation and Recombination Noise.** When a generation or recombination event occurs at $x = x'$, the charge neutrality is still preserved in the entire region between $x = 0$ and $x = w$ so that there is no majority-carrier relaxation. However, there is a perturbation in the minority-carrier distribution that can be removed by minority-carrier relaxation flow, and therefore, results in current noise.

The initial action of this generation and recombination noise is the respective instantaneous appearance and disappearance of an electron. The generation corresponds to a flow $-q\,\delta(t)$ from nowhere to the $x = x'$ plane. The departure from the unperturbed electron distribution is illustrated in Figure 11.12 for this case. Perturbed electron distribution $n''(j\Omega)$ is similarly obtained from continuity Eq. (11.92), having boundary conditions $n''(j\Omega) = 0$ at $x = 0$ and $x = w$, and $n''(j\Omega) = n_1''$ at $x = x'$.

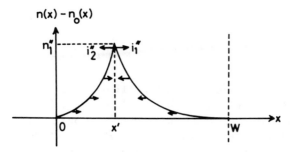

**Figure 11.12.** Perturbed electron distribution and relaxation currents due to a generation event.

Relaxation current densities $i_1''(j\Omega)$ at $x = x' - 0$ and $i_2''(j\Omega)$ at $x = x' + 0$ are obtained as the relationships

$$i_1''(j\Omega) = qD_n \frac{\partial n''(j\Omega)}{\partial x}\bigg|_{x=x'-0} = k_1 n_1'' \qquad (11.113)$$

and

$$i_2''(j\Omega) = qD_n \frac{\partial n''(j\Omega)}{\partial x}\bigg|_{x=x'+0} = -k_2 n_1'' \qquad (11.114)$$

Constants $k_1$ and $k_2$ are given by Eqs. (11.95) and (11.98). The current continuity at $x = x'$ imposes the equation

$$i_1''(j\Omega) - i_2''(j\Omega) - q = 0 \qquad (11.115)$$

From this relationship, constant $n_1''$ is found by

$$n_1'' = \frac{q}{k_1 + k_2} \qquad (11.116)$$

The external current pulse is the difference between the two Fourier-transformed current densities at $x = 0$ and $x = w$ in

$$i_T''(j\Omega) = i_0''(j\Omega) - i_w''(j\Omega) \qquad (11.117)$$

where

$$i_0''(j\Omega) = q\frac{k_0}{k_1 + k_2} \qquad (11.118)$$

and

$$i_w''(j\Omega) = -q\frac{k_w}{k_1 + k_2} \tag{11.119}$$

Constants $k_0$ and $k_w$ are given by Eqs. (11.107) and (11.108).

As indicated by Eqs. (11.118) and (11.119), $i_0''(j\Omega)$ and $i_w''(j\Omega)$ are caused mainly by the generation–recombination events occurring in the vicinity of $x = 0$ and $x = w$. Currents $i_0''(j\Omega)$ and $i_w''(j\Omega)$ flow in the opposite direction so that they constructively contribute to external current $i_T''(j\Omega)$. On the other hand, currents $i_0'(j\Omega)$ and $i_w'(j\Omega)$, due to thermal fluctuations (diffusion), flow in the same direction so that they cancel each other out in external current $i_T'(j\Omega)$. When $w \gg L_n$, the contribution of $i_w'(j\Omega)[i_w''(j\Omega)]$ is much smaller than that of $i_0'(j\Omega)[i_0''(j\Omega)]$ because the excess electron density concentrates in the vicinity of the edge of the depletion layer where $x = 0$.

The average number of recombination events per second in the small volume $A \Delta x$ is given by

$$\gamma_R = \frac{nA \Delta x}{\tau_n} \tag{11.120}$$

while the average number of generation events is

$$\gamma_G = \frac{n_{p0} A \Delta x}{\tau_n} \tag{11.121}$$

The power spectrum of the current noise due to the two events in this small volume follows the relationship

$$\Delta P_{i''}(\Omega) \equiv 2(\gamma_R + \gamma_G)|i_T''(j\Omega)|^2 = 2\frac{[n(x') + n_{p0}]A \Delta x}{\tau_n}q^2\left[\frac{k_0 + k_w}{k_1 + k_2}\right]^2 \tag{11.122}$$

Then, integrating Eq. (11.122) from $x = 0$ to $x = w$ results in

$$\begin{aligned}P_{i''}(\Omega) &\equiv 2\frac{Aq^2}{\tau_n}\int_0^\infty [n(x) + n_{p0}]\left[\frac{k_0 + k_w}{k_1 + k_2}\right]^2 dx \\ &\simeq \frac{2Aq^2 D_n}{L_n}\left[\frac{n_p - n_{p0}}{3} + n_{p0}\right] \quad (w \gg L_n) \tag{11.123}\end{aligned}$$

**11.4.1.4. Thermal Fluctuation of Majority-Carrier Flow.** Random thermal motions of the majority carriers in both the $p^+$ and $N$ semiconductors

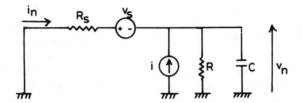

**Figure 11.13.** Noise-equivalent circuit of a $p^+$–$N$ junction diode.

result in current noise in the external circuit. These noise components are known to obey the Nyquist formula for the thermal noise associated with the resistance in each part. If diode series resistance $R_S$ is assigned to express all of the resistive parts, the power spectrum of the current noise complies with the relationship [40]

$$P_{i''}(\Omega) = \frac{2\hbar\Omega}{R_S} \coth\left[\frac{\hbar\Omega}{2k_BT}\right] \simeq \frac{4k_BT}{R_S} \qquad (11.124)$$

The second equality results from the assumption $\hbar\Omega \ll k_BT$ and indicates that the quantum zero-point fluctuation is not important for the electrical current fluctuation.

**11.4.1.5. Constant Voltage Operation versus Constant Current Operation.** The noise equivalent circuit of a $p^+$–$N$ single-heterojunction diode is shown in Figure 11.13. Here, current noise $i'''$, due to the thermal fluctuation of majority-carrier flow, is transformed into voltage noise source $v_s$ by using the Thevenin theorem. Its power spectrum is given by

$$P_{v_s}(\Omega) = 4k_BTR_S \leftarrow \text{thermal fluctuation of majority-carrier flow}$$

$$(11.125)$$

The terms $R_S$, $R$, and $C$ are the diode's series resistances, which also include the source resistance, the diode's differential resistance [Eq. (11.90)] and the diffusion capacitance [Eq. (11.91)]. Current noise source $i$ includes the two noise mechanisms, and its power spectrum is given by

$$P_i(\Omega) = \frac{4Aq^2 D_n}{L_n}\left[\frac{n_p}{3} + \frac{n_{p0}}{6}\right]$$

$$\uparrow$$

thermal fluctuation of minority-carrier flow

$$+ \frac{2Aq^2 D_n}{L_n}\left[\frac{n_p}{3} + \frac{2}{3}n_{p0}\right]$$

$$\uparrow$$

generation–recombination noise $\qquad (11.126)$

Next we will confirm that this noise-equivalent circuit is adequate to explain the experimental results under the three following operating conditions.

1. *Zero Applied Voltage (V = 0)*. When the applied voltage is zero, the electron density at $x = 0$ is equal to thermal equilibrium $n_p = n_{p0}$. Equation (11.126) then reduces to

$$P_i(\Omega) = \frac{4Aq^2 D_n n_{p0}}{L_n} = \frac{4k_B T}{R(V = 0)} \qquad (11.127)$$

where $R(V = 0)$ is the differential resistance [Eq. (11.90)] when $V$ is 0. The current noise source is equal to the Nyquist formula, as it should be, because the diode is in thermal equilibrium when $V$ is 0. It is interesting to note that this thermal noise is equally contributed to by diffusion noise $i'$ and generation–recombination noise $i''$.

2. *Constant-Voltage-Driving Case*. When $R_S$ is much smaller than $R$, junction voltage noise $v_n$ is suppressed, as shown by

$$v_n = \frac{-v_s + R_S i}{1 + R_S(1/R + j\omega C)} \to 0 \quad (R_S \ll R) \qquad (11.128)$$

This situation corresponds to the thermal fluctuations of majority-carrier flow in the resistive parts being negligibly small. Junction-current noise $i_n$ becomes equal to current noise source $i$ indicated in the relationship

$$i_n = \frac{i + v_s(1/R + j\omega C)}{1 + R_S(1/R + j\omega C)} \to -i \quad (R_S \ll R) \qquad (11.129)$$

The power spectrum of $i_n$ becomes equal to Eq. (11.126), shown in

$$P_{i_n}(\Omega) = \frac{2Aq^2 D_n}{L_n}(n_p + n_{p0}) = 2q(I + 2I_s) \simeq 2qI \quad (11.130)$$

where $I = (Aq D_n/L_n)(n_p - n_{p0})$ is the total current and $I_s = (Aq D_n/L_n)n_{p0}$ is the reverse saturation current. The current noise reduces to the shot-noise formula. This is a "theorist's shot noise." In most cases, this shot noise cannot be measured because the measurement circuit breaks the constant-voltage operation. Two-thirds of the full-shot noise originates from the diffusion noise, and one-third stems from the generation and recombination noise.

3. *Constant-Current-Driving Case.* When $R_S$ is much larger than $R$, junction-voltage noise $v_n$ is not suppressed. In such a case, however, junction-current noise $i_n$ is suppressed by the source and reduces to the thermal-noise-limited value represented by

$$i_n \rightarrow \frac{v_s}{R_S} \quad (R_S \gg R) \tag{11.131}$$

and

$$P_{i_n}(\Omega) \rightarrow \frac{4k_BT}{R_S} \tag{11.132}$$

Power spectrum $P_{i_n}(\Omega)$ can be much smaller than the shot-noise level $2qI$ if the condition

$$R_S > \frac{2V_T}{I} = 2R \tag{11.133}$$

is satisfied. This condition is usually satisfied under a strong forward-biased condition, because the differential resistance monotonically decreases with the junction current, and the bulk resistance and contact resistance of the diode are finite.

Junction voltage noise $v_n$ can be represented by the relationships

$$v_n \rightarrow \frac{i}{1/R + j\omega C} \quad (R_S \gg R) \tag{11.134}$$

and

$$P_{v_n}(\Omega) = \frac{R^2}{1 + \Omega^2\tau^2}P_i(\Omega) \simeq 2qIR^2 \quad \left(\Omega \ll \frac{1}{\tau}\right) \tag{11.135}$$

where $\tau = 1/RC$. In fact, the normalized junction-voltage noise $P_{v_n}(\Omega)/R^2$ is equal to the shot-noise formula. Most experimental studies on noise in a $p$–$n$ junction diode [41] correspond to this situation, which is sometimes erroneously referred to as "the observation of the shot-noise-limited junction current." However, what is actually measured is not the shot-noise-limited junction-current noise but the junction-voltage noise. This is an "experimentalist's shot noise" and is entirely different from the "theorist's shot noise."

The important observation made in this section is that the junction-current noise is suppressed to below the ordinary shot-noise level when $R_S$ is greater than $2R$ ("high-impedance suppression"). This condition is automatically satisfied under a strong forward-biased condition. It is also important to note that shot-noise-limited current fluctuation or voltage fluctuation does not originate from the pump source, but is the consequence of the internal noise mechanisms of the junction itself. The noise generated in a pump source is the thermal noise [Eq. (11.132)] under the condition $\hbar\Omega \ll k_B T$.

### 11.4.2.  Current Noise of a Double-Heterojunction Diode

**11.4.2.1.  Current–Voltage Characteristic.** Consider next the $P-_p-N$ double heterojunction diode shown in Figure 11.14. In the double-heterojunction diode, injected electrons from the wide-band-gap $N$ semiconductor to the narrow-band-gap $p$-type active layer cannot diffuse freely toward the $p$-side metal contact. This is because of the conduction-band discontinuity between the $p$-type wide-band-gap semiconductor and the $p$-type narrow-band-gap active layer ($P-_p$ isotype heterojunction).

When thickness $d$ of the active layer is much smaller than electron diffusion length $L_n$, the electron density becomes constant along the entire active layer. For such a case, the electron and hole densities in the active layer can be described uniquely by quasi-Fermi levels $\phi_n$ for

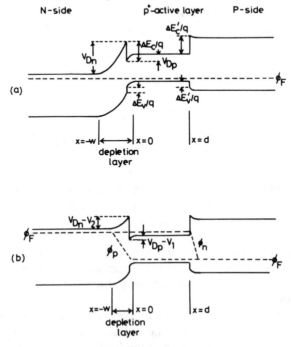

**Figure 11.14.**  Band diagrams of a $P-p^+-N$ double-heterostructure laser diode. ($a$) Zero-bias case $V = 0$; ($b$) forward-bias case $V > 0$.

the conduction band and $\phi_p$ for the valence band, as in

$$n = n_i \exp\left[\frac{\psi - \phi_n}{V_T}\right] \tag{11.136}$$

and

$$p = p_i \exp\left[\frac{\phi_p - \psi}{V_T}\right] \tag{11.137}$$

where $n_i$ and $p_i$ are the respective intrinsic electron and hole densities and $\psi$ is the intrinsic Fermi level of the active layer. For low injection, where $n \ll (N_{A1}^- - N_{D1}^+)$, and high injection, where $n \gg (N_{A1}^- - N_{D1}^+)$, electron density $n$ can be expressed in terms of the junction voltage by

$$n = \begin{cases} n_{p0} \exp\left[\dfrac{V}{V_T}\right], & \text{low injection} \\[2ex] n_i \exp\left[\dfrac{V}{2V_T}\right], & \text{high injection} \end{cases} \tag{11.138}$$

This result is based on the Boltzmann approximation (assumption at the beginning of Section 11.4.1).

In the double-heterojunction diode, the junction current is not carried by the diffusion process, but crosses the imaginary plane between the conduction and valence bands by radiative or nonradiative recombination processes. Therefore, it is expressed by

$$I = q\frac{N_c}{\tau_n} \tag{11.139}$$

where

$$\tau_n = \left[\frac{1}{\tau_{nr}} + \frac{1}{\tau_{sp}}\right]^{-1} \tag{11.140}$$

Here, $N_c = A\,dn$ is the total electron number in the active layer, $d$ is the active-layer thickness, $\tau_{sp}$ is the electron lifetime due to spontaneous emission, and $\tau_{nr}$ is the nonradiative lifetime due to surface state recombination, Auger recombination, and other factors.

The diode's differential resistance is given by

$$R \equiv \left[\frac{dI}{dV}\right]^{-1} = \frac{mV_T}{I} = \frac{mV_T \tau_n}{qN_c} \tag{11.141}$$

Here, $m$ is unity for the low-injection case and is 2 for the high-injection case. Diffusion capacitance $C$ is defined by $dQ_c$, the incremental increase in charge for an incremental change in junction voltage $dV$. This can be represented by

$$C \equiv \frac{dQ_c}{dV} = q \frac{dN_c}{dV} = \frac{qN_c}{mV_T} \qquad (11.142)$$

From this equation it can be seen that total electron number fluctuation $\Delta N_c$ is uniquely related to junction-voltage fluctuation $v_n$,

$$\Delta N_c = \frac{C}{q} v_n = \frac{N_c}{mV_T} v_n \qquad (11.143)$$

### 11.4.2.2. Thermal Fluctuation of Minority Carrier Flow.

In a double-heterojunction diode driven by a constant-voltage source, electron densities at $x = 0$ ($p$–$N$ anisotype heterojunction) and $x = d$ ($p$–$P$ isotype heterojunction) are fixed constants. These two junctions maintain the boundary conditions and also act as the sources of minority carriers and majority carriers for neutralizing potentials in the active layer. Here, the effect of a finite bulk resistance and electrode contact resistance is neglected.

Relaxation current densities at $x = 0$ and $x = d$ due to thermal fluctuations of electrons in the active layer are given by Eqs. (11.105) and (11.106), if $w$ is replaced by $d$. When $d \ll L_n$, the two Fourier-transformed current densities become equal, as in the relationship

$$i_0'(j\Omega) \simeq i_d'(j\Omega) \simeq \frac{ql_f}{d} \qquad (11.144)$$

Therefore, total outflow $i_T'(j\Omega) = i_0'(j\Omega) - i_d'(j\Omega)$ due to the thermal fluctuation event is equal to zero.

The thermal fluctuation of minority-carrier flow does not contribute to the external current noise. At each junction, however, the power spectra of noise current flow are provided by

$$P_{i_0} = P_{i_d'}(\Omega) \equiv \int_0^d \frac{2n(x)A}{\tau_f} \left[ \frac{ql_f}{d} \right]^2 dx = \frac{2q^2 N_c}{\tau_f} \left[ \frac{l_f}{d} \right]^2 = 4qI \left[ \frac{L_n}{d} \right]^2 \qquad (11.145)$$

These noise currents are larger by the factor $2(L_n/d)^2$ than the shot-noise level, but they exactly cancel each other out because of the correlation between $i_0'$ and $i_d'$.

**11.4.2.3.  Generation and Recombination Noise.** Relaxation current densities at $x = 0$ and $x = d$ due to the generation and recombination events in the active layer are given by Eqs. (11.118) and (11.119), if $w$ is replaced by $d$. When $d \ll L_n$, the two current densities are given by

$$i_0''(j\Omega) \simeq -\left[\frac{x'}{d}\right]q \tag{11.146}$$

and

$$i_d''(j\Omega) \simeq -\left[1 + \frac{x'}{d}\right]q \tag{11.147}$$

Therefore, total outflow $i_T''(j\Omega) = i_0''(j\Omega) - i_d''(j\Omega)$ due to either a generation or recombination event is exactly equal to the charge of one carrier.

The average number of generation events per second in the entire active layer is found by

$$v_G = N_{c0}\left[\frac{1}{\tau_{sp}} + \frac{1}{\tau_{nr}}\right] \tag{11.148}$$

where

$$N_{c0} = n_{p0}Ad \tag{11.149}$$

The average number of recombination events is given by

$$v_R = N_c\left[\frac{1}{\tau_{nr}} + \frac{1}{\tau_{sp}}\right] \tag{11.150}$$

From Carson's theorem, the power spectrum of the current noise due to generation and recombination events is denoted by

$$P_i(\Omega) = 2q^2(N_c + N_{c0})\left[\frac{1}{\tau_{nr}} + \frac{1}{\tau_{sp}}\right] \simeq 2qI \tag{11.151}$$

Here, Eq. (11.139) is used in Eq. (11.151), and it is assumed that $N_{c0} \ll N_c$. The current noise is again the full-shot noise, but in this case, it is contributed only by the generation–recombination noise. The noise-equivalent circuit shown in Figure 11.13 holds also for a double-heterojunction diode, where $R$ is given by Eq. (11.141) and $C$ by Eq. (11.142), instead of by Eqs. (11.90) and (11.91).

**11.4.2.4. Effect of Stimulated Emission and Absorption.** In a double-heterojunction diode laser, the junction current mainly crosses the imaginary plane between the conduction and valence bands by stimulated emission and absorption processes in an active layer. For such a case, the electron lifetime is

$$\tau_n = \left( \frac{1}{\tau_{nr}} + \frac{1}{\tau_{sp}} + \frac{1}{\tau_{st}} \right)^{-1} \tag{11.152}$$

and the generation and recombination rates are

$$\nu_G = N_{c0} \left( \frac{1}{\tau_{nr}} + \frac{1}{\tau_{sp}} \right) + E_{vc} S_0 \tag{11.153}$$

$$\nu_R = N_c \left( \frac{1}{\tau_{nr}} + \frac{1}{\tau_{sp}} \right) + E_{cv} S_0 \tag{11.154}$$

Here, $E_{cv}$ and $E_{vc}$ are stimulated emission and absorption coefficients

$$E_{cv} = \sum_{kk'} \frac{|g_{kk'}|^2 \gamma_{kk'} n_{kc} (1 - n_{k'v})}{\gamma_{kk'}^2 + (\omega_{kk'} - \omega)^2} \tag{11.155}$$

$$E_{vc} = \sum_{k,k'} \frac{|g_{kk'}|^2 \gamma_{kk'} n_{k'v} (1 - n_{kc})}{\gamma_{kk'}^2 + (\omega_{kk'} - \omega)^2} \tag{11.156}$$

where $g_{kk'}$ is the transition matrix element and $\omega_{kk'}$ is the frequency difference, both between the conduction electron with wavenumber $k$ and the valence hole with wavenumber $k'$, $\gamma_{kk'}$ is the dephasing time constant, and $n_{kc}$ and $n_{k'v}$ are the conduction and valence electron numbers; $S_0$ is the total photon number inside the active region.

The power spectrum of the current noise is

$$P_i(\Omega) = 2q^2 \left[ (N_c + N_{c0}) \left( \frac{1}{\tau_{nr}} + \frac{1}{\tau_{sp}} \right) + (E_{cv} + E_{vc}) S_0 \right]$$

$$\simeq 2qI + 4q^2 E_{vc} S_0 \tag{11.157}$$

The current noise is higher than the shot-noise level, because the junction current is given by the difference between the stimulated emission and absorption rates, $I = q[N_c(1/\tau_{nr} + 1/\tau_{sp}) + (E_{cv} - E_{vc})S_0]$, but the current noise is contributed independently by these two processes [42].

### 11.4.3. Noise-Equivalent Circuit of a Semiconductor Laser

The following facts are discovered in the discussions in Sections 11.4.1 and 11.4.2.

1. The shot-noise-limited current noise in a $p-n$ junction diode is not the noise introduced by the pump source, but is the result of the thermal fluctuation of minority carrier flow (diffusion noise) and generation–recombination noise inside the diode.

2. This shot-noise-limited current noise exists only when the diode is biased by a constant-voltage source (negligible source resistance).

3. The pump noise for a semiconductor laser is the Johnson–Nyquist (thermal) noise generated in the source resistance. In the strong forward-bias condition, the diode's differential resistance becomes smaller than the source resistance, and in such a case the (thermal) pump noise becomes smaller than the shot-noise level (high-impedance suppression).

We can now construct a complete and self-consistent noise-equivalent circuit for a semiconductor laser, including the mutual coupling between the pump source and the junction. The noise-equivalent circuit discussed in this section is different from that presented in Figure 11.3, which

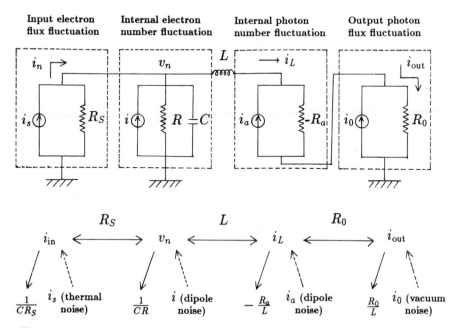

**Figure 11.15.** Noise-equivalent circuit of a semiconductor laser and the dissipation–fluctuation relations.

describes both the amplitude and phase of a photon field. Here, we are treating only the photon-number fluctuation (not at an optical frequency $\omega$, but at a radio frequency $\Omega$). This is because the purpose of this subsection is to demonstrate the conservation of input electron flux and output photon flux, even at a quantum level.

The noise-equivalent circuit shown in Figure 11.15 consists of the four parts: pump (input electron flux) fluctuation, population inversion (internal electron number) fluctuation, internal photon number fluctuation, and output photon flux fluctuation, respectively.

### 11.4.3.1. Fluctuation–Dissipation Theorem for Internal Electron Number.
Electron number fluctuation $\Delta N_c$ is uniquely determined by junction voltage fluctuation $v_n$ via Eq. (11.143). Thus, the electron-number fluctuation inside the active region corresponds to the fluctuation of the charge stored in capacitance $C$ in Figure 11.15. Electron-number fluctuation $\Delta N_c$ has two decay processes, one via source resistance $R_S$ and the other via differential resistance $R$. The circuit equation for the first process (assuming $R \to \infty$ and $i \to 0$) is

$$\left( \frac{1}{R_S} + j\omega C \right) v_n = i_s \tag{11.158}$$

or equivalently

$$\frac{d}{dt} \Delta N_c = - \frac{1}{CR_S} \Delta N_c + \frac{i_s}{q} \tag{11.159}$$

The power spectral density of the noise source $i_s/q$

$$P_{i_s/q}(\Omega) = \frac{4k_B T}{q^2 R_S} = 2 \frac{N_c}{CR_S} \tag{11.160}$$

is equal to twice of the electron-number decay rate via the source resistance. This is the fluctuation–dissipation theorem for the $CR_S$ decay process. The pump source functions to restore the electron number to its average value. This relaxation process with decay rate $1/CR_S$ accompanies noise [Eq. (11.160)], which is the true origin of the pump noise for a semiconductor laser. The noise current is the thermal noise of the source resistance, since $\hbar\Omega \ll k_B T$.

The circuit equation for the second process (assuming $R_S \to \infty$ and $i_s \to 0$) is

$$\left( \frac{1}{R} + j\omega C \right) v_n = i \tag{11.161}$$

or equivalently

$$\frac{d}{dt}\Delta N_c = -\frac{1}{CR}\Delta N_c + \frac{i}{q} \qquad (11.162)$$

As expected, decay rate $1/CR$ is equal to the inverse of effective electron lifetime $1/\tau_n = (1/\tau_{nr} + 1/\tau_{sp} + 1/\tau_{st})$, indicating that the second process is the relaxation process of the electron number fluctuation due to internal radiative and nonradiative recombination. The power spectral density of the noise source $i/q$

$$P_{i/q}(\Omega) = 2\left[\frac{N_c}{CR} + 2E_{vc}S_0\right] \qquad (11.163)$$

is equal to twice of the electron-number decay rate via the differential resistance (full-shot noise), plus a spurious stimulated absorption contribution. This is the fluctuation–dissipation theorem for the $CR$ decay process. The origin of this noise current is dipole moment noise sources $\hat{S}_\Sigma$ and $\hat{S}_\tau$ as indicated in Eq. (11.50). Combining Eqs. (11.159) and (11.162) gives

$$\frac{d}{dt}\Delta N_c = -\left(\frac{1}{CR_S} + \frac{1}{\tau_n}\right)\Delta N_c + \frac{1}{q}(i_s + i) \qquad (11.164)$$

### 11.4.3.2. Fluctuation–Dissipation Theorem for Internal Photon Number.
The average junction current carried by the lasing photons is $I_L = q\gamma S_0$. The photon-number fluctuation is thus determined by photon current fluctuation $i_L$ via $\Delta S = i_L/q\gamma$. The photon-number fluctuation inside the active region is proportional to the fluctuation of the current flowing in inductance $L$ in Figure 11.15. We assume that the electron energy is stored in capacitance $C$, that the photon energy is stored in inductance $L = \tau_{st}/C\gamma$, and that $\omega_r \equiv 1/\sqrt{LC} = \sqrt{\gamma/\tau_{st}}$ is the relaxation oscillation frequency. Current fluctuation $i_L$ is amplified in negative resistance $-R_a$ (stimulated emission gain) and is attenuated in positive resistance $R_0$ (output coupling loss). The circuit equation for the first process (assuming $R_0 \to 0$) is

$$(j\omega L - R_a)i_L = R_a i_a \qquad (11.165)$$

or equivalently

$$\frac{d}{dt}\Delta S = \frac{R_a}{L}\Delta S + \frac{i_a}{q} \qquad (11.166)$$

Here, the relation $R_a/L = E_{cv} - E_{cv} \simeq \gamma$ is used. The power spectral density of the noise source $i_a/q$ is

$$P_{i_a/q}(\Omega) = 2\left(\frac{R_a}{L}S_0 + 2E_{vc}S_0\right) \tag{11.167}$$

which satisfies the fluctuation–dissipation theorem for the amplification process and also includes the spurious noise due to stimulated absorption. The origin of this noise current is the dipole moment noise source $\hat{S}_\Sigma$, as indicated in Eq. (11.47). Therefore, $i_a$ is correlated with noise current $i$. The power spectrum of the mutual correlation function is

$$P_{ii_a/q^2}(\Omega) = -2(E_{cv} + E_{vc})S_0$$

The circuit equation for the second process (assuming $R_a \to 0$) is

$$(j\omega L + R_0)i_L = -R_0 i_0 \tag{11.168}$$

or equivalently

$$\frac{d}{dt}\Delta S = -\frac{R_0}{L}\Delta S - \frac{i_0}{q} \tag{11.169}$$

Here, relation $R_0/L = \gamma$ is used. The power spectral density of noise source $i_0/q$ is

$$P_{i_0/q}(\Omega) = 2\gamma S_0 \tag{11.170}$$

which satisfies the fluctuation–dissipation theorem for the decay process. The origin of this noise current is the incident vacuum field fluctuation, as indicated in Eq. (11.46). Combining Eqs. (11.166) and (11.169) gives

$$\frac{d}{dt}\Delta S = -[\gamma - (E_{cv} - E_{vc})]\Delta S + \frac{1}{q}(i_a - i_0) \tag{11.171}$$

### 11.4.3.3. Coupling of Internal Electron-Number Fluctuation and Internal Photon Number Fluctuation.
Junction voltage fluctuation $v_n$ and photon current fluctuation $i_L$ are coupled via capacitance $C$ and inductance $L$. The circuit equations for $v_n$ and $i_L$ are

$$v_n = j\Omega L i_L + (i_L + i_a)(-R_a) + (i_L + i_0)R_0 \tag{11.172}$$

and

$$i_L = -\left(\frac{1}{R_S} + \frac{1}{R} + j\Omega C\right)v_n + i_s + i \tag{11.173}$$

The Langevin equations for the internal electron-number fluctuation and internal photon-number fluctuation are obtained by replacing $j\Omega$ with $d/dt$ and by using relations $v_n = (q/C)\Delta N_c$ and $i_L = q\gamma\Delta S$ to give

$$\frac{d}{dt}\Delta N_c = -\left(\frac{1}{CR_S} + \frac{1}{\tau_n}\right)\Delta N - \gamma\Delta S + \frac{1}{q}(i_s + i) \qquad (11.174)$$

$$\frac{d}{dt}\Delta S = -[\gamma - (E_{cv} - E_{vc})]\Delta S + \frac{\Delta S}{\tau_{st}} + \frac{1}{q}(i_a - i_0) \quad (11.175)$$

Equation (11.174) is different from the conventional Langevin equation [Eq. (11.48)] for the electron-number fluctuation, as can be stated with the following two points: there is a new decay rate of the electron-number fluctuation, which represents the junction voltage pinning effect by the source; and the pump noise is not proportional to the pump rate, but is given by thermal noise current $i_s$. When $R_S$ is very large, decay rate $1/CR_S$ of the electron-number fluctuation becomes small, and accordingly, the pump noise becomes small.

**11.4.3.4. Second Threshold for Amplitude Squeezing.** The input electron flux fluctuation, that is, the external circuit current fluctuation $i_{in}$, is given by the Kirchhoff law

$$i_{in} = -\frac{v_n}{R_S} + i_s \qquad (11.176)$$

Junction voltage fluctuation $v_n$ is partly caused by noise current $i_s$ from the (reservoir) pump source, as shown in Eq. (11.158). Therefore, $v_n$ and $i_s$ are correlated. Junction voltage fluctuation $v_n$, in turn, affects external current $i_{in}$ flowing in the source resistance. This is the boundary condition at the energy input plane and represents the back action of system $v_n$ on the (reservoir) pump source.

The output photon flux fluctuation, i.e., the external photon current fluctuation, is given by the Kirchhoff law

$$i_{out} = i_L + i_0 \qquad (11.177)$$

Photon current fluctuation $i_L$ is partly caused by noise current $i_0$ from the (reservoir) external photon fields (vacuum fluctuations), as shown in Eq. (11.169). Therefore, $i_L$ and $i_0$ are correlated. Photon current fluctuation $i_L$, in turn, affects external current $i_{out}$ flowing in the load resistance. This is the boundary condition at the energy output plane and represents the back action of system $i_L$ on the (reservoir) external photon fields.

The external photon current fluctuation is related to output field fluctuation $\Delta \hat{r}$ as

$$i_{\text{out}} \leftrightarrow q \cdot 2r_0 \, \Delta\hat{r} \tag{11.178}$$

and the internal photon current fluctuation is related to internal field fluctuation $\Delta\hat{A}$ as

$$i_L \leftrightarrow q \cdot \gamma \cdot 2A_0 \, \Delta\hat{A} \tag{11.179}$$

Using relation $r_0 = \sqrt{\gamma} A_0$, boundary condition Eq. (11.177) is reduced to Eq. (11.17). Here, the fact that the power spectral densities of $\hat{S}_A$ and $i_0/2qr_0$ are both equal to $\frac{1}{2}$ is used.

From circuit equations Eqs. (11.172) and (11.173), $v_n$ and $i_L$ are obtained in a low-frequency limit as

$$v_n \simeq (R_0 - R_a)i_s + (R_0 - R_a)i + R_0 i_0 - R_a i_a \tag{11.180}$$

$$i_L \simeq i_s + i + \left(\frac{1}{R_S} + \frac{1}{R}\right)R_a i_a - \left(\frac{1}{R_S} + \frac{1}{R}\right)R_0 i_0 \tag{11.181}$$

Using the boundary conditions [Eqs. (11.176) and (11.177)], $i_{\text{in}}$ and $i_{\text{out}}$ are

$$i_{\text{in}} = \left(1 - \frac{R_0 R_a}{R_S}\right)i_s - \frac{R_0 - R_a}{R_S}i + \frac{R_0}{R_S}i_0 - \frac{R_a}{R_S}i_a \tag{11.182}$$

$$i_{\text{out}} = i_s + i + \left(\frac{1}{R_S} + \frac{1}{R}\right)R_a i_a + \left[1 - \left(\frac{1}{R_S} + \frac{1}{R}\right)R_0\right]i_0 \tag{11.183}$$

Here, threshold condition $R_0 \simeq R_a$ is used. As $R_S$ approaches infinity, $i_{\text{in}} = i_s \to 0$ and

$$i_{\text{out}} = \left(1 - \frac{R_0}{R}\right)i_0 + i + \frac{R_a}{R}i_a \simeq -xi_0 + i + (1+x)i_a \tag{11.184}$$

where $x = [n_{\text{sp}}(P/P_{\text{th}} - 1)]^{-1} = (n_{\text{sp}}R_P)^{-1}$ and $R_0/R \simeq R_a/R \simeq \tau_{\text{st}}/\tau_n \simeq 1 + x$ are used. The power spectrum of the output photon flux is

$$P_{i_{\text{out}}/q} = 2\gamma S_0\left[2n_{\text{sp}} \cdot x^2 + x\right] \tag{11.185}$$

Since $2\gamma S_0$ is the shot-noise-limited output photon flux fluctuation, Eq. (11.185) can be reduced to below the shot-noise level when pump rate $R_P \equiv P/P_{\text{th}} - 1$ satisfies

$$R \geq \frac{4}{\sqrt{8n_{\text{sp}} + 1} - 1} \tag{11.186}$$

When $n_{\text{sp}} = 1$ (complete population inversion), pump rate $P/P_{\text{th}} \geq 3$ is required to produce a number-phase squeezed state. This "second thresh-old pump rate" decreases with the population inversion parameter. As $x$ approaches infinity, $i_{out} \to 0$. Thus, the input electron flux and the output photon flux do not fluctuate. This is reasonable because a fixed number of electrons are injected by the high-impedance constant current source, and all the injected electrons are converted to lasing photons by the dominant stimulated emission process and are extracted from the output coupling mirror.

**11.4.3.5. Physical Interpretation.** When source resistance $R_S$ is very large, the pump source does not have the junction voltage pinning effect; in other words, the $CR_S$ time constant is long compared to carrier lifetime $\tau_n$. Since the dissipation rate is small, the noise from the pump reservoir is also small. Junction voltage fluctuation $v_n$ and photon current fluctuation $i_L$ are coupled and self-stabilize themselves (macroscopic self-organiza-tion). However, there remain residual fluctuations of $v_n$ and $i_L$, which coherently counteract incident and reflected fluctuations $i_s$ and $i_0$, respec-tively, to cancel them out of the external circuits. In this sense, the quiet input electron flux and output photon flux are realized because of the residual noisy internal systems.

When $R_S$ is very small, on the other hand, the pump source has the junction voltage pinning effect; that is, the $CR_S$ time constant is short compared to carrier lifetime $\tau_n$. Each time the electron number fluctuates, the steady-state value is restored by the impulsive relaxation current flowing in the external circuit. The amplitude stabilization due to the coupling between the photon number fluctuation and the electron-number fluctuation does not work. As a result, the photon-number fluctuation is only amplified instead of attenuated.

The situation is similar to the number-state-like Block oscillation of the superconductor tunnel junction with a large source resistance and the coherent-state-like Josephson oscillation with a small source resistance [43]. It has been recently demonstrated that the regularity of electron tunneling events in an ultrasmall tunnel junction depends on the relation between the source resistance and the junction tunnel resistance [44].

### 11.4.4. Numerical Examples

Circuit elements $R$, $R_{se} = R_0 - R_a$, $L$, and $C$ in the noise-equivalent circuit (Figure 11.15) versus normalized pump level $R_P = P/P_{th} - 1$ are shown in Figure 11.16 for a typical semiconductor laser. The following numerical parameters are assumed: spontaneous electron lifetime $\tau_{sp} = 3$ ns, nonradiative electron lifetime $\tau_{nr} = \infty$, stimulated electron lifetime $\tau_{st} \simeq \tau_{sp}/n_{sp}R_P$, population inversion parameter $n_{sp} = E_{cv}/(E_{cv} - E_{vc})$ $= 2$, photon lifetime $\tau_p = 2$ ps (or the photon decay rate, $\omega/Q = 5 \times n^{11}$ s$^{-1}$), stimulated emission rate $E_{cv} = n_{sp}(E_{cv} - E_{vc}) \simeq n_{sp}(\omega/Q) = 10^{12}$ s$^{-1}$, absorption rate $E_{vc} \simeq 5 \times 10^{11}$ s$^{-1}$, thermal voltage $V_T = 26$ mV,

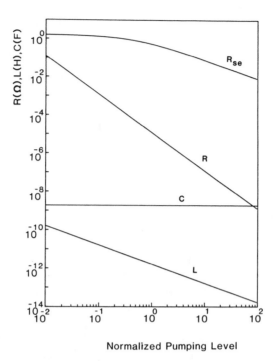

**Figure 11.16.** The diode's differential resistance, $R$, diffusion capacitance $C$, effective resistance $R_{se}$, and effective inductance $L$ versus normalized pump level $R_P = P/P_{th} - 1$.

parameter $m = 1.5$, spontaneous emission coefficient $\beta = 1 \times 10^{-5}$, total photon number $S_0 = R_P n_{sp}/\beta$, active-layer volume $V = 3 \times 10^{-10}$ cm$^3$, and threshold electron density $N_c/V = 1.5 \times 10^{18}$ cm$^{-3}$.

The power spectra for the external field amplitude fluctuation in the optically pumped semiconductor laser with shot-noise-limited pump fluctuation and in the injection current-driven semiconductor laser with suppressed-pump fluctuation are compared in Figure 11.17. The amplitude fluctuation spectrum is coincident with the standard quantum limit at a high pump level for the optically pumped semiconductor laser. On the other hand, that for the injection current-driven semiconductor laser is reduced to below the standard quantum limit in the frequency region below the cavity bandwidth.

The amplitude fluctuation power spectral densities in the low-frequency region are shown in Figure 11.18 as a function of series resistance $R_S$. Since diode differential resistance $R$ decreases monotonically with the pump level, as shown in Figure 11.16, criterion $2R < R_S$, wherein the pump fluctuation becomes smaller than the shot-noise level, is always satisfied for any finite $R_S$ value. Figures 11.18a and 11.18b correspond to

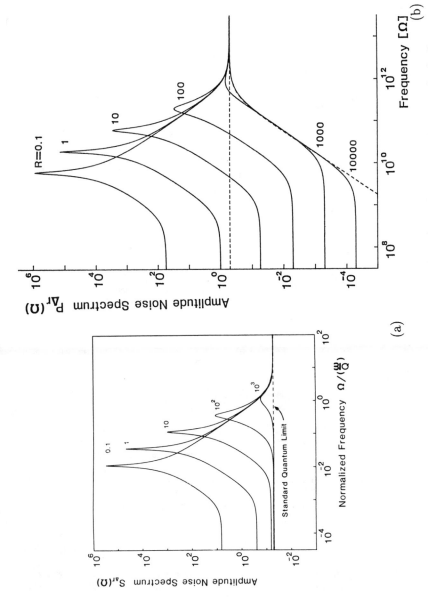

**Figure 11.17.** Amplitude noise spectra as a function of a normalized pump level $R_P$; (a) with shot-noise-limited pump amplitude fluctuation; (b) with suppressed pump noise.

515

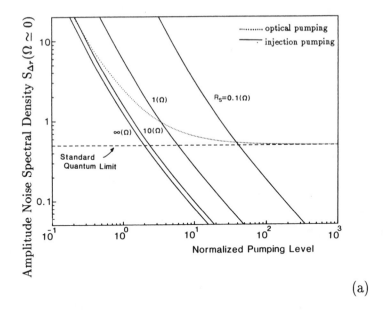

(a)

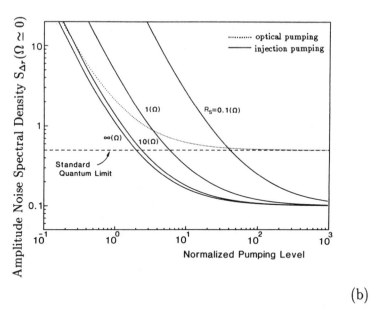

(b)

**Figure 11.18.** Amplitude noise spectral densities in the low-frequency region, $\Omega \ll \omega/Q$ versus normalized pump level $R_P$. The dotted line and solid line correspond to the optical pumping with shot-noise-limited pump amplitude fluctuation and the injection pumping with a source resistance $R_s$: (a) with no internal loss $\omega/Q_0 = 0$; (b) with a finite internal loss $\omega/Q_0 = 1 \times 10^{11}$ s$^{-1}$.

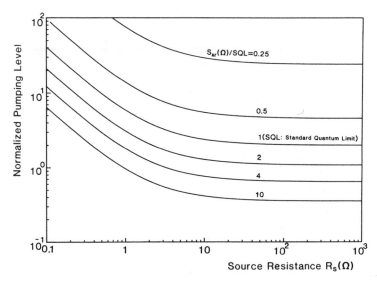

**Figure 11.19.** Contours for the equal amplitude noise spectral densities in the normalized pump level and source resistance plane.

the respective cases of no internal loss ($\omega/Q_0 = 0$ and $\omega/Q_e = \omega/Q = 5 \times 10^{11}$ s$^{-1}$) and finite internal loss ($\omega/Q_0 = 1 \times 10^{11}$ s$^{-1}$ and $\omega/Q_e = 4 \times 10^{11}$ s$^{-1}$). The amplitude squeezing is reduced by the presence of cavity internal loss.

The contours for the equal amplitude noise level are plotted in Figure 11.19 for the pump level and the source resistance. The numerical parameters are the same as those for Figure 11.18b. A higher pump level, larger source resistance, and smaller cavity internal loss are preferable for obtaining a large reduction of amplitude noise.

## 11.5. SQUEEZED-STATE GENERATION EXPERIMENTS

### 11.5.1. Thermal Noise Squeezing Experiment at 20 GHz

As demonstrated in Figure 11.3, there is one-to-one correspondence between a microwave oscillator and a laser oscillator. The amplitude noise at below the cavity cutoff frequency is squeezed to below the external input noise level in both oscillators, if the internal noise is smaller than the external noise, as shown in Figure 11.4. Thus, the principle of amplitude squeezing can be checked not only by a laser but also by a microwave oscillator. In nearly all microwave oscillators, however, the dominant noise source is internal active device noise. A definite exception is a Josephson junction oscillator, but it is known to be very sensitive to external noise and unstable.

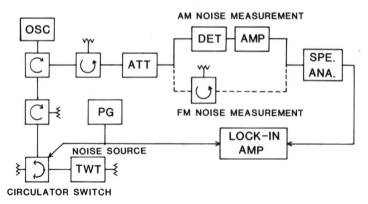

**Figure 11.20.** Experimental setup for AM noise spectrum and field power spectrum measurements of a Gunn diode oscillator.

To verify the theoretical predictions, a Gunn diode oscillator at 20 GHz was used as a negative resistance oscillator in circulator configuration, and a large external broadband and nearly white noise from a traveling-wave tube amplifier was injected into the Gunn diode oscillator, as shown in Figure 11.20. The cold cavity cutoff frequency was 70 MHz. The traveling-wave tube amplifier with 35-dB signal gain and 25-dB noise figure generated a large thermal noise power density of about $10^6$ $k_BT$, which was much higher than the internal noise of the Gunn diode. This was fed into the Gunn oscillator through a switching circulator, an isolator, and an input circulator. The outgoing wave from the Gunn oscillator was fed into a Schottky barrier diode to measure the amplitude noise spectrum, and was also fed directly into a spectrum analyzer to observe the field power spectrum. The amplitude noise spectrum caused only by the external noise from the traveling-wave tube amplifier can be measured using on–off modulation of the external noise source at the switching circulator and phase-sensitive lock-in amplifier detection.

Figure 11.21 shows the amplitude noise spectra caused by the external thermal noise and the internal noise source. The amplitude noise spectrum without an external noise source was found to be Lorentzian with a cutoff frequency of 70 MHz, except for the $1/f$ noise at low frequencies below several hundred kilohertz. The internal noise was about 20 dB higher than thermal background noise $4k_BT$. The amplitude noise caused by the external thermal noise was, on the other hand, suppressed at below the cold cavity cutoff frequency by $(\Omega/\Omega_c)^2$ as expected from Figure 11.3. The classical (thermal-noise-limited) amplitude noise was thus squeezed by 25 dB, which fully confirms the argument of Section 11.3.2. But this is not the experimental evidence for the quantum noise squeezing. The quantum vacuum field fluctuation was far below the measured noise level.

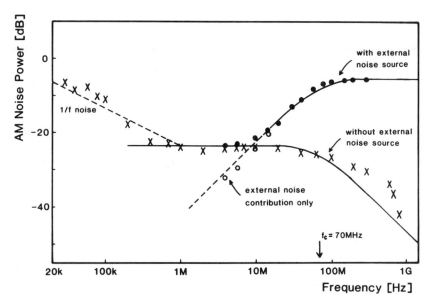

**Figure 11.21.** Normalized AM noise spectra of a Gunn diode with and without an external noise source.

### 11.5.2. Evidence for High-Impedance Suppression of Pump Current Noise in a Semiconductor Laser

A semiconductor laser has parasitic elements, as shown in Figure 11.22. The chip, consisting of $p$- and $n$-type bulk semiconductors and metal electrodes, has a series resistance of 5–10 $\Omega$ and a shunt capacitance of 10–20 pF. The bonding wire and package have an inductance of 1–2 nH, as well as a small capacitance and resistance. In a low fluctuation frequency below 100 MHz, however, those parasitic elements can be safely disregarded. Figure 11.23 shows the differential resistance and light output of a typical semiconductor laser as a function of the pump rate. At the pump rate not far from the threshold, the series resistance is still smaller than the diode's differential resistance. Therefore, the semiconductor laser can be biased by either a high-impedance constant current source or a low-impedance constant voltage source.

The two bias circuits were used to demonstrate the principle of high-impedance suppression of pump current, as shown in Figure 11.22. One circuit is the $LC$ parallel circuit. The pump source features a high impedance at $LC$ circuit resonant frequency $f_r \simeq 11$ MHz and a low impedance at all other frequencies. The amplitude noise (photocurrent noise) spectrum from the InGaAsP conventional Fabry–Perot cavity semiconductor laser driven by this bias circuit is shown in Figure 11.24$a$. The

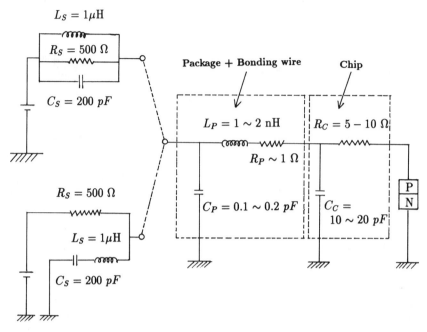

**Figure 11.22.** Typical parasitic elements and the two bias circuits for a semiconductor laser.

laser is operated at 77 K, and the pump rate is not far above the threshold. The amplitude noise spectrum is reduced at $LC$ circuit resonant frequency $f_r$. The other circuit is the $LC$ series circuit with a 500-$\Omega$ parallel resistance. This pump source features a low impedance at the same $LC$ circuit resonant frequency $f_r \simeq 11$ MHz, and a high impedance at all other frequencies. The amplitude noise spectrum from the same laser driven by this bias circuit is shown in Figure 11.24$b$. The amplitude noise is reduced except at the $LC$ circuit resonant frequency.

The external current noise level measured via the voltage fluctuation across source resistance $R_S$ was found to be much smaller than the shot-noise level, when $R_S$ is much greater than the diode's differential resistance. From these experimental results, the principle of high-impedance suppression of pump current noise in a semiconductor laser was fully confirmed.

### 11.5.3. Shot-Noise Calibration by Balanced Detectors and Light-Emitting Diodes

Calibration of the shot-noise level is important for the squeezed-state generation experiment. One can measure DC photocurrent $I$ and calculate the shot-noise level by the Schottky formula $P_s(\Omega) = 2qI$. This

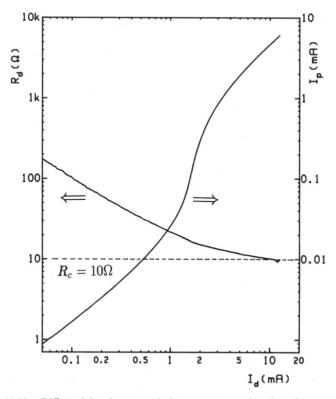

**Figure 11.23.** Differential resistance and photocurrent as a function of pump current.

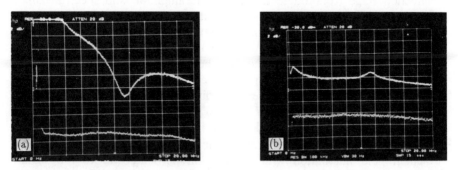

**Figure 11.24.** Amplitude noise (photocurrent noise) spectra of the InGaAsP semiconductor laser driven by the *LC* parallel circuit (*a*) and by the *LC* series circuit (*b*).

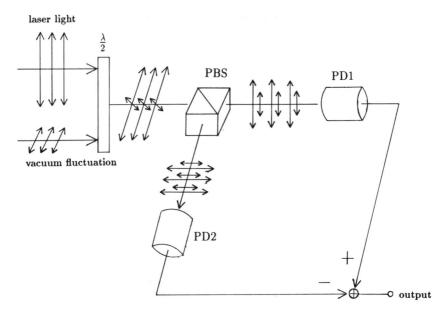

**Figure 11.25.** Dual-balanced detectors using a polarization beamsplitter.

method cannot be used in real experiments because the photodetector quantum efficiencies at DC and at $\Omega \neq 0$ are not necessarily the same and because the electronic amplifier gain and the spectrum analyzer resolution bandwidth are difficult to measure precisely. The saturation at any of the detection stages (photodetector, electronic amplifier, and spectrum analyzer) affects the final result. The dual-balanced detectors shown in Figure 11.25 are specifically suitable for calibrating the shot-noise level [45]. The laser output is split precisely into the two arms by a half-wave plate and a polarization beam splitter. The two beams are detected by the two photodiodes with identical response. The two output currents are subtracted by a differential amplifier. The differential amplifier output corresponds to $\hat{I}_1 - \hat{I}_2$, where $\hat{I}_1$ and $\hat{I}_2$ are the photodetector current operators. The quantum theory of the dual-balanced detectors [45–47] shows that the fluctuation of $\hat{I}_1 - \hat{I}_2$ corresponds to the quantum noise of an incident vacuum field fluctuation with polarization perpendicular to the signal wave, and that the current fluctuation density is exactly equal to shot-noise level $2q(\langle \hat{I}_1 \rangle + \langle \hat{I}_2 \rangle)$.

Figure 11.26 shows the differential amplifier output for the directly intensity modulated semiconductor laser. When one of the two beams are blocked, the noise spectral density is higher by 20–30 dB, which indicates the common-mode suppression factor of the balanced detectors is 30 dB at near DC and 20 dB at a modulation frequency of 140 MHz. This means

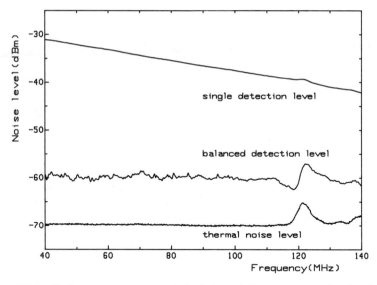

**Figure 11.26.**   Single detector output and the balanced detectors output for the directly intensity modulated semiconductor laser.

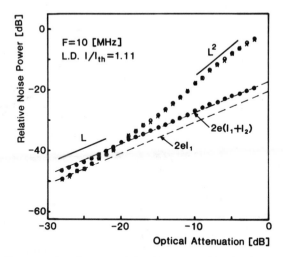

**Figure 11.27.**   Photocurrent noise spectral density measured by the single detector and the balanced detectors as a function of the optical loss.

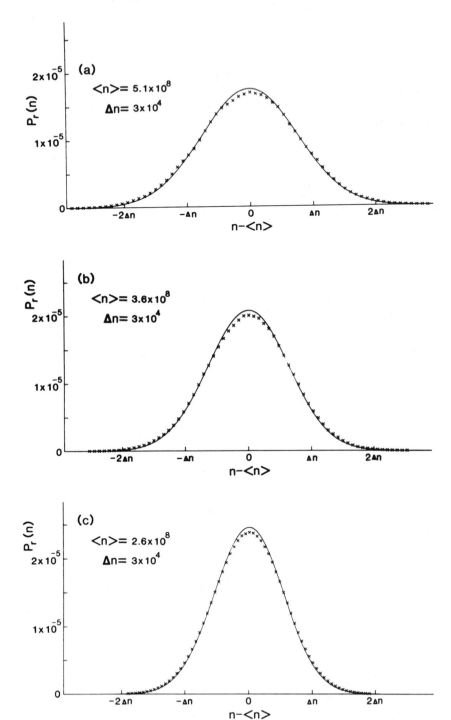

**Figure 11.28.** Photocurrent distributions produced by the LED and the Poissonian distributions.

the unbalance of the two photodetectors' quantum efficiencies and of the differential amplifier gain for the two input channels is less than 0.1–1%.

Figure 11.27 shows the amplitude noise (photocurrent noise) spectral density at 10 MHz of the semiconductor laser biased at near the threshold $R_P = 0.1$ as a function of optical loss between the laser and balanced detectors. When the laser output is measured by the single detector (by adjusting the polarization direction in Figure 11.25), the laser noise itself can be measured, and it is decreased in proportion to the square of optical loss $L^2$. This is the characteristic of the excess-noise-limited light [47]. When the optical loss is greater than 20 dB, the noise is decreased in proportion to optical loss $L$, which is the characteristic of the shot-noise-limited light [47]. When the laser output is measured by the balanced detectors, the noise is decreased in proportion to optical loss $L$, even in a small loss case. Thus, it is clear that the balanced detectors suppress the excess noise and always output the shot-noise level corresponding to the total photocurrent.

The shot-noise level is also calibrated by a light-emitting diode (LED). Figure 11.28 shows the photocurrent distributions (photoelectron statistics) for the LED output [48]. The solid lines are Poissonian distributions for average photoelectron numbers. The slightly broader distribution of the photocurrent, compared to the Poissonian distribution, is due to the thermal noise of the electronic amplifiers. Figure 11.29 shows photocurrent distributions for the semiconductor laser and for the LED by the single detector. The photocurrent distribution for the semiconductor laser is broader than that for the LED, but the balanced detectors output for the semiconductor laser agrees well with that for the LED.

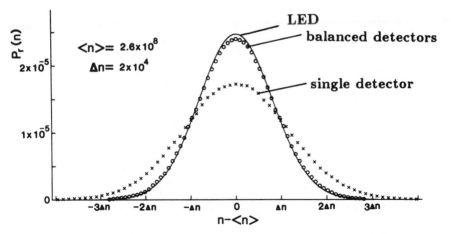

**Figure 11.29.** Photocurrent distributions of the laser noise measured by the single detector and the balanced detectors, with that of the LED.

### 11.5.4. Observation of Amplitude Squeezing by Conventional Balanced Detectors

The experimental setup is shown in Figure 11.30 [16, 17]. The semiconductor laser, the high-impedance bias circuit, and the collimating microlens are mounted inside a cryostat. The AlGaAs/GaAs transverse junction stripe (TJS) semiconductor laser (Mitsubishi, ML-2308) with a $0.81$-$\mu$m oscillation wavelength was operated at 77 K. Stable single-longitudinal-mode operation was obtained at a wide pump rate range of $R_P = 1 \sim 10$. The threshold current was about 1 mA, and the differential quantum efficiency at the front facet above the threshold was about 40%. The output coupling loss from a high-reflection-coated rear facet and the internal absorption–scattering loss are responsible for the relatively small differential quantum efficiency. The diode's differential resistance is 20 $\Omega$ at the threshold and is inversely proportional to the current. The diode's series resistance is 7 $\Omega$. The bias circuit has a series resistance of 1 k$\Omega$, which is high enough to suppress the pump current noise to well below the shot-noise level.

In this procedure $S_i$ photodiodes with a 240-$\mu$m diameter and 1-ns response time (NEC model NDL-2102) are used as photodetectors. GaAs FETs with an input impedance of 240 $\Omega$ are used as a front-end differential amplifier. The quantum efficiency of a $S_i$ photodiode is about 93%. The overall detection quantum efficiency, including collimating-lens loss, cryostat window loss, gold mirror reflection loss, isolator insertion loss, and focusing-lens loss, is about 55%.

The single-detector photocurrent noise spectrum for the laser pumped at $R_P = 8.5$ is shown by curve $A$ in Figure 11.31. Curve $D$ is the balanced detectors output for the same laser. Another semiconductor laser (Hitachi model HLP-1400, CSP laser) with the same emission wavelength is oper-

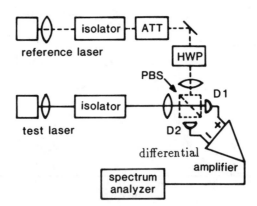

**Figure 11.30.** Experimental setup for measuring the amplitude noise of a pump-noise-suppressed semiconductor laser.

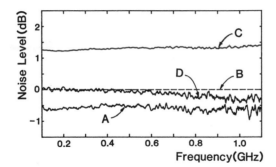

**Figure 11.31.**   Photocurrent noise spectra of the TJS laser and the CSP laser measured by the single detector and the balanced detectors.

ated at room temperature. The threshold current was about 30 mA, and the output power is much higher than that of the TJS laser. After attenuation of about 10 dB, the CSP laser output is detected by the single and balanced detectors. The results are shown by curves $C$ and $B$ in Figure 11.31. The ordinate is normalized by the balanced detectors output for the attenuated CSP laser light. The photocurrent noise spectrum of the CSP laser light is 1.2–1.4 dB higher than the shot-noise level, even after the optical loss of about 10 dB, because the original laser output biased at levels near the threshold has much excess noise. The excess noise included in the detected CSP laser light is smaller by $-5$ dB than the shot-noise level. Since the balanced detectors have the common-mode suppression factor of about 30 dB, the excess noise included in the curve B is smaller than the shot-noise level by $-34$ dB, and so the calibration error of the shot-noise level is less than 0.2%. The TJS laser light features amplitude squeezing of 0.4–0.7 dB (8–19%) below the shot-noise level in a frequency region from 100 MHz to 1.1 GHz. Balanced detectors output $C$ for the TJS laser light retraces shot-noise level $B$ obtained by the attenuated CSP laser light, as was expected. In a high-frequency region, however, a deviation of 0.2–0.4 dB exists, mainly as a result of the smaller common-mode rejection factor. The excess noise is not suppressed enough in curve $B$, and the squeezed noise is not eliminated sufficiently in curve $D$. The true shot-noise level may exist between curves $B$ and $D$. It is obvious, in spite of the small ambiguity for the shot-noise level calibration, that amplitude squeezing was observed in the entire detection bandwidth from 100 MHz to 1.1 GHz.

### 11.5.5.   Observation of Amplitude Squeezing by Balanced Detectors with a Delay Line

In the above mentioned experiment, two measurement steps are required, and the DC photocurrent levels of the laser noise measurement and

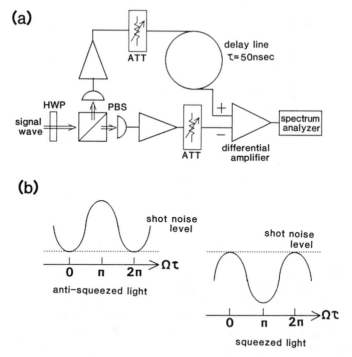

**Figure 11.32.** (*a*) Balanced direct detectors with a delay line and attenuators (ATT). (*b*) Current noise spectra of an amplitude antisqueezed light and an amplitude-squeezed light.

shot-noise calibration differ by a factor of 2. The noise level usually saturates at a DC current level beyond a few milliamperes. To eliminate the error introduced by the two-step measurements and the photodetector saturation effect, the balanced detectors with a delay line shown in Figure 11.32 are developed [18]. One detector output is delayed (by $\tau = 50$ ns in our measurement), but the other detector output is not. The difference in these two outputs is produced by a differential amplifier. A coaxial cable delay line has a loss coefficient proportional to $\sqrt{\Omega}$, where $\Omega$ is the fluctuation frequency. This frequency-dependent attenuation imposed on the delayed signal is compensated for by introducing the same attenuation to the other signal.

For a fluctuation frequency $\Omega_{in}$ satisfying inphase delay condition $\Omega_{in}\tau = 2N\pi$, where $N$ is an integer, the differential amplifier output measures $\hat{I}_1 - \hat{I}_2$, where $\hat{I}_1$ and $\hat{I}_2$ are the two photodetector currents. The current-fluctuation spectral density is exactly equal to the shot-noise level. For a fluctuation frequency $\Omega_{out}$ satisfying out-of-phase delay condition $\Omega_{out}\tau = (2N + 1)\pi$, the differential amplifier output measures $\hat{I}_1 + \hat{I}_2$. The quantum-mechanical theory of a balanced detector [36–38] shows that $\hat{I}_1 + \hat{I}_2$ corresponds to the quantum noise of the laser itself with polarizations parallel to the coherent excitation. Thus, the detector output simultaneously displays the laser-noise level and the corresponding shot-noise level on a spectrum analyzer with a frequency period of $\Delta\Omega = 2\pi/\tau$.

When the signal wave is amplitude-antisqueezed (super-Poissonian), the current fluctuation at $\Omega_{in}$ is smaller than that at $\Omega_{out}$, as shown in Figure 11.32$b$. On the other hand, when the signal wave is amplitude-squeezed (sub-Poissonian), the current fluctuation at $\Omega_{in}$ is larger than that at $\Omega_{out}$, as shown in Figure 11.32$b$. This inverted modulation in the current-fluctuation spectrum is an unmistakable mark of amplitude-squeezed light. This is a single-step measurement; therefore, the ambiguity in the shot-noise level owing to photodetector saturation can be eliminated.

A short-cavity AlGaAs/GaAs transverse-junction stripe semiconductor laser with antireflection coating on the front facet and high-reflection coating on the rear facet was used at 77 K. The threshold current was 1 mA, and the differential quantum efficiency above the threshold was 70%, which was better than the 40% value of the laser used in the preceding experiment. To eliminate a minute optical reflection feedback to the laser from the measurement optical elements, an optical isolator was used, and all optical elements were antireflection-coated and slanted with respect to the beam direction. The overall detection quantum efficiency, including losses from the laser collimating lens, cryostat window, gold mirror, optical isolator, half-wave plate, photodetector focusing lens, polarization beam splitter, and photodetector quantum efficiency, was 60%.

Figure 11.33 shows current noise spectra at two different bias levels. The current noise spectrum for bias level $R_P = 0.03$, which is shown by curve $A$, features lower noise power at $\Omega_{in}$ than at $\Omega_{out}$. This indicates that the field is amplitude anti-squeezed (super-Poissonian). The amplifier noise level is shown by curve $B$. The front-end amplifier is an AC-coupled bipolar transistor (NEC 2SC3358) with a noise figure of 1.1 dB and a load resistance of 220 $\Omega$. The current noise spectrum for bias level $R_P = 12.6$, which is shown by curve $C$, features higher noise power at $\Omega_{in}$ than at $\Omega_{out}$. This indicates that the field is amplitude squeezed (sub-Poissonian). The total DC photocurrents are 15 $\mu$A and 6.12 mA for $R_P = 0.03$ and 12.6, respectively. Curves $D$ and $E$ are the current noise spectra when one of the two incident signal waves for $R_P = 12.6$ is blocked. The modulation disappears, as expected, in a low-frequency region. At high frequencies, however, the noise power is reflected back at the differential amplifier input, and so the modulation due to the round trip in a delay line appears. Curve $F$ is the sum of the current noise spectra indicated by curves $D$ and $E$. The noise level of curve $F$ is not equal to a 3-dB noise rise from the noise level indicated by curve $D$ or $E$ because of the amplifier thermal noise (curve $B$). Note that the current noise spectrum indicated by curve $F$

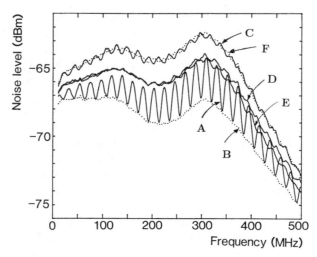

**Figure 11.33.**   Current noise spectra for bias levels $I/I_{th} = 1.03$ (curve $A$) and $I/I_{th} = 13.6$ (curve $C$). Curve $B$ is the amplifier thermal noise. Curves $D$ and $E$ are obtained when one of the two signal beams is blocked for $I/I_{th} = 13.6$. Curve $F$ is the sum of noise curves $D$ and $E$.

is between the shot-noise level at $\Omega_{in}$ and the reduced noise level at $\Omega_{out}$. This is because the noise level of amplitude-squeezed light increases to approach the shot-noise level when the amplitude is attenuated.

Figure 11.34 shows the current noise spectra normalized by the shot-noise level for the two bias levels. The shot-noise level calibrated by the measurement of $\hat{I}_1 - \hat{I}_2$ is compared with the shot-noise level generated by the light-emitting diode with the same wavelength as the laser. The difference is smaller than 0.1 dB. The current noise spectrum at $R_P = 0.03$ shows an enhanced noise peak at the relaxation–oscillation frequency. The current noise spectrum at $R_P = 12.6$ shows a noise level reduced to $-1.3$ dB below the shot-noise level. The observed 0.6-dB squeezing becomes 1.3-dB squeezing in Figure 11.4, because the effect of the amplifier thermal noise is subtracted. This noise reduction is much larger than the error bar of the shot-noise-level calibration. If the increase in the noise level owing to the detection quantum efficiency of 60% is corrected, the measured noise level corresponds to the squeezing of $-3$ dB (50%) at the laser output.

### 11.5.6.   Observation of Amplitude Squeezing by Face-to-Face Coupled Balanced Detectors

In the experiment using the polarization beam splitter, the polarization noise of the laser output, if it exists, is converted to the correlated intensity noise in the two arms. The balanced detectors output does not

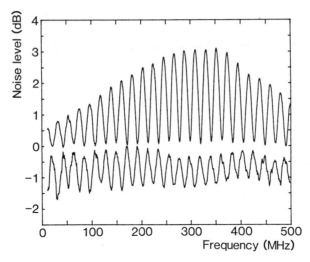

**Figure 11.34.** Current noise spectra normalized by the shot-noise level for bias levels $I/I_{th} = 1.03$ (upper curve) and $I/I_{th} = 13.6$ (lower curve). The amplifier thermal noise is subtracted in the normalization process.

cancel but simply adds this correlated noise, and so the error is introduced in the shot-noise-level calibration. Theoretical analysis [49] also showed that the squeezed noise is easily destroyed by a small amount of reflected light. The effect of external optical reflection feedback can be relaxed if the reflection point is close to the laser [49]. Such observation motivated us to construct the balanced detectors directly coupled to the semiconductor laser and to the LED with separations of less than 1 mm, as shown in Figure 11.35. The whole system including the driving circuits, laser, LED, balanced detectors, and differential amplifiers is put inside the cryostat. It is confirmed by the common-mode rejection experiment shown in Figure 11.26 that the difference between the two photodiode quantum efficiencies and the two differential amplifier input gains is less than 1%. The laser

Noise measurement setup

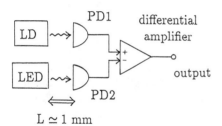

**Figure 11.35.** The face-to-face coupled balanced detectors.

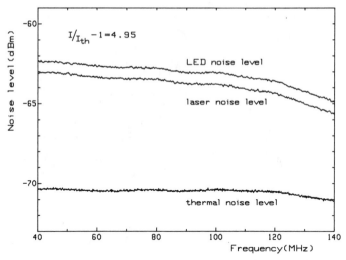

**Figure 11.36.** The photocurrent noise spectra of the TJS laser and the LED measured by the face-to-face coupled balanced detectors.

noise can be measured by driving only the laser, and the shot-noise level can be calibrated by driving only the LED.

The photocurrent noise spectra for the GaAs TJS laser and the LED are shown in Figure 11.36. Amplitude squeezing up to $-1.5$ dB below the shot-noise level was directly observed in this experiment [49].

### 11.5.7.  Degree of Squeezing Versus Optical Loss

The photocurrent noise spectral density at 800 MHz at pump rate $R_P = 10.4$ versus optical loss $L$ put in front of the balanced detectors is shown in Figure 11.37. The experimental setup shown in Figure 11.30 was used. The ordinate is normalized by the corresponding shot-noise level. The noise level increases to the shot-noise level as the optical loss increases, and the amplitude squeezing is eventually lost in the limit of infinite loss. This is because the optical loss process couples the vacuum field fluctuation to the laser light. In the limit of large optical loss, the original quantum noise of the laser is entirely replaced by the vacuum field fluctuation, and thus, the shot-noise emerges. This apparent increase of normalized photocurrent noise is an unmistakable mark of amplitude squeezing.

The overall quantum efficiency from the laser injection-current increment to photodetector current increment is 22% in this experiment, when artificial optical attenuation is eliminated. The overall detection quantum efficiency of 55% consists of a photodetector quantum efficiency of 0.93, focusing-lens loss of 0.90, isolator insertion loss of 0.81, mirror reflection

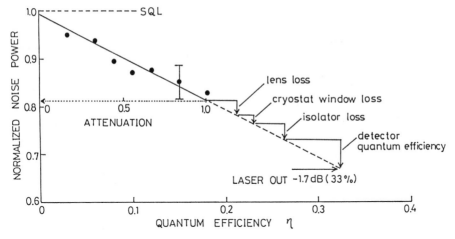

**Figure 11.37.**   Degree of squeezing versus optical loss.

loss of 0.95, cryostat window loss of 0.93, and laser collimating-lens loss of 0.93. The amplitude noise levels corrected for these factors are shown in figure 11.37. The observed noise level corresponds to amplitude squeezing of $-1.7$ dB (33%) below the SQL at the output of the laser facet, as shown in Figure 11.37. This is the noise level that the laser actually produced at the output mirror. The laser output coupling efficiency is

$$\eta_c = \frac{L^{-1} \ln\left(R_1^{-1/2}\right)}{\alpha + L^{-1} \ln\left[\left(R_1 R_2\right)^{-1/2}\right]}$$

where $L = 250$ $\mu$m is the cavity length, $R_1 = 0.32$ and $R_2 = 0.6$ are power reflectivities of the front and rear facets, respectively, and $\alpha = 22$ cm$^{-1}$ is the internal absorption loss. If the laser output-coupling efficiency due to nonideal rear-facet reflectivity $\eta_M \simeq 0.57$ and that due to internal loss $\eta_A = 0.70$ are also corrected, the observed noise level corresponds to amplitude squeezing of $-7$ dB (80%) below the shot-noise level. This is the intrinsic noise level achievable if the rear-facet reflectivity is increased to 100% and internal absorption loss is eliminated.

### 11.5.8.   Degree of Squeezing versus Laser Pump Rate

The amplitude noise level corrected for the detection quantum efficiency and also for the laser quantum efficiency, that is, assuming the unity output coupling efficiency, is plotted in Figure 11.38 as a function of laser pump rate $R_P$. The solid line represents the theoretical result for the suppressed pump current noise case; and the dashed line, that for the

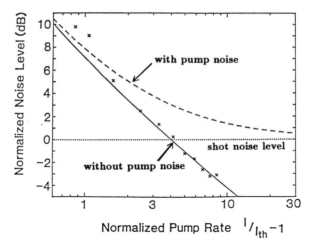

**Figure 11.38.**    Normalized amplitude noise level versus pump rate.

shot-noise-limited pump current noise case. The experimental results fully confirm the principle of squeezed-state generation by a high-impedance constant current driven semiconductor laser.

Similar experimental results have been reported by Richardson and Shelby, using the GaAs TJS laser at 4 K [50].

## 11.6. CONCLUSION

For many years, the quantum-statistical properties of a semiconductor laser were thought to be essentially the same as those of other optical pumped laser systems. It was also believed that the quantum limit of the amplitude noise was equal to the standard quantum limit of a coherent state.

The main conclusion of this chapter is that a constant-current-driven semiconductor laser produces a number-phase squeezed state in the frequency region below the cavity cutoff frequency. The principle is based on the pump current noise suppression by the high-impedance constant current source, the dipole moment noise suppression by the strong laser-field-induced gain saturation, and the incident vacuum field noise suppression by the quantum-mechanical destructive interference.

The advantages of this scheme, compared to other squeezed-state generation schemes, are as follows:

1. The degree of squeezing is not limited by optical nonlinear constants, but is determined only by the laser quantum efficiency. Since a quantum efficiency higher than 90% is readily achieved in a

semiconductor laser, noise reduction of more than 10 dB is possible without requiring a high-intensity pump wave.

2. The squeezing bandwidth is ultimately determined by a photon lifetime on the order of 1 psec. This broadband property suggests the potential capability of the production of a squeezed state of very short pulse duration.

3. A squeezed state can be generated at any wavelength between 0.7 and 10 $\mu$m by choosing the appropriate semiconductor material system.

On the other hand, the disadvantage of the scheme is one common to almost all squeezed state generation schemes:

1. The squeezing is quickly destroyed by a small amount of optical loss.

This may limit applications of the squeezed state generated by a semiconductor laser to low-loss systems. Optical systems such as optical memory readout, optical signal processing, optical pumping of other laser systems, and short-distance optical communications are examples that meet the above requirement.

## REFERENCES

1. Y. Yamamoto, S. Machida, and O. Nilsson, *Phys. Rev.* **A34**, 4025 (1986).

2. Y. Yamamoto and S. Machida, *Phys. Rev.* **A35**, 5114 (1987).

3. D. Bohm, *Quantum Theory*, Prentice-Hall, Englewood Cliffs, NJ, 1951.

4. A. Messiah, *Quantum Mechanics*, McGraw-Hill, New York, 1961.

5. H. Takahashi, *Adv. Commun. Syst.* **1**, 227 (1965).

6. D. Stoler, *Phys. Rev.* **D4**, 1925 (1971).

7. H. P. Yuen, *Phys. Rev.* **A13**, 2226 (1976).

8. R. J. Glauber, *Phys. Rev.* **131**, 2766 (1963).

9. H. P. Yuen and J. H. Shapiro, *IEEE Trans. Inform. Theory* **IT-26**, 78 (1980).

10. For a recent review, see the special issues on squeezed state of the electromagnetic field, *J. Opt. Soc. Am.* **B4** (October 1987) D. F. Walls and S. E. Slusher, eds., and *J. Mod. Optics* (June 1987) R. Loudon and P. L. Knight, eds.

11. R. E. Slusher, L. W. Hollberg, B. Yurke, J. C. Mertz, and J. F. Valley, *Phys. Rev. Lett.* **55**, 2409 (1985); R. M. Shelby, M. D. Levenson, S. H. Perlmutter, R. G. DeVoe, and D. F. Walls, *Phys. Rev. Lett.* **57**, 691 (1986); L. A. Wu, H. J. Kimble, J. L. Hall, and H. Wu, *Phys. Rev. Lett.* **57**, 2520 (1986); M. W. Maeda, P. Kumar, and J. H. Shapiro, *Opt. Lett.* **12**, 161 (1987).

12. R. Jackiw, *J. Math. Phys.* **9**, 339 (1968).

13. M. Kitagawa and Y. Yamamoto, *Phys. Rev.* **A34**, 3974 (1986); M. Kitagawa, N. Imoto, and Y. Yamamoto, *Phys. Rev.* **A35**, 5270 (1987); K. Watanabe and Y. Yamamoto, *Phys. Rev.* **A38**, 3556 (1988).

14. P. Carruthers and M. M. Neito, *Rev. Mod. Phys.* **40**, 411 (1968).

15. Y. Yamamoto and H. A. Haus, *Rev. Mod. Phys.* **58**, 1001 (1986).

16. S. Machida, Y. Yamamoto, and Y. Itaya, *Phys. Rev. Lett.* **58**, 1000 (1987).

17. S. Machida and Y. Yamamoto, *Phys. Rev. Lett.* **60**, 792 (1988).

18. S. Machida and Y. Yamamoto, *Opt. Lett.* **14**, 1045 (1989).

19. M. Lax, *Phys. Rev.* **145**, 110 (1966); M. Lax, in *Physics of Quantum Electronics*, P. L. Kelley, et al., eds., McGraw-Hill, New York, 1966; M. Lax, *Phys. Rev.* **160**, 290 (1967); M. Lax, *Phys. Rev.* **157**, 213 (1967); M. Lax and W. H. Louisell, *Phys. Rev.* **185**, 568 (1969); M. Lax and W. H. Louisell, *IEEE J. Quantum Electron.* **QE-3**, 47 (1967); M. Lax and M. Zwanziger, *Phys. Rev.* **A7**, 750 (1973).

20. H. Haken, *Light and Matter*, Vol. XXV of *Hendbuch der Physics*, Springer-Verlag, Berlin, 1970; H. Haken, *Light*, North-Holland, Amsterdam, 1981, Vols. 1 and 2.

21. M. Sargent III, M. O. Scully, and W. E. Lamb, Jr., *Laser Physics*, Addison-Wesley, Reading, MA, 1974).

22. K. Kurokawa, *Proc. IEEE* **61**, 1386 (1973).

23. C. W. Gardinar and M. J. Collet, *Phys. Rev.* **A31**, 3761 (1985).

24. H. A. Haus and Y. Yamamoto, *Phys. Rev.* **A29**, 1261 (1984).

25. Y. Yamamoto and N. Imoto, *IEEE J. Quantum Electron.* **QE-22**, 2032 (1986).

26. O. Nilsson, Y. Yamamoto, and S. Machida, *IEEE J. Quantum Electron.* **QE-22**, 2043 (1986).

27. Yu. M. Golubev and I. V. Sokolov, *Sov. Phys.* **JETP 60**, 234 (1984).

28. A. W. Hull and N. H. Williams, *Phys. Rev.* 25, 147 (1925); B. J. Thompson, D. O. North, and W. A. Harris, *RCA Rev.* **4**, 269 (1939).

29. M. Teich and B. E. Saleh, *J. Opt. Soc. Am.* **2**, 275 (1985).

30. K. K. Likharev and A. B. Zorin, *J. Low Temp. Phys.* **59**, 347 (1985).

31. T. A. Fulton and G. J. Dolan, *Phys. Rev. Lett.* **59**, 109 (1987); L. S. Kuzmin and K. K. Likharev, *JETP Lett.* **45**, 496 (1987).

32. H. A. Haus and Y. Yamamoto, *Phys. Rev.* **A34**, 270 (1987); Y. Yamamoto and H. A. Haus, *Phys. Rev.* **A41**, 5164 (1990).

33. J. Bergon, L. Davidovich, M. Orszag, C. Benkert, M. Hillery, and M. O. Scully, *Phys. Rev.* **A40**, 5073 (1989).

34. F. Haake, D. F. Walls, and M. J. Collet, *Phys. Rev.* **A39**, 3211 (1989).

35. M. J. Buckingham, *Noise in Electronic Devices and Systems*, Wiley, New York, 1983.

36. S. M. Sze, *Physics of Semiconductor Devices*, Wiley, New York, 1969.

37. R. L. Anderson, *Solid-State Electron.* **5**, 341 (1962).

38. S. O. Rice, *Bell Syst. Tech. J.* **23**, 282 (1944); **24**, 46 (1945).

39. W. Shockley, *Electrons and Holes in Semiconductors*, Van Nostrand, New York, 1963.

40. H. B. Callern and T. A. Welton, *Phys. Rev.* **83**, 34 (1951).

41. B. Schneider and M. J. O Strutt, *Proc. IRE* **47**, 546 (1959).

42. H. Hang, *Z. Phys.* **206**, 163 (1967); *Z. Phys.* **200**, 57 (1967); *Phys. Rev.* **184**, 338 (1969).

43. D. V. Avelin and K. K. Likharev, *J. Low Temp. Phys.* **62**, 345 (1986).

44. M. Ueda and Y. Yamamoto, *Phys. Rev.* **B41**, 3082 (1990).

45. H. P. Yuen and V. W. S. Chan, *Opt. Lett.* **8**, 177 (1983).

46. G. L. Abbas, V. W. S. Chan, and T. K. Yee, *IEEE J. Lightwave Technol.* **LT-3**, 1110 (1985).

47. S. Machida and Y. Yamamoto, *IEEE J. Quantum Electron.* **QE-22**, 6017 (1986).

48. Y. Yamamoto, N. Imoto, and S. Machida, *Phys. Rev.* **A33**, 3243 (1986); S. Machida and Y. Yamamoto, *Opt. Commun.* **57**, 290 (1986).

49. Y. Yamamoto and S. Machida, talk presented at the Winter Colloquium on Quantum Electronics (January 1989, Snowbird).

50. W. Richardson and R. M. Shelby, *Phys. Rev. Lett.* **64**, 400 (1990).

# 12

# Generation of Photon-Number-Squeezed Light by Semiconductor Incoherent Light Sources

MALVIN C. TEICH
*Columbia Radiation Laboratory, Columbia University, New York, New York*

BAHAA E. A. SALEH
*Department of Electrical and Computer Engineering, University of Wisconsin, Madison, Wisconsin*

FEDERICO CAPASSO
*AT&T Bell Laboratories, Murray Hill, New Jersey*

## 12.1. INTRODUCTION

Photon-number-squeezed light, by definition, exhibits a photon-number uncertainty that is squeezed below the minimum classical value, which is associated with the Poisson distribution [1, 2]. Such light is therefore also called sub-Poisson light. Photon-number-squeezed light is expected to find use in a variety of applications, ranging from lightwave communications [3, 4] to biology [5], where the capacity of light to carry information is limited by photon-number uncertainty. Indeed, the use of a fixed number of photons to represent a bit of information can, in principle, provide noise-free direct-detection lightwave communications [3, 4]. The noise is squeezed into the phase fluctuations, which are not registered by the process of direct detection.

Photon-number-squeezed light may be generated in many ways. When it is desired to impart information on the phase of a light beam, the use of

*Coherence, Amplification, and Quantum Effects in Semiconductor Lasers*, Edited by Yoshihisa Yamamoto.
ISBN  0-471-512494  © 1991 John Wiley & Sons, Inc.

quadrature-squeezed light [6, 7] that is mixed with coherent light at a beamsplitter [8] is particularly useful. The homodyne process converts the quadrature-squeezed light into photon-number-squeezed light [9]. The distinction between quadrature-squeezed and photon-number-squeezed light has been elucidated elsewhere [2, 10, 11].

Techniques in which photon-number-squeezed light is directly generated are sometimes preferable because of their simplicity. Such techniques are useful when information is to be imparted directly to the photon number. Yamamoto and his colleagues have considered several schemes that in principle permit the synthesis of light with a particular quantum state [10]; these include unitary transformation from a coherent state, nonunitary state reduction by measurement, the combination of measurement and feedback, and lasing with suppressed pump-noise fluctuations. The latter approach is considered in the context of semiconductor injection lasers in the previous chapter of this book [11].

An alternative approach, initially used by us [1, 12] and considered in this chapter, focuses on the point process that characterizes the generation and detection of photons in terms of their arrival times. It is most readily applied to a description of incoherent photon-number-squeezed light. This approach is meritorious in the physical intuition that it provides and the fact that it includes time dynamics, but it does not provide a framework that allows for the synthesis of light of a particular quantum state.

Most sources of laser light produce a statistically independent stream of photons represented by the Poisson point process. The generation of photon-number-squeezed light requires that anticorrelations be introduced into the photon stream. These anticorrelations may be manifested in the times at which the radiators emit photons (excitation control) or they may be derived from the emitted photons themselves (photon control) [1, 2].

In this chapter, we focus on the generation of photon-number-squeezed light by techniques that rely on excitation control (Section 12.2). Excitation control may be imparted by mechanisms that rely either on a physical process (Section 12.3), or on an externally provided feedback control signal (Section 12.4). Both of these techniques can be used in conjunction with semiconductor light sources. Two proposed applications of photon-number-squeezed light are briefly considered in Section 12.5. An analysis of the generation of photon-number-squeezed light from a stochastic point process point of view is detailed in the Appendix.

In principle, these techniques can be used to generate ideal photon-number-state light (which has no uncertainty in its photon number). However, it is important to note that photon-number squeezing is a fragile effect. Once produced it is readily diluted by the ever-present random loss of photons and by contamination arising from the presence of (unsqueezed) background photons [1, 13].

## 12.2. EXCITATION CONTROL

The generation of photon-number-squeezed light by excitation control may be visualized in terms of the schematic representation shown in Figure 12.1. Two key effects regulate the photon-number-squeezing possibilities for light generated by a two-step process of excitation and emission: (1) the statistical properties of the excitations themselves and (2) the statistical properties of the individual emissions. The role of these two factors is heuristically illustrated in Figure 12.1 and analytically examined in the Appendix (which is drawn from Section 3.4 of Ref. 1).

In Figure 12.1a, we show an excitation process that is Poissonian. Consider each excitation as generating photons independently. Now if each excitation instantaneously produces a single photon, and if we ignore the effects of interference, the outcome is a Poisson stream of photons, which is obviously not sub-Poisson. This is the least random situation that we could hope to produce, given the Poisson excitation statistics. If interference is present, it will redistribute the photon occurrences, leading to the results for chaotic light [14]. On the other hand, the individual nonstationary emissions may consist of multiple photons or random numbers of photons. In this case, we encounter two sources of randomness, one associated with the excitations and another associated with the emissions. The outcome will then be super-Poisson; that is, it will exhibit photon-number fluctuations greater than those associated with the Poisson distribution.

In particular, if the emissions are also described by Poisson statistics, and the counting time is sufficiently long, the result turns out to be the Neyman Type-A counting distribution, as has been discussed in detail elsewhere [15, 16]. Even if the individual emissions comprise deterministic numbers of photons, the end result is the fixed-multiplicative Poisson distribution, which is super-Poisson [16]. Related results have been obtained when interference is permitted [14]. It is quite clear, therefore, that if the excitations themselves are Poisson (or super-Poisson), there is no hope of generating photon-number-squeezed light by such a two-step process.

In Figure 12.1b we consider a situation in which the excitations are more regular than Poisson. For illustration and concreteness, we choose the excitation process to be produced by deleting every other event of a Poisson pulse train. The outcome is the gamma-2 (or Erlang-2) renewal process, whose analytical properties are well understood (see Appendix). Single-photon emissions, in the absence of interference, result in sub-Poisson photon statistics. Poisson emissions, on the other hand, result in super-Poisson light statistics. Of course, the presence of interference can introduce additional randomness. Clearly, a broad variety of excitation processes can be invoked for generating many different kinds of light. A

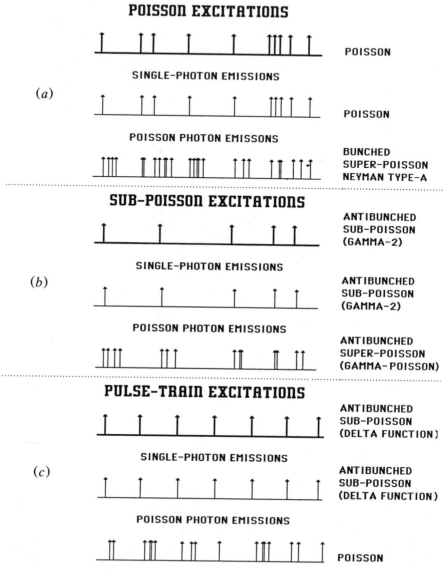

**Figure 12.1.** Schematic representation of a two-step process for the generation of light, illustrating stochastic excitations (first line) with either single-photon emissions (second line) or Poisson multiple-photon emissions (third line). Interference effects are ignored in this simple representation. (*a*) Poisson excitations; (*b*) sub-Poisson, antibunched excitations (gamma-2); (*c*) pulse-train excitations (random phase). After Teich et al. [12].

process similar to the gamma-2, and for which many analytical results are available, is the nonparalyzable dead-time-modified Poisson process (see Appendix). Resonance fluorescence radiation from a single atom is described by a process of this type since, after emitting a single photon, the atom decays to the ground state, where it remains for a period of time and cannot radiate. Short and Mandel used such a scheme to produce conditionally photon-number-squeezed emissions from isolated atoms [17].

Finally, in Figure 12.1c, we consider the case of pulse-train excitations (with random initial time). This is the limiting result both for the gamma family of processes and for the dead-time-modified Poisson process. In the absence of interference, single-photon emissions in this case yield ideally sub-Poisson photon statistics. Interference does not destroy the sub-Poisson nature in the long-counting-time limit. Poisson emissions give rise to Poisson photon statistics.

The illustration presented in Figure 12.1 is intended to emphasize the importance of the excitation and emission statistics as determinants of the character of the generated light. To produce sub-Poisson photons by direct generation, both sub-Poisson excitations and sub-Poisson emissions are required.

The statistical properties of light generated by sub-Poisson excitations, with each excitation leading to a single-photon emission, has been examined in considerable detail by Teich et al. [12]. These authors also addressed the effects of different locations for the different emissions and the rates of photon coincidence at pairs of positions in the detection plane. Some of the results are summarized in the Appendix.

The sub-Poisson excitations are characterized by a time constant $\tau_e$ that represents the time over which excitation events are anticorrelated (antibunched). The single-photon emissions, on the other hand, are characterized by a photon excitation–emission lifetime $\tau_p$. The detected light will be photon-number-squeezed provided $T \gg \tau_e, \tau_p$; $A \gg A_c$, where $T$ is the detector counting time, $A$ is the detector counting area, and $A_c$ is the coherence area. Different methods of sub-Poisson excitation result in different values of $\tau_e$ whereas different mechanisms of photon generation result in different values of $\tau_p$ and $A_c$.

Invoking these limits assures that all memory of the fields from individual emissions lie within the detector counting time and area, in which case the randomization of photon occurrences associated with interference does not extend beyond these limits. Consequently, the photon-counting statistics are determined by the only remaining source of variability, which is the randomness in the excitation occurrences. In this limit the photons behave as classical particles.

To recapitulate, a stationary stream of photon-number-squeezed light can be generated by a two-step process if sub-Poisson statistics are obeyed both by the excitations and by the individual emissions. For sufficiently large counting times and large detection areas, interference effects are washed out and the photons behave as classical particles. If the emissions are single photons, the overall photon statistics then directly mimic the statistics of the excitations.

This result is applicable even for physical processes that operate on the basis of nonindependent emissions as long as the counting time and area are sufficiently large in comparison with the time and area over which dependent emissions occur. Thus our approach is useful for understanding the generation of photon-number-squeezed light by incoherent sources, as described in this chapter, and by devices such as semiconductor lasers as described in the previous chapter [11].

## 12.3.  EXCITATION CONTROL PROVIDED BY A PHYSICAL PROCESS

We now consider several excitation control methods that make use of the sub-Poisson excitations inherent in space-charge-limited current flow. Such a current is naturally sub-Poisson as a result of the intrinsic Coulomb repulsion of the flowing electrons. In this case the excitation control results from a physical process.

Coulomb repulsion, which is the underlying physical process responsible for space-charge-limited current flow, is ubiquitous when excitations are achieved by means of charged particle beams. The single-photon emissions may be obtained in any number of ways. Three examples are provided: In Section 12.3.1 (which serves as the prototype for the following subsections) they arise from spontaneous fluorescence emissions in Hg vapor, in Section 12.3.2 they represent spontaneous recombination photons in a semiconductor, and in Section 12.3.3 they are stimulated recombination photons in a semiconductor. These methods all generate photon-number-squeezed light by transferring the anticlustering properties of the electrons, ultimately arising from Coulomb repulsion, directly to the photons.

### 12.3.1.  Space-Charge-Limited Franck–Hertz Experiment

The space-charge-limited Franck–Hertz effect [18–20] provided the first source of unconditionally photon-number-squeezed light. The essential element of this experiment is a collection of Hg atoms excited by inelastic collisions with a low-energy space-charge-limited ("quiet") electron beam. The space-charge reduction of the usual shot noise associated with thermionically emitted electrons can be substantial [21, 22]. A convenient measure of the noise reduction is provided by the Fano factor, which is defined as the ratio of the variance $\text{Var}(m)$ to the mean $\langle m \rangle$ of a random variable: $F_m \equiv \text{Var}(m)/\langle m \rangle$. Fano factors for the electron stream with values $F_e < 0.1$ are typical, and values as low as 0.01 are possible. After

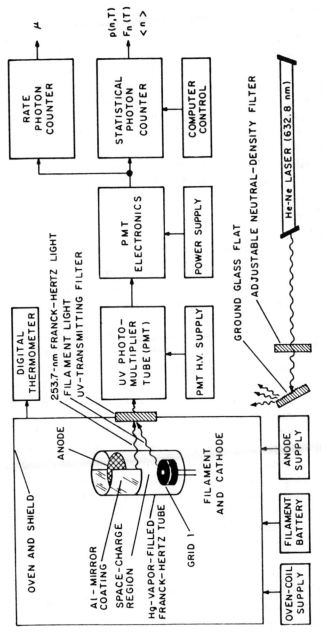

**Figure 12.2.** Block diagram of the space-charge-limited Franck–Hertz experiment. The use of Hg resulted in ultraviolet photon-number-squeezed light at a wavelength of 253.7 nm. After Teich and Saleh [20].

excitation, each atom emits a (sub-Poisson) single photon via the Franck–Hertz (FH) effect.

A block diagram of apparatus originally used to carry out this experiment is shown in Figure 12.2. The light was generated in a specially constructed 25-mm-diameter UV-transmitting Franck–Hertz tube, filled with 0.75 g of Hg. The radiation impinged on a UV photon-counting photomultiplier tube (PMT) in a special base that provided preamplification, discrimination, and pulse shaping. The output of this circuitry was fed to electronic photon-counting equipment that measured the probability distribution $p(n, T)$ for the detection of $n$ photoelectrons in time $T$. The mean count $\langle n \rangle$ and the Fano factor $F_n(T)$ were calculated from $p(n, T)$. The photon count was only slightly sub-Poisson, but the result was statistically significant. The small degree of photon-number squeezing resulted principally from optical losses in the experimental apparatus. The details of the experiment and the experimental results have been described elsewhere [19, 20].

### 12.3.2. Space-Charge-Limited Excitation of Recombination Radiation

A useful source of photon-number-squeezed light should exhibit a photon Fano factor that is substantially below unity while producing a large photon flux, preferably in a directed beam. It should also be small in size and rapidly switchable.

This has led us to propose a semiconductor device structure in direct analogy with the Franck–Hertz experiment described above. Sub-Poisson electron excitations are attained through space-charge-limited current flow and single-photon emissions are achieved by means of recombination radiation [23]. A device of this nature will emit incoherent photon-number-squeezed recombination radiation and should be far more efficient than its vacuum-tube cousin. The energy-band diagram for such a space-charge-limited light-emitting device (SCLLED) is illustrated in Figure 12.3. Sub-Poisson electrons are directly converted into sub-Poisson photons, as in the space-charge-limited Franck–Hertz experiment, but these are now recombination photons in a semiconductor. In designing such a device, carrier and photon confinement should be optimized and optical losses should be minimized. The basic structure of the device is that of a $p^+-i-n^+$ diode. Recombination radiation is emitted from the LED-like region.

The current noise in such a space-charge-limited diode [24] can be quite low. It has a thermal (rather than shot-noise) character [25–27]. The current noise spectral density $S_e(\omega)$ for a device in which only electrons participate in the conduction process is given by [23]

$$\frac{S_e(\omega)}{2e\langle I_e \rangle} = \frac{8k\theta}{e\langle V_e \rangle} \qquad (12.1)$$

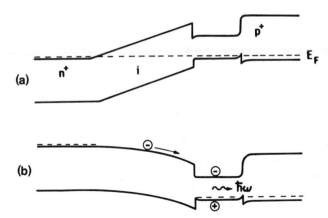

**Figure 12.3.** Energy-band diagram of a specially designed solid-state space-charge-limited light-emitting device under (*a*) equilibrium conditions and (*b*) strong forward-bias conditions. The curvature of the intrinsic region under forward-bias conditions indicates the space-charge potential. After Teich et al. [23].

where $\langle I_e \rangle$ is the average forward current in the device, $\langle V_e \rangle$ is the applied forward-bias voltage, $k$ is Boltzmann's constant, $\theta$ is the device temperature in kelvins, $\omega$ is the angular frequency, and $e$ is the electronic charge.

The degree of photon-number squeezing of the detected photons is then expected to be given by [23]

$$F_n(T) = 1 + \eta \left( \frac{8k\theta}{e\langle V_e \rangle} - 1 \right) \qquad (12.2)$$

provided that background light is absent. For a space-charge-limited diode, such as that shown in Figure 12.3, it is estimated that $8k\theta/e\langle V_e \rangle \approx$ 0.1 when $\theta = 300$ K and $\langle V_e \rangle = 2$ V (corresponding to $\langle I_e \rangle \approx 33$ mA). This ratio can be further reduced by cooling the device. If a dome-shaped surface-emitting GaAs/AlGaAs configuration and a Si $p$–$i$–$n$ photodetector are used, the overall quantum efficiency is estimated to be $\eta \approx 0.113$, yielding an overall estimated postdetection Fano factor $F_n(T) \approx 0.90$. A commercially available standard LED should provide $F_n(T) \approx 0.97$. In both cases, $T$ can be as short as $\approx 1$ ns. In principle, the degree of photon-number squeezing of the recombination radiation from the SCLLED is limited only by the geometrical collection efficiency.

### 12.3.3. Sub-Poisson Excitations and Stimulated Emissions

The properties of the light generated by the SCLLED could be improved if stimulated emissions were permitted. The advantages include improved beam directionality, switching speed, spectral properties, and coupling to an optical fiber. This could be achieved by use of an edge-emitting (rather than surface-emitting) LED configuration, with its waveguiding geometry and superfluorescence properties (single-pass stimulated emission). The theoretical results associated with the simple model presented in Section 12.2 (see Appendix) will apply provided that $T \gg \tau_e, \tau_c$; $A \gg A_c$, where $\tau_c$ and $A_c$ are now the coherence time and coherence area of the superfluorescent emission, respectively. The effect of the stimulated emissions is to extend $\tau_p$ into $\tau_c$ and to reduce the coherence area $A_c$. From a physical point of view, the photons still behave as classical particles in this regime since each electron gives rise to a single photon and there is no memory beyond the counting time $T$.

There will likely be further advantage in combining space-charge-limited current injection with a semiconductor laser structure rather than with the LED structure considered above. This could provide increased emission efficiency as well as additional improvement in beam directionality, switching speed, spectral properties, and coupling. This will be beneficial when the laser can be drawn into a realm of operation in which it produces a state that exhibits photon-number squeezing [28], such as a number-phase minimum uncertainty state [10, 11, 29]. Yamamoto and his colleagues [30, 31] have shown that this mode of operation can be attained in a semiconductor laser oscillator, within the cavity bandwidth and at high photon-flux levels, but in their case the pump fluctuations are suppressed below the shot-noise level by means of externally provided excitation control (see Section 12.4.4). Suggestions of this kind have also been made by Smirnov and Troshin [32] and by Carroll [33].

## 12.4. EXTERNALLY PROVIDED EXCITATION CONTROL

External mechanisms can also be used to ensure that the current flowing in a circuit is sub-Poisson. These include both optoelectronic and current-stabilization schemes. In Section 12.4.1 we discuss the use of two schemes that rely on the use of a light source and photodetector in a negative-feedback loop. The use of a beamsplitter to extract a portion of these in-loop photons is not useful for producing photon-number-squeezed light, as discussed in Section 12.4.2.

In Section 12.4.3 we discuss the possibility of generating sub-Poisson photons from sub-Poisson electrons by making use of an externally provided feedback control signal and an in-loop auxiliary optical source. Sub-Poisson electrons flow through the auxiliary source and produce sub-Poisson photons en route. Finally, in Section 12.4.4 we discuss the generation of sub-Poisson electrons by means of a purely electronic scheme, external current stabilization.

### 12.4.1. Optoelectronic Generation of Sub-Poisson Electrons

Two optoelectronic experiments incorporating externally provided excitation control have been used to generate sub-Poisson *electrons*. One of

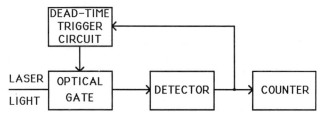

**Figure 12.4.**   Generation of sub-Poisson and antibunched electrons by external feedback, as studied by Walker and Jakeman [34].

these was carried out by Walker and Jakeman [34]; the other, by Machida and Yamamoto [35, 36]. The simplest form of the experiment carried out by Walker and Jakeman is illustrated in Figure 12.4. The registration of a photoevent at the detector operates a trigger circuit that causes an optical gate to be closed for a fixed period of time $\tau_d$ following the time of registration. During this period, the power $P_t$ of the (He–Ne) laser illuminating the detector is set precisely equal to zero so that no photo-events are registered. This is a dead-time optical gating scheme. Sub-Poisson photoelectrons were observed.

Machida and Yamamoto's experiment [35] has a similar thrust, al-though it is based on rate compensation. They used a single-longitudinal-mode GaAs/AlGaAs semiconductor injection laser to generate light (LD) and a Si $p$–$i$–$n$ photodiode (PD) to detect it, as shown in Figure 12.5$a$. Negative feedback from the detector was provided to the current driving the laser diode. A sub-shot-noise spectrum and sub-Poisson photoelectron counts were observed.

The similarity in the experimental results reported by Walker and Jakeman and by Machida and Yamamoto can be understood from a physical point of view [37]. In the configuration used by the latter authors, the injection-laser current (and therefore the injection-laser light output) is reduced in response to peaks of the in-loop photodetector current $i_t$. This rate compensation is essentially the same effect as that produced in the Walker–Jakeman experiment where the He–Ne laser light output is reduced (in their case to zero) in response to photoevent registrations at the in-loop photodetector. The feedback acts like a dead time, suppressing the emission of light in a manner that is correlated with photoevent occurrences at the in-loop detector.

### 12.4.2.   Extraction of In-Loop Photons by a Beamsplitter

These simple configurations cannot generate usable sub-Poisson *photons* since the feedback current is generated from the annihilation of the in-loop photons. Any attempt to remove in-loop photons by means of a beamsplitter (BS), such as that illustrated in Figure 2.5$b$, will lead to

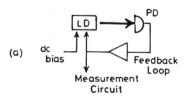

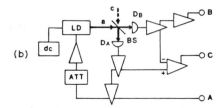

**Figure 12.5.** (*a*) Generation of sub-Poisson and antibunched electrons by external feedback using rate compensation, as investigated by Machida and Yamamoto [35]. (*b*) The removal of in-loop photons by a beamsplitter leads to super-Poisson light at the out-of-loop detector ($D_B$). After Machida and Yamamoto [35].

super-Poisson light. This result can be understood in terms of the arguments of Walker and Jakeman [34] and Shapiro and coworkers [37, 38] and is confirmed by the experiments of Walker and Jakeman [34].

Nevertheless, under special circumstances, and when components other than beamsplitters are used, the feedback technique can be useful in generating photon-number-squeezed light. These involve the use of quantum-nondemolition measurements and correlated photon pairs [1, 38].

### 12.4.3. Use of an In-Loop Auxiliary Optical Source

One of the more direct ways of producing photon-number-squeezed light from a system making use of external feedback is to insert an auxiliary optical source in the path of the sub-Poisson electron stream, as suggested by Capasso and Teich [39]. Two alternative configurations are shown in Figure 12.6. The character of the photon emitter is immaterial; it has been chosen to be an LED for simplicity but it could be a laser. In Figure 12.6*a* the photocurrent derived from the detection of light is negatively fed back to the LED input. It has been established both experimentally [35] and theoretically [37] that, in the absence of the block labeled "source," sub-Poisson electrons (i.e., a sub-shot-noise photocurrent) will flow in a circuit such as this. This conclusion is also valid in the presence of this block, which in this case acts simply as an added impedance to the electron flow.

Incorporating this element into the system offers access to the loop and permits the sub-Poisson electrons flowing in the circuit to be converted into sub-Poisson photons by means of electronic transitions. This is achieved by replacing the detector used in the feedback configurations of Machida and Yamamoto [35] and Walker and Jakeman [34] with a structure that acts simultaneously as a detector and a source. The sub-Poisson electrons emit sub-Poisson photons and continue on their way. The

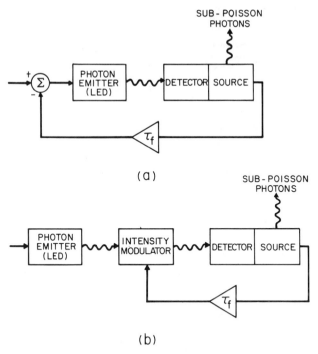

**Figure 12.6.** Generation of photon-number-squeezed light by insertion of an auxiliary source into the path of a sub-Poisson electron stream, as proposed by Capasso and Teich [39]. Wavy lines represent photons; solid lines represent the electron current ($\tau_f$ signifies the feedback time constant). The schemes represented in (a) and (b) make use of the sub-Poisson electron production methods illustrated in Figures 12.5 and 12.4, respectively. After Capasso and Teich [39].

configuration presented in Figure 12.6b is similar except that the (negative) feedback current gates the light intensity at the output of the LED in the manner of Walker and Jakeman, rather than the current at its input in the manner of Machida and Yamamoto. Any similar scheme, such as selective deletion [1, 40], could be used instead.

Two possible solid-state detector–source configurations have been suggested [39]. The scheme shown in Figure 12.7a makes use of sequential resonant tunneling [41] and single-photon dipole electronic transitions between the energy levels of a quantum-well heterostructure. The device consists of a reverse-biased $p^+$–$i$–$n^+$ diode where the $p^+$ and $n^+$ heavily doped regions have a wider band gap than the high-field, light-absorbing/emitting $i$ region. This arrangement ensures both high quantum efficiency at the incident photon wavelength (to which the $p^+$ window layer is transparent) and high collection efficiency (due to the waveguide geometry) for the light generated by the electrons drifting in the $i$ layer. An edge-emitting geometry is therefore appropriate. To maximize the

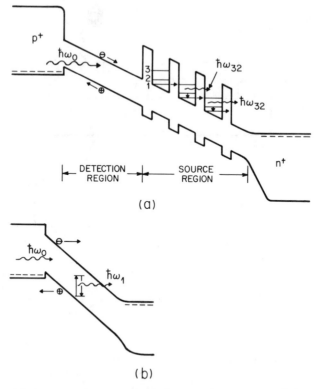

**Figure 12.7.** (*a*) Representative energy-band diagram of a quantum-well detector–source device (see Figure 12.6). The energy of the incident photon emitted by the LED is denoted $\hbar\omega_0$. Detection and source regions are shown. Photons of energy $\hbar\omega_{32}$ are emitted via electronic quantum-well transitions. (*b*) Representative energy-band diagram of a detector–source device with electroluminescent centers impact-excited by energetic photo-electrons, emitting photons with energy $\hbar\omega_1$. After Capasso and Teich [39].

collection efficiency, some of the facets of the device could be reflectively coated. The scheme shown in Figure 12.7*b* is similar except that it uses the impact excitation of electroluminescent centers in the *i* region by drifting electrons. Of course, the ability of configurations such as these to generate photon-number-squeezed light requires the usual interrelations among the various characteristic times associated with the system.

An estimate of the degree to which this mechanism will give rise to photon-number-squeezed light is provided by the Fano factor. The relevant relations are similar to those for the Franck–Hertz source. However, in this situation a single electron may give rise to multiple photons since there are *u* stages of the device (see Appendix). Numerical estimates for the Fano factor turn out to be similar for both structures illustrated in Figure 12.7, namely, $F_n \approx 0.97$ (under the assumption that the photodetector has an efficiency of 0.8). This is not as good as the value attainable by the SCLLED, principally because of low radiative efficiency in the tunneling scheme. Furthermore the external feedback mechanism may well be slower than the internal feedback scheme of the SCLLED.

### 12.4.4. Use of a Current Source with External Compensation

Steady-state current stabilization can be achieved by use of a constant voltage source in series with an external resistor $R$ [29], or in series with some other optoelectronic component with suitable $I-V$ characteristic.

Strong photon-number-squeezed light has been generated in two experiments that make use of external compensation. Tapster *et al.* [42] carried out an elegantly simple experiment, using a high-efficiency commercial GaAs LED fed by a Johnson-noise-limited high-impedance current source. They achieved a Fano factor $F_n \approx 0.96$ over a bandwidth of about 100 kHz, with a current transfer efficiency in excess of 11%. Machida *et al.* [30] fed an InGaAsP/InP single-longitudinal-mode distributed-feedback laser oscillator, operating at a wavelength of 1.56 $\mu$m, with a current source whose fluctuations were suppressed by the use of an external high-impedance element. In their first experiments, these authors obtained an average Fano factor $F_n \approx 0.96$ over a bandwidth of about 100 MHz, with a minimum Fano factor $F_n \approx 0.93$. They calculated that the radiation produced by their device is in a near number-phase minimum-uncertainty state, in the frequency range below the cavity bandwidth (which is in excess of 100 GHz for a typical semiconductor laser). Further results, which are indeed impressive, have been reported more recently [11, 31].

### 12.5. APPLICATIONS

We consider two specific examples where the use of photon-number-squeezed light might prove beneficial. In an idealized direct-detection lightwave communication system, errors (misses and false alarms) can be caused by noise from many sources, including photon noise intrinsic to the light source [1, 3]. If photon noise is the limiting factor, the use of photon-number-squeezed light in place of coherent light can bring about a reduction in this noise, and thereby the probability of error. For a coherent source each pulse of light (which carries a bit of information) contains a Poisson number of photons so that the photon-number standard deviation $\sigma_n = \langle n \rangle^{1/2}$. For photon-number-squeezed light, each pulse contains a sub-Poisson number of photons so that $\sigma_n < \langle n \rangle^{1/2}$. This noise reduction results in a decrease of the error probability. In a simple binary on–off-keying system whose only source of noise is assumed to be binomial photon counts (with Fano factor $F_n$), the mean number of photons per bit $\langle n' \rangle$ required to achieve an error probability of $10^{-9}$ decreases below its coherent-light "quantum limit" of 10 photons/bit as $F_n$ decreases below unity [1, 3, 43]. The "quantum limit" of a lightwave communication system should therefore more properly be designated as the "shot-noise limit."

The use of photon-number-squeezed light in visual science [5] could serve to clarify the functioning of ganglion cells in the mammalian retina. These cells transmit signals to higher visual centers in the brain via the optic nerve. In response to light, the ganglion cell generates a neural signal that takes the form of a sequence of nearly identical electrical events occurring along the time axis. The statistical nature of this neural signal is generally assumed to be governed by two nonadditive elements of stochasticity: the incident photons (which are Poisson-distributed in all experiments to date) and a randomness intrinsic to the cell itself [44]. If the statistical fluctuations of the photons could be controlled by exciting the retina with photon-number-squeezed light, the essential nature of the randomness intrinsic to the cell could be isolated and unambiguously determined. The use of photon-number-squeezed light as a stimulus in visual psychophysics experiments could also be helpful in clarifying the nature of seeing at threshold [5].

## APPENDIX　　GENERATION OF PHOTON-NUMBER-SQUEEZED AND ANTIBUNCHED LIGHT FROM INDEPENDENT RADIATORS

Consider an arbitrary (in general non-Poisson) excitation point process. Let each event of this process $\{t_1, t_2, \ldots, t_k, \ldots\}$ initiate a statistically independent and identical emission, so that the radiated light is the superposition of these emissions [12]. Even though the individual emissions typically take the form of pulses lasting a short time, and are therefore nonstationary, the overall radiation is stationary because of the assumed stationarity of the excitation process.

### Characterization of the Excitation Point Process

Two important descriptors of a stationary point process are the rate $\mu$ (events per unit time) and the rate of coincidence $\mu^2 g_e^{(2)}(\tau)$ of pairs of events at times separated by $\tau$. These descriptors are not sufficient to characterize an arbitrary point process completely [45, 46]; in general knowledge of the probability of multicoincidences of events at $k$ points, for $k = 1, 2, \ldots, \infty$, is required. If $m$ is the number of events that occur in a time interval $[0, T]$, then its mean is

$$\langle m \rangle = \mu T \qquad (12.\text{A1})$$

and its Fano factor (ratio of variance to mean) is

$$F_m(T) \equiv \frac{\text{Var}(m)}{\langle m \rangle} = 1 + \frac{\langle m \rangle}{M_e} \qquad (12.\text{A}2)$$

where

$$M_e^{-1} = \left(\frac{2}{T}\right) \int_0^\infty \left(1 - \frac{\tau}{T}\right) [g_e^{(2)}(\tau) - 1] \, d\tau \qquad (12.\text{A}3)$$

The simplest example is the Poisson point process, for which $g_e^{(2)}(\tau) = 1$ and $F_m(T) = 1$. If $g_e^{(2)}(0) < 1$, the excitation process is said to be anti-bunched or anticorrelated, whereas if $g_e^{(2)}(0) > 1$ it is said to be bunched or correlated. The characteristic time associated with the function $[g_e^{(2)} - 1]$ is denoted $\tau_e$. Similarly, if $F_m(T) < 1$, the excitation counts are said to be sub-Poisson (for this counting time $T$), whereas if $F_m(T) > 1$, the counts are said to be super-Poisson. The Poisson point process has neither memory nor aftereffects.

For the self-exciting point process (SEPP), on the other hand, the probability of occurrence of an event at a particular time depends on the times and numbers of previous occurrences [46]. Renewal point processes (RPPs) form an important subclass of SEPPs for which the rate $\mu$ and the normalized coincidence rate $g_e^{(2)}(\tau)$ do characterize the process completely [45]. These are processes for which the interevent time intervals are statistically independent and identically distributed. The following are important examples of renewal point processes that exhibit antibunched events and sub-Poisson counts:

1. *The Gamma-$\mathcal{N}$ Process.* This process is obtained from a Poisson process by decimation, that is, by selecting every $\mathcal{N}$th event and discarding all others [45, 47], as illustrated in Figure 12.1b for $\mathcal{N} = 2$. The process is so named because the interevent time distribution $P(\tau)$ is a gamma distribution of order $\mathcal{N}$. For the particular case when $\mathcal{N} = 2$, it turns out that [12]

$$g_e^{(2)}(\tau) = 1 - \exp(-4\mu|\tau|) \qquad (12.\text{A}4)$$

$$F_m(T) \approx \tfrac{1}{2} \qquad (12.\text{A}5)$$

2. *The Nonparalyzable Dead-Time-Modified Poisson Process.* This process is obtained from a Poisson process by deleting events that fall within a specified dead time $\tau_d$ following the registration of an event [45–51]. It is

characterized by [12]

$$g_e^{(2)}(\tau) = \frac{1}{\mu} \sum_{l=1}^{\infty} \frac{[\lambda(\tau - l\tau_d)]^{l-1}}{(l-1)!} \exp[-\lambda(\tau - l\tau_d)]U(\tau - l\tau_d) \quad (12.\text{A}6)$$

$$F_m(T) \approx (1 - \mu\tau_d)^2 \quad (12.\text{A}7)$$

with

$$\lambda = \frac{\mu}{1 - \mu\tau_d} \quad (12.\text{A}8)$$

where $U(t)$ is the unit step function, $\lambda$ is the initial rate of the process, and $\mu$ is the rate after dead-time modification. [Equation (12.A6) replaces Eq. (3.43) in Ref. 1, which has a typographical error.] Its interevent-time density function is a decaying exponential function displaced to the minimum permissible interevent time $\tau_d$.

Another example is a pulse train with random time of initiation [12, 52].

## Photon Statistics for Emissions at Antibunched Times

When the underlying excitation process has known rate $\mu$ and normalized coincidence rate $g_e^{(2)}(\tau)$, but is otherwise arbitrary, what can be said about the statistics of the radiation? Because of their finite lifetime, emissions overlap and interfere. To determine their bunching properties it is necessary to know not only the rate of coincidence of the excitation process at pairs of time instants but also the coincidence rates at triple points, and so on. If such information is not available, the bunching properties of the superposed radiation cannot be determined.

However, in the limit in which the counting time $T$ is much longer than the lifetime $\tau_p$ of the individual emissions, interference has a negligible effect on the total number of collected photocounts. The total number of photons $n$ is then simply the sum of the number of photons emitted *independently* by the individual emissions. If $m$ is the number of emissions and $\alpha_k$ is the number of photoevents associated with the $k$th emission, then $n = \sum_{k=1}^{m} \alpha_k$. Using the fact that the $\{\alpha_k\}$ are statistically independent and identical, it is not difficult to show that the mean and variance of $n$ are

$$\langle n \rangle = \langle \alpha \rangle \langle m \rangle \quad (12.\text{A}9)$$

$$\text{Var}(n) = \langle \alpha \rangle^2 \, \text{Var}(m) + \langle m \rangle \text{Var}(\alpha) \quad (12.\text{A}10)$$

from which it follows that the corresponding Fano factors are related by

$$F_n = \langle \alpha \rangle F_m + F_\alpha \qquad (12.\text{A}11)$$

or

$$F_n = 1 + [F_\alpha - 1 + \langle \alpha \rangle] + \langle \alpha \rangle (F_m - 1) \qquad (12.\text{A}12)$$

Equation (12.A10) is known as the *cascade variance formula* [53–55].

Equation (12.A12) shows that the Fano factor comprises three contributions. The first term is that of a Poisson process. The second term (in square brackets) represents excess noise due to randomness in the number of photons per emission (if $\alpha = 1$, then $F_\alpha = 0$ and it vanishes). The third term admits the possibility of noise reduction due to anticorrelations in the excitation process. This term vanishes if the excitation process is Poisson (since $F_m = 1$), whereupon Eq. (12.A11) assumes the well-known form

$$F_n = \langle \alpha \rangle + F_\alpha \qquad (12.\text{A}13)$$

We now consider an example in which each of the individual emissions is described by a one-photon number state (i.e., single-photon emissions with $\alpha = 1$). This corresponds to $\langle \alpha \rangle = 1$ and $F_\alpha = 0$, from which Eq. (12.A11) yields

$$F_n = F_m \qquad (12.\text{A}14)$$

This is to be expected. For single-photon emissions, the number of photons counted over a long time interval is approximately equal to the number of excitations (assuming there are no losses). If the excitation point process is sub-Poisson, the photons will also be sub-Poisson. It is of interest to note that we need not go outside the domain of linear (one-photon) optics to see such uniquely quantum-mechanical effects.

Equations (12.A11) and (12.A12) reveal the key to obtaining sub-Poisson photons from sub-Poisson excitations. In order to have $F_n < 1$, $F_\alpha$ must be $< 1$, as is apparent from Eq. (12.A11). Furthermore, a necessary condition for $F_n < 1$ is that $F_m < 1$ (because the term in square brackets in Eq. (12.A12) is nonnegative). It follows that for $F_n$ to be less than unity, both $F_\alpha$ and $F_m$ must be less than unity. Therefore, the generation of a stationary stream of sub-Poisson photons from a superposition of independent emissions requires that both the excitation process and the photons of the individual emissions be sub-Poisson.

Physical mechanisms that provide control of the excitation point process, and that are well described by the model presented here, have been discussed in Sections 12.3 and 12.4.

### Bunching/Antibunching Properties of Emissions Initiated'
### at Antibunched Times

Determination of the short-time behavior of the photoevents requires knowledge of the normalized photocoincidence rate $g^{(2)}(\tau)$. This is not possible unless the excitation point process is completely specified (higher-order multicoincidence rates specified). Teich et al. [12] examined this problem under the assumption that the excitation point process is a renewal point process. Using the assumption of single-mode individual emissions, they showed that

$$g^{(2)}(\tau) = 1 + |g^{(1)}(\tau)|^2 + \left(\frac{1}{\mu\tau_p}\right)\bar{g}^{(2)}(\tau) + r(\tau) \quad (12.A15)$$

The first three terms on the right-hand side of Eq. (12.A15) emerge from a Poisson excitation process. The fourth term, which is given by

$$r(\tau) = \int_0^\infty \psi(\tau,t)\left[g_e^{(2)}(t) - 1\right]dt \quad (12.A16)$$

with

$$\psi(\tau,t) = \int_0^\infty \left[I_0(t')I_0(t' + \tau - t) + V_0^*(t')V_0(t' + \tau)\right.$$

$$\left. \times V_0^*(t + t')V_0(t + t' + \tau)\right]dt' \quad (12.A17)$$

represents the effects of deviation of the excitation process from Poisson. When the excitation point process is antibunched, this term is negative, thereby introducing anticorrelations into the photon process. If it is sufficiently strong, it can counterbalance the bunching effects due to wave interference [second term in Eq. (12.A15)] and due to the randomness of the individual emissions [third term in Eq. (12.A15)].

With the availability of Eq. (12.A15), the Fano factor for the photon counts in a time interval of arbitrary duration can be determined. The result can be put in the form [12]

$$F_n(T) = 1 + \frac{\langle n\rangle}{M} + \frac{F_\alpha(\infty) - 1 + \langle\alpha\rangle}{\mathscr{M}} + \frac{\langle n\rangle}{\mathscr{M}'} \quad (12.A18)$$

where $M$, $\mathscr{M}$, and $\mathscr{M}'$ are degrees-of-freedom parameters, the latter associated with the term $r(\tau)$. The parameter $\mathscr{M}'$ depends, in a complex way, on the relation between the counting time $T$, the emission lifetime $\tau_p$, and the excitation point process memory time $\tau_e$ (which is the width of the function $[g_e^{(2)}(\tau) - 1]$). For counting times that are long ($T \gg \tau_p, \tau_e$), however, it turns out that $M = \infty$ and wavelike (interference) noise is washed out; $\mathcal{M} = 1$ so that the role of noise in the individual emissions is enhanced; and $\mathcal{M}'$ is given by the degrees-of-freedom parameter for the excitation process $M_e$ given in Eq. (12.A3). It then follows that Eq. (12.A18) reduces to Eq. (12.A12), which was directly obtained by use of the cascade variance formula.

## ACKNOWLEDGMENT

This work was supported in part by the Joint Services Electronics Program through the Columbia Radiation Laboratory.

## REFERENCES

1. M. C. Teich and B. E. A. Saleh, "Photon bunching and antibunching," in *Progress in Optics*, Vol. 26, E. Wolf, ed., North-Holland, Amsterdam, 1988, pp. 1–104.
2. M. C. Teich and B. E. A. Saleh, *Quantum Opt.* **1**, 153 (1989); *Phys. Today* **43** (6), 26 (1990).
3. B. E. A. Saleh and M. C. Teich, *Phys. Rev. Lett.* **58**, 2656 (1987).
4. Y. Yamamoto and H. A. Haus, *Rev. Mod. Phys.* **56**, 1001 (1986).
5. M. C. Teich, P. R. Prucnal, G. Vannucci, M. E. Breton, and W. J. McGill, *Biol. Cybern.* **44**, 157 (1982).
6. R. E. Slusher, L. W. Hollberg, B. Yurke, J. C. Mertz, and J. F. Valley, *Phys. Rev. Lett.* **55**, 2409 (1985).
7. L.-A. Wu, M. Xiao, and H. J. Kimble, *J. Opt. Soc. Am. B* **4**, 1465 (1987).
8. R. Loudon and P. L. Knight, *J. Mod. Opt.* **34**, 709 (1987).
9. L. Mandel, *Phys. Rev. Lett.* **49**, 136 (1982).
10. Y. Yamamoto, S. Machida, N. Imoto, M. Kitagawa, and G. Björk, *J. Opt. Soc. Am. B* **4**, 1645 (1987).
11. Y. Yamamoto and S. Machida, Chapter 11, this volume.
12. M. C. Teich, B. E. A. Saleh, and J. Peřina, *J. Opt. Soc. Am. B* **1**, 366 (1984).
13. M. C. Teich and B. E. A. Saleh, *Opt. Lett.* **7**, 365 (1982).
14. B. E. A. Saleh, D. Stoler, and M. C. Teich, *Phys. Rev. A* **27**, 360 (1983).
15. B. E. A. Saleh and M. C. Teich, *Proc. IEEE* **70**, 229 (1982).
16. M. C. Teich, *Appl. Opt.* **20**, 2457 (1981).
17. R. Short and L. Mandel, *Phys. Rev. Lett.* **51**, 384 (1983).
18. M. C. Teich, B. E. A. Saleh, and D. Stoler, *Opt. Commun.* **46**, 244 (1983).
19. M. C. Teich, B. E. A. Saleh, and T. Larchuk, "Observation of sub-Poisson Franck–Hertz light at 253.7 nm," in *Digest XIII Int. Quant. Electron. Conf.*, Anaheim, CA, 1974 (Optical Society of America, Washington, DC, 1984), Paper PD-A6.

20. M. C. Teich and B. E. A. Saleh, *J. Opt. Soc. Am. B* **2**, 275 (1985).

21. S. K. Srinivasan, *Nuovo Cimento* **38**, 979 (1965).

22. S. K. Srinivasan, *Opt. Acta* **33**, 207 (1986).

23. M. C. Teich, F. Capasso, and B. E. A. Saleh, *J. Opt. Soc. Am. B* **4**, 1663 (1987).

24. S. Sze, *Physics of Semiconductor Devices*, 1st ed., Wiley, New York, 1969, p. 421, Eq. (95).

25. M. A. Lampert and A. Rose, *Phys. Rev.* **121**, 26 (1961).

26. M. A. Nicolet, H. R. Bilger, and R. J. J. Zijlstra, *Phys. Stat. Sol.* **B70**, 9 (1975).

27. M. A. Nicolet, H. R. Bilger, and R. J. J. Zijlstra, *Phys. Stat. Sol.* **B70**, 415 (1975).

28. P. Filipowicz, J. Javanainen, and P. Meystre, *Phys. Rev. A* **34**, 3077 (1986).

29. Y. Yamamoto, S. Machida, and O. Nilsson, *Phys. Rev. A* **34**, 4025 (1986).

30. S. Machida, Y. Yamamoto, and Y. Itaya, *Phys. Rev. Lett.* **58**, 1000 (1987).

31. S. Machida and Y. Yamamoto, *Phys. Rev. Lett.* **60**, 792 (1988).

32. D. F. Smirnov and A. S. Troshin, *Opt. Spektrosk.* **59**, 3 (1985) [transl. in *Opt. Spectrosc.* (USSR) **59**, 1 (1985)].

33. J. E. Carroll, *Opt. Acta* **33**, 909 (1986).

34. J. G. Walker and E. Jakeman, *Proc. Soc. Photo-Opt. Instrum. Eng.* **492**, 274 (1985).

35. S. Machida and Y. Yamamoto, *Opt. Commun.* **57**, 290 (1986).

36. Y. Yamamoto, N. Imoto, and S. Machida, *Phys. Rev. A* **33**, 3243 (1986).

37. J. H. Shapiro, M. C. Teich, B. E. A. Saleh, P. Kumar, and G. Saplakoglu, *Phys. Rev. Lett.* **56**, 1136 (1986).

38. J. H. Shapiro, G. Saplakoglu, S.-T. Ho, P. Kumar, B. E. A. Saleh, and M. C. Teich, *J. Opt. Soc. Am. B* **4**, 1604 (1987).

39. F. Capasso and M. C. Teich, *Phys. Rev. Lett.* **57**, 1417 (1986).

40. B. E. A. Saleh and M. C. Teich, *Opt. Commun.* **52**, 429 (1985).

41. F. Capasso, K. Mohammed, and A. Y. Cho, *Appl. Phys. Lett.* **48**, 478 (1986).

42. P. R. Tapster, J. G. Rarity, and J. S. Satchell, *Europhys. Lett.* **4**, 293 (1987).

43. K. Yamazaki, O. Hirota, and M. Nakagawa, *Trans IEICE Japan* **71**, 775 (1988).

44. B. E. A. Saleh and M. C. Teich, *Biol. Cybern.* **52**, 101 (1985).

45. D. R. Cox, *Renewal Theory*, Methuen, London, 1962.

46. D. L. Snyder, *Random Point Processes*, Wiley, New York, 1975.

47. E. Parzen, *Stochastic Processes*, Holden-Day, San Francisco, 1962.

48. L. M. Ricciardi and F. Esposito, *Kybernetik* (*Biol. Cybern.*) **3**, 148 (1966).

49. J. W. Müller, *Nucl. Instrum. Meth.* **117**, 401 (1974).

50. B. I. Cantor and M. C. Teich, *J. Opt. Soc. Am.* **65**, 786 (1975).

51. M. C. Teich and G. Vannucci, *J. Opt. Soc. Am.* **68**, 1338 (1978).

52. R. Loudon, *Rep. Progr. Phys.* **43**, 913 (1980).

53. W. Shockley and J. R. Pierce, *Proc. IRE* **26**, 321 (1938).

54. L. Mandel, *Brit. J. Appl. Phys.* **10**, 233 (1959).

55. R. E. Burgess, *J. Phys. Chem. Solids* **22**, 371 (1961).

# 13

# Controlled Spontaneous Emission in Microcavity Semiconductor Lasers

Y. YAMAMOTO, S. MACHIDA, AND K. IGETA
*NTT Basic Research Laboratories, Nippon Telegraph and Telephone Corporation, Musashino-shi, Tokyo, Japan*

G. BJÖRK
*Department of Microwave Engineering, The Royal Institute of Technology, Stockholm, Sweden*

## 13.1. INTRODUCTION

It is known that spontaneous emission of an atom is not an immutable property of an atom, but can be altered by a cavity wall [1, 2]. The standard interpretation for this phenomenon is that the cavity wall modifies the vacuum field fluctuations that "stimulate" a spontaneous emission [3]. Recently, several experiments have demonstrated inhibition and enhancement of spontaneous emission, at both microwave frequency [4–6] and optical frequency [7–9]. Such capability is no less important in semiconductors, for which spontaneous emission plays a decisive role in the physics [10] and in device performance [11].

This chapter discusses the theoretical and experimental studies on enhanced and inhibited spontaneous emission in microcavity semiconductor lasers. In Section 13.2, the physics of spontaneous emission of an atom in free space is briefly reviewed. The difference between vacuum-field-induced Rabi oscillation and the irreversible spontaneous decay described by Weisskopf–Wigner theory is pointed out. The radiation pattern is calculated in polar coordinates. Section 13.3 treats the spontaneous emission of an atom between two (hypothetical) ideal planar mirrors. The principle of enhanced and inhibited spontaneous emission is delineated.

*Coherence, Amplification, and Quantum Effects in Semiconductor Lasers*, Edited by Yoshihisa Yamamoto.
ISBN 0-471-512494 © 1991 John Wiley & Sons, Inc.

Section 13.4 discusses two practical microcavity structures: a one-dimensional periodic structure with a half-wavelength central cavity layer, one with a one-wavelength central cavity layer. The radiation pattern, the coupling efficiency into a cavity resonant mode, the spontaneous lifetime, and the cavity bandwidth (photon lifetime) are calculated as functions of various parameters. In Section 13.5, experimental evidence for enhanced and inhibited spontaneous emission in GaAs quantum wells in microcavities is presented. The characteristics of microcavity semiconductor lasers with controlled spontaneous emission are finally discussed in Section 13.6. These characteristics include an extremely low threshold pump rate, very broad modulation bandwidth, and amplitude squeezing at all pump rates.

## 13.2. BRIEF REVIEW OF SPONTANEOUS EMISSION

### 13.2.1. Vacuum-Field-Induced Rabi Oscillation

Suppose an atom in an excited state couples with a single electromagnetic field mode with no photon. This simplest situation was recently experimentally demonstrated by injecting an excited Rydberg atom into a single-mode superconductor high-Q cavity [12]. During the time of flight, the atom couples with only one electromagnetic field mode in the vacuum state. The interaction Hamiltonian is [13]

$$\hat{H}_I = \hbar g (\hat{a}\hat{\sigma}^\dagger + \hat{a}^\dagger\hat{\sigma}) \tag{13.1}$$

where $\hat{a}$ and $\hat{a}^\dagger$ are the annihilation and creation operators of the field mode, $\hat{\sigma}$ and $\hat{\sigma}^\dagger$ are the lowering and raising operators (Pauli's spin operators) of the two-level atom, and $g$ is the coupling coefficient proportional to the atomic dipole transition matrix element $\langle e|\hat{p}|g\rangle$, in which $\hat{p}$ is the dipole operator. The wavefunction for the combined atom–field system is expressed as

$$|\psi\rangle = C_{e0}|e\rangle_a|0\rangle_f + C_{g1}|g\rangle_a|1\rangle_f \tag{13.2}$$

in a Schrödinger picture. In Eq. (13.2), $C_{e0}$ is the probability amplitude for the state with the excited atom and the vacuum field, and $C_{g1}$ is that for the state with the unexcited atom and the single photon field. When the atomic transition frequency and the field frequency are equal (in resonance), the time evolution of the two probability amplitudes is periodic, that is

$$C_{e0} = \cos(gt) \tag{13.3}$$

and

$$C_{g1} = -i\sin(gt) \tag{13.4}$$

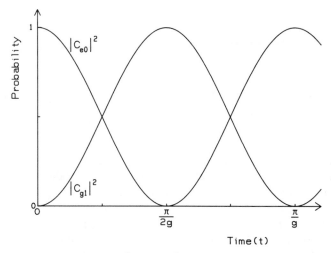

**Figure 13.1.**   Time evolution of $|C_{e0}|^2$ and $|C_{g1}|^2$ in a vacuum-field-induced Rabi oscillation.

Figure 13.1 shows the time evolution of $|C_{e0}|^2$ and $|C_{g1}|^2$. The atom and the field pass energy back and forth. This is a vacuum-field-induced Rabi oscillation [13].

Since the field is initially in a vacuum state and the atom emits a photon after $t = \pi/2g$, this process is considered spontaneous emission. However, ordinary spontaneous emission is an irreversible process with an exponential decay. The vacuum-field-induced Rabi oscillation discussed here is a reversible process with periodic energy exchange.

### 13.2.2.   Weisskopf–Wigner Theory of Spontaneous Emission

Next let us consider an atom in an excited state coupled with infinite and continuous electromagnetic field modes with no photons. The new interaction Hamiltonian and system wavefunction are

$$\hat{H}_I = \hbar \sum_k g_k \left( \hat{a}_k \hat{\sigma}^\dagger + \hat{a}_k^\dagger \hat{\sigma} \right) \tag{13.5}$$

and

$$|\psi\rangle = C_{e0} |e\rangle_a |0\rangle_f + \sum_k C_{gk} |g\rangle_a |1_k\rangle_f \tag{13.6}$$

Here $|0\rangle_f$ indicates all field modes are in vacuum states; $|1_k\rangle_f$ means the mode $k$ is in a single photon state and all the other modes are in vacuum states. The equation of motion for $C_{e0}$ after eliminating the coefficients

$C_{gk}$ is

$$\frac{d}{dt}C_{e0} = -\sum_k g_k^2 \int_0^t dt' \exp[-i(\omega_k - \omega_0)(t - t')]C_{e0}(t')$$

$$= -\int D(\omega_k) g^2(\omega_k) d\omega_k \int_0^t dt' \exp[-i(\omega_k - \omega_0)(t - t')]C_{e0}(t')$$

$$\simeq -\pi g^2(\omega_0) D(\omega_0) C_{e0} \qquad (13.7)$$

Here the summation $\Sigma_k$ is replaced by the integral $\int D(\omega_k) d\omega_k$, in which $D(\omega_k)$ is the density of the electromagnetic field modes and $\omega_0$ is the atomic transition frequency. It is assumed that both $D(\omega_k)$ and $g(\omega_k)$ are constant in the frequency region where the time integral has an appreciable value. The equation $\gamma = 2\pi g^2(\omega_0)D(\omega_0)$ expresses the spontaneous decay rate of the excited atom. This is "Fermi's golden rule."

Irreversible spontaneous decay emerges when the atom couples with infinite and continuous field modes. The atomic dipole couples to many field modes and decays to the ground state. However, many field modes couple to vacuum field fluctuations, which are independent of each other, and to the atom, in turn. This is the fluctuation–dissipation theorem [14]. Since each vacuum field fluctuation oscillates at a different frequency, the atomic dipole and the field modes lose their mutual phase coherence and periodic exchange of energy becomes impossible.

### 13.2.3. Spontaneous Emission Patterns of an Atom in Free Space

Let us consider a dipole moment along the $x$ axis in a polar coordinate $(r, \psi, \gamma)$ shown in Figure 13.2. The dipole emits electromagnetic fields having polarizations both perpendicular ($s$ wave) and parallel ($p$ wave) to the emission plane. The radiation intensity at the point $(r, \psi, \gamma)$ is [2]

$$dI(r, \psi, \gamma) = \eta \frac{P_{12}^2 E_0^2}{r^2} \times \begin{cases} \sin^2 \psi, & s \text{ wave} \\ \cos^2 \psi \cos^2 \gamma & p \text{ wave} \end{cases} \qquad (13.8)$$

Here $\eta$ is a constant and $E_0^2$ is the vacuum field intensity at the location $(r = 0)$ of a dipole moment. In free space, $E_0^2$ is independent of $\psi$ and $\gamma$ (isotropic). The total emitted intensity is given by

$$I \equiv \int_0^{2\pi} d\psi \int_0^{\pi} d\gamma \, r^2 \sin \gamma \, dI(r, \psi, \gamma) = \tfrac{8}{3}\pi \eta P_{12}^2 E_0^2 \qquad (13.9)$$

Three-fourths of Eq. (13.9) is radiated as the $s$ wave and the one-fourth as the $p$ wave. When the dipole moment is along the $y$ axis, the result is exactly the same.

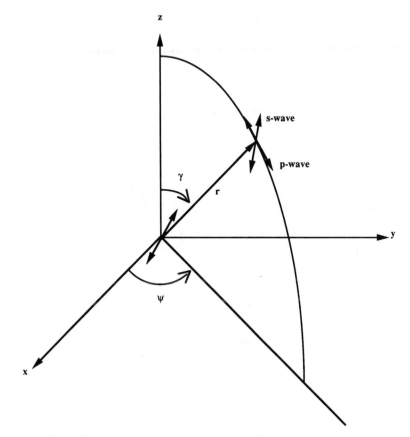

**Figure 13.2.**   Polar coordinates $(r, \psi, \gamma)$ for radiation pattern calculation.

If the dipole moment is along the $z$ axis, the radiation intensity at the point $(r, \psi, \gamma)$ is [2]

$$dI(r, \psi, \gamma) = \eta \frac{P_{12}^2 E_0^2}{r^2} \times \begin{cases} 0, & s \text{ wave} \\ \sin^2 \gamma & p \text{ wave} \end{cases} \qquad (13.10)$$

The total emitted intensity,

$$I \equiv \int_0^{2\pi} d\psi \int_0^{\pi} d\gamma \, r^2 \sin \gamma \, dI(r, \psi, \gamma) = \tfrac{8}{3}\pi \eta P_{12}^2 E_0^2 \qquad (13.11)$$

is equal to Eq. (13.9) for the dipole moment along the $x$ axis, as it should be, but in this case, all the emitted power is carried by the $p$ wave. The choice of the emission plane including the $z$ axis employed here, thus the definition of the $s$ wave and the $p$ wave, has no physical significance in free space. However, as discussed in subsequent sections, the vacuum field intensity $E_0^2$ in one-dimensional periodic structures becomes a function of not only the angle $\gamma$ but also the polarization direction ($s$ or $p$).

## 13.3.  SPONTANEOUS EMISSION OF AN ATOM BETWEEN IDEAL MIRRORS

We consider the spontaneous emission of an atom placed between two hypothetical ideal mirrors in this section. Assume that a mirror has constant reflection and transmission coefficients and phase shift, which are all independent of incident angle $\gamma$. Even though such a mirror is not realistic, it still gives a simple physical picture for the enhanced and inhibited spontaneous emission realized by one-dimensional periodic structures treated below.

### 13.3.1.  Local Vacuum Field Fluctuation Coupled with an Atom

Let us consider the modification of the vacuum field intensity $E_0^2$ by the presence of the two planar mirrors shown in Figure 13.3. The two field

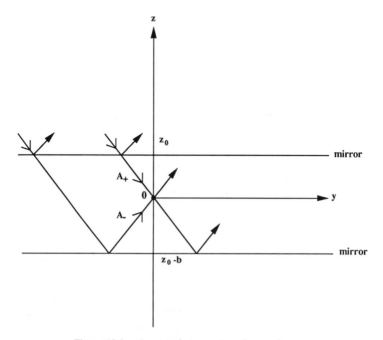

**Figure 13.3.**   An atom between two planar mirrors.

amplitudes $A_+$ and $A_-$ at the location of the atom ($z = 0$) are

$$A_+ = \frac{\sqrt{T} \exp(ikz_0 \cos \gamma)}{1 - R \exp[i(2kb \cos \gamma + 2\phi_r)]} E_0 \qquad (13.12)$$

and

$$A_- = \frac{\sqrt{RT} \exp[i\{k(2b - z_0)\cos \gamma + \phi_r\}]}{1 - R \exp[i(2kb \cos \gamma + 2\phi_r)]} E_0 \qquad (13.13)$$

where $R$ and $T$ are the reflection and transmission coefficients, $\phi_r$ is the phase shift due to reflection, $b$ is the mirror spacing, and $z_0$ is the distance between the upper mirror and the atom.

If the dipole moment is along the $z$ axis, the vacuum field intensity coupled with the dipole is modified according to

$$E_z^2 = |A_+ - A_-|^2 = \{|A_+|^2 + |A_-|^2 - 2|A^+||A^-|\cos \phi\}E_0^2 \quad (13.14)$$

where

$$\phi = 2k(b - z_0)\cos \gamma + \phi_r \qquad (13.15)$$

If the dipole moment is along the $x$ axis or $y$ axis, the vacuum field intensity coupled with the dipole is modified according to

$$E_x^2 = E_y^2 = |A_+ + A_-|^2 = \{|A_+|^2 + |A_-|^2 + 2|A_+||A_-|\cos \phi\}E_0^2 \quad (13.16)$$

### 13.3.2. Modified Radiation Pattern and Lifetime in a Half-Wavelength Cavity

If the phase shift $\phi_r$ due to reflection is $\pi$, the horizontal dipole ($P_x$ or $P_y$) couples only with odd modes and the vertical dipole ($P_z$) couples only with even modes. Suppose the two mirrors are separated by a distance equal to half a wavelength, $b = \lambda/2$, and the atom is located midway between the two mirrors, $z_0 = \lambda/4$. The atom is located at the antinode of the horizontal field of a resonant standing wave ($N = 1$ odd mode) propagating along the $z$ axis, so that the spontaneous emission in the $z$ axis by the horizontal dipole should be enhanced. The horizontal component of the vacuum field intensity $E_x^2(E_y^2)$ propagating in the direction $\gamma$ for the dipole moment along the $x$ axis ($y$ axis) is

$$E_x^2 = E_y^2 = \frac{T\{1 + R + 2\sqrt{R} \cos[\pi(\cos \gamma + 1)]\}}{(1 - R)^2 + 4R \sin^2[\pi(\cos \gamma + 1)]} E_0^2 \qquad (13.17)$$

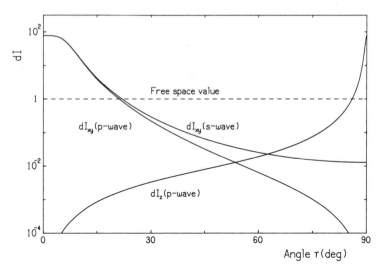

**Figure 13.4.** Spontaneous radiation patterns from an atom between two ideal mirrors with $d = \lambda/2$, $R = 0.95$, and $\phi_r = \pi$.

The radiation intensity at the point $(r, \psi, \gamma)$ is calculated by using Eq. (13.17) for $E_0^2$ in Eq. (13.8):

$$dI(r,\psi,\gamma) = \eta \frac{P_{12}^2 E_0^2}{r^2} \cdot \frac{(1-R)\{1 + R + 2\sqrt{R}\cos[\pi(\cos\gamma + 1)]\}}{(1-R)^2 + 4R\sin^2[\pi(\cos\gamma + 1)]}$$

$$\times \begin{cases} \sin^2\psi, & s \text{ wave} \\ \cos^2\psi\cos^2\gamma & p \text{ wave} \end{cases} \quad (13.18)$$

Numerical examples of the radiation pattern in Eq. (13.18) for $T = 0.05$ and $R = 0.95$ are plotted in Figure 13.4. The radiation patterns in both polarizations concentrate on $\gamma \simeq 0$, as expected. In a small $\gamma$ value where the radiation intensity is appreciable, Eq. (13.18) is approximated by

$$dI(r,\psi,\gamma) = \eta \frac{P_{12}^2 E_0^2}{r^2} \cdot \frac{4(1-R)}{(1-R)^2 + \pi^2 R\gamma^4} \times \begin{cases} \sin^2\psi, & s \text{ wave} \\ \cos^2\psi, & p \text{ wave} \end{cases}$$

$$(13.19)$$

The peak intensity at $\gamma = 0$ is enhanced by a factor of $4/(1-R)$ above the free-space value and the diverging angle of the main lobe is $\sqrt{(1-R)}/\pi$. The total intensity of emission in all directions is given by

$$I \simeq 2 \int_0^{2\pi} d\psi \int_0^{\pi/2} d\gamma \, dI(r,\psi,\gamma) r^2 \sin\gamma \simeq 4\pi\eta P_{12}^2 E_0^2 \quad (13.20)$$

Here the integral formula $\int_0^\infty x/(1 + x^4)\,dx = \pi/4$ and $\sin\gamma \simeq \gamma$ are used. The total emitted power is 1.5 times larger than the free-space value, which means that the spontaneous lifetime is reduced by a factor of $\frac{2}{3}$. This change in the spontaneous lifetime is independent of the reflection coefficient $R$. The spontaneous emission is carried equally by the $p$ wave and the $s$ wave.

The vertical component of the vacuum field intensity $E_z^2$ propagating in the direction $\gamma$ for the dipole moment along the $z$ axis is, on the other hand, given by

$$E_z^2 = \frac{T\{1 + R - 2\sqrt{R}\cos[\pi(\cos\gamma + 1)]\}}{(1 - R)^2 + 4R\sin^2[\pi(\cos\gamma + 1)]} E_0^2 \qquad (13.21)$$

The radiation intensity at the point $(r, \psi, \gamma)$ is given by Eqs. (13.21) and (13.10):

$$dI(r, \psi, \gamma) = \eta\frac{P_{12}^2 E_0^2}{r^2} \cdot \frac{(1 - R)\{1 + R - 2\sqrt{R}\cos[\pi(\cos\gamma + 1)]\}}{(1 - R)^2 + 4R\sin^2[\pi(\cos\gamma + 1)]}$$

$$\times \begin{cases} 0 & s \text{ wave} \\ \sin^2\gamma & p \text{ wave} \end{cases} \qquad (13.22)$$

The numerical example of the radiation pattern in Eq. (13.22) is plotted in Figure 13.4. The radiation pattern concentrates on $\gamma \simeq \pi/2$. This is because the vertical dipole couples with an $N = 0$ even mode that propagates within the $x$–$y$ plane. In a $\gamma$ value close to $\pi/2$ where the radiation intensity is appreciable, Eq. (13.22) is approximated by

$$dI(r, \psi, \gamma) = \eta\frac{P_{12}^2 E_0^2}{r^2} \cdot \frac{4(1 - R)}{(1 - R)^2 + 4\pi^2 R(\pi/2 - \gamma)^2}$$

$$\times \begin{cases} 0 & s \text{ wave} \\ \sin^2\gamma & p \text{ wave} \end{cases} \qquad (13.23)$$

The peak intensity at $\gamma = \pi/2$ is enhanced by a factor $4/(1 - R)$ above the free-space value and the diverging angle centered at $\gamma = \pi/2$ is $(1 - R)/2\pi$. The total intensity of emission in all directions is given by

$$I = 2\int_0^{2\pi} d\psi \int_0^{\pi/2} d\gamma\, dI(r, \psi, \gamma)r^2 \sin\gamma \simeq 4\pi\eta P_{12}^2 E_0^2 \quad (13.24)$$

Here the integral formula $\int_0^\infty dx/(ax^2 + b) = \pi/2\sqrt{ab}$ is used. The total emitted power is again 1.5 times larger (the spontaneous lifetime is shorter by a factor $\frac{2}{3}$) than the free-space value. The spontaneous emission is carried only by the $p$ wave.

### 13.3.3.  Modified Radiation Pattern and Lifetime in a One-Wavelength Cavity

Next let us consider the case where the two mirrors are separated by a distance equal to one wavelength, $b = \lambda$, and the atom is again located midway between the two mirrors, $z_0 = \lambda/2$. If the phase shift $\phi_r$ due to reflection is zero, the horizontal dipole couples with even modes and the vertical dipole couples with odd modes. The atom is located at the antinode of the horizontal field component of a resonant standing wave ($N = 2$ even mode) propagating along the $z$ axis. Therefore, the spontaneous emission in the $z$ direction by the horizontal dipole is expected to be enhanced in this case also. The vacuum field intensity $E_x^2(E_y^2)$ for the dipole moment along the $x$ axis ($y$ axis) is

$$E_x^2 = E_y^2 = \frac{T\{1 + R + 2\sqrt{R}\cos(2\pi\cos\gamma)\}}{(1 - R)^2 + 4R\sin^2(2\pi\cos\gamma)}E_0^2 \qquad (13.25)$$

The radiation intensity at the point $(r, \psi, \gamma)$ is

$$dI(r,\psi,\gamma) = \eta\frac{P_{12}^2 E_0^2}{r^2} \cdot \frac{(1 - R)\{1 + R + 2\sqrt{R}\cos(2\pi\cos\gamma)\}}{(1 - R)^2 + 4R\sin^2(2\pi\cos\gamma)}$$

$$\times \begin{cases} \sin^2\psi & s\text{ wave} \\ \cos^2\psi\cos^2\gamma & p\text{ wave} \end{cases} \qquad (13.26)$$

The radiation pattern concentrates on $\gamma \simeq 0$ and also on $\gamma \simeq \pi/2$, as shown in Figure 13.5. The radiation peak near $\gamma = \pi/2$ is due to the $N = 0$ even mode propagating within the $x$–$y$ plane. In a small $\gamma$ value where the radiation intensity is appreciable, Eq. (13.26) is approximated by

$$dI(r,\psi,\gamma) = \eta\frac{P_{12}^2 E_0^2}{r^2} \cdot \frac{4(1 - R)}{(1 - R)^2 + 4\pi^2 R\gamma^4} \times \begin{cases} \sin^2\psi & s\text{ wave} \\ \cos^2\psi & p\text{ wave} \end{cases}$$

$$(13.27)$$

Comparing Eq. (13.27) with Eq. (13.19), the peak intensity is enhanced by the same factor $4/(1 - R)$ above the free-space value, but the diverging angle of the main lobe is reduced by a factor of $\sqrt{2}$, that is, to $\sqrt{(1 - R)/2\pi}$ for the one-wavelength cavity from $\sqrt{(1 - R)/\pi}$ for the

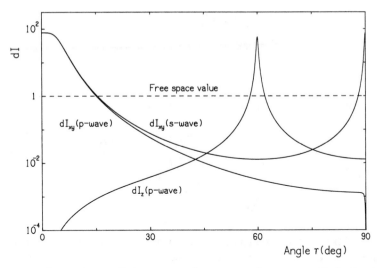

**Figure 13.5.** Spontaneous radiation patterns from an atom between two ideal mirrors with $d = \lambda$, $R = 0.95$, and $\phi_r = 0$.

half-wavelength cavity. The integrated emission intensity is given by

$$I \simeq 2\pi\eta P_{12}^2 E_0^2 \tag{13.28}$$

In a $\gamma$ value region close to $\pi/2$ where the radiation intensity is also appreciable, Eq. (13.26) is approximated by

$$dI(r,\psi,\gamma) = \eta \frac{P_{12}^2 E_0^2}{r^2} \cdot \frac{4(1-R)}{(1-R)^2 + 16\pi^2 R[\pi/2 - \gamma]^2}$$

$$\times \begin{cases} \sin^2\psi & s \text{ wave} \\ 0 & p \text{ wave} \end{cases} \tag{13.29}$$

The peak intensity in $s$ polarization at $\gamma = \pi/2$ is enhanced by a factor of $4/(1 - R)$ above the free-space value and the diverging angle is $(1 - R)/4\pi$. The integrated emission intensity is

$$I \simeq \pi\eta P_{12}^2 E_0^2 \tag{13.30}$$

The total emitted power is $I \simeq 3\pi\eta P_{12}^2 E_0^2$, which is larger than the free-space value by a factor of $\frac{9}{8}$; the spontaneous lifetime is decreased by a factor of $\frac{8}{9}$. This factor is again independent of the reflection coefficient $R$.

The vacuum field intensity $E_z^2$ for the dipole moment along the $z$ axis is

$$E_z^2 = \frac{T\{1 + R - 2\sqrt{R}\cos(2\pi\cos\gamma)\}}{(1 - R)^2 + 4R\sin^2(2\pi\cos\gamma)}E_0^2 \qquad (13.31)$$

The radiation intensity at the point $(r, \psi, \gamma)$ is

$$dI(r,\psi,\gamma) = \eta\frac{P_{12}^2 E_0^2}{r^2} \cdot \frac{(1 - R)\{1 + R + 2\sqrt{R}\cos(2\pi\cos\gamma)\}}{(1 - R)^2 + 4R\sin^2(2\pi\cos\gamma)}$$

$$\times\begin{cases} 0 & s \text{ wave} \\ \sin^2\gamma & p \text{ wave} \end{cases} \qquad (13.32)$$

The radiation pattern concentrates on $\gamma = \pi/3$ as shown in Figure 13.5. This is because the $N = 1$ odd mode propagates a zigzag path with $\gamma = \pi/3$ in the $\lambda$ layer. For $\gamma$ values close to $\gamma = \pi/3$ where the radiation intensity is appreciable, Eq. (13.32) is approximated by

$$dI(r,\psi,\gamma) = \eta\frac{P_{12}^2 E_0^2}{r^2} \cdot \frac{4(1 - R)}{(1 - R)^2 + 12\pi^2 R(\gamma - \pi/3)^2}$$

$$\times\begin{cases} 0 & s \text{ wave} \\ \sin^2\gamma & p \text{ wave} \end{cases} \qquad (13.33)$$

The intensity of the total emission is given by

$$I \simeq 3\pi\eta P_{12}^2 E_0^2 \qquad (13.34)$$

The total emitted power is larger than the free-space value by a factor of $\frac{9}{8}$; the spontaneous lifetime is decreased by a factor of $\frac{8}{9}$.

As shown above, the spontaneous emission from the horizontal dipole moment along the $x$ axis or $y$ axis can be concentrated on $\gamma \simeq 0$ by the half-wavelength cavity with $\phi_r = \pi$. This is a very useful feature for an efficient surface emitting laser. The spontaneous emission from the vertical dipole moment along the $z$ axis can be concentrated on $\gamma \simeq \pi/2$ by the half-wavelength cavity with $\phi_r = \pi$. This is a very useful feature for an efficient edge emitting laser. The one-wavelength cavity with $\phi_r = 0$, however, has either two emission peaks (horizontal dipole) or one peak emission in the oblique (60°) direction, and so it is less interesting for those device applications.

A real mirror does not have a constant reflection coefficient independent of incident angle $\gamma$. In the next section, we will discuss the half-wavelength cavity and one-wavelength cavity surrounded by one-dimensional

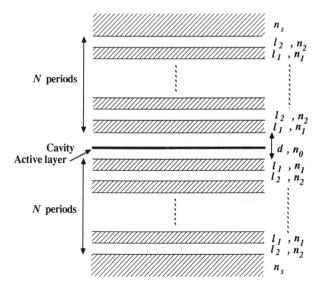

**Figure 13.6.** A microcavity structure with one-dimensional distributed Bragg reflectors.

distributed Bragg reflectors. The structure is leaky so that the spontaneous emission is not only in the vertical or horizontal directions, but also in oblique angles where the mirror reflection coefficient is small.

## 13.4. SPONTANEOUS EMISSION OF AN ATOM BETWEEN DISTRIBUTED BRAGG REFLECTORS

The microcavity structure with one-dimensional distributed Bragg reflectors (DBRs) shown in Figure 13.6 is studied in this section. One structure is the half-wavelength cavity, in which the central optical cavity layer has a lower refractive index and the first layer of the DBRs facing the central layer has a higher refractive index. Thus, the phase shift $\phi_r$ for a vertically propagating wave is $\pi$, and this microcavity is expected to have a performance similar to that discussed in Section 13.3.2. The other structure is the one-wavelength cavity, in which the central optical cavity layer has a higher refractive index and the first layer of the DBRs has a lower refractive index. Thus, the phase shift $\phi_r$ for a vertically propagating wave is zero, and this microcavity is expected to feature a performance similar to that discussed in Section 13.3.3. Exact analysis for the electromagnetic wave propagation in such structures can be done by the computer-aided transfer matrix method [15]. We assume that the outer medium has the highest refractive index, so that there are no guided modes propagating within the layer. All vacuum field fluctuations coupled with an atom can be excited from the outer medium. This makes the theoretical analysis easier, and basic performance is not lost by the assumption.

### 13.4.1.  Reflection Characteristics

The reflection coefficients for a vertically propagating wave in the half-wavelength cavity are shown in Figure 13.7 as a function of optical wavelength. Here, $n_0 = n_2 = 2.96$ (AlAs) and $n_e = n_1 = 3.6$(GaAs) are assumed. Owing to a Fabry–Perot cavity effect, the reflection coefficient at a resonant wavelength $\lambda_0 = 2\,dn_0$ decreases to zero, where $d$ is the central layer thickness and $n_0$ is the refractive index. As the number of reflector pairs increases, the reflection coefficient within the stopband also increases. On the other hand, the stopband width and the resonant transmission bandwidth decrease. In Figure 13.7, the cavity resonant wavelength $\lambda_0$ and the center wavelength $\lambda_B = 4l_1n_1 = 4l_2n_2$ of the DBRs are assumed to be equal.

Figures 13.8a and 13.8b show the reflection coefficients for the $s$ wave and the $p$ wave at the cavity resonant wavelength $\lambda_0$ versus the incident angle $\gamma$. With increasing $\gamma$ from zero to $\pi/2$, the cavity resonant wavelength and the center wavelength of the DBRs shift to a shorter wavelength according to $\lambda_0' = 2\,dn_0 \cos\gamma$ and $\lambda_B' = 4l_1n_1 \cos\gamma = 4l_2n_2 \cos\gamma$, respectively. Therefore, the reflection characteristics at a fixed wavelength $\lambda_0$ features first a resonant transmission window near $\gamma = 0$, which is followed by the stopband with the cutoff angle $\gamma_s = \cos^{-1} 2n_2/(n_1 + n_2)$

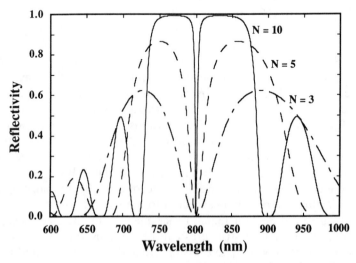

**Figure 13.7.**   Reflection coefficients for a normally incident light versus optical wavelength of a half-wavelength microcavity. $\lambda_0 = \lambda_B = 800$ nm, $n_0 = n_2 = 2.96$, $n_e = n_1 = 3.6$.

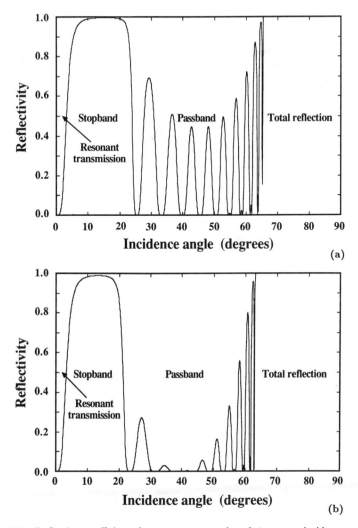

**Figure 13.8.** Reflection coefficients for a resonant wavelength $\lambda_0$ versus incident angle $\gamma$ of a half-wavelength microcavity: (a) $n_0 = n_2 = 2.96$ (s wave); (b) $n_0 = n_2 = 2.96$ (p wave); (c) $n_0 = n_2 = 2.0$ (s wave); (d) $n_0 = n_2 = 2.0$ (p wave).

$\simeq 27°$ and the passband. Since it is assumed that the external medium has highest refractive index $n_e$, the reflection coefficient increases to one at incident angles that are considerably larger than the critical angle of total reflection $\gamma_t' = \sin^{-1} n_2/n_e \simeq 55°$. The effective critical angle of total reflection is $\gamma_t = \sin^{-1} n_{\text{eff}}/n_e = 65°$, where $n_{\text{eff}} = 2n_1 n_2/(n_1 + n_2)$ is the effective refractive index of the DBRs. The difference in the s and p polarizations is due to the Brewster angle $\gamma_B = \tan^{-1} n_2/n_e = 40°$ for the p wave.

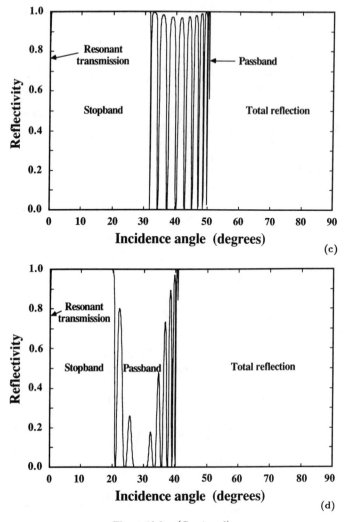

**Figure 13.8.** *(Continued)*

Comparing Figures 13.8c and 13.8d with 13.8a and 13.8b, we see that the open window decreases as the refractive index difference $n_1 - n_2$ of the DBRs increases, where $n_1 = 3.6$ and $n_2 = 2.0$. In Figure 13.8c, the stopband cutoff angle $\gamma_s \simeq 58°$ and the total reflection critical angle $\gamma_t \simeq 32°$ overlap. However, there are several resonant open windows between $\gamma = 30°$ and $\gamma = 50°$. Those resonances are due to leaky guided modes supported by the periodic DBR structure. In Figure 13.8d, a large open window corresponds to the Brewster angle $\gamma_B \simeq 28°$. Even though the open windows due to leaky guided modes and the Brewster phenomenon survive at some angles, they can be decreased by increasing the refractive index difference. In this way, the ideal half-wavelength cavity

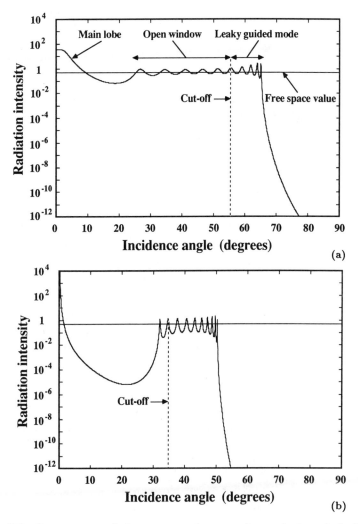

**Figure 13.9.** Spontaneous radiation patterns of $s$ wave from a horizontal dipole in a half-wavelength microcavity, $n_1 = n_e = 3.6$: ($a$) $n_0 = n_2 = 2.96$; ($b$) $n_0 = n_2 = 2.0$; ($c$) $n_0 = n_2 = 1.3$.

discussed in Section 13.3.2 is approximately realized by the dielectric DBRs.

### 13.4.2.  Modified Spontaneous Emission in a Half-Wavelength Cavity

**13.4.2.1.  Radiation Patterns.** The radiation intensities of the $s$ wave and the $p$ wave emitted by the horizontal dipole integrated with respect to $\phi$ versus the emission angle $\gamma$ are shown in Figures 13.9 and 13.10. In addition to the main lobe near $\gamma = 0$, there is a side lobe corresponding to

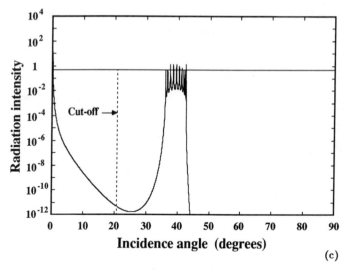

Figure 13.9. (*Continued*)

the passband between the stopband edge angle $\gamma_s$ and the total reflection critical angle $\gamma_t'$ (shown by the "cutoff") in the case of Figures 13.9*a* and 13.10*a*. The side lobe is also caused by leaky guided modes, specifically when the refractive index difference increases (Figures 13.9*b*, 13.9*c*, 13.10*b*, 13.10*c*). The side lobe decreases as the refractive index difference increases, as expected.

The radiation intensity of the *p* wave emitted by the vertical dipole integrated with respect to $\phi$ versus the emission angle $\gamma$ is shown in Figure 13.11. The spontaneous emission is mainly radiated into leaky guided modes propagating in the *x–y* plane. The peak emission angle in Figure 13.11*a* is approximately given by $\gamma_P \simeq \sin^{-1}[n_{\text{eff}}/n_e]$. This means that the spontaneous emission is carried by the leaky guided modes with the effective mode index $n_{\text{eff}}$ and is leaked into the GaAs layer with an angle $\gamma_P$.

### 13.4.2.2. Coupling Efficiency of Spontaneous Emission into a Main Lobe.
The diverging angle (FWHM) of the main lobe and the cavity resonant transmission bandwidth (FWHM) monotonically decrease as the number of DBR layers increases (Figure 13.12). This is because the reflection coefficient $R$ of the DBR increases with the number of DBR layers. However, the coupling efficiency of the spontaneous emission from the horizontal dipole coupled into the main lobe propagating along the *z* axis is kept constant. The coupling efficiency $\beta$ for the horizontal dipole is given by

$$\beta = \frac{\int_0^{2\pi} d\psi \int_{\text{main lobe}} d\gamma \left[ dI^P(r,\psi,\gamma) + dI^S(r,\psi,\gamma) \right] r^2 \sin \gamma}{\int_0^{2\pi} d\psi \int_0^{\pi/2} d\gamma \left[ dI^P(r,\psi,\gamma) + dI^S(r,\psi,\gamma) \right] r^2 \sin \gamma} \quad (13.35)$$

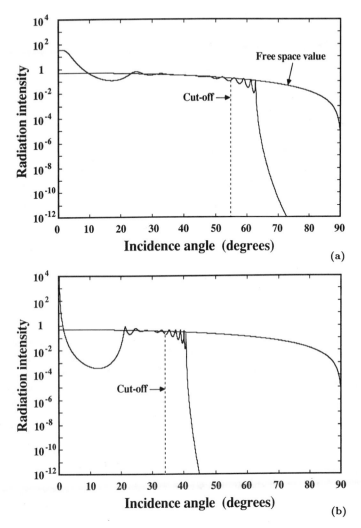

**Figure 13.10.** Spontaneous radiation patterns of a $p$ wave from a horizontal dipole in a half-wavelength microcavity, $n_1 = n_e = 3.6$: (a) $n_0 = n_2 = 2.96$; (b) $n_0 = n_2 = 2.0$; (c) $n_0 = n_2 = 1.3$.

where $dI^P(r, \psi, \gamma)$ and $dI^S(r, \psi, \gamma)$ are the $p$-polarized and $s$-polarized spontaneous emission intensities from the horizontal dipole at the point $(r, \psi, \gamma)$. The coupling efficiency is determined only by the refractive index difference. This is because the peak spontaneous emission intensity increases, but the diverging angle decreases, with the number of DBR layers so that the integrated spontaneous emission power is independent of the number of DBR layers.

The coupling efficiency $\beta$ of spontaneous emission into the main lobe propagating along the $z$ axis is plotted in Figure 13.13 as a function of the

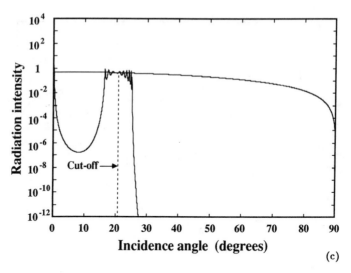

**Figure 13.10.** *(Continued)*

refractive index $n_2$ of the DBR material, where $n_e = n_1 = 3.6$ is fixed constant. The solid line represents the coupling efficiency for the horizontal dipole, $P_x^2 = P_y^2 = P_{12}^2/2$ and $P_z^2 = 0$. As the refractive index difference increases, the coupling efficiency approaches unity, as expected. On the other hand, the coupling efficiency for the randomly oriented dipoles, $P_x^2 = P_y^2 = P_z^2 = P_{12}^2/3$, saturates at $\beta = \simeq \frac{2}{3}$. The coupling efficiency $\beta$ for the random dipoles is given by

$$\beta = \frac{\int_0^{2\pi} d\psi \int_{\text{main lobe}} [2\, dI^P(r, \psi, \gamma) + 2\, dI^S(r, \psi, \gamma)] r^2 \sin\gamma}{\int_0^{2\pi} d\psi \int_0^{\pi/2} d\gamma [2\, dI^P(r, \psi, \gamma) + 2\, dI^S(r, \psi, \gamma) \atop \qquad\qquad + dI^V(r, \psi, \gamma)] r^2 \sin\gamma} \quad (13.36)$$

Here $dI^V(r, \psi, \gamma)$ is the $p$-polarized spontaneous emission intensity from the vertical dipole. This is because the vertical dipole emits spontaneously mainly in the $x$–$y$ plane.

**13.4.2.3. Spontaneous Emission Lifetime.** The modification in the spontaneous emission lifetime for the horizontal dipole, that is, $P_x^2 = P_y^2 = P_{12}^2/2$ and $P_z^2 = 0$, is given by

$$\frac{\tau_{\text{sp}}}{\tau_{\text{sp}}^0} = \frac{\int_0^{2\pi} d\psi \int_0^{\pi/2} d\gamma [dI_0^P(r, \psi, \gamma) + dI_0^S(r, \psi, \gamma)] r^2 \sin\gamma}{\int_0^{2\pi} d\psi \int_0^{\pi/2} d\gamma [dI^P(r, \psi, \gamma) + dI^S(r, \psi, \gamma)] r^2 \sin\gamma} \quad (13.37)$$

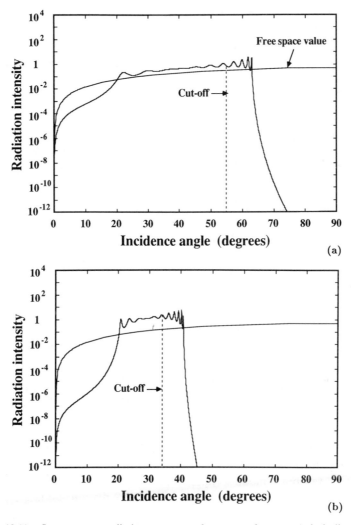

**Figure 13.11.** Spontaneous radiation patterns of $p$ wave from a vertical dipole in a half-wavelength microcavity, $n_1 = n_e = 3.6$: ($a$) $n_0 - n_2 = 2.96$; ($b$) $n_0 = n_2 = 2.0$; ($c$) $n_0 = n_2 = 1.3$.

where $\tau_{sp}^0$ is the spontaneous emission lifetime in free space, $dI_0^P = \eta(P_{12}^2 E_0^2/r^2)\cos^2 \psi \cos^2 \gamma$ and $dI_0^S = \eta(P_{12}^2 E_0^2/r^2)\sin^2 \psi$ are the $p$-polarized and $s$-polarized spontaneous emission intensities in free space, and $dI^P$ and $dI^S$ are those in the half-wavelength cavity and are given by Eq. (13.18). Figure 13.14 shows the modification factor of the spontaneous emission lifetime as a function of the refractive index $n_2$ of the DBR material.

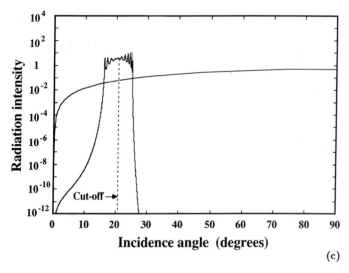

(c)

**Figure 13.11.** *(Continued)*

If the dipole is randomly oriented, that is, $P_x^2 = P_y^2 = P_z^2 = P_{12}^2/3$, the modification factor is given by

$$\frac{\tau_{sp}}{\tau_{sp}^0} = \frac{\int_0^{2\pi} d\psi \int_0^{\pi/2} d\gamma \left[2 \, dI_0^P + 2 \, dI_0^S + dI_0^V\right] r^2 \sin \gamma}{\int_0^{2\pi} d\psi \int_0^{\pi/2} d\gamma \left[2 \, dI^P + 2 \, dI^S + dI^V\right] r^2 \sin \gamma} \quad (13.38)$$

The modification factor is also plotted in Figure 13.14. Even though the radiation pattern and the coupling efficiency of spontaneous emission into the cavity resonant mode propagating along the $z$ axis are modified drastically by the microcavity, the spontaneous emission lifetime is hardly altered.

### 13.4.3.  Modified Spontaneous Emission in a One-Wavelength Cavity

Figures 13.15 and 13.16 show the spontaneous emission intensities of the $s$ wave and $p$ wave versus the emission angle from the horizontal dipole in the one-wavelength cavity. The radiation peak near $\gamma \simeq 75°$ in Figure 13.15a corresponds to the peak at $\gamma = \pi/2$ in Figure 13.5. The spontaneous emission coupled to the $N = 0$ even mode is leaked into this angle. Figure 13.17 shows the spontaneous emission intensity of the $p$ wave versus the emission angle from the vertical dipole in the one-wavelength cavity. Overall, the characteristics of the one-wavelength cavity are similar to those of the half-wavelength cavity.

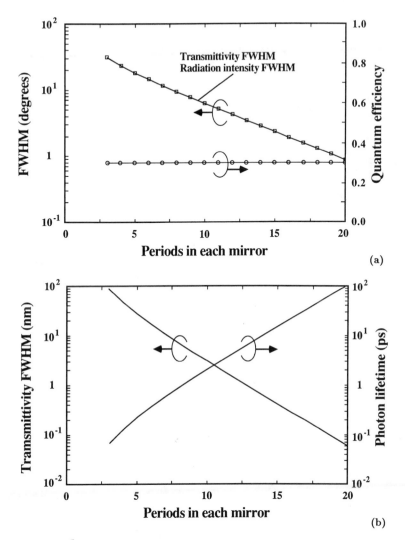

**Figure 13.12.** ($a$) A diverging angle (FWHM) of a main lobe and coupling efficiency of spontaneous emission into a main lobe versus number of periods of DBRs, $n_1 = n_e = 3.6$ and $n_0 = n_2 = 2.96$. ($b$) Resonant transmission bandwidth (FWHM) and effective photon lifetime versus number of periods of DBRs.

The coupling efficiency of the spontaneous emission into the main lobe versus the refractive index difference is plotted in Figure 13.18 for the two cases of $P_x^2 = P_y^2 = P_{12}^2/2$, $P_z^2 = 0$ and $P_x^2 = P_y^2 = P_z^2 = P_{12}^2/3$. The modification in the spontaneous emission lifetime versus the refractive index difference is plotted in Figure 13.19.

In the microcavity having one-dimensional distributed Bragg reflectors, the integrated intensity of emission in the main lobe is constant for

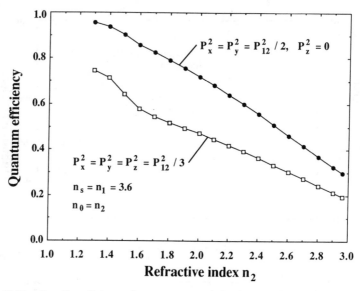

**Figure 13.13.** Coupling efficiency of spontaneous emission into a main lobe versus refractive index $n_2$ of DBR material.

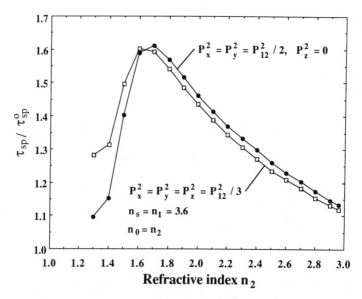

**Figure 13.14.** Normalized spontaneous emission lifetime versus refractive index $n_2$ of DBR material.

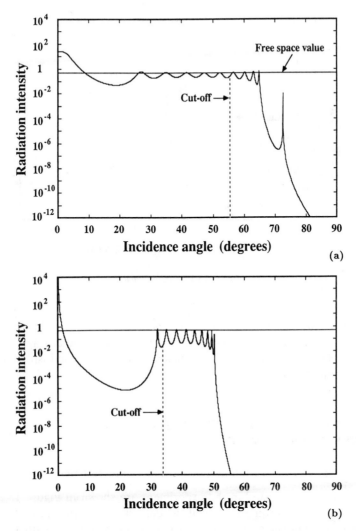

**Figure 13.15.** Spontaneous radiation patterns of $s$ wave from a horizontal dipole in a one-wavelength microcavity, $n_0 = n_2 = n_e = 3.6$: (a) $n_1 = 2.96$; (b) $n_1 = 2.0$; (c) $n_1 = 1.3$.

different number of DBR layers, as shown in Figure 13.12$a$. This is because the peak intensity at $\gamma = 0$ increases but the diverging angle (FWHM) of the main lobe decreases as the number of DBR layers increases. To increase the integrated intensity of emission in the main lobe, the refractive index difference of DBRs must be increased. In this way, the coupling efficiency into the main lobe can be increased to unity. However, the spontaneous emission lifetime cannot be decreased, because the increased intensity of emission in the main lobe is canceled out by the

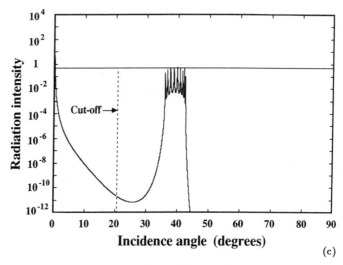

**Figure 13.15.**  *(Continued)*

decreased intensity of emission in all the other directions. In Appendix 13.A, we will discuss the metal clad optical waveguide microcavity, which realizes the increased coupling efficiency and decreased spontaneous lifetime simultaneously.

## 13.5.   EXPERIMENTAL EVIDENCE FOR MODIFIED SPONTANEOUS EMISSION

### 13.5.1.   Device Preparation and Reflection Characteristics

The three samples tested in the experiment are shown in Figure 13.20. A 7.5-nm-thick GaAs quantum well is embedded at the center of an $Al_{0.2}Ga_{0.8}As$ optical cavity layer having either half-wavelength ($\lambda/2 \sim$ 130 nm) or one-wavelength ($\lambda \sim 260$ nm) thickness. The optical cavity is sandwiched between distributed Bragg reflectors consisting of 10 pairs of quarter-wavelength layers of $Al_{0.2}Ga_{0.8}As$ and AlAs. The whole structure is fabricated on a GaAs substrate by a molecular beam epitaxy. The thickness ($L_z = 7.5$ nm) of the GaAs quantum well is much smaller than the optical wavelength $\lambda \simeq 260$ nm, so the emitting dipole is well localized at the node and antinode positions of the horizontal field component of the standing wave vacuum field fluctuations in the $\lambda/2$ cavity and $\lambda$ cavity, respectively. Note that the phase shift due to reflection is $\pi$ for both cavities. Thus, it is expected that spontaneous emission in a normal direction is enhanced in the $\lambda$ cavity, but suppressed in the $\lambda/2$ cavity. The third sample is just the GaAs quantum well embedded in a thick

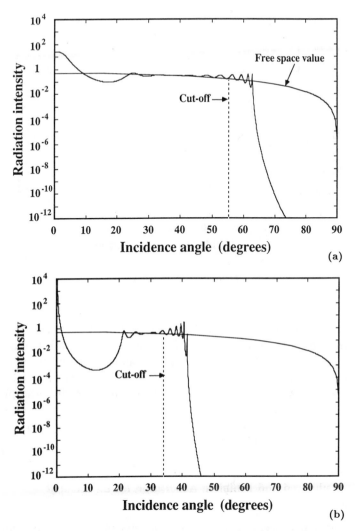

**Figure 13.16.** Spontaneous radiation patterns of $p$ wave from a horizontal dipole in a one-wavelength microcavity, $n_0 = n_2 = n_e = 3.6$: (a) $n_1 = 2.96$; (b) $n_1 = 2.0$; (c) $n_1 = 1.3$.

$Al_{0.2}Ga_{0.8}As$ layer without any cavity configurations. This sample is used as the standard spontaneous emission from the GaAs quantum well.

The measured reflection coefficient $R$ of the $\lambda$-cavity sample as a function of the probe wavelength $\lambda_p$ is compared in Figure 13.21 with the theoretical result. The cavity resonant wavelength $\lambda_0$ is 720 nm. The power reflectivity of the DBR is $\sim 0.97$ within the stopband between 680 and 770 nm. The theoretical transmission window linewidth (FWHM) is given by $\Delta\lambda_{1/2} = (1 - R)/(\pi\sqrt{R})\lambda_0 \simeq 7$ nm, which is in fairly good agree-

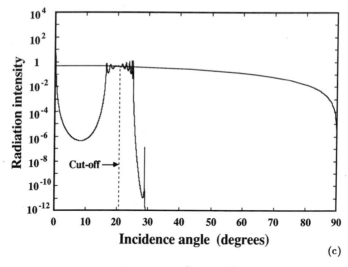

**Figure 13.16.** *(Continued)*

ment with the measured values. The reflection coefficient at $\lambda_0$ does not decrease to zero, because the cavity is surrounded by outer media having different refractive indices, air on one side and GaAs on the other. The measured reflection coefficient is larger than the theoretical result at the cavity resonant wavelength $\lambda_0$. This is mainly because the absorption in the GaAs quantum well is not taken into account in the theory. The measured reflection characteristics of the $\lambda/2$ cavity are similar to those of the $\lambda$ cavity shown in Figure 13.21.

### 13.5.2. Enhanced and Inhibited Absorption for an Incident Coherent Light

The absorption process is the reverse process of spontaneous emission. Since it is expected that the spontaneous emission is enhanced and inhibited by modifying the vacuum field fluctuation with a cavity, it is also expected that the absorption rate is enhanced and inhibited by modifying the externally incident coherent light with a cavity. The measured absorption rate of the $\lambda$ cavity for a normal incident light is shown in Figure 13.22 as a function of the pump laser wavelength detuning $\Delta\lambda = \lambda_p - \lambda_0$ from the cavity resonant wavelength $\lambda_0$. The cavity resonant wavelength is 720 nm, which is well above the absorption edge of the GaAs quantum well; the spontaneous emission wavelength is 820 nm, which is outside the stopband. A wavelength-tunable dye laser (or Ti:Al$_2$O$_3$ laser) illuminates the sample at 77 K from a normal direction. The laser wavelength is changed in the region of $720 \pm 30$ nm. The absorption rate is measured

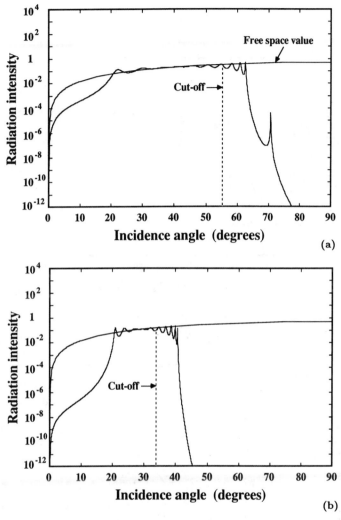

**Figure 13.17.** Spontaneous radiation patterns of $p$ wave from a vertical dipole in a one-wavelength microcavity, $n_0 = n_2 = n_e = 3.6$: ($a$) $n_1 = 2.96$; ($b$) $n_1 = 2.0$; ($c$) $n_1 = 1.3$.

indirectly by the spontaneous emission intensity from the GaAs quantum well at 820 nm. The intrinsic absorption coefficient of the GaAs quantum well is constant within the scanned wavelength region of 720 ± 30 nm, so the modulation in the absorption rate (Figure 13.22) is a pure cavity effect. The absorption rates are different by a factor of 4000 between the on-resonance and off-resonance cases. If we assume the absorption rate in free space shown by the dashed line in Figure 13.22, the absorption rate is enhanced at on-resonance by a factor of $4/(1 - R) \simeq 130$ and is sup-

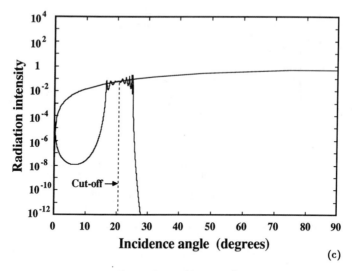

**Figure 13.17.** *(Continued)*

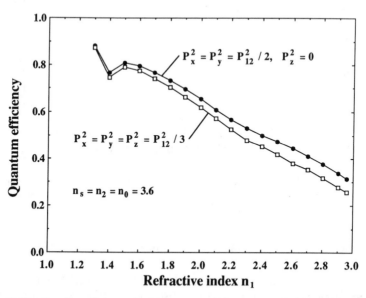

**Figure 13.18.** Coupling efficiency of spontaneous emission into a main lobe versus refractive index $n_1$ of DBR material.

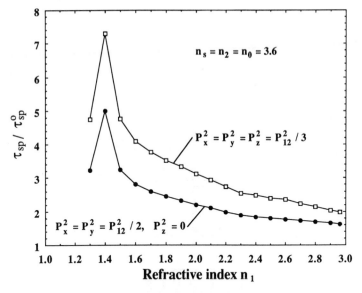

**Figure 13.19.** Normalized spontaneous emission lifetime versus refractive index $n_1$ of DBR material.

pressed at off-resonance by a factor of $1 - R \simeq \frac{1}{30}$, as the theory predicts. The measured linewidth (FWHM) of the enhanced absorption rate is also in good agreement with the theoretical result, $[(1 - R)/\pi\sqrt{R}]\lambda_0 \simeq 7$ nm.

### 13.5.3. Excitonic Spontaneous Emission

To observe the controlled spontaneous emission by a microcavity, the emission linewidth must be much narrower than the cavity resonant transmission linewidth. The spontaneous emission spectrum from the GaAs quantum well without a cavity at 77 K is shown in Figure 13.23. The main peak at 800 nm is the free excitonic transition between the lowest conduction band and the lowest heavy hole band (1e–1hh) and the side peak at 795 nm is that between the lowest conduction band and the lowest light hole band (1e–1lh). The spectral linewidth, 0.5–0.7 nm, is broader than the homogeneous linewidth ($\sim 0.2$ nm) at 77 K [16], but is still much narrower than the cavity transmission linewidth ($\sim 7$ nm). It is known that the dipole moment for the 1e–1hh transition is formed mainly in the plane of the quantum well (horizontal dipole) and the dipole for the 1e–1lh transition is formed mainly in the perpendicular direction (vertical dipole) [17]. Thus, we can observe enhanced and inhibited spontaneous emission more effectively at the 1e–1hh transition with the horizontal dipole.

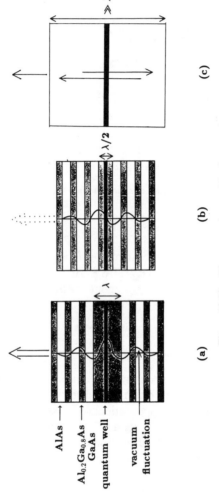

**Figure 13.20.** Three samples used in the experiment.

AlAs →

$Al_{0.2}Ga_{0.8}As$ →

GaAs →

quantum well →

vacuum
fluctuation

(a)

(b)

(c)

$\lambda$

$\lambda/2$

$\lambda$

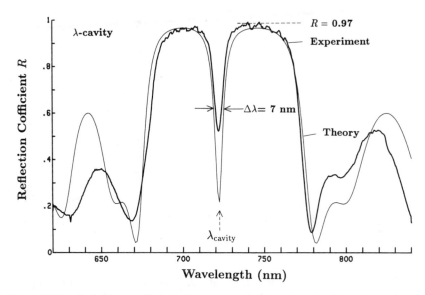

**Figure 13.21.** Reflection coefficient $R$ versus optical wavelength of a one-wavelength microcavity.

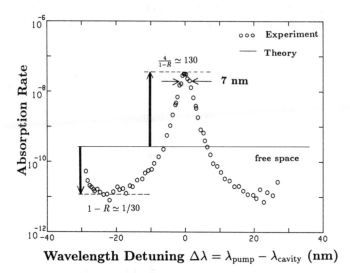

**Figure 13.22.** Absorption rate versus optical wavelength detuning $\Delta\lambda = \lambda_p - \lambda_0$ from cavity resonant wavelength of a one-wavelength microcavity.

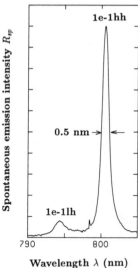

**Figure 13.23.** Spontaneous emission spectrum from GaAs quantum well at 77 K.

The spontaneous emission intensity $R_{sp}$ at the 1e–1hh transition is plotted in Figure 13.24 as a function of a pump power $P$. In a low pump region, the spontaneous emission intensity is proportional to the square of the pump power. This is a characteristic of the nonradiative recombination dominant case. In a high-pump region, however, the spontaneous emission intensity is proportional to the pump power. This is a characteristic of the

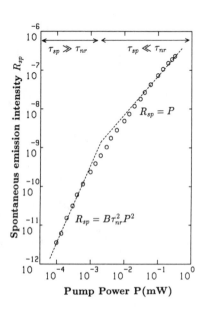

**Figure 13.24.** Spontaneous emission intensity at 1e–1hh transition versus pump power from GaAs quantum well at 77 K.

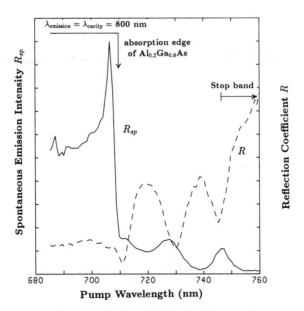

**Figure 13.25.** Spontaneous emission intensity at 1e–1hh transition versus pump laser wavelength $\lambda_p$ from a one-wavelength microcavity sample.

radiative spontaneous recombination dominant case. In this range of pump rate, the amplified spontaneous emission and the superradiance effects are neglected.

### 13.5.4. Pump Wavelength Dependence of Spontaneous Emission Intensity

The measured spontaneous emission intensity $R_{sp}$ at $\lambda_e = 800$ nm versus the pump wavelength $\lambda_p$ between 680 nm and 760 nm is shown in Figure 13.25, together with the reflection coefficient for the pump wavelength $\lambda_p$. If the pump wavelength is within the stopband, the spontaneous emission intensity is low, because most of the pump power is simply reflected. There is modulation of the spontaneous emission intensity according to the modulation of the reflection coefficient outside of the stopband. The strong peak of the spontaneous emission intensity at 710 nm is due to the excitonic absorption of the $Al_{0.2}Ga_{0.8}As$ DBR layers. The pump light at wavelengths below 710 nm is once absorbed by the DBRs and re-emitted as a broad spectral light which penetrates into the GaAs quantum well and is absorbed there.

### 13.5.5. Modified Radiation Patterns

The experimental and theoretical radiation patterns of the spontaneous emission from the GaAs quantum well in the uniform multi-wavelength

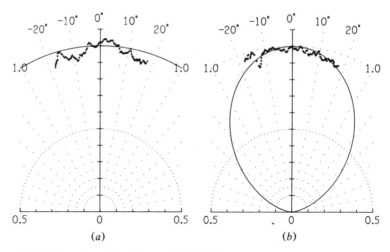

**Figure 13.26.** Experimental and theoretical radiation patterns from GaAs quantum well in a thick $Al_{0.2}Ga_{0.8}As$ layer: (*a*) *s* wave; (*b*) *p* wave.

$Al_{0.2}Ga_{0.8}As$ layer are shown in Figure 13.26. The theoretical radiation patterns from the horizontal dipole for the two polarizations perpendicular (*s* wave) and parallel (*p* wave) to the emission plane are given by integrating Eq. (13.8) with respect to $\psi$. They are shown by the solid curves in Figures 13.26*a* and 13.26*b*. The experimental radiation patterns are measured in liquid $N_2$ and are transformed to those values in the $Al_{0.2}Ga_{0.8}As$ layer by taking into account Shell's law and Fresnel reflection coefficients. Within the critical angle of total internal reflection at an $Al_{0.2}Ga_{0.8}As$/liquid $N_2$ interface, $\gamma_c = 23°$, the measured radiation intensity of the *s* wave is independent of $\gamma$, as it should be. The radiation pattern of the *p* wave also agrees with the theoretical curve.

In the $\lambda$-cavity sample with the DBRs, however, the vacuum field intensity is no longer isotropic. Within the stopband, the modified vacuum field intensity is calculated by simply assuming two lumped reflectors separated by one wavelength, and is given by Eq. (13.25). The stopband edge angle for the present sample is $\gamma_S \simeq 27°$. Therefore, Eq. (13.25) is a good approximate formula for the measured angle region ($\gamma \leq 23°$). The spontaneous emission intensity is enhanced in the normal direction by a factor of $4/(1 - R) \simeq 130$. The cavity resonant transmission angle for the present sample is $\gamma_P = [(1 - R)/2\pi\sqrt{R}]^{1/2} \simeq 4°$.

The experimental and theoretical radiation patterns of the spontaneous emission from the $\lambda$-cavity sample are shown in Figure 13.27. The spontaneous emission wavelength $\lambda_e = 800$ nm is coincident with the cavity resonant wavelength $\lambda_0$ in this sample. The pump light at 720 nm strikes the sample at an incident angle of 5°. Both polarizations have identical radiation patterns for such a small radiation angle ($\gamma \leq 5°$).

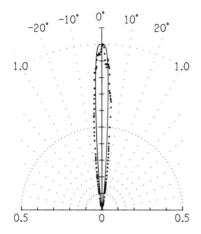

**Figure 13.27.** Experimental and theoretical radiation patterns from GaAs quantum well in a one-wavelength microcavity, $\lambda_e = \lambda_0 = 800$ nm.

When the spontaneous emission wavelength is shorter than the cavity resonant wavelength, $\lambda_e < \lambda_0$, the resonance condition is still satisfied at a small offset angle, $\gamma_r = \cos^{-1}(\lambda_e/\lambda_0)$. This was confirmed by experiments. Figure 13.28a shows the experimental radiation pattern for $\lambda_e = 800$ nm and $\lambda_0 = 815$ nm. The theoretical resonance angle is 11°, which is in good agreement with the experimental result. When $\lambda_e > \lambda_0$, however, the resonance condition is not satisfied by any angle. In fact, the experimental radiation pattern for $\lambda_e > \lambda_0$ does not feature such a conical emission, but features a simply attenuated single peak centered at $\gamma = 0$, as shown in Figure 13.28b.

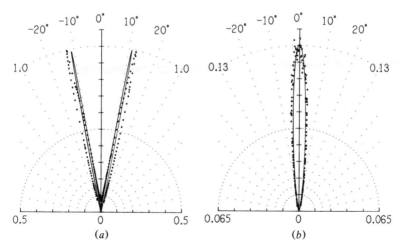

**Figure 13.28.** Experimental and theoretical radiation patterns from GaAs quantum well in a one-wavelength microcavity: (a) $\lambda_e = 800$ nm and $\lambda_0 = 815$ nm; (b) $\lambda_e = 800$ nm and $\lambda_0 = 790$ nm.

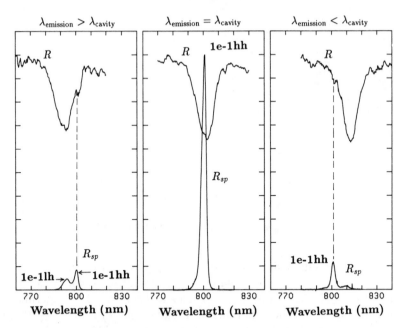

**Figure 13.29.** Spontaneous emission intensities at $\gamma = 0$ and reflection coefficients for three cases ($\lambda_e < \lambda_0$, $\lambda_e = \lambda_0$ and $\lambda_e > \lambda_0$) in a one-wavelength microcavity.

### 13.5.6. Enhanced and Inhibited Spontaneous Emission

The spontaneous emission spectra from the GaAs quantum well in the $\lambda$ cavity are shown in Figure 13.29 for three cases. When the spontaneous emission wavelength at the 1e–1hh transition coincides with the cavity resonant wavelength, spontaneous emission is strongly enhanced. When the spontaneous emission wavelength is detuned from the cavity resonant wavelength, spontaneous emission is suppressed.

The vacuum field intensity is modulated by the spontaneous emission wavelength detuning $\Delta\lambda = \lambda_e - \lambda_0$ from the cavity resonant wavelength and by the position offset $\Delta z$ of the quantum well from the center of the optical cavity layer. Within the stopband, the modified vacuum field intensity is given by

$$E_0^2 \rightarrow E_0^2 \cdot \frac{(1-R)\{1 + R \pm 2\sqrt{R}\cos(\phi_1 - 2\phi_2)\}}{(1-R)^2 + 4R\sin^2(\phi_1/2)} \tag{13.39}$$

where the normal emission direction $\gamma = 0$ is assumed. Here, $\phi_1 = 2\pi\,\Delta\lambda/\lambda_e$, $\phi_2 = 2\pi\,\Delta z/\lambda_e$, and the plus and minus signs correspond to the $\lambda$-cavity and $\lambda/2$-cavity samples. When $\phi_1 = \phi_2 = 0$, the vacuum field

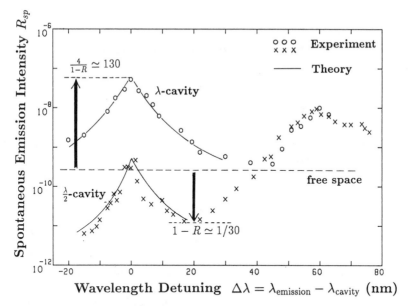

**Figure 13.30.** Spontaneous emission intensities at $\gamma = 0$ versus emission wavelength detuning $\Delta\lambda$ from cavity resonant wavelength in a one-wavelength microcavity and half-wavelength microcavity.

intensity $E_0^2$ is enhanced by a factor of $4/(1 - R)$ in the $\lambda$ cavity and is suppressed by a factor of $(1 - \sqrt{R})^2/(1 - R) \simeq (1 - R)/4$ in the $\lambda/2$ cavity. At off-resonance ($\phi_1 \simeq \pi$) in the $\lambda/2$ cavity, on the other hand, the vacuum field intensity is suppressed by a factor of $1 - R$.

The measured spontaneous emission intensities at $\gamma = 0$ in the $\lambda$-cavity sample and the $\lambda/2$ cavity sample are plotted in Figure 13.30 as a function of $\Delta\lambda$. In the samples used in this particular experiment, the two embedded GaAs quantum wells are separated by 8 nm and are not localized exactly at the antinode or node position, specifically, $\phi_2 \simeq 8°$. The solid lines are the theoretical results calculated using Eq. (13.39). The cavity resonant wavelength is modulated over $\pm 60$ nm centered at the emission wavelength $\lambda_e = 800$ nm. This was done by tapering the thickness of all layers within the sample and irradiating the pump light onto a different position. If we assume the spontaneous emission intensity in free space as indicated by the dashed line, the spontaneous emission intensity from the $\lambda$ cavity at $\Delta\lambda = 0$ is enhanced above the free-space value by a factor of $4/(1 - R) \simeq 130$ and the spontaneous emission from the $\lambda/2$ cavity at off-resonance is suppressed by a factor of $1 - R \simeq \frac{1}{30}$, as expected. The spontaneous emission intensity in the $\lambda/2$ cavity at $\Delta\lambda \simeq 0$ is not suppressed by a factor of $(1 - R)/4 \simeq \frac{1}{130}$, but is comparable to the free-space value. This is because the active dipoles "see" leakage vacuum field fluctuations at $\phi_2 \simeq 8°$. The modification of spontaneous emission intensity by the microcavity is well confirmed by these experiments.

## 13.6.  CHARACTERISTICS OF MICROCAVITY SEMICONDUCTOR LASERS

In a conventional semiconductor laser, a very small part of the spontaneous emission is coupled into a single lasing mode. The reason is threefold: (1), the radiation pattern of spontaneous emission is isotropic, so a substantial part of the spontaneous emission radiates in oblique directions; (2), the spontaneous emission spectral linewidth is much broader than the resonant cavity mode linewidth, so a substantial part of the spontaneous emission is coupled to nonresonant modes; and (3), the active volume size is much larger than the optical wavelength, so a substantial part of the spontaneous emission radiates in nonlasing cavity resonant modes. The coupling efficiency $\beta$ of spontaneous emission into a single lasing mode is often referred to as a "spontaneous emission coefficient" and is on the order of $10^{-5}$ in a conventional semiconductor laser (see Appendix 13.B). As discussed already, this coupling efficiency = spontaneous emission coefficient $\beta$ can be increased to close to one by a microcavity structure. The characteristics of such a microcavity semiconductor laser are very different from a conventional laser when the spontaneous emission coefficient $\beta$ is close to one.

### 13.6.1.  Basic Equations

The quantum Langevin equation for the (total) electron number operator $\hat{N}_c$ is written as

$$\frac{d}{dt}\hat{N}_c = P - \left(\frac{1-\beta}{\tau_{sp}} + \frac{\beta}{\tau_{sp}}\right)\hat{N}_c - (\hat{E}_{cv} - \hat{E}_{vc})\hat{n} + \hat{\Gamma}_p + \hat{\Gamma}_{sp} + \hat{\Gamma} \quad (13.40)$$

where $P$ is pump rate, $\beta/\tau_{sp}\langle\hat{N}_c\rangle$ is the spontaneous emission rate into the lasing mode, $(1 - \beta)/\tau_{sp}\langle\hat{N}_c\rangle$ is the spontaneous emission rate into all modes except the lasing mode, $n_{sp} = \langle\hat{E}_{cv}\rangle/(\langle\hat{E}_{cv}\rangle - \langle\hat{E}_{vc}\rangle)$ is the population inversion parameter, and $\hat{n}$ is the (total) photon number operator.

The spontaneous emission rate and the stimulated emission rate by one photon are equal, so that we have

$$\frac{\beta}{\tau_{sp}}\langle\hat{N}_c\rangle = \langle\hat{E}_{cv}\rangle \quad (13.41)$$

The noise operators associated with the pump process, the spontaneous emission process and the stimulated emission–absorption process satisfy

$$\left\langle \hat{\Gamma}_p(t)\hat{\Gamma}_p(u) \right\rangle = 0 \tag{13.42}$$

$$\left\langle \hat{\Gamma}_{sp}(t)\hat{\Gamma}_{sp}(u) \right\rangle = \delta(t - u) \cdot \frac{\langle \hat{N}_c \rangle}{\tau_{sp}} \tag{13.43}$$

$$\left\langle \hat{\Gamma}(t)\hat{\Gamma}(u) \right\rangle = \delta(t - u) \cdot (\langle \hat{E}_{cv} \rangle + \langle \hat{E}_{vc} \rangle)\langle \hat{n} \rangle \tag{13.44}$$

Here we assume pump noise suppression [Eq. (13.42)] by a constant current source [18].

The quantum Langevin equation for the (total) photon number operator $\hat{n}$ is written as

$$\frac{d}{dt}\hat{n} = -\left[\gamma - (\hat{E}_{cv} - \hat{E}_{vc})\right]\hat{n} + \frac{\beta}{\tau_{sp}}\hat{N}_c + \hat{F} + \hat{F}_e \tag{13.45}$$

where $\gamma$ is photon decay rate. The noise operators associated with the stimulated emission–absorption process and the photon decay process satisfy

$$\left\langle \hat{F}(t)\hat{F}(u) \right\rangle = \delta(t - u) \cdot \left[(\langle \hat{E}_{cv} \rangle + \langle \hat{E}_{vc} \rangle)\langle \hat{n} \rangle + \frac{\beta\langle \hat{N}_c \rangle}{\tau_{sp}}\right] \tag{13.46}$$

$$\left\langle \hat{F}_e(t)\hat{F}_e(u) \right\rangle = \delta(t - u) \cdot \gamma\langle \hat{n} \rangle \tag{13.47}$$

Here $\hat{F}_e = 2\sqrt{\gamma\langle \hat{n} \rangle}\hat{f}$, and $\hat{f}$ is the vacuum field fluctuation amplitude incident on an output coupling mirror. Since $\hat{\Gamma}' \equiv \hat{\Gamma} + \hat{\Gamma}_{sp}$ and $\hat{F}$ are originally from the same dipole moment noise operator, they are correlated as

$$\left\langle \hat{\Gamma}'(t)\hat{F}(u) \right\rangle = \left\langle \hat{F}(t)\hat{\Gamma}'(u) \right\rangle$$
$$= -\delta(t - u) \cdot \left[(\langle \hat{E}_{cv} \rangle + \langle \hat{E}_{vc} \rangle)\langle \hat{n} \rangle + \langle \hat{E}_{cv} \rangle\right] \tag{13.48}$$

Here Eq. (13.41) is used.

The photon flux operator $\hat{N}$ emanating from the output coupling mirror is [18]

$$\hat{N} = \gamma\hat{n} - \hat{F}_e \tag{13.49}$$

The first term represents the transmitted internal photon number, and the second term represents the quantum interference between the transmitted internal coherent excitation $\sqrt{\gamma}\langle\hat{n}\rangle$ and the reflected vacuum field fluctuation $f$.

### 13.6.2. Reduction of Threshold Pump Rate

Let us consider the steady-state (average) solutions of Eqs. (13.40) and (13.45). By using the linearized solutions, $\hat{N}_c = N_{c0} + \Delta\hat{N}_c$ and $\hat{n} = n_0 + \Delta\hat{n}$, in Eqs. (13.40) and (13.45), we obtain

$$P - \frac{N_{c0}}{\tau_{sp}} - \frac{\beta N_{c0}}{\tau_{sp}n_{sp}} \cdot n_0 = 0 \tag{13.50}$$

$$-\left[\gamma - \frac{\beta N_{c0}}{\tau_{sp}n_{sp}}\right]n_0 + \frac{\beta N_{c0}}{\tau_{sp}} = 0 \tag{13.51}$$

At pump rates well above the threshold, the photon decay rate $\gamma$ is equal to the net gain $E_{cv} - E_{vc}$ ($=$ stimulated emission gain-stimulated absorption loss); thus

$$\gamma = \frac{\beta N_{c0,\,th}}{\tau_{sp}n_{sp}} \tag{13.52}$$

where $N_{c0,\,th}$ is the clumped (total) electron number. The threshold condition is usually written as

$$\exp[(g_{th} - \alpha)L] = \frac{1}{R} \tag{13.53}$$

where $g_{th}$ is the gain constant per unit length, $\alpha$ is the absorption coefficient, $L$ is the active layer length, and $R$ is the power reflection coefficient of the end reflectors. Using $g_{th} = (1/v)(E_{cv} - E_{vc})_{th}$ and $(1/v)(\alpha + (1/L)\ln(1/R)) = \gamma$ in Eq. (13.53), we can obtain the threshold condition

$$\gamma = (E_{cv} - E_{vc})_{th} = \frac{E_{cv,\,th}}{n_{sp}} \tag{13.54}$$

At the threshold pump rate, all the pump electrons are recombined via the spontaneous emission (pump rate = total spontaneous emission rate) so that

$$E_{cv,\,th} = \beta P_{th} \tag{13.55}$$

From Eqs. (13.54) and (13.55), the threshold pump rate is calculated as

$$P_{th} = \frac{\gamma n_{sp}}{\beta} \qquad (13.56)$$

Note that the threshold pump rate is inversely proportional to the spontaneous emission coefficient $\beta$.

From Eq. (13.51), the average photon number $n_0$ is

$$n_0 = \frac{\beta N_{c0}/\tau_{sp}}{\gamma - (\beta N_{c0}/\tau_{sp} n_{sp})} \qquad (13.57)$$

It is obvious from this equation that the real electron number $N_{c0}$ never reaches the threshold value $N_{c0,\,th}$ as long as the spontaneous emission coefficient $\beta$ is nonzero. From Eq. (13.51), the average electron number $N_{c0}$ is given by

$$N_{c0} = \begin{cases} N_{c0,\,th} \cdot \dfrac{(r + 1) - \sqrt{(r + 1)^2 - 4(1 - \beta)r}}{2(1 - \beta)} & (\beta \neq 1) \\[4mm] N_{c0,\,th} \cdot \dfrac{r}{1 + r} & (\beta = 1) \end{cases} \qquad (13.58)$$

where $r = P/P_{th}$ is the normalized pump rate and $N_{c0,\,th} = \gamma \tau_{sp} n_{sp}/\beta$. Using Eq. (13.57) in Eq. (13.58), the average photon number $n_0$ is

$$n_0 = \begin{cases} n_{sp} \cdot \dfrac{\left[(r + 1) - \sqrt{(r + 1)^2 - 4(1 - \beta)r}\right]/2(1 - \beta)}{\left[1 - [(r + 1) - \sqrt{(r + 1)^2 - 4(1 - \beta)r}\right]/2(1 - \beta)\right]} & (\beta \neq 1) \\[6mm] n_{sp} \cdot r & (\beta = 1) \end{cases}$$

$$(13.59)$$

Figure 13.31 shows the internal photon number $n_0$ versus the pump current $I$ as a function of the spontaneous emission coefficient $\beta$. It is assumed that $\gamma = 10^{12}$ s$^{-1}$ and $n_{sp} = 1$. The threshold pump current $I_{th} \equiv qP_{th}$ is on the order of 10 mA for a conventional semiconductor laser with $\beta \simeq 10^{-5}$. The threshold pump current decreases with increasing $\beta$ and becomes less than 1 $\mu$A for a microcavity semiconductor laser with $\beta \simeq 0.1 \sim 1$. In the limit of $\beta = 1$, the jump in the quantum efficiency from $\eta_D = \beta$ to $\eta_D = 1$ disappears and the quantum efficiency is irrespective of the pump rate and is always unity. The threshold pump rate defined by Eq. (13.56), at which the average internal photon number is unity,

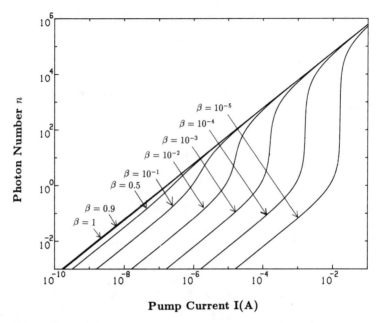

**Figure 13.31.** Internal photon number $n_0$ versus pump current $I$ as a function of spontaneous emission coefficient $\beta$. $n_{sp} = 1$ and $\gamma = 10^{12}$ s$^{-1}$.

corresponds to the threshold pump current $I_{th} = 160$ nA. At pump rates below this value, spontaneous emission is dominant. At pump rates above this value, stimulated emission is dominant. Figure 13.32 shows the average electron number $N_{c0}$ versus the pump current, where $\tau_{sp} = 1$ ns is assumed.

### 13.6.3.   Increase in Modulation Bandwidth

When the pump rate is sinusoidally modulated as $P_0 + \Delta Pe^{i\Omega t}$, the response of the internal photon number modulation is evaluated by Eqs. (13.40) and (13.45). The power spectrum of $\Delta n$ is

$$S_{\Delta n}(\Omega) = \frac{[(1 + r)/\tau_s P]^2 (\Delta P)^2}{\Omega^4 + \Omega^2 \{[\gamma/(1 + r)]^2 + [(1 + r)/\tau_{sp}]^2 - 2r\gamma/\tau_{sp}\} + \gamma^2[(1 + r)/\tau_{sp}]^2} \quad (13.60)$$

Figure 13.33 shows the normalized power spectrum as a function of the normalized pump rate. The modulation response is dependent on $\gamma$, $\tau_{sp}$, and $r$, but is independent of $\beta$. When $r \equiv P/P_{th} > \gamma\tau_{sp}$, Eq. (13.60) is

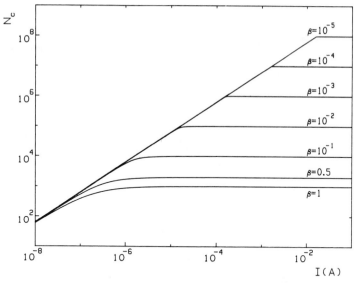

**Figure 13.32.** Electron number $N_{c0}$ versus pump current $I$ as a function of spontaneous emission coefficient $\beta$. $n_{sp} = 1$. $\gamma = 10^{12}$ s$^{-1}$ and $\tau_{sp} = 10^{-9}$ s.

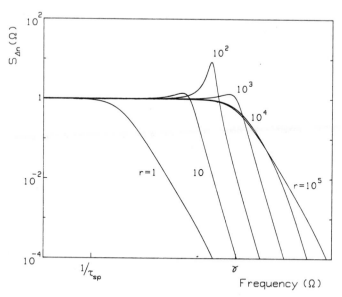

**Figure 13.33.** Normalized intensity modulation response $S_{\Delta n}(\Omega)$ as a function of pump rate $r = P/P_{th}$.

reduced to

$$S_{\Delta n}(\Omega) \simeq \frac{(\Delta P)^2}{\gamma^2} \cdot \frac{1}{1 + 1/\gamma^2\Omega^2 + n_{\text{sp}}^2/\gamma^2 r^2 \Omega^4} \qquad (13.61)$$

The relaxation oscillation is shifted beyond the cavity cutoff frequency $\Omega_c = \gamma$, and the modulation bandwidth is equal to the cavity cutoff frequency. Since $P_{\text{th}}$ is given by Eq. (13.56), the pump rate satisfying this ultimate modulation bandwidth is

$$P > P_{\text{th}}\gamma\tau_{\text{sp}} = \frac{\gamma^2 \tau_{\text{sp}} n_{\text{sp}}}{\beta} \qquad (13.62)$$

When $\gamma = 10^{13}$ s$^{-1}$, $\tau_{\text{sp}} = 10^{-9}$ s, $n_{\text{sp}} = 1$, and $\beta = 1$, pump rate $P$ must be greater than $10^{17}$ (s$^{-1}$), which corresponds to the pump current of 16 mA. If the pump rate satisfies this condition, the 3-dB down-modulation bandwidth is as high as $f_c = \gamma/2\pi \simeq 1.6$ THz.

The physical reason for such enormous increase in the modulation bandwidth of a microcavity laser is that the electron lifetime is shortened greatly by the stimulated emission process. The internal field strength produced by a single photon increases with decreasing cavity size.

### 13.6.4.  Amplitude Squeezing

The small fluctuation operators $\Delta \hat{N}_c$ and $\Delta \hat{n}$ obey

$$\frac{d}{dt}\Delta\hat{N}_c = -\frac{1}{\tau_{\text{sp}}}\left(1 + \frac{\beta n_0}{n_{\text{sp}}}\right)\Delta\hat{N}_c - \frac{\beta N_{c0}}{\tau_{\text{sp}} n_{\text{sp}}}\Delta\hat{n} + \hat{\Gamma}_p + \hat{\Gamma}_{\text{sp}} + \hat{\Gamma} \quad (13.63)$$

$$\frac{d}{dt}\Delta\hat{n} = -\left(\gamma - \frac{\beta N_{c0}}{\tau_{\text{sp}} n_{\text{sp}}}\right)\Delta\hat{n} + \frac{\beta}{\tau_{\text{sp}}}\left(1 + \frac{n_0}{n_{\text{sp}}}\right)\Delta\hat{N}_c + \hat{F} + \hat{F}_e \quad (13.64)$$

The Fourier transformed photon number fluctuation operator is

$$\Delta\tilde{n} = \frac{A_4\left(\tilde{\Gamma}_p + \tilde{\Gamma}_{\text{sp}} + \tilde{\Gamma}\right) - (j\Omega + A_1)\left(\tilde{F} + \tilde{F}_e\right)}{A_2 A_4 - (j\Omega + A_1)(j\Omega + A_4)} \qquad (13.65)$$

Here the tilde ($\sim$) stands for the Fourier transformed operator and the coefficients are

$$A_1 = \frac{1}{\tau_{sp}}\left(1 + \frac{\beta n_0}{n_{sp}}\right) \tag{13.66}$$

$$A_2 = \frac{\beta N_{c0}}{\tau_{sp} n_{sp}} \tag{13.67}$$

$$A_3 = \gamma - \frac{\beta N_{c0}}{\tau_{sp} n_{sp}} \tag{13.68}$$

$$A_4 = -\frac{\beta}{\tau_{sp}}\left(1 + \frac{n_0}{n_{sp}}\right) \tag{13.69}$$

Using the boundary condition [Eq. (13.49)], the Fourier-transformed photon flux fluctuation operator becomes

$$\Delta \tilde{N} = \left\{ \frac{\gamma}{A_4}\left[\tilde{\Gamma}_p + \tilde{\Gamma}_{sp} + \tilde{\Gamma}\right] - (j\Omega + A_1)\gamma\tilde{F} \right.$$

$$\left. - \left[\Omega^2 - (A_1 A_3 - A_2 A_4 + \gamma A_1) - j\Omega(A_1 + A_3 - \gamma)\right]\tilde{F}_e \right\} \bigg/$$

$$\left\{ \Omega^2 - (A_1 A_3 - A_2 A_4) - j\Omega(A_1 + A_3) \right\} \tag{13.70}$$

The power spectrum is

$$S_{\Delta \tilde{N}}(\Omega) = 2 \times \left\{ \gamma^2 A_4^2 \left[\frac{\gamma n_{sp}}{\beta} + \gamma(2n_{sp} - 1)n_0\right] \right.$$

$$+ \gamma^2(\Omega^2 + A_1^2 + 2A_1 A_4)\left[\gamma n_{sp} + \gamma(2n_{sp} - 1)n_0\right]$$

$$+ \left(\left[\Omega^2 - (A_1 A_3 - A_2 A_4 + \gamma A_1)\right]^2 \right.$$

$$\left. + \Omega^2(A_1 + A_3 - \gamma)^2\right)\gamma n_0 \bigg/$$

$$\left\{ \left[\Omega^2 - (A_1 A_3 - A_2 A_4)\right]^2 + \Omega^2(A_1 + A_4)^2 \right\} \tag{13.71}$$

Figure 13.34 shows the normalized power spectral density $S_{\Delta \tilde{N}}(\Omega)/2N_0$ in the limit $\Omega \to 0$ as a function of the normalized pump rate $r$ for various $\beta$ values. When the spontaneous emission coefficient $\beta$ is on the order of $10^{-5}$, the output photon flux fluctuation is reduced to below the shot-noise level only at high pump rates. However, if the spontaneous emission coefficient $\beta$ is close to one, the output photon flux fluctuation is reduced

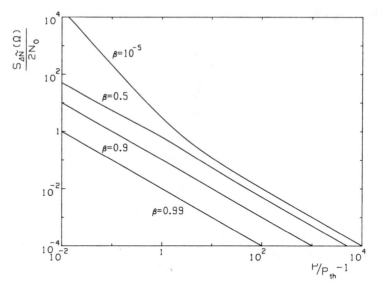

**Figure 13.34.** Normalized photon flux spectral density as a function of normalized pump rate $P/P_{th} - 1$.

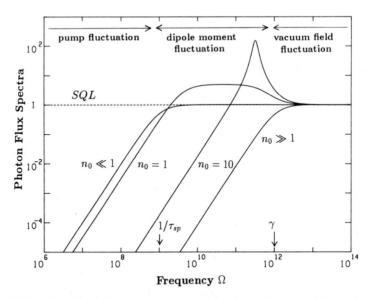

**Figure 13.35.** Normalized photon flux spectra for $\beta = 1$ as a function of internal photon number $n_0$.

to below the shot-noise level at all pump rates including below, near, and above the threshold.

The physical reason for this change is obvious from the energy conservation argument. If the spontaneous emission coefficient $\beta$ is unity, all the injected electrons are converted to photons in a single mode and are extracted from the cavity sooner or later. Therefore, if the output photon number is counted for a time interval much longer than the delay time involved in the electron–photon conversion process and in the photon escaping process, the photon number is constant because of the constant electron number. The time constant is the spontaneous lifetime $\tau_{sp}$ at a pump rate far below the threshold, and is the photon lifetime $1/\gamma$ at a pump rate far above the threshold. This is clearly demonstrated in Figure 13.35, in which the normalized photon flux fluctuation spectra are plotted as a function of the pump rate. When the spontaneous emission coefficient $\beta$ is much smaller than unity, on the other hand, the conversion efficiency from the injected electrons to the emitted photons jumps suddenly at the threshold, where the output photon number is fluctuating.


## 13.7.  CONCLUSION

The microcavity semiconductor laser can concentrate all the spontaneous emission power into the single cavity resonant mode and realize a near-unity quantum efficiency of spontaneous emission, even though it is a one-dimensional periodic structure. The characteristics of a semiconductor laser can be altered drastically by increasing the coupling efficiency of spontaneous emission in this way. The microcavity semiconductor laser is a high-quantum-efficiency, low-threshold pump-rate, high-speed, and low-noise surface-emitting light source. Potential applications include ultra-high-speed optical communication, optical connection in LSI, and two-dimensional optical information processing.

It is essential that the spontaneous emission linewidth be much narrower than the cavity resonant linewidth. A possible "exciton laser" [19] is one candidate for meeting the above requirement. A "quantum wire or dot laser" [20] is another candidate.


## ACKNOWLEDGMENT

The authors wish to thank Dr. Yoshiji Horikoshi of the NTT Basic Research Laboratories for preparing the GaAs quantum well samples.

## APPENDIX 13.A.   MODIFIED SPONTANEOUS EMISSION IN A METAL CLAD OPTICAL WAVEGUIDE MICROCAVITY

Let us consider a metal clad optical waveguide, in which a semiconductor core region (post) has the same layered structure as the half-wavelength or one-wavelength microcavity discussed in Section 13.4 and a dielectric buffer layer exists between semiconductor core and metal clad. It is possible to make only the fundamental mode of the waveguide be resonant for the microcavity by satisfying the conditions

$$\beta_0 d = \pi \tag{13.A1}$$

and

$$\beta_0' l_1 = \beta_0'' l_2 = \frac{\pi}{2} \tag{13.A2}$$

where $\beta_0$, $\beta_0'$, and $\beta_0''$ are the propagation constants of the fundamental mode in the optical cavity layer and in the two DBR layers. It is also possible to make all the other guided modes be off-resonant (either in the stopband or outside of the stopband) if the waveguide is not too large. A low refractive index buffer layer is inserted between semiconductor core and metal clad to decrease the absorption loss of the fundamental mode due to metal surface. The vacuum field intensity of the fundamental mode is enhanced by the microcavity, as long as the decreased absorption loss $\alpha_0$ satisfies

$$\alpha_0 < \frac{1}{L_{\text{eff}}} \ln \frac{1}{R} \tag{13.A3}$$

where $L_{\text{eff}}$ is the effective cavity length of the microcavity and $R$ is the reflection coefficient of the DBR. On the other hand, the vacuum field intensities of all higher-order modes are not enhanced by the microcavity, because these modes are either inside or outside the stopband of the DBRs. Therefore, the structure realizes increased coupling efficiency of spontaneous emission into the fundamental mode and decreased spontaneous emission lifetime simultaneously.

The propagation constant $\beta_0$ of the fundamental mode is $\beta_0 = 2\pi\sqrt{\nu^2 - \nu_0^2}/c$, where $\nu_0$ is the fundamental mode cutoff frequency and is given by $\nu_0 = 0.58(c/a)$ for circular and by $\nu_0 = 0.71(c/a)$ for square metal clad waveguides. Here $c$ is the light velocity in core medium and $a$ is core diameter. If the core medium does not have a microcavity structure but rather has a uniform refractive index, the density of states for the fundamental mode is given by [3]

$$\rho_0(\nu) = \frac{4}{cA_g} \cdot \frac{\nu}{\sqrt{\nu^2 - \nu_0^2}} \simeq \frac{16\nu_0^2}{\xi c^3} \cdot \frac{\nu}{\sqrt{\nu^2 - \nu_0^2}} \qquad (13.A4)$$

Here $A_g$ is the cross section of the waveguide and $\xi = 1$ for circular and $\xi = 2$ for square waveguides. The density of states normalized by the free-space value, $\rho_{fs}(\nu) = 8\pi\nu^2/c^3$, is given by

$$\frac{\rho_0(\nu)}{\rho_{fs}(\nu)} = \frac{2}{\pi\xi}\left(\frac{\nu_0}{\nu}\right)^2 \frac{\nu}{\sqrt{\nu^2 - \nu_0^2}} \qquad (13.A5)$$

The normalized density of states is enhanced enormously just above its cutoff $\nu \simeq \nu_0$ [3]. This is because the fundamental mode near its cutoff does not propagate along the waveguide but is almost "trapped" at same position as a resonator mode. This is the case for microwave superconductor waveguides. However, the fundamental mode near its cutoff has fairly large absorption loss due to metal surface at optical frequencies. Consequently, this resonant enhancement of spontaneous emission is not practical at optical frequencies.

If the fundamental mode is well above its cutoff, the absorption loss is decreased but the normalized density of states is also decreased according to Eq. (13.A5). Moreover, there are higher-order guided modes, and the normalized density of states for higher-order modes is given by [3]

$$\sum_{j \geq 1} \frac{\rho_j(\nu)}{\rho_{fs}(\nu)} = \frac{2}{\pi\xi}\left(\frac{\nu_0}{\nu}\right)^2 \sum_j \frac{\nu}{\sqrt{\nu^2 - \nu_j^2}} \qquad (13.A6)$$

Here $\nu_j$ is the cutoff frequency of the mode $j$. In general, Eq. (13.A6) is larger than Eq. (13.A5). By choosing the waveguide cross section appropriately, Eq. (13.A6) can be made smaller than one. This becomes possible unless the highest order guided mode is just above its cutoff. In this way, the total density of states for all guided modes except the fundamental mode can be made smaller than the free-space value:

$$\sum_{j \geq 1} \frac{\rho_j(\nu)}{\rho_{fs}(\nu)} < 1 \qquad (13.A7)$$

The inequality Eq. (13.A7) is unchanged even if the microcavity structure is introduced in semiconductor core, because all the higher-order modes are either in the stopband or outside of the stopband of the DBRs and so the vacuum field intensity modulation by the microcavity is weak.

The vacuum field intensity for the fundamental mode is, on the other hand, enhanced efficiently by the microcavity. When the effective reflection coefficient of DBRs is $R$, the normalized density of states is given by

$$\frac{\rho_0'(\nu)}{\rho_{fs}(\nu)} = \frac{8}{\pi\xi(1-R)}\left(\frac{\nu_0}{\nu}\right)^2 \frac{\nu}{\sqrt{\nu^2 - \nu_0^2}} \qquad (13.A8)$$

Here the resonantly enhanced vacuum field intensity is expressed as the increased density of states for convenience. If AlAs/GaAs core-metal clad square waveguide with $a = 0.5$ $\mu$m is assumed, $\nu_0/\nu \simeq 0.32$ for $\lambda = 0.8$ $\mu$m and Eq. (13.A8) reduces to $\rho_0'(\nu)/\rho_{fs}(\nu) \simeq 0.1/(1-R)$. When $R = 0.95$, $R = 0.99$, and $R = 0.999$, $\rho_0'(\nu)/\rho_{fs}(\nu) \simeq 2$, 10, and 100. The coupling efficiency of spontaneous emission into the fundamental mode is approximately given by $\beta = \rho_0'(\nu)/[\rho_0'(\nu) + \sum_{j\geq 1}\rho_j(\nu)]$ and is about 0.67, 0.91, and 0.99, respectively. Note that large refractive index difference is not required to achieve high quantum efficiency. The spontaneous emission lifetime normalized by the free-space value is approximately given by $\tau_{sp}/\tau_{sp}^0 = \rho_{fs}(\nu)/[\rho_0'(\nu) + \sum_{j\geq 1}\rho_j(\nu)]$ and is about 0.33, $10^{-1}$ and $10^{-2}$, respectively.

The absorption loss of the fundamental mode is calculated by the Maxwell equation taking into account the complex refractive index of metal [21, 22]. For a 0.5-$\mu$m-thick AlGaAs slab waveguide cladded by Au, the absorption loss of $TE_0$ and $TM_0$ modes are about 80 and 800 cm$^{-1}$, respectively. The absorption loss of the fundamental mode of the square waveguide is approximately given by the sum of the absorption coefficients of $TE_0$ and $TM_0$ mode in a slab waveguide. If a dielectric buffer layer with refractive index $n_b$ and thickness $b$ is inserted between AlGaAs core and Au clad, the absorption loss is decreased by a factor

$$\exp\left(-2\sqrt{\beta_0^2 - k_0^2 n_b^2}\, b\right) \qquad (13.A9)$$

where $k_0 = 2\pi/\lambda$ is a free-space wavenumber. When $\beta_0 \simeq 3.5k_0$, $n_b = 1.45$ (for SiO$_2$), and $b = 0.1$ $\mu$m, the factor is about $10^{-2}$. Therefore, the absorption loss of the fundamental mode is reduced to about 9 cm$^{-1}$ by the buffer layer. Since the effective cavity length of AlAs/GaAs microcavity $L_{eff}$ is about 1 $\mu$m, the inequality Eq. (13.A2) is satisfied for the reflection coefficient $R$ of 0.999. Therefore, the coupling efficiency $\beta \simeq 0.99$ and the normalized spontaneous emission lifetime $\tau_{sp}/\tau_{sp}^0 \simeq 10^{-2}$ can be realized simultaneously by the metal clad optical waveguide microcavity.

## APPENDIX 13.B.   COUPLING EFFICIENCY $\beta$ OF SPONTANEOUS EMISSION IN CONVENTIONAL SEMICONDUCTOR LASERS

Let us consider an active volume $V$ enclosed by "perfect reflectors." The number of modes per unit frequency interval is given by $8\pi\nu^2 V/c^3$. If we assume that the active dipoles are distributed uniformly in the volume $V$ and are randomly oriented, the electric dipole coupling constant $|g|^2$ is constant for all the modes. Therefore, the total spontaneous rate is given by

$$R_{\text{sp}} = \int \frac{8\pi\nu^2 V}{c^3}|g|^2 \frac{2\Gamma N_c}{4\pi^2(\nu - \nu_0)^2 + \Gamma^2} \, d\nu = \frac{8\pi\nu^2 V|g|^2 N_c}{c^3} \quad (13.\text{B}1)$$

Here $2\Gamma$ is the spontaneous emission linewidth in radians per second (FWHM). The spontaneous emission rate $E_{cv}$ into the lasing mode at $\nu = \nu_0$ is

$$E_{cv} = \frac{2|g|^2 N_c}{\Gamma} \quad (13.\text{B}2)$$

From Eqs. (13.B1) and (13.B2), the spontaneous emission coefficient $\beta$ is calculated as

$$\beta \equiv \frac{E_{cv}}{R_{\text{sp}}} = \frac{c^3}{4\pi\nu^2 V\Gamma} = \frac{\lambda^4}{4\pi^2 V\Delta\lambda n^3} \simeq 0.025\frac{\lambda^4}{n^3\,\Delta\lambda V} \quad (13.\text{B}3)$$

Here $\Delta\lambda = (\lambda^2/\pi c_0)\Gamma$ is the spontaneous emission linewidth in meters (FWHM) and $c = c_0/n$.

This model is only an approximate one for a real semiconductor laser. A more accurate model for estimating the spontaneous emission coefficient is described below. The spontaneous emission is partly coupled to guided modes in an active waveguide and is partly coupled to radiation continuum modes. The spontaneous emission radiated at angles $\gamma$ larger than the critical angle $\gamma_c$ of the total internal reflection is trapped in guided modes, so that the coupling efficiency into the guided modes is

$$\eta_g = \frac{\int_{\gamma_c}^{\pi/2}\left[\sin^3\gamma + \sin\gamma + \sin\gamma\cos^2\gamma\right] d\gamma}{\int_0^{\pi/2}\left[\sin^3\gamma + \sin\gamma + \sin\gamma\cos^2\gamma\right] d\gamma}$$

$$\begin{array}{ccc} \nearrow & \nearrow & \nearrow \\ p \text{ wave} & s \text{ wave} & p \text{ wave} \\ \text{vertical} & \text{horizontal} & \text{horizontal} \\ \text{dipole} & \text{dipole} & \text{dipole} \end{array}$$

$$= \cos\gamma_c \quad (13.\text{B}4)$$

Of the spontaneous emission coupled to guided modes, only a small fraction is actually coupled to the lasing mode. Suppose the active waveguide supports only the fundamental transverse mode. The number of guided modes within the spontaneous emission linewidth is then given by

$$M = 2\frac{\Delta\lambda}{\Delta\lambda_c} = \frac{4nL\,\Delta\lambda}{\lambda^2} \tag{13.B5}$$

where a factor of 2 accounts for the two polarization directions and $\Delta\lambda_c = \lambda^2/2nL$ is the longitudinal-mode separation. The single-mode waveguide has a cross-sectional area on the order of $A_{\text{eff}} \sim \lambda^2/4n^2$, so that the active layer length $L$ is related to the active volume $V$ via $L = 4n^2V/\lambda^2$. The coupling efficiency into the lasing mode from all the guided modes is thus given by

$$\eta_l \equiv \frac{1}{M} = \frac{\lambda^4}{16n^3\,\Delta\lambda V} \tag{13.B6}$$

From Eqs. (13.B4) and (13.B5), the spontaneous emission coefficient $\beta$ is

$$\beta \equiv \eta_g \times \eta_l = \frac{\lambda^4 \cos\gamma_c}{16n^3\,\Delta\lambda\,V} \tag{13.B7}$$

If the refractive index ratio $n_2/n_1$ of the active waveguide and the cladding medium is $\sim 0.92$, which is typical in a semiconductor laser, $\cos\gamma_c \approx 0.4$, and Eq. (13.B7) reduces to

$$\beta \approx 0.025\frac{\lambda^4}{n^3\,\Delta\lambda\,V} \tag{13.B8}$$

The two different models result in the same spontaneous emission coefficient [Eqs. (13.B3) and (13.B8)]. If we use the numerical parameters of a typical semiconductor laser, $\lambda = 8 \times 10^{-7}$ m, $n = 3.5$, $\Delta\lambda = 5 \times 10^{-8}$ m, and $V = 4 \times 10^{-16}$ m$^3$, the spontaneous emission coefficient is on the order of $\beta \approx 10^{-5}$.

## REFERENCES

1. E. M. Percell, *Phys. Rev.* **69**, 681 (1946).
2. K. H. Drexhage, *Progress in Optics*, Vol. 12, E. Wolf, ed., North-Holland, New York, 1974, p. 165.
3. D. Kleppner, *Phys. Rev. Lett.* **47**, 233 (1981).
4. P. Goy, J. M. Raimond, M. Gross, and S. Haroche, *Phys. Rev. Lett.* **50**, 1903 (1983).

5. G. Gabrielse and H. Dehmelt, *Phys. Rev. Lett.* **55**, 67 (1985).

6. R. G. Hulet, E. S. Hilfer, and D. Kleppner, *Phys. Rev. Lett.* **55**, 2137 (1985).

7. W. Jhe, A. Anderson, E. A. Hinds, D. Meschede, L. Moi, and S. Haroche, *Phys. Rev. Lett.* **58**, 666 (1987).

8. D. Heinzen, J. J. Childs, J. E. Thomas, and M. S. Feld, *Phys. Rev. Lett.* **58**, 1320 (1987).

9. F. DeMartini, G. Innocenti, G. R. Jacobovitz, and P. Mataloni, *Phys. Rev. Let.* **59**, 2955 (1987).

10. E. Yablonovitch, *Phys. Rev. Lett.* **58**, 2059 (1987).

11. T. Kobayashi, T. Segawa, A. Morimoto, and T. Sueta, *Technical Digest of 43rd Fall Meeting of Japanese Applied Physics Society*, 29a-B-6 (1982).

12. D. Meshede, H. Walther, and G. Müller, *Phys. Rev. Lett.* **54**, 551 (1985).

13. J. J. Sanchez-Mondragon, N. B. Narozhny, and J. H. Eberly, *Phys. Rev. Lett.* **51**, 550 (1983).

14. I. R. Senitzky, *Phys. Rev. Lett.* **31**, 955 (1973); I. R. Senitzky, *Phys. Rev.* **A6**, 1175 (1972).

15. G. Björk and O. Nilsson, *IEEE J. Lightwave Technol.* **LT-5**, 140 (1987).

16. A. Honold, L. Schultheis, J. Kuhl, and C. W. Tu, *XVIth International Conf. Quantum Electron.* ThB-6, Tokyo (July 1988).

17. M. Yamanishi and I. Suemune, *Jpn. J. Appl. Phys.* **23**, L35 (1984).

18. Y. Yamamoto, S. Machida, and O. Nilsson, *Phys. Rev.* **A34**, 4025 (1986); Y. Yamamoto and S. Machida, *Phys. Rev.* **A35**, 5114 (1987); S. Machida, Y. Yamamoto, and Y. Itaya, *Phys. Rev. Lett.* **58**, 1000 (1987); S. Machida and Y. Yamamoto, *Phys. Rev. Lett.* **60**, 792 (1988).

19. T. T. J. M. Berendschot, H. A. J. M. Reinen, and H. J. A. Bluyssen, *Appl. Phys. Lett.* **54**, 1827 (1989).

20. Y. Arakawa and H. Sakaki, *Appl. Phys. Lett.* **40**, 939 (1982).

21. T. Takano and J. Hamasaki, *IEEE J. Quantum Electron.* **QE-8**, 206 (1972).

22. Y. Yamamoto, T. Kamiya, and H. Yanai, *IEEE J. Quantum Electron.* **QE-11**, 729 (1975).

# Index

617